Integrated Science

Volume 19

Editor-in-Chief

Nima Rezaei, Tehran University of Medical Sciences, Tehran, Iran

The **Integrated Science** Series aims to publish the most relevant and novel research in all areas of Formal Sciences, Physical and Chemical Sciences, Biological Sciences, Medical Sciences, and Social Sciences. We are especially focused on the research involving the integration of two of more academic fields offering an innovative view, which is one of the main focuses of Universal Scientific Education and Research Network (USERN), science without borders.

Integrated Science is committed to upholding the integrity of the scientific record and will follow the Committee on Publication Ethics (COPE) guidelines on how to deal with potential acts of misconduct and correcting the literature.

Juan Carlos Castilla · Juan J. Armesto ·
María José Martínez-Harms · David Tecklin
Editors

Conservation in Chilean Patagonia

Assessing the State of Knowledge, Opportunities, and Challenges

EDICIONES UC

Editors

Juan Carlos Castilla
Departamento de Ecología
Pontificia Universidad Católica de Chile
Santiago, Chile

Juan J. Armesto
Departamento de Ecología
Pontificia Universidad Católica de Chile
Santiago, Chile

María José Martínez-Harms
Center for Research and Innovation
in Climate Change
Universidad Santo Tomás
Santiago, Chile

David Tecklin
Programa Austral Patagonia
Universidad Austral de Chile
Valdivia, Chile

ISSN 2662-9461   ISSN 2662-947X (electronic)
Integrated Science
ISBN 978-3-031-39407-2   ISBN 978-3-031-39408-9 (eBook)
https://doi.org/10.1007/978-3-031-39408-9

Jointly published with EdicionesUC
ISBN of the Co-Publisher's edition: 978-9-561-42820-1

Translation from the Spanish language edition: "CONSERVACIÓN EN LA PATAGONIA CHILENA"
by Juan Carlos Castilla et al., © Ediciones UC 2021. Published by Ediciones UC. All Rights Reserved.

The translation was done with the help of artificial intelligence (machine translation by the service
DeepL.com). A subsequent human revision was done primarily in terms of content.

This Springer imprint is published by the registered company Springer Nature Switzerland AG
The registered company address is: Gewerbestrasse 11, 6330 Cham, Switzerland

Paper in this product is recyclable.

Foreword by Mary T. Kalin Arroyo

A Brief Vision of the Past, Present, and Future of Chilean Patagonia

Patagonia is a biologically unique region of the world. This is confirmed by irrefutable facts for any visitor. Its western slope (Chilean Patagonia) harbors forests whose tree species are evidence of biogeographic connections with distant regions of the planet, such as New Zealand, Australia, and New Guinea. Such relationships attest to an ancient terrestrial connection through the Antarctic continent linking these southern territories and South America [4]. In particular, the coastal forests of Aysén are remarkable reserves of biodiversity comparable to the Valdivian forests. Developed on vertical soils close to the sea, they are nourished in part thanks to the great richness of mosses and bryophytes, which in turn are fed by abundant and endless rains.

The flora along the coast is rich in species with bright red flowers, in addition to the centuries-old coihues and canelos of the northernmost forests of Aysén, a true treasure. These forests are known for their high number of endemic species and genera [1]. Although the angiosperms and gymnosperms that inhabit these forests are well known, our knowledge of the multitudinous diversity of mosses, liverworts, lichens, and fungi that cover the forest floors and shelter the tree trunks is still precarious. On the other hand, the Patagonian steppe, under a drier climate, contains younger ecosystems than the forests, whose origin in Patagonia is related to the Andean uplift, which produced the rain shadow that gave rise to the present semi-arid climate of the eastern slope (see in this book Radic *et al.* 2023). In addition to the gradual and more recent appearance of the steppe, the uplift of the Cordillera promoted the emergence of the so-called high Andean belt above the tree line, with a unique and fragile flora due to the severe climatic conditions. All these factors give the Patagonian biota diverse ecological, landscape, and biogeographical values, many of which are highlighted in the chapters of this book.

Chilean Patagonia is an extraordinary natural laboratory, where ecosystems of different ages and characteristics coexist in a relatively small physiographic space, which we should undoubtedly appreciate and care for. Chilean Patagonia constitutes a treasure for the development of the nature-based tourism industry, an industry that requires the incorporation of people and professionals from different specialties and is therefore ideal for Patagonian regional development (see in this book Guala *et al.* 2023). Nature-based tourism requires experts in flora and fauna trained in digital photography, leaders in gastronomy, and transportation, as well as the incorporation of Patagonian citizens in general, such as the "gauchos" who from my own experience, know much about the history and geography of the area.

The evolutionary history of this remarkable ecological system is only part of the story. The forests of Chilean Patagonia provide local societies with valuable ecosystem services. The mosses and bryophytes of the forests fertilize soils that have been excavated and washed into the rivers and oceans by ancient glaciers. In the face of the Holocene climate warming [5], the forests advanced from the north, colonizing the mountains and rocky soils as the glaciers retreated. The forests typical of the southern Patagonian region with lenga, ñirre (deciduous), and Magellanic coigue as protagonists, are some of the southernmost terrestrial ecosystems in the world [3]. They develop on very thin soils where large trees must stoically resist hurricanes and Patagonian winds, which can flatten them at any time. The inhabitants of this extreme physical space, including humans, survive permanent catastrophes and maintain their populations in the face of adversity. The integrity of the Patagonian mountain flora and the Andean belt with its great diversity of plants and animals is essential for the protection of soils and the supply of water to all lower altitudes. Patagonia's riches go beyond its biodiversity, ecosystem functions, and spectacular landscapes. They have global values as an area that preserves elements of pre-industrial environments, which today's humans can use as references in the face of climate change and anthropogenic disturbances that are transforming our planet. This region can be seen as an incredible time machine, through which we can witness the cooling action of the ice age (which prevailed in the last million years); places where trees still coexist with the great glaciers, as well as the past periods when extensive wetlands or peatlands, surrounded by ice, dominated the landscape of southern Chile. Today, thanks to the Andean uplift, the territory presents strong altitudinal gradients that provide a privileged setting for understanding how plant and animal species adapt and survive abrupt changes in environmental conditions. In some of the chapters of this book, the study of these gradients and their biota provides us with useful lessons to face the uncertain future of the planet (see in this book Marquet *et al.* 2023).

Several of the book's chapters emphasize the need to recognize that our knowledge and wisdom regarding Chilean Patagonia has deep roots in its inhabitants (see in this book Aylwin *et al.* 2023). Sea-faring peoples who, for centuries preceding the European invasion, traveled and knew the territory, its flora and fauna in great detail. Our Patagonia is known by many Chileans more for its spectacular landscapes (e.g., glaciers and mountains such as Torres del Paine) than for its biological values and cultural heritage, which requires an ability to look and appreciate beyond the most grandiose landscape elements. Many Chileans still have the impression of Patagonia as a barren land of low productivity and extreme climates, with little mineral wealth, which is the way we have long assessed the value of our territories. Undoubtedly, the great riches of Chilean Patagonia lie in its unknown landscapes steep mountain ranges, hidden lakes, and extensive peat bogs, which have hardly been traveled by humans. These places represent a gigantic pool of environmental wealth for the world and for a country like Chile. Unfortunately, in just a few decades, many of these fragile environments have already suffered severe anthropogenic degradation [2] (see in this book Marquet *et al.* 2023).

How to protect Patagonia and all its splendor? It is not an easy task, but with well-planned and integrated work, as proposed in this book, it is possible to move forward (see in this book Tacón *et al.* 2023; Tecklin *et al.* 2023).

The book highlights the need for a comprehensive conservation strategy at different levels, ranging from the proper management of large terrestrial and marine parks and reserves to the implementation of goals that ensure that its inhabitants and visitors admire and respect the nature that surrounds them. Thanks to a long tradition in Chile of creating protected areas, with support from the State and from visionary, passionate, and generous individuals, we have advanced to a regional panorama of large protected areas. However, with more people visiting the ecosystems of these remote national parks and reserves, fires, the introduction of harmful invasive species, and the extraction of plants and animals become a greater danger.

It is essential that Chile implements concrete actions to contain the advance of such impacts which, in turn, threaten the welfare of Chileans who live in the area and depend on the tourism and ecotourism industries, thus creating complex situations in terms of social welfare. For terrestrial conservation, it is urgent to change the design of highways and roads, minimizing the amount of area affected, so as to prevent impacts on the natural environment, and monitor the advance of invasive species that enter along roads.

As highlighted in the initial chapters of the book, serious thought should be given to regulating the number of visitors to Patagonian parks and access areas, thus reducing the possibility of large and disastrous fires, recovery from which takes decades, and which have large economic costs for the country. It goes without saying that parks as extensive as the Patagonian parks must have an adequate number of park rangers, both men and women, well trained in these matters and adequately remunerated. These are matters that the nascent National Biodiversity and Protected Areas Service should keep in mind. However, we should not underestimate the importance of preserving the biodiversity of Chilean Patagonia outside of its large parks. The conservation of Patagonia's biodiversity requires the collaboration of all stakeholders. Environmental education programs aimed at the many visitors and society in general are essential.

On a last note, in the 1980s, I had the opportunity to work for several summers, in the high and cold peaks of Torres del Paine National Park and the Sierra de los Baguales with the help of local people and the staff of the National Forestry Corporation (CONAF). I can honestly say that the ecosystems of Chilean Patagonia and the stories of the inhabitants were a great inspiration for my scientific research. I thank the local people and CONAF for their kindness in showing me the beauty and secrets of these wild areas and invite all young scientists in the region to recognize the enormous value of these remote ecosystems in advancing universal knowledge of how the biosphere works.

My congratulations to the editors of the book for bringing together a wealth of expertise and valuable information to support a renewed Patagonian conservation strategy. One that has the capacity to improve the lives of the inhabitants and promote future scientific work.

Mary T. Kalin Arroyo
National Science Award, 2010
Full Professor, Universidad de Chile
Researcher Institute of Ecology
and Biodiversity (IEB), Santiago,
Chile

References

1. Arroyo MTK, Marquet PA, Marticorena C, Simonetti JA, Cavieres L, Squeo F, Rozzi R (2004) Chilean winter rainfall-valdivian forests. In: Mittermeier RA, Gil PR, Hoffmann M, Pilgrim J, Brooks T, Lamoreux J, Da Fonseca y GAB, (eds) Hotspots revisited: Earth's biologically wealthiest and most threatened ecosystems, CEMEX, México D. F. p 99–103
2. Relva MA, Damascos MA, Macchi P, Mathiasen P, Premoli AC, Quiroga MP, Radovani NI, Raffaele E, Sackmann P, Speziale K, Svriz M, Vigliano PH (2013) Impactos humanos en la Patagonia. In: Raffaele E, de Torres-Curth M, Morales CL, Kitzberger yT (eds) Ecología e historia natural de la Patagonia andina: un cuarto de siglo de investigación en biogeografía, ecología y conservación. Fundación de Historia Natural Félix Azara, Buenos Aires, pp 157–181
3. Rozzi R, Armesto JJ, Gutiérrez J, Massardo F, Likens G, Anderson CB, Poole A, Moses K, Hargrove G, Mansilla A, Kennedy JH, Willson M, Jax K, Jones C, Callicott JB, Kalin MT (2012) Integrating ecology and environmental ethics: Earth stewardship in the southern end of the Americas. BioSci 62(3):226–236
4. Segovia RA, Armesto JJ (2015) The Gondwanan legacy in South American biogeography. J Biogeograp 42(2):209–217
5. Villagrán C (2018) Biogeografía de los bosques subtropical-templados del sur de Sudamérica. Hipótesis históricas. Magallania (Punta Arenas) 46(1):27–48

Foreword by Humberto E. González

Chilean Patagonia, a View from the Ocean

Chilean Patagonia, with its magnificent geomorphology, sculpted by the gouging of glaciers, is one of the world's great freshwater reservoirs and a southern bastion of regional and global climate stability. Both the dynamics of the atmosphere (and hence climate) and ocean circulation (and hence the distribution of heat, gases, and nutrients) are partially modulated from Patagonia. This part of the planet currently offers us an enormous volume of ecosystem services that are described throughout this book, and that have been provided to Indigenous peoples for their well-being for millennia [1] (see in this book Aylwin *et al.* 2023). This is what we want to conserve, and this is the vision of our Chilean Patagonia from the ocean. Why is it important to conserve it? Because Patagonia is a central part of the history and culture of our country, it is necessary to keep nature as healthy as possible to fulfill its function of climate, ecological, environmental, and cultural regulation, and because it is one of the areas most vulnerable to Global and Climate Change (GC and CC), to anthropogenic impacts/pollution and to the misuse of its enormous potential for economic-productive activities (see in this book Buschmann *et al.* 2023).

In this sense, it is necessary to highlight the lesser-known ecosystem services of the Chilean Patagonian marine territory, such as its capacity to capture, export, and sequester carbon during the productive period [11]. In addition, several functional groups of plankton (such as euphausiids) and/or cetaceans (such as whales), make significant contributions in ocean fertilization and/or carbon export through their metabolic waste (i.e., fecal pellets) or behavior (vertical migrations) [4], [10]. All these actions are part of nature-based solutions that contribute, at no cost to us, to the great challenge of protecting and maintaining "healthy" Patagonia. How? Using legal tools such as marine protected areas, marine spatial planning, municipal conservation areas, rational use of the coastline, and above all, integrating the communities that live in these areas (i.e., Indigenous Peoples Coastal Marine Spaces) into conservation (see in this book Hucke-Gaete *et al.* 2023; Tecklin *et al.* 2023; Haussermann *et al.* 2023).

This book highlights the need for urgent conservation actions throughout Chilean Patagonia. Commitments to environmental protection, sustainable and responsible use of its goods and services in a Patagonia that is conformed by "Hybrid Terrestrial-Marine Systems" (SHT-M), where the Patagonian marine territory not only interacts with terrestrial systems but also with adjacent oceanic ecosystems [8] (see in this book Rozzi *et al.* 2023). This reality determines a very complex governance, as it includes (i) different SHT-M type ecosystems; (ii) threats to ecosystem functions and services that are transversal to these systems; (iii) socio-economic and cultural activities that include and influence them [9]. Chilean Patagonia with its system dominated by fjords, channels, bays, etc. hosts a very productive, quasi-pristine system that receives nutrients, particulate and dissolved matter (both organic and inorganic), and abundant freshwater discharge from terrestrial systems for its functioning [5]. These flows project into the marine territory as "plumes" of brackish surface water and partially into deeper areas (saline subantarctic currents). The influx of freshwater is what, in part, sustains the functioning of the marine environment, but in turn, makes it more vulnerable to acidification. This, together with the loss of glacier mass and the warming of its waters, is gradually changing the physical and chemical conditions, where a rich biodiversity and endemism of plankton and benthic organisms coexist with birds, cetaceans, and other marine mammals that make use of their ecosystem services and that could serve as "focal points and sentinels" for climate change (see in this book Hucke-Gaete *et al.* 2023).

Chilean Patagonia is characterized by high seasonal variability, including aquatic productivity that transitions between primarily sunlight and photoperiod control in winter to primarily nutrient control in summer [3], [4]. The impacts of GC and CC (natural and anthropogenic) in Patagonia are multifactorial and given that we have a single large ocean (with different names) and a single large atmosphere, the potential threats come from Chile and the world. This obliges us to make efforts and commitments at the national level with all the actors involved: academia, public and private sectors, government agencies, NGOs, communities, and Indigenous peoples. The great commitment is to protect and conserve Chilean Patagonia from the threats described throughout this book. What to do? Marine-protected areas are a first step; making them effective so that they fulfill their function is the second and greatest challenge.

The nations that are leading the CC efforts in the Ocean Panel (including Chile) have indicated the urgency of five actions: (i) sustainable management of marine resources; (ii) CC mitigation; (iii) assessing the possibilities of economic recovery; (iv) integrated ocean management; and (v) halting biodiversity loss [7]. All these efforts are represented in Chilean Patagonia and discussed in the book by specialists in various areas. We should add the interface between natural and social sciences and the changes generated by stressors (environmental, extractive, and productive activities such as tourism, aquaculture, fishing, etc.) and the contribution of Indigenous peoples in the conservation of Patagonian aquatic systems (see in this book Aylwin *et al.* 2023).

This book describes the threats to aquatic biodiversity, the introduction of exotic species (i.e., salmonids, *Dydimo*), and the risk of frequent algal blooms, many of which are harmful and affect productive activities (artisanal and industrial) and public health. In recent years, the expansion of toxic species to the north (*Alexandrium catenella*) and south (*Pseudochattonella* cf. *verruculosa*) of Chilean Patagonia has been reported [12]. These processes could be related to an expansion of anthropogenic activity and/or changes in chemical and physical conditions in the marine environment due to CC and operate at various spatial and temporal scales. The range of involved processes is very broad, including large-scale atmospheric (i.e., wind regime) and oceanographic (i.e., upwelling of water masses) factors, to more local freshwater inputs that change the stratification and stoichiometry of fjords and channels. The marine environment receives a large freshwater incursion from a low-lying mountain range with sectors covered with ice fields and glaciers, most of which are in the process of ice mass loss (see in this book Rivera *et al.* 2023) ranging between 20 and 30 Gt a-1 over the last two decades [2]. The wide extension of the coastal zone below 10 m in Chilean Patagonia makes it susceptible to extreme events such as floods, landslides, avalanches, and tidal waves, with negative effects on the provision of ecosystem services [6] (see in this book Rivera *et al.* 2023).

In summary, the Chilean Patagonian marine environment is a very complex and diverse system with very conspicuous gradients, both east–west and north–south. The Andes Mountain range, with an N–S orientation throughout South America, changes direction (90°) in the Darwin mountain range, becoming E–W oriented (i.e., Beagle Channel), as a result of the pressure of the Antarctic plate on the South American continent. These orographic changes have opened bi-oceanic routes to which the Strait of Magellan is added, enhancing the geographic, climatic, biogeographic, and physicochemical singularities (see in this book Rozzi *et al.* 2023).

Finally, our Patagonia is a hybrid between land and sea, Pacific and Atlantic, Antarctic and Sub-Antarctic, with physical influences from the atmosphere, the cryosphere, and the ocean. The challenge is to achieve an efficient conservation of the ecosystems, and their cultural, social, and ecological legacies which underlie the diverse ecosystem services that Chilean Patagonia provides us, for the well-being of the current population and those who come after us.

Humberto E. González
Full Professor, Institute of Marine
and Limnological Sciences,
Universidad Austral de Chile
and Centro de Investigación en
Dinámica de Ecosistemas Marinos de
Altas Latitudes (FONDAP-IDEAL),
Valdivia, Chile

Acknowledgements To the FONDAP 15150003 program and to the editors (JCC, JJA and MJM-H) and DT (UAch) for their great leadership, commitment, and effort in bringing this great work to completion.

References

1. Dillehay TD, Ramírez C, Pino M, Collins J, Rossen, Pino-Navarro J (2008) Monte Verde: seaweed, food, medicine, and the peopling of South America. Science, 320:784–786
2. Dussaillant IE, Berthier F, Brun M, Masiokas, R, Hugonnet, Favier V, Rabatel A, Pitte P, Ruiz L (2019) Two decades of glacier mass loss along the Andes. Nature Geosci 12:802–808
3. González H.G., M. J. Calderón, L. Castro, A. Clement, L. Cuevas, G. Daneri, J. L. Iriarte, L. Lizárraga, R. Martínez, E. Menschel, N. Silva, C. Carrasco, C. Valenzuela, C. A. Vargas, C. Molinet (2010) Primary production and its fate in the pelagic food web of the Reloncaví Fjord and plankton dynamics of the Interior Sea of Chiloé, Northern Patagonia, Chile.- Mar. Ecol Prog Ser. 402: 13–30
4. González HE, Graeve M, Kattner G, Silva N, Castro L, Iriarte JL, Osmán L, Daneri G, Vargas C (2016) Carbon flow through the pelagic food web in southern Chilean Patagonia: relevance of *Euphausia vallentini* as key species. Mar Ecol Prog Ser 557:91–110
5. González HE, Nimptsch J, Giesecke R, Silva N (2019) Organic matter distribution, composition, and its possible fate in the Chilean north-Patagonian estuarine system. Sci Total Environ 657:1419–1431
6. Iriarte JL, González HE, Nahuelhual L (2010) Patagonian fjord ecosystems in southern Chile as a highly vulnerable region: problems and needs. Ambio 39(7):463–466
7. Lubchenco J, Haugan P, Pangestu E (2020) Five priorities for a sustainable ocean economy. Nature 588:30–32
8. Pavés H, González HE, Castro L, Iriarte JL (2015) Carbon flows through the pelagic sub-food web in two basins of the Chilean Patagonian coastal ecosystem: the significance of coastal-ocean connection on ecosystem parameters. Estuaries and Coast 38:179–191
9. Pittman J, Armitage D (2016) Governance across the land-sea interface: a systematic review. Environ Sci Policy 64:9–17
10. Ratnarajah L, Nicol S, Bowie AR (2018) Pelagic iron recycling in the Southern Ocean: exploring the contribution of marine animals. Front Mar Sci 5:109
11. Torres R, Pantoja S, Harada N, González HE, Daneri G, Frangopulos M, Rutllant JA, Duarte C, Ruiz-Halpern S, Mayol E, Fukasawa M (2011) Air-sea fluxes along the coast of Chile: from CO_2 outgassing in central-northern upwelling waters to CO_2 uptake in southern Patagonia fjords. J Geophys Res Oceans 116, C09006
12. Trainer VL, Moore S, Hallegraeff G, Kudela R, Clement A, Mardones J, Cochlan W (2020) Pelagic harmful algal blooms and climate change: lessons from nature's experiments with extremes. Harmful Algae 91:101591

Introduction and Acknowledgments

Chilean Patagonia offers an exceptional opportunity, both nationally and globally, for the comprehensive protection of relatively undisturbed landscapes, seascapes, and freshwater systems. With vast areas still wild, including watersheds draped in primary forests, extensive steppes, peatlands, large rivers, glaciers, inland seas, estuaries formed by the Patagonian archipelagos, and a diversity of ancestral cultures and Indigenous peoples, the region is a reservoir of socio-natural heritage on a global scale. Despite its importance, threats to the region are rapidly increasing. These include climate change; introduced invasive terrestrial, freshwater, and marine species; and socio-economic drivers, such as aquaculture that replicate and extend patterns of high impact, indiscriminate natural resource use.

Notwithstanding the significant national and international scientific interest in the conservation and sustainable use of Chilean Patagonia's ecosystems and biodiversity, until now there has been no integrated compilation and evaluation of the scientific evidence on the conservation status of its ecosystems or the needs and priorities for their effective protection. In general, the scientific information and data available on the conservation and management of the region's ecosystems remains fragmented and has not been integrated or analyzed at a regional scale. For example, the conservation of terrestrial and marine ecosystems has generally been treated in separate literatures. In a region such as Chilean Patagonia, whose most outstanding characteristic is the vast extent of the land-marine interface, such segregation represents an essential limitation for both understanding and action. Added to this is the fact that many of the studies that address issues of planning, management, local development, and Patagonian tourism lie within a current of grey literature that is dispersed among many institutions and not always easily accessible to the scientific community.

An integrated perspective on the state of knowledge, the conservation status of Chilean Patagonian ecosystems, and the drivers of change that threaten them is essential for developing innovative, proactive, and operational conservation practices to effectively manage and protect the region's biodiversity and ecosystem services. Sound scientific evidence on terrestrial, marine, and land-sea interface biodiversity, ecosystem processes, and knowledge of the human dimensions of socio-environmental problems in the region is a requirement for informed decision-making on conservation and sustainability in Chilean Patagonia. This is due to the fact that the main drivers of degradation in Patagonian ecosystems, such as climate change, exotic invasive species, habitat loss, overfishing, and the impacts of aquaculture, occur at the sea-land-society interface. Addressing these complex environmental problems requires an interdisciplinary approach including diverse sources of information, especially those based on Indigenous and local knowledge to understand the diversity of perspectives related to relationships with nature.

This book represents an effort to address this major challenge and seeks to provide a synthesis of the most salient local, regional, and global attributes of conservation in Chilean Patagonia through an integrated compilation and analysis of the available information by thematic experts at the regional scale. The hypothesis underlying this book is that there is a much higher level of scientific information in the region than has historically been used for conservation decision-making. Compiling and critically analyzing this knowledge and putting it at the service of actions to improve the conservation of the most threatened ecosystems, as well as the sustainability and resilience of Chilean Patagonia was the central motivation for this initiative.

The book also represents a collaborative effort involving 68 authors who over the course of 18 chapters cover the following subjects: terrestrial, marine, and freshwater ecosystems and biodiversity; the accelerating pressures of global and local changes on ecosystems; impacts of aquaculture; the dynamics of the land-sea interface; the conservation of glaciers, grasslands, peatlands, and intact primary forests; conservation led by Indigenous communities as well as the management of protected areas; and socioeconomic trends in the region, among other topics. Acknowledging that the Chilean Patagonia region has no technical boundaries, for the purposes of this book, it was geographically defined as the area located between the Reloncaví Sound—where the fjords and inland sea begin—and the Diego Ramírez Islands, at the southern limit of the continental shelf. The authors compiled the information available in the scientific and grey literatures, critically reviewed it, analyzed the gaps and opportunities, and formulated recommendations for better conservation and management of Chilean Patagonian ecosystems. As gaps were identified on some topics, authors contributed additional original research to the existing literature. In addition, a synthesis is presented in the first chapter that comprehensively analyzes the state of knowledge on the conservation and management of ecosystems and the drivers of global change that threaten the region and provides cross-cutting recommendations.

The book is the product of almost three years of collaborative work between the Austral Patagonia Program (ProAP) of the Universidad Austral de Chile and a team of scientists from the Institute of Ecology and Biodiversity (IEB), with the participation of researchers from many Chilean universities who contributed as authors, advisors, or reviewers of the chapters. ProAP was formed in 2018 with financial support from The Pew Charitable Trusts with the goal of improving the conservation status of marine and terrestrial ecosystems in Chilean Patagonia, in particular, through improving the management of public protected areas by generating and disseminating information, and capacity building. ProAP relies on a network of researchers from diverse disciplines at the Universidad Austral de Chile and professionals specializing in conservation issues and local development in alliance with social organizations and local communities in Patagonia. The IEB is a non-profit institution, established in Chile in 2008, whose main objective is to carry out cutting-edge scientific research in biodiversity sciences and contribute to the country's sustainable development, linking the efforts of scientists and professionals from various academic institutions throughout Chile.

Conservation in Chilean Patagonia: Assessing the State of Knowledge, Challenges, and Opportunities was originally published in Spanish with the goal of providing an accessible interdisciplinary resource for interested researchers, conservation practitioners, and decision-makers in Chile and the broader Latin American region. We also conceived this as a first step towards a continuing effort to integrate knowledge at the regional scale with the hope that it will be followed by successive phases of work carried out by researchers in the region and worldwide, and in this way to contribute to a renewed, scientifically informed, and reinvigorated conservation agenda for the region.

This English language edition seeks to bring the compilation and analysis to the broader international public particularly given the rapidly growing interest in the Patagonian region, as well as in interdisciplinary conservation assessment and new approaches to science-policy dialogues. We trust that this work will contribute to growing the field of researchers and institutions working toward understanding and addressing the conservation of Chilean Patagonia's critically important ecosystems.

Acknowledgments

We would like to thank, first of all, the authors of the chapters and the other collaborators in the generation of the book, including the peer reviewers and administrative and technical support professionals, who contributed their knowledge and experience. For the original Spanish version of the book, we are grateful for the invaluable help of Claudia Papic in the coordination of the project and of Taryn Fuentes-Castillo in technical aspects of project management of the manuscripts. Aldo Farías and the Geographic Information Systems laboratory of ProAP collaborated with the systematization and homogenization of the cartographic information and the production of maps used in several of the chapters.

Our deep appreciation is also due to the team who contributed to the arduous effort to produce an English version of what is a lengthy and varied text. Eaton Lafayette provided crucial copy editing of texts suffering from translation difficulties. María Paz Peña contributed vital editorial assistance, including reworking all of the figures for this book. Camilo Ruiz and the ProAP executive team played an essential supporting role and ensured the process continued through many challenging moments. Finally, we thank Gonzalo Cordova our counterpart at Springer for his inspiration and support in bringing out the English version of the book.

In addition, many people and institutions generously contributed to the underlying assessment and integration of information leading up to this book. We acknowledge the valuable guidance and multiple contributions of the scientific panel for the regional conservation assessment of Chilean Patagonia that gave rise to this volume, which, in addition to the editors, was formed by Mg. María Victoria Castro, Professor Emeritus, Department of Anthropology, Faculty of Social Sciences, Universidad de Chile; Dr. Giovanni Daneri, Executive Director and Researcher at the Center for Research on Patagonian Ecosystems; Dr. Míriam Fernández, Full Professor, Department of Ecology, Faculty of Biological Sciences, Pontificia Universidad Católica de Chile; Dr. Humberto González, Full Professor, Institute of Marine and Limnological Sciences, Universidad Austral de Chile; Dr. Stefan Gelcich, Associate Professor, Department of Ecology, Faculty of Biological Sciences, Pontificia Universidad Católica de Chile; Dr. Rodrigo Hucke, Assistant Professor, Institute of Marine and Limnological Sciences, Universidad Austral de Chile; Dr. Pablo Marquet, Full Professor, Department of Ecology, Faculty of Biological Sciences, Pontificia Universidad Católica de Chile; Dr. Ricardo Rozzi, Full Professor, Universidad de Magallanes and Full Professor at the University of North Texas in the United States.

The assessment included three workshops beginning in Santiago in 2018 with the meeting, "Strategic partnership to form a high-level scientific panel and conduct a regional conservation assessment for Chilean Patagonia." There, the project was planned, the methodology for information synthesis was discussed, and the writing of 10 technical reports was commissioned. The second workshop, "Regional conservation assessment project meeting for Chilean Patagonia", was held in Coyhaique, in September 2018 and was attended by the lead authors of reports that had been commissioned and a review of work by the broader panel. The third workshop, "Review of the science panel's synthesis report," was held in Punta Arenas in March 2019, where a summary of the full collection of reports was presented. The subsequent discussion identified key gaps in scientific topics of importance and recommendations for broadening and strengthening the analysis. Based on this feedback the IEB and ProAP team agreed to commission specialists to provide additional reports and to consolidate and publish all of this work as a single volume.

We sincerely thank the leaders of the Indigenous peoples of Patagonia, who attended the Punta Arenas workshop and contributed to a productive dialogue to seek bridges between scientific-technical knowledge and traditional local knowledge. We also appreciate the contributions of professionals from the National Forestry Corporation, the Ministry of the Environment, and other public agencies, as well as technicians and professionals from civil society organizations with long experience in Chilean Patagonia who participated in this workshop. We are also grateful for the collaboration of research institutions in Patagonia and in particular the Center for Research in Patagonian Ecosystems in Aysén, the Universidad de

Magallanes and the Pontificia Universidad Católica de Chile, Faculty of Biological Sciences, who generously supported us in conducting workshops in Santiago, Aysén, and Punta Arenas. We are also grateful for the financial support provided by The Pew Charitable Trusts, which made possible the preparation and printing of this book.

Juan Carlos Castilla
David Tecklin
María José Martínez-Harms
Juan J. Armesto
César Guala

Contents

Editors and Contributors

About the Editors

Juan Carlos Castilla (Ph.D., Marine Biology, Bangor University, UK) is Professor Emeritus at the Pontificia Universidad Católica de Chile. Over the course of his career, he has specialized in experimental marine ecology, marine conservation, coastal benthic marine resource management, coastal pollution, and marine education. In 2010, he received Chile's National Award in Applied Sciences and Technology, in 2011 the Ramon Margalef Award in Ecology, and in 2012 the Mexico Award in Science and Technology. He is a foreign member of the US National Academy of Sciences.

Juan J. Armesto (Ph.D., Botany and Plant Physiology, Rutgers University) is a Full Professor at Pontificia Universidad Católica de Chile, an Adjunct Scientist at the Cary Institute of Ecosystem Studies, and a Visiting Professor at the Universidad de Concepción. His research interests focus on forest ecosystems, in particular biogeochemical cycles and the relationship between biodiversity and ecosystem functions. He is President and Researcher at the Senda Darwin Foundation, Chiloé, and Director of the Institute of Ecology and Biodiversity, Chile.

María José Martínez-Harms (Ph.D., Biodiversity Conservation Science, University of Queensland, Australia) is a researcher at the Center for Research and Innovation in Climate Change, Universidad Santo Tomas, Santiago, Chile. Her research focuses on landscape ecology, ecosystem services, evidence-based conservation, and spatial planning for biodiversity conservation and ecosystem services. In 2019, she received the L'Oreal Chile-UNESCO Award for Women in Science.

David Tecklin (Ph.D., Geography, University of Arizona) is a Research Associate at the Austral Patagonia Program, Universidad Austral de Chile, and Principal Officer for South American Land and Fresh Water Conservation at The Pew Charitable Trusts. His work as a conservation practitioner has included development of multiple conservation programs and collaborative initiatives for forests, freshwater, and marine ecosystems. His research focuses on the analysis of institutions, policies, and governance.

Contributors

David Alday Yagán Community, Bahía Mejillones, Chile

Ricardo Álvarez School of Archaeology, Universidad Austral de Chile, Balneario Pelluco, Los Pinos S/N, Puerto Montt, Chile

Juan Carlos Aravena GAIA-Antarctic Research Center, Universidad de Magallanes, Punta Arenas, Chile;
Network for Extreme Environments Research (NEXER), Universidad de Magallanes, Punta Arenas, Chile;
Cape Horn International Center, Universidad de Magallanes, Puerto Williams, Chile;
Centro de Investigación Gaia Antártica (CIGA), Universidad de Magallanes, Punta Arenas, Chile

Lorena Arce Biodiversity and Alternatives to Development Program, Observatorio Ciudadano; Southern Cone Coordinator, TICCA Consortium, Territories and Areas Conserved By Indigenous People and Local Communities, Temuco, Chile

Juan J. Armesto Estación Biológica Senda Darwin, Ancud, Chile;
Departamento de Ecología, Pontificia Universidad Católica de Chile, Santiago, Chile;
Facultad de Ciencias Naturales y Oceanográficas, Universidad de Concepción, Concepción, Chile;
Instituto de Ecología y Biodiversidad, Universidad de Concepción, Concepción, Chile;
Institute of Ecology and Biodiversity, Santiago, Chile

Anna Astorga Roine Centro de Investigaciones en Ecosistemas de La Patagonia (CIEP), Coyhaique, Chile;
Centro Internacional Cabo de Hornos (CHIC), Universidad de Magallanes, Puerto Williams, Chile

José Aylwin Globalization and Human Rights Program, Observatorio Ciudadano, Faculty of Legal and Social Sciences, Universidad Austral de Chile, Temuco, Chile

Alejandro H. Buschmann i-mar Research Center, Universidad de Los Lagos, Puerto Montt, Chile;
Center for Biotechnology and Bioengineering (CeBiB), Universidad de Los Lagos, Puerto Montt, Chile

Nicolás Butorovic Universidad de Magallanes, Punta Arenas, Chile

Alejandra Carmona Prospectiva Local Consultores, Valdivia, Chile;
Centro de Educación Continua, Universidad Austral de Chile, Valdivia, Chile

Leticia Caro Community Grupos Familiares Nómades del Mar, Vina del Mar, Chile

Juan Carlos Castilla Department of Ecology, Faculty of Biological Sciences and Interdisciplinary Center for Global Change. Pontificia Universidad Católica de Chile, Santiago, Chile

Cristián Chiguay Mon Fen de Yaldad Community, Quellón, Chile

Tamara Contador Centro Internacional Cabo de Hornos (CHIC), Universidad de Magallanes, Puerto Williams, Chile

Derek Corcoran Departmento de Ecología, Facultad de Ciencias Biológicas, Pontificia Universidad Católica de Chile, Santiago, Chile;
Centro de Modelamiento Matemático (CMM), Universidad de Chile, International Research Laboratory 2807, CNRS, Santiago, Chile

Cristián Correa Facultad de Ciencias Forestales y Recursos Naturales, Instituto de Conservación, Biodiversidad y Territorio (ICBTe), Universidad Austral de Chile, Valdivia, Chile;
Centro de Humedales Río Cruces (CEHUM), Universidad Austral de Chile, Valdivia, Chile

Paulo Corti Universidad Austral de Chile, Los Ríos, Chile

Erwin Domínguez Instituto de Investigaciones Agropecuarias (INIA), Punta Arenas, Chile

Patricio Andrés Díaz i-Mar Center, Universidad de Los Lagos, Puerto Montt, Chile

Aldo Farías Austral Patagonia Program, Faculty of Economics and Administrative Sciences, Universidad Austral de Chile, Valdivia, Chile

Günter Försterra Facultad de Recursos Naturales, Escuela de Ciencias del Mar, Universidad Católica de Valparaíso, Valparaíso, Chile

Taryn Fuentes-Castillo Wildland Ecobenefit Conservancy, WEConserv Foundation, Santiago, Chile;
Faculty of Forest Sciences and Nature Conservation, University of Chile, Santiago, Chile;
Instituto de Geografía, Facultad de Historia, Geografía y Ciencia Política, Pontificia Universidad Católica de Chile, Santiago, Chile;
Faculty of History, Geography and Political Science, Institute of Geography, Pontificia Universidad Católica de Chile, Santiago, Chile

Magdalena García Department of Geography, University of Montreal, Montréal, Québec, Canada

René Garreaud Departamento de Geofísica, Facultad de Ciencias Físicas y Matemáticas, Universidad de Chile, Santiago, Chile;
Centro de Ciencia del Clima y la Resiliencia (CR)2, Santiago, Chile

Xiomara Gélvez Austral Patagonia Program, Faculty of Economics and Administrative Sciences, Universidad Austral de Chile, Valdivia, Chile

César Guala Austral Patagonia Program, School of Economic and Administrative Sciences, Institute of Tourism, Universidad Austral de Chile, Valdivia, Chile

Felipe Guerra Legal Area, Observatorio Ciudadano, Faculty of Legal and Social Sciences, Universidad Austral de Chile, Temuco, Chile

Vreni Häussermann Escuela de Ingeniería en Gestión de Expediciones y Ecoturismo, Facultad de Ciencias de la Naturaleza, Universidad San Sebastián, Puerto Montt, Chile

Juan Marcos Henríquez Instituto de la Patagonia, Universidad de Magallanes, Punta Arenas, Chile

Jorge Hoyos-Santillán Network for Extreme Environments Research (NEXER), Universidad de Magallanes, Punta Arenas, Chile;
School of Biosciences, University of Nottingham, Loughborough, UK;
Environmental Biogeochemistry in Extreme Ecosystems Laboratory, Universidad de Magallanes, Punta Arenas, Chile;
Center for Climate and Resilience Research (CR)2, Santiago, Chile

Rodrigo Hucke-Gaete Institute of Marine and Limnological Sciences, Universidad Austral de Chile, Valdivia, Chile;
NGO Centro Ballena Azul, Valdivia, Chile

Carolina Huenucoy Kawésqar Community Resident, Puerto Edén, Chile

Jürgen Laudien Alfred-Wegener-Institute, Helmholtz Zentrum für Polar Und Meeresforschung, Am Handelshafen 12, Bremerhaven, Germany

Roy Mackenzie Cape Horn International Center, Universidad de Magallanes, Puerto Williams, Chile;
Programa de Conservación Biocultural Subantártica, Universidad de Magallanes, Puerto Williams, Chile;
Instituto de Ecología y Biodiversidad, Puerto Williams, Chile

Isaí Madriz Fulbright–National Geographic Fellowship, Coyhaique, Chile

Andrés Mansilla Cape Horn International Center (CHIC) and Parque Etnobotánico Omora, Universidad de Magallanes, Puerto Williams, Chile;
Laboratorio de Macroalgas Subantárticas y Antárticas, Universidad de Magallanes, Punta Arenas, Chile

Claudia A. Mansilla GAIA-Antarctic Research Center, Universidad de Magallanes, Punta Arenas, Chile;
Instituto de la Patagonia, Universidad de Magallanes, Punta Arenas, Chile;
Network for Extreme Environments Research (NEXER), Universidad de Magallanes, Punta Arenas, Chile;

Cape Horn International Center, Universidad de Magallanes, Puerto Williams, Chile

Pablo Mansilla Institute of Geography, Pontificia Universidad Católica de Valparaíso, Valparaíso, Chile

Pablo A. Marquet Departmento de Ecología, Facultad de Ciencias Biológicas, Pontificia Universidad Católica de Chile, Santiago, Chile;
Centro de Modelamiento Matemático (CMM), Universidad de Chile, International Research Laboratory 2807, CNRS, Santiago, Chile;
Centro de Cambio Global UC, Pontificia Universidad Católica de Chile, Santiago, Chile

María José Martínez-Harms Institute of Ecology and Biodiversity, Santiago, Chile;
Center for Research and Innovation in Climate Change of Universidad Santo Tomás, Santiago, Chile;
Department of Ecology, Faculty of Biological Sciences, Pontificia Universidad Católica de Chile, Santiago, Chile;
Center for Applied Ecology and Sustainability (CAPES), Universidad Católica de Chile, Santiago, Chile

Francisca Massardo Cape Horn International Center (CHIC) and Parque Etnobotánico Omora, Universidad de Magallanes, Puerto Williams, Chile

Carlos Molinet Instituto de Acuicultura, Universidad Austral de Chile, Puerto Montt, Chile;
Programa de Investigación Pesquera, Universidad Austral de Chile-Universidad de Los Lagos, Puerto Montt, Chile

Paulo Moreno-Meynard Centro de Investigaciones en Ecosistemas de la Patagonia, Coyhaique, Chile

René Muñoz-Arriagada Universidad de Magallanes, Punta Arenas, Chile

Laura Nahuelhual Departament of Social Sciences, Universidad de los Lagos, Osorno, Chile;
Centro de Investigación Dinámica de Ecosistemas Marinos de Altas Latitudes (IDEAL), Valdivia, Chile;
Instituto Milenio en Socioecología Costera (SECOS), Santiago, Chile

Edwin J. Niklitschek Programa de Investigación Pesquera, Universidad Austral de Chile-Universidad de Los Lagos, Puerto Montt, Chile;
Centro i~Mar, Universidad de los Lagos, Puerto Montt, Chile;
i-mar Research Center, Universidad de Los Lagos, Puerto Montt, Chile

David Núñez NGO Poloc, Providencia, Santiago, Chile

Mariela Núñez-Ávila Estación Biológica Senda Darwin, Ancud, Chile; Instituto de Ecología y Biodiversidad, Universidad de Concepción, Concepción, Chile

Sandra V. Pereda i-mar Research Center, Universidad de Los Lagos, Puerto Montt, Chile

María Paz Peña Austral Patagonia Program, Faculty of Economics and Administrative Sciences, Universidad Austral de Chile, Valdivia, Chile; Programa Austral Patagonia, Faculty of Economic and Administrative Sciences, Universidad Austral de Chile, Valdivia, Chile

Patricio Pliscoff Centro de Modelamiento Matemático (CMM), Universidad de Chile, International Research Laboratory 2807, CNRS, Santiago, Chile; Instituto de Geografía, Facultad de Historia, Geografía y Ciencia Política, Pontificia Universidad Católica de Chile, Santiago, Chile; Department of Ecology, Faculty of Biological Sciences, Pontificia Universidad Católica de Chile, Santiago, Chile; Faculty of History, Geography and Political Science, Institute of Geography, Pontificia Universidad Católica de Chile, Santiago, Chile; Center for Applied Ecology and Sustainability (CAPES), Universidad Católica de Chile, Santiago, Chile

Sergio Radic-Schilling Universidad de Magallanes, Punta Arenas, Chile

Brian Reid Centro de Investigaciones en Ecosistemas de La Patagonia (CIEP), Coyhaique, Chile; Centro Internacional Cabo de Hornos (CHIC), Universidad de Magallanes, Puerto Williams, Chile

Andrés Rivera Department of Geography, Universidad de Chile, Santiago, Chile

Paulina Rojas R. University of California, Davis, CA, USA

Sebastián Rosenfeld Cape Horn International Center (CHIC) and Parque Etnobotánico Omora, Universidad de Magallanes, Puerto Williams, Chile; Facultad de Ciencias, Millennium Institute Biodiversity of Antarctic and Subantarctic Ecosystems (BASE), Universidad de Chile, Santiago, Chile; Laboratorio de Macroalgas Subantárticas y Antárticas, Universidad de Magallanes, Punta Arenas, Chile; Estación Biológica Senda Darwin, Ancud, Chile

Ricardo Rozzi Cape Horn International Center (CHIC) and Parque Etnobotánico Omora, Universidad de Magallanes, Puerto Williams, Chile; Departament of Philosophy and Religion and Department of Biological Sciences, University of North Texas, Denton, TX, USA

Alejandro Salazar Instituto de Geografía, Facultad de Historia, Geografía y Ciencia Política, Pontificia Universidad Católica de Chile, Santiago, Chile

Laura Sánchez-Jardón Universidad de Magallanes, Punta Arenas, Chile

Fernanda Sariego Austral Patagonia Program, School of Economic and Administrative Sciences, Institute of Tourism, Universidad Austral de Chile, Valdivia, Chile

Maximiano Sepúlveda The Pew Charitable Trusts, Santiago, Chile

Alejandro Simeone Department of Ecology and Biodiversity, Faculty of Life Sciences, Universidad Andrés Bello, Santiago, Chile

Alberto Tacón Programa Austral Patagonia, Faculty of Economic and Administrative Sciences, Universidad Austral de Chile, Valdivia, Chile;
Cooperativa Calahuala, Servicios para la Conservación, Valdivia, Chile

David Tecklin Austral Patagonia Program, Faculty of Economics and Administrative Sciences, Universidad Austral de Chile, Valdivia, Chile;
Programa Austral Patagonia, Faculty of Economic and Administrative Sciences, Universidad Austral de Chile, Valdivia, Chile

Alejandra Urra Rio Cruces Wetland Centre, Universidad Austral de Chile, Valdivia, Chile

Katerina Veloso Austral Patagonia Program, School of Economic and Administrative Sciences, Institute of Tourism, Universidad Austral de Chile, Valdivia, Chile

Francisco A. Viddi Blue Whale Center, Universidad Austral de Chile, Valdivia, Chile ;
Institute of Marine and Limnological Sciences, Universidad Austral de Chile, Valdivia, Chile;
NGO Centro Ballena Azul, Valdivia, Chile

Rodrigo Villa-Martínez GAIA-Antarctic Research Center, Universidad de Magallanes, Punta Arenas, Chile;
Cape Horn International Center, Universidad de Magallanes, Puerto Williams, Chile

Part I

Synthesis

An Integrated Conservation Vision for Chilean Patagonia

Juan J. Armesto, María José Martínez-Harms,
Juan Carlos Castilla, and Taryn Fuentes-Castillo

Abstract

Chilean Patagonia is a globally outstanding region notable for the current extent of its protected areas, which account for 51% of the terrestrial area and 41% of coastal waters, even if not entirely in terms of effective management. The remoteness of many of its vast landscapes, some of which remain untransformed by humans, the value of its spectacular mountain and island settings for recreation and nature-based tourism, and its highly endemic biota make this region unique for nature protection. The chapters in this book document recent human impacts on Patagonian ecosystems, including the challenges posed by climate change, changing use of sea and land, invasive non-native species, increasing tourist visitation, and expansion of salmon farming. These chapters underscore the critical need of protecting the region's exceptional values, both for regional

J. J. Armesto · J. C. Castilla
Department of Ecology, Faculty of Biological Sciences, P, Universidad Católica de Chile, Santiago, Chile

J. J. Armesto · M. J. Martínez-Harms (✉)
Institute of Ecology and Biodiversity, Santiago, Chile
e-mail: mmartinez-harms@ieb-chile.cl

J. J. Armesto
Faculty of Natural Sciences and Oceanography, Universidad de Concepción, Concepción, Chile

M. J. Martínez-Harms
Center for Research and Innovation in Climate Change of Universidad Santo Tomás, Santiago, Chile

J. C. Castilla
Interdisciplinary Center On Global Change, P. Catholic University of Chile, Santiago, Chile

T. Fuentes-Castillo
Wildland Ecobenefit Conservancy, WEConserv Foundation, Santiago, Chile

Faculty of Forest Sciences and Nature Conservation, University of Chile, Santiago, Chile

© Pontificia Universidad Católica de Chile 2023
J. C. Castilla et al. (eds.), *Conservation in Chilean Patagonia*, Integrated Science 19,
https://doi.org/10.1007/978-3-031-39408-9_1

and global nature conservation efforts. Scientific interest in Chilean Patagonia has increased greatly over the last decades. Through a review of the literature, in this chapter, we discuss the state of knowledge of biodiversity and the conservation status of coastal, marine, and freshwater ecosystems in Chilean Patagonia. We identify important gaps in knowledge of the ancestral history of human occupation, the impact of present socioeconomic systems on Patagonian environments, the biodiversity and characterization of freshwater systems, and the interconnections of land–ocean-human systems. The review of the literature identifies promising avenues to advance in the prevention and mitigation of current and future human impacts on protected areas. It underscores the necessity of interdisciplinary approaches to bolster conservation, from the planning and implementation of marine and terrestrial protected areas to their ongoing management in Chilean Patagonia. Finally, we summarize specific recommendations based on the analysis of each type of ecosystem presented in the chapters of this book and propose overarching policy recommendations that aim to foster a comprehensive, integrated conservation perspective that considers the intricate connections between land, ocean, and human systems throughout the Chilean Patagonia region.

Keywords

Patagonia • Chile • Conservation • Integrated land–ocean-human conservation • Socio-ecological systems

1   Introduction

1.1   Context

Chilean Patagonia extends for approximately 1,600 km along the southwestern margin of South America, from the Reloncaví Sound to the Diego Ramírez Islands (41° 42'S 73° 02'W; 56° 29'S 68° 44'W), occupying a continental territory intensely fragmented by glacial activity and tectonic phenomena that occurred during the Pleistocene (the last 1.5 million years). It is the largest system of estuaries and fjords in the Southern Hemisphere and one of the largest extensions of land-sea area remaining wild in the world. Its total area is 452,204 km^2, including the inland sea and terrestrial landscape. The coastal zone is rugged, with steep gradients between 0 and 3,000 m altitude, with the presence of a relatively shallow inland sea (between 100 and 1,000 m), separated from the Pacific Ocean by island chains [48]. It is in the protected inland sea, in the channels and fjords where extraordinary marine biodiversity is concentrated, as well as the main flow of matter and energy, and where the highest primary productivity has been recorded [19, 24].

The southern tip of South America is an area of climatic contrasts, from hyper-humid conditions on the western margin to semi-arid on the eastern margin; there is spatial contiguity between marine, freshwater, and terrestrial environments in a system of gulfs, fjords, and estuaries, and the most extensive latitudinal continuity of forests and wetlands in the entire Southern Hemisphere (41°–56° S). It is undoubtedly one of the most exceptional landscapes in the world, with its unique scenic beauty [22] and diversity of ecosystems, where numerous remote enclaves that have been scarcely transformed by human activity remain [5, 31, 37, 42, 49, 50, 59].

The persistence of these remote areas is of special scientific interest because they are important reservoirs of pre-industrial ecological processes and constitute enclaves for buffering and counteracting the effects of global change on the planet [14, 33, 72, 73]. The integrity of ecosystem functions in Chilean Patagonia is strengthened by the large land area dedicated to parks and reserves, which cover 51% of the territory, equivalent to 71% of the total area protected in Chile [65]. Official public conservation of the Patagonian marine-coastal systems reaches 41%, including 11 Marine Parks and Marine Reserves, Multiple-Use Marine and Coastal Marine Protected Areas and Nature Sanctuaries, with 11,218 km^2 (6% of the Patagonian marine territory), and the marine-coastal space of 7 National Parks and National Reserves of the National Protected Area System (SNASPE in Spanish), with 63,703 km^2 (35% of the marine territory; [68]). Unfortunately, the recognition of the SNASPE's coastal marine areas by public institutions and their management has been highly variable.

The lack of protection and limited knowledge of the biological and physical characteristics of Patagonian freshwater systems is notable. They are represented in southwestern Patagonia by a diversity of lake basins, among the most transparent and deepest in the world, as well as the largest and most torrential rivers in Patagonia and Chile [55]. The extensive continental ice fields (Fig. 1), the largest outside of Antarctica [56], are important regional and global water reserves, whose flows feed numerous rivers and wetlands. The extensive coastal wetlands dominated by *Sphagnum* moss cover deep strata of soils rich in organic carbon, of high relevance for climate regulation [35]. The region has one of the most continuous (120,000 km^2) and still sparsely modified forest covers [21], which represents important carbon storage that contributes to climate change mitigation [5]. This synthesis argues that understanding and safeguarding the exceptional values of these vast southern ecosystems requires an integrated conservation vision, leading to the protection and management of the marine-terrestrial interface and the local livelihoods of the Chilean Patagonia inhabitants.

1.2 Conservation Vision in Southwestern Patagonia

The contributions in this book have developed a vision of conservation that considers the territory of Chilean Patagonia in a unified manner, covering the region under temperate to cold climatic conditions, from the Reloncaví Sound, *ca.* 41° S, to the Diego Ramírez Islands, *ca.* 56° S. Although several authors have subdivided the region on the basis of topographic and ecological differences, and history of

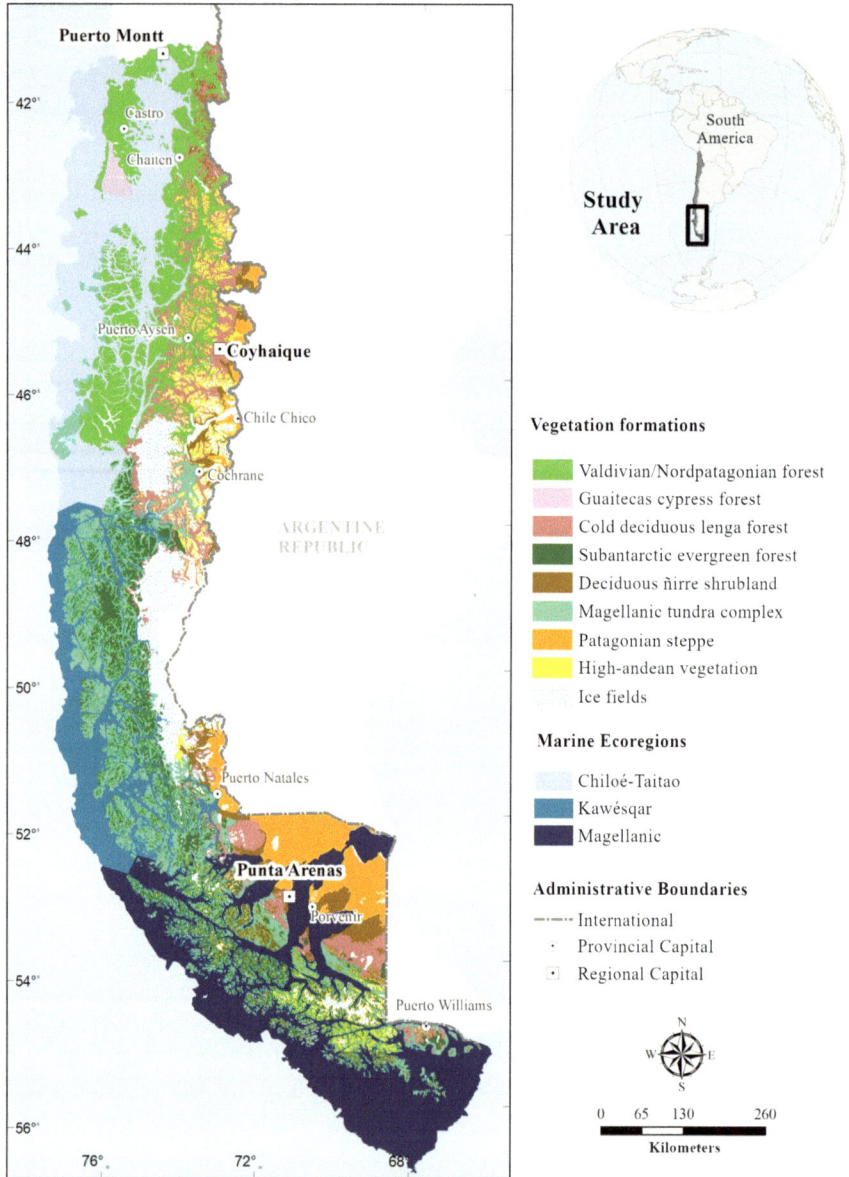

Fig. 1 Marine ecoregions of Chilean Patagonia according to [57] and map of terrestrial plant formations, compiled from various sources, including descriptions and maps published by [3, 34, 39, 51, 69]

human occupation, our integrated vision is based on current and historical processes that cut across all of Chilean Patagonia and that identify it regionally and globally. One of the physical processes that has affected the entire region over millennia is the repeated cycles of glacial advances that fragmented and modeled the territory [56], creating an extensive system of islands, archipelagos, channels, and fjords.

The process of prehistoric settlement and the establishment of diverse cultures of native peoples of navigators, hunter-gatherers, and fishers is also a common element throughout the region [6] that is very different from the advance of European colonization and Chilean settlers from southern Chile in the nineteenth and twentieth centuries. This colonization had devastating effects on native peoples and their cultures throughout the Patagonian territory [6]. The historical patterns of settler migration through the Patagonian region were spontaneous or state-sponsored advances, commonly originating from Chiloé, culturally and socially connecting much of Chilean Patagonia. The environmental impacts of this colonization process were often devastating for the Patagonian territory. Large, forested areas were lost by fires and by the expansion of plagues of rabbits, hares, and other exotic animals, including the widespread impact of domestic livestock and wild animals such as beavers.

There are ocean processes common to the entire region, such as the mixing of fresh and saltwater in the numerous estuaries, in addition to the contributions of meltwater from continental glacial fields, many of which are currently retreating [56]. These processes have generated unique conditions for the fauna of the Patagonian marine territory, which tolerates wide ranges of salinity and nutrients. The study area is subject to the direct influence of westerlies, originated by the atmospheric circulation that prevails in these latitudes, as well as marine currents derived from the circum-Antarctic system, which bifurcates on reaching Patagonia between *ca.* 41°–46° S, and gives rise to the cold Humboldt current, that flows north along the Pacific coastline, and the Cape Horn current, which flows south (for details of seasonal variations see [63]. These ocean–atmosphere interaction systems maintain the hyper-humid condition of Patagonia's western edge, their variation on a millennial scale has influenced the characteristics of glacial and interglacial periods which affected southern South America during the Pleistocene. This climatic pattern also generated the drying of the opposite sector of the continent to the east of the Andes, which produced steppe vegetation, well represented in Argentina and in bordering sectors of Chile [54].

More than half of the region's continental territory is currently incorporated into National Parks and National Reserves, in contrast to the situation in the central region of the country where ecosystem protection is scarce [2, 16, 52, 65]. Chilean Patagonia includes two of the largest terrestrial-marine protected areas (PAs) in the world, the Bernardo O'Higgins National Park, with an area of *ca.* 39,000 km^2, and the Kawésqar National Park and Reserve, of *ca.* 52,000 km^2 [65, 68]. The effective conservation of these vast Patagonian ecosystems, defined as when the conservation actions and strategies implemented contribute to improving the status of biodiversity and ecosystem services, is of global relevance because they are some of the best-preserved systems since the beginning of the industrial era. However, most of these terrestrial and marine ecosystems are now threatened by large-scale anthropogenic processes such as the southward expansion of the salmon farming industry [13], increased tourism, the construction of roads and other infrastructure, and the advance of invasive non-native species. To mitigate these trends of accelerated change [36], it is necessary to strengthen conservation governance, management, and monitoring and enforcement systems, particularly with regard to established PAs [68].

A network of interconnected and effective PAs in Chilean Patagonia would be a conservation strategy conducive to reducing biodiversity losses, increasing ecosystem resilience to industrialization processes, and mitigating and adapting to the effects of climate change. Many drivers of global change originate beyond the boundaries of conservation areas. For example, anthropogenic activities on the continents have important consequences for coastal marine ecosystems [18]. Biogeochemical and ecological connections between terrestrial and marine systems support numerous trophic chains through energy and nutrient flow [1]. The effects of human intervention on the management and disproportionate extraction of resources from coastal terrestrial systems are transmitted through watersheds to the ocean, affecting marine biodiversity [62]. A limitation of the current system of PAs in Chilean Patagonia is that the extensive marine-terrestrial interface adjacent to the PAs has not been fully integrated into conservation design and management, nor has there been an internalization of the close link between terrestrial and marine ecosystems and society in the context of the current Anthropocene [17]. Because of its geographic configuration and history, conservation in Chilean Patagonia requires to incorporate the links between sea, land and society in conservation governance and planning. This is undoubtedly a great challenge that requires coordinating efforts of many actors with the environmental commitments of the region, the country, and the world.

The cross-cutting recommendations proposed at the end of this chapter are intended to implement the following vision of integrated conservation in Patagonia: strengthen the ecosystem protection system and its ecosystem services, integrating land and sea and incorporating the development expectations of local inhabitants and the rights of Indigenous peoples, based on the scientific evidence and traditional knowledge of local communities. We propose adopting an approach that explicitly considers energy flows and ecological connections between marine and terrestrial systems, to identify and analyze threats to design mitigation and adaptation actions to global change.

We recommend here that the overall conservation policy for Patagonian ecosystems be focused on human well-being and the conservation of the livelihoods of its inhabitants, consistent with the theoretical framework proposed by the Intergovernmental Science-Policy Platform on Biodiversity and Ecosystem Services [15, 29], which highlights the interdependency between inhabitants and ecosystems, as well as the need to reconcile the influence and perceptions of diverse knowledge systems and forms of habitation on changes in the natural world. This view is shared with the socio-ecological proposal for the sustainability of the oceans in the next decade (2020–2030), which aims to develop new forms of cooperation based on a multicultural ethic [9] and the newly adopted Kunming-Montreal Global Biodiversity Framework that guides international nature conservation efforts until 2030 [77]. Both visions are consistent with an inclusive approach to conservation that reinforces the link between human society and natural systems.

2   Scope and Objectives

This synthesis is based on the premise that a systematic review of published scientific, socio-ecological, and anthropological studies relevant to the integrated conservation of Chilean Patagonia can help to identify and overcome deficiencies in governance, planning, and management currently carried out by governmental, private, and civil society entities. The chapter has the following purposes: (i) to review, based on the material presented in this book and a systemic analysis of the published scientific evidence on the region [38], the state of knowledge on Chilean Patagonian ecosystems, including terrestrial, marine, freshwater, cryosphere, and sea-land interface connections; (ii) to identify scientific, socio-environmental and global change opportunities and challenges that Chilean Patagonia faces; (iii) to synthesize and highlight the major cross-cutting recommendations (theoretical and practical) that emerge from the chapters of this book and from our vision, both in terms of conservation in action and in relation to public policies.

3   Study Area: Chilean Patagonia and Its Singularities

Chilean Patagonia, with a land area of 148,000 km^2, a marine territory of 183,087 km^2 and 100,627 km of linear coastline and with more than 40,000 islands [27, 68], is a region with its own biophysical, political, and cultural identity, extending across a territory that has major climatic, biotic and ethnic distinctions that have been used to define a diversity of sub-regions, biomes, ecosystems and terrestrial and marine ecoregions.

For this synthesis, the Chilean Patagonian region comprises the area between Reloncaví Sound (41° 42' S, 73° 02' W) and the Diego Ramírez Islands (56° 29' S, 68° 44' W), which are located approximately 100 km southwest of Cape Horn and are the southernmost point of the South American continent. The area includes archipelagos covered by temperate and Subantarctic forests [4, 58], dry steppes in the eastern zone with rain shadow [54], peatlands and other wetlands [35] located mainly in western Patagonia, as well as high Andean vegetation above the tree line (Fig. 1). Large ice fields [56] are also found in the continental area and in Tierra del Fuego, extending to the coast.

3.1   Description

Due to its complex geography and topography, Chilean Patagonia is home to different terrestrial, marine, and freshwater ecosystems and ecoregions, which are very significant areas because they are feeding, reproductive, life cycle development, and migratory routes for a great diversity of organisms [24, 27, 55]. The south-central zone of Chilean Patagonia (47°–55° S) is a refuge for numerous animal and plant species with an endemic gene pool, a large global freshwater reserve [55], and an area that contributes to mitigating global climate change.

Patagonia is bounded tectonically by three oceanic plates (Nazca, South American, and Antarctic) that meet at the so-called Linquiñe-Ofqui fault in front of the Taitao Peninsula (47° S). This fault extends for more than 1000 km along the Andes, generating numerous volcanoes. At the southern end of America, the movement of the Antarctic plate has given the Darwin Cordillera an E-W orientation (Beagle Channel); the western half (higher altitude) has large glacier systems [56] and the eastern half (lower altitude) has forest systems, scrublands, steppes, and peatlands [35, 54]. This heterogeneity of environments harbors a remarkable biodiversity of terrestrial and aquatic organisms, as documented in the chapters of this book [4, 24, 27, 55, 58].

The southern tip of South America, where the continent narrows with latitude, is the most ice-free land mass in the Southern Ocean, extending 22° farther south than the southern tip of Africa, 14° farther south than Tasmania, and 9–10° farther south than the southern tip of New Zealand. It is a unique and formidable natural obstacle to westerly drifting wind systems and the Antarctic Circumpolar Current, which move from west to east, affecting oceanographic systems, wind circulation, and climate. It extends South American terrestrial ecosystems to latitudes with no equivalent in other continents of the Southern Hemisphere [59].

The continental coastline of Chilean Patagonia has been fragmented and modeled for millennia by glacial advances and retreats [56]. The terrestrial and marine landscapes are the product, on one hand of the subsidence of the Central Valley of Chile at Reloncaví Sound (Puerto Montt) and, on the other, of the powerful erosive forces of the glaciers, which covered the area during the entire Pleistocene, until *ca.* 15,000 years ago [70]. These effects have produced an irregular and fissured coastline with numerous channels, straits, fjords, sounds, estuaries, and islands extending between 41° and 56° S.

The Andes divides Patagonia between the eastern slope, with extensive, relatively dry plains, and the much narrower western slope, with steep slopes, estuaries, and coastal wetlands. The slopes rise up to 4000 m in altitude on Mount San Valentín, to 3600 m on Mount Murallón, and 3400 m on Mount Fitz Roy, where there are large permanent ice fields between Aysén and Puerto Natales and in the Darwin Range, with projections that flow into lakes or directly into Patagonian fjords. The main rivers have a torrential snow-pluvial regime and short hydrographic basins with high flow [55, 74]. Due to the barrier effect of the Andes and the elevation of the Patagonian mountain ranges, on its western slope Chilean

Patagonia has rainfall that can reach over >6000 mm per year [35, 36]. The circulation dynamics of the fjords are influenced by rivers and freshwater runoff. Horizontal circulation of surface water (<30 m, with low salinity) occurs from the interior of the fjords towards the mouth of the gulfs and the ocean, while salty sub-surface water masses enter through the mouths of the gulfs, due to strong westerly winds and large tides, producing mixing processes (Sobarzo, 2009) [49, 50]. However, knowledge of oceanographic processes in Chilean Patagonia is still incipient [20, 31]. Pickard and Stanton [50] described the existence of three zones in the Chilean Patagonian maritime territory (approximate latitudes): (i) northern Patagonia, 41°–47° S, (ii) central Patagonia, 47°–53° 30' S; (iii) southern Patagonia, 53° 30'–56° S. [57], based on a comprehensive literature review, propose to distinguish three ecoregions in Chilean Patagonia: (i) Chiloé-Taitao, 42°–47° S, (ii) Kawésqar, 47°–54° S; and (iii) Magallanes, 54°–56° S (Fig. 1). This classification of three Patagonian marine ecoregions is used by different authors (including some in this book) as equivalent to Patagonian marine biophysical macrozones or macro-sectors, calling them northern, central, and southern Patagonia; with boundaries similar to those used by [57] for ecoregions (see [24, 27, 43, 68]. Previously, [61, 64] had proposed the recognition of only two major marine ecoregions for Chilean Patagonia: (i) Chiloense, 41°–47° S, and (ii) Channels and Fjords of Southern Chile, 47°–56° S. In this book different authors use these terminologies to distinguish ecoregions and/or macro-geographic zones, in each case providing new biological/ecological background.

4   Conservation Based on Scientific Evidence in Chilean Patagonia

For informed conservation decision-making in Chilean Patagonia, it is essential to compile and synthesize evidence on biodiversity distribution, ecological processes, and knowledge of the human dimensions of the most pressing environmental problems [59]. This is because the main causes of ecosystem degradation, from climate change, invasive species, habitat loss, overfishing, and salmon farming [4, 13, 36, 43, 55, 58], are most evident at the interface between coastal fjord, channel, and inland sea ecosystems and human communities. Addressing these problems requires an interdisciplinary framework supported by different sources of information. Of particular importance are disciplines such as ecology, conservation, fisheries, economics, political science, environmental law, geography, anthropology, and psychology to understand fully the diversity of people's relationships with nature, especially those based on traditional and local knowledge [11, 59, 66].

Panoramic view over Puerto Aguirre and Huichas Islands, Aysén Region. Photograph by Javier Godoy

To contribute to integrated conservation in Chilean Patagonia, we analyzed and synthesized the evidence available in the literature on the region. To do so, we compiled and analyzed published studies applying a systematic mapping approach [32], which is defined as a reliable synthesis of the quantity and quality of evidence in relation to a research question of broad relevance [23]. We investigated the state of knowledge on conservation and management of Chilean Patagonian ecosystems. This process facilitated describing and cataloging the evidence available in published regional conservation studies, covering the breadth of science needed to address questions that impact public policy. The team for this study was led by two senior experts, who were supported by a technical secretariat, who had the role of systematically collecting, compiling, and cataloguing the evidence, using the systematic mapping method [38]. A national scientific panel made up of an interdisciplinary group of 8 experts supervised the thematic and geographic review of the region.

The 17 other chapters of this book analyze the marine, terrestrial, and freshwater biodiversity of Chilean Patagonia, the accelerating pressures of global and local changes on ecosystems, the impacts of aquaculture and fisheries, the interrelationship of the land-sea interface, the conservation of glaciers, peatlands, steppes, and primary forests, Indigenous-led conservation, evidence-based conservation, as well as the management of protected areas and socioeconomic trends in the region. The chapters compile available information from the literature, critically review key conservation issues, and formulate specific recommendations for integrated management of Patagonian conservation. Evidence was coded with semantic analysis using R Bibliometrix software and each publication was classified into each of the five study systems: (i) terrestrial; (ii) marine; (iii) freshwater; (iv) social; (v) other [40]. We considered the five direct drivers of change in biodiversity and ecosystem services identified by the Millennium Ecosystem Assessment [29, 41]: i climate change, ii habitat change; iii invasive species; iv overexploitation; v pollution [40].

The publications were grouped into one or several drivers; to validate their classification, we manually inspected the classification of the articles (n = 986) for each ecological system. One hundred percent of them were classified in one of the five study systems and 56% of the articles were classified by their focus on one or more of the drivers of change.

4.1   Time Trend

We compiled a database on Chilean Patagonia that clearly documents an increase in the number of publications over the last decade [38]. Most of the compiled publications refer to the Terrestrial and marine ecological systems (Fig. 2). The systematic map showed an exponential increase in evidence during the last 10 years, distributed in marine systems (325 articles; 33%), terrestrial systems (282; 29%), social systems (205; 21%), freshwater (148; 15%), and others (26; 3%). A growing number of recent publications include social variables and human dimensions of conservation (Fig. 2).

4.2   Distribution of Evidence by Drivers of Change

The classification by global change drivers (Fig. 3) showed that more studies have focused on climate change (191 studies; 19%), followed by studies of invasive species (131; 13%), especially addressing the impacts of salmon farming and beavers; followed by studies of pollution (102; 10%), habitat changes (79; 8%) and overexploitation of marine and terrestrial resources (53; 5%). The publications on terrestrial systems are mainly concerned with climate change (56; 6%), invasive species (49; 5%) and habitat changes (34; 3.4%).

4.3   Spatial Distribution

Georeferencing the studies in the database publications (2,059 sampling sites for 986 records) found that 72% of the sites analyzed are in the terrestrial system and only 28% in the marine system (Fig. 2). Coding the compiled evidence by drivers of change (Fig. 3) by their spatial distribution in Chilean Patagonia in the 11 current administrative provinces and the three marine ecoregions [57], we obtained the results described below (Fig. 4).

Climate change has been the most studied topic in the provinces of Última Esperanza (108 sites; 5%) and Capitán Prat (101 sites; 5%) where the ice fields

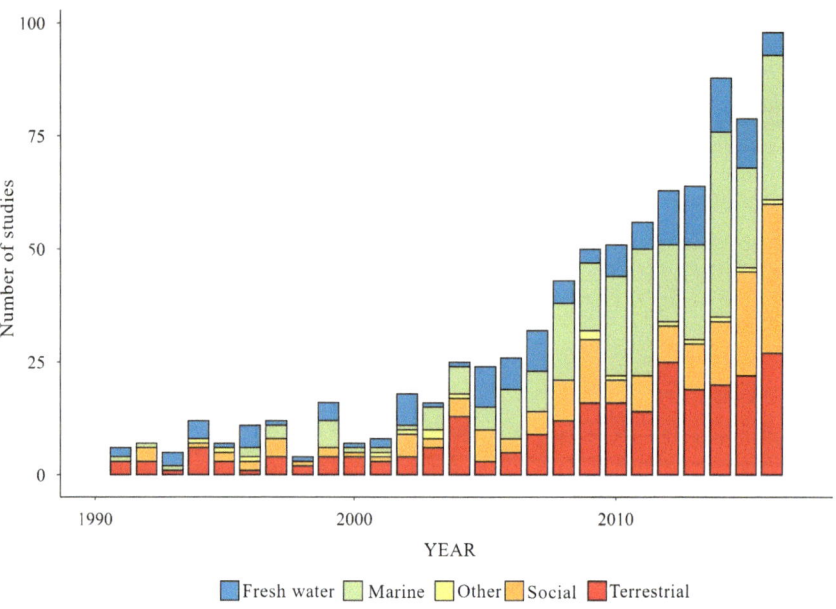

Fig. 2 Cumulative number of studies on the Chilean Patagonia region (published between 1980–2017) classified by study system. The X axis represents the years, and the Y axis represents the number of publications per year. The relatively low number of papers on freshwater systems during the period stands out, in addition to the growth in recent decades in the number of papers that include social variables

are located, with studies documenting glacial retreat. The largest number of publications on invasive species is concentrated in Coyhaique (115 sites; 6%), Palena (60 sites; 3%), and Aysén (53 sites; 3%). The effects of pollution have been little studied in the terrestrial provinces due to their relatively low impact in the region; however, the largest number of studies is concentrated in Llanquihue, with 33 sites. Habitat change and overexploitation of natural resources have also been little studied in the literature referring to terrestrial systems (Fig. 4).

Climate change has been addressed most frequently in Patagonian marine ecoregions [57] in the Chiloé-Taitao ecoregion (42 sites), followed by the Kawésqar ecoregion (37 sites), and Magallanes (24 sites). Invasive species studies are equally concentrated in Chiloé-Taitao (30 sites) and Magallanes (30 sites), with a smaller number in the Kawésqar ecoregion (10 sites). Industrial pollution has been addressed almost entirely in the Chiloé-Taitao ecoregion (60 sites), with very few studies in the Kawésqar and Magallanes ecoregions. Habitat change has been scarcely addressed in the different marine ecoregions, while studies of resource overexploitation have been concentrated mainly in the Chiloé-Taitao ecoregion (41 sites), with few studies in Kawésqar and Magallanes. After spatializing the study sites from the database records (2,059 sampling sites for 986 records), we overlaid the map of terrestrial (Tacon et al., 2023) and marine [68] protected areas with the

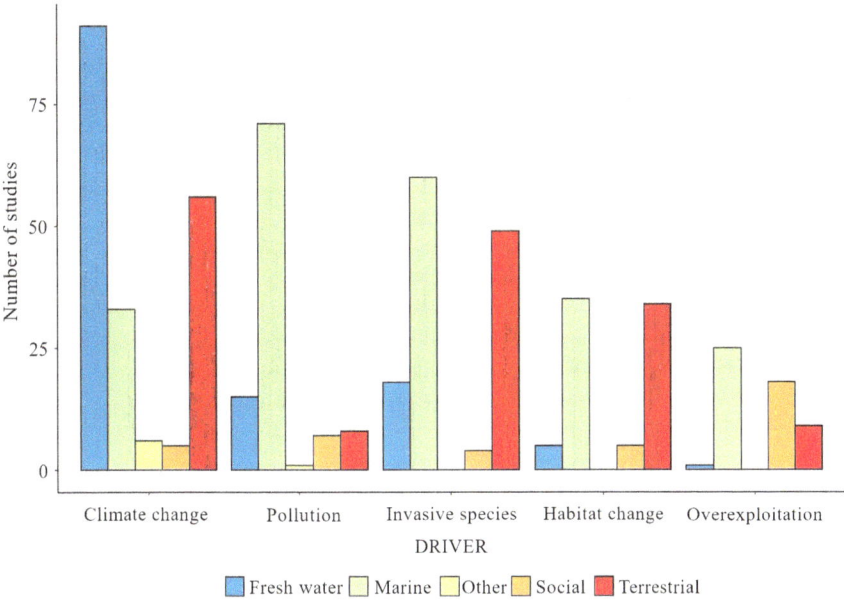

Fig. 3 Distribution of the number of publications by direct drivers of change in the different ecosystems (X-axis) and by ecosystem of interest (bar colors). The Y-axis represents the numbers of publications accumulated over the period 1980–2017

evidence map for the Chilean Patagonia region. We found that less than 27% of the evidence is based on information collected within protected areas. Most studies concentrated only on the three largest national parks: Bernardo O'Higgins, Laguna San Rafael, and Torres del Paine, revealing that a substantial fraction of this region remains poorly explored.

5   Opportunities and Recommendations for the Conservation of Chilean Patagonia

The holistic and inclusive approach to managing terrestrial, marine, freshwater, and sea-land interface environments represents a unique opportunity to promote a distinctive process of land use in Chilean Patagonia, setting it apart from the rest of Chile. The environmental liabilities left by the extractive development model in other regions of Chile could be avoided in Chilean Patagonia with the promotion of a new proposal for integrated conservation of the sea, land, and society [18, 44].

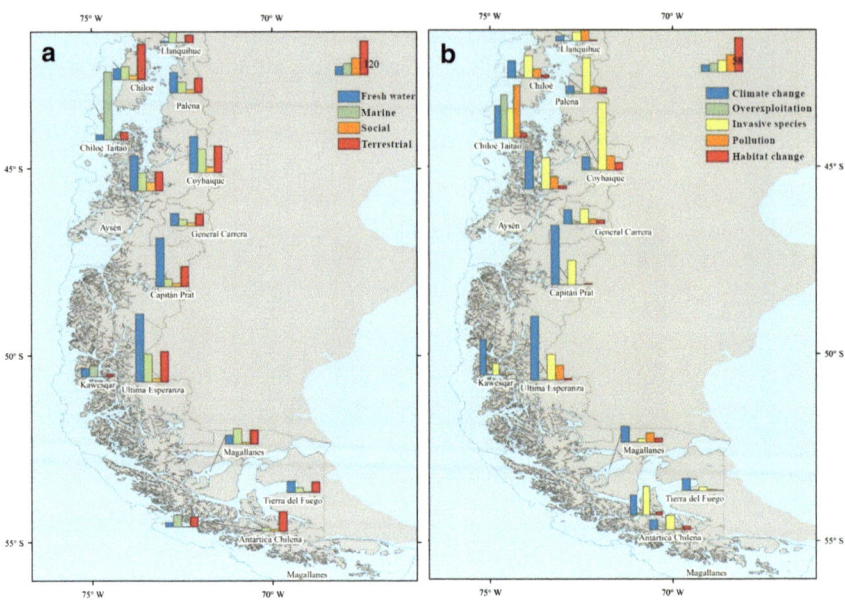

Fig. 4 Spatial distribution of the number of publications by administrative province and by marine ecoregion [57] of Chilean Patagonia for evidence coded by study system (**a**) and by direct change forcing (**b**)

5.1 Bases for an Integrated System of Protected Areas in Chilean Patagonia

Chilean Patagonia has National Parks, National Reserves, and Natural Monuments (SNASPE), many of which are adjacent to coastal systems, and which cover approximately half of the land area [53, 65, 68]. This conservation platform represents a unique opportunity for Chile, and unusual worldwide, to integrate the conservation of large terrestrial and marine ecosystems. The numerous terrestrial and marine protected areas [53, 65, 68] have low levels of implementation, particularly the marine ones, and a total absence in freshwater systems, such deficits generally include the absence of well-developed management plans, insufficient monitoring, and limited financial and human resources to achieve real protection. To advance and overcome the current conservation situation in Chilean Patagonia, in the face of fragmented and competitive research and conservation schemes, in this chapter (see below we consider of high relevance the creation of an Interdisciplinary Center for Conservation in Chilean Patagonia, with a public–private orientation, and including the development of incentives for binational Chile-Argentina collaboration.

The development of mechanisms to achieve an adequate balance between the protection of marine and terrestrial systems in Chilean Patagonia is both an opportunity and an important challenge that can also contribute to the global conservation proposal for the future of the oceans [33]. An integrated and inclusive sea-land-society conservation vision will allow anticipating new and growing challenges, such as the expansion of aquaculture, new mining interest in the ocean floor, the development of coastal renewable energy projects, the regional expansion of tourism, and the expansion of use rights and productive activities in coastal marine territories. These challenges are present today in Chilean Patagonia and call for strengthening an integrated regional conservation system that goes beyond the current situation of parks and reserves, mostly on paper without real support [68].

Tacon [65] note that the SNASPE of the Chilean Patagonian region establish legal protection for ca. 83% of the surface area of snow and glaciers (29,784 km^2), 40% of the surface area of native forests (36,168 km^2) and scrublands, and 68% of the surface area of peatlands (22,042 km^2). These proportions of the different environments suggest that there are still relevant conservation challenges, especially in less intervened areas, which have been identified both inside and outside PAs [4, 53, 55]. An important omission from public conservation is the vast areas of wetlands and peatlands, which are particularly fragile to the impacts of climate change and human intervention [35].

Most of the studies of ecosystem services or benefits to society conducted in Chile focus on inland waters, but there is little information on the value of freshwater ecosystems in the Patagonian region [55]. The few studies on the subject in Southwest Patagonia come from the Aysén basin and Isla Navarino/Tierra del Fuego [76]. To understand the relationship between water provision for humans and well-preserved forests better, a network of weather stations in headwater streams (which also provide drinking water to many rural communities) needs to be supplemented, distributed along the bioclimatic gradient of western Patagonia. Along with flow monitoring, it is important to protect these headwater streams with some legal designation that avoids their mismanagement and degradation [4].

Various private conservation initiatives (PCIs) have contributed to improving the representation, coverage, and connectivity between terrestrial and aquatic ecosystems in Chilean Patagonia. As of 2014, 47 PCIs were identified between the regions of Los Lagos and Magallanes, covering an approximate area of 9,640 km^2, equivalent to ca. 57% of the total nationally [47]. Some of the largest PCIs in the country have been established in western Patagonia, Tantauco Park in insular Chiloé in 2003, with 1,180 km^2 and the Karukinka Nature Reserve in Tierra del Fuego in 2004, with 2,700 km^2. Even though the PCIs have been de facto consolidating as a complementary category of national conservation, these territories still remain in an uncertain official position [67], as progress in the matter has been slow. Only in 2020 was there a formal proposal for standards for private conservation in Chile [75].

Another type of Patagonian conservation area that has attempted to integrate research, education, and social participation in management is the Biosphere Reserves (BR). The first two BRs in Chilean Patagonia were declared in 1978: Torres del Paine and Laguna San Rafael. The BRs declared after 2000 were expanded to follow the logic proposed by UNESCO to constitute conservation landscapes, including core areas, buffer zones, biological corridors, and natural resource management areas, with scientific support in decision making, and citizen participation in management. However, in most of Chile's BRs these proposals have not materialized and their real insertion in the national system of protected areas is not clear.

In our opinion, BRs could become management models for the entire Patagonian region under the sea-land-society integrated conservation paradigm if their theoretical objectives are fulfilled, because they focus on landscapes with highly complex environments and uses and because their inhabitants play a central role. The adoption of this BR model to connect the northern and southern ends of the Chilean Patagonian region has been considered [59], where the area currently dedicated to public and private conservation reaches the highest relative proportion in the country and where it is a priority to extend conservation from the coastal edge towards the oceans that make up the archipelagos and channels environment. This scheme should consider that all PAs in Chilean Patagonia are home to tourism activities and enterprises that promote the development of neighboring towns and communities [22, 44].

5.2 Summary of the Main Cross-Cutting Recommendations for the Conservation of Chilean Patagonia

The following is a synthesis of the main regional cross-cutting recommendations for Chilean Patagonia at different levels of analysis, from the most urgent and general to those that are more specific or require more gradual implementation. The specific recommendations by Patagonian ecosystems are detailed in each of the 17 chapters of the book:

- **Recommendations to prevent losses of biodiversity and ecosystem services.**
  It is urgent to complete the assessment of the state of conservation in Chilean Patagonia and to clearly specify the baselines with respect to threats, opportunities, challenges, and connectivity priorities of the different Patagonian ecosystems. The most urgent actions are: (i) protection of threatened biodiversity in freshwater systems; (ii) concrete measures to prevent the impact of massive salmon farming in the region, especially in Magallanes; (iii) limiting and regulating the impact of intensive tourism in remote areas; (iv) reducing the risk from wildfires. We propose an integrative analysis of the structures and dynamics of all Patagonian socio-ecosystems in order to connect land-sea-society interactions better with the protection of ecosystem services. Such an assessment has not been used for the design and selection of PAs in Chile. Its implementation is an urgent need In Chilean Patagonia, due to its intricate geography, singularities, and multiple productive activities [44].

  A latent threat to the coastal zone of Chilean Patagonia is the rapid advance of the salmon farming industry south of the Chiloé-Taitao ecoregion, with a growing number of concessions in channel zones, archipelagos, and fjords. It is urgent to legislate and implement a system of environmental liability that regulates and penalizes environmental damage caused by massive salmon escapes from net pens, applying preventive measures and technologies [13]. This model will need to internalize the environmental costs of nutrient discharges and implement mitigation measures, such as integrated farming with algae and/or filter feeders. Under a precautionary approach, it would be reasonable to freeze the advance of salmon farming in Magallanes until mitigation measures and a regulatory system are in place to prevent the impacts of growth in biomass and nutrient discharges.

- **Recommendations for sea-land-society conservation planning.** It is urgent to promote integrated planning and management of marine-terrestrial-freshwater ecosystems, to optimize conservation efforts and transfer the capacities already installed in Chilean Patagonia. One of the problems of the current PA system is that the analysis of costs and conservation actions related to sea-land interfaces have not been integrated into management or conservation priorities [8]. The current model of marine and terrestrial reserves usually assumes that each site is an independent ecological system. Human intervention in the management and extraction of resources from coastal terrestrial systems (forests and wetlands) can severely alter watersheds connected to the ocean, affecting marine biodiversity [1, 58]. It is important that the recognition of connectivity between water, forests, and soils be integrated into the watershed concept, as an instrument for public policies and/or conservation planning and regional land use changes. It is recommended that a system of incentives be designed for landowners to conserve the most pristine or valuable areas, along with improved management. The old-growth forests of Chilean Patagonia occupy important watersheds. Consideration of the interactions (water, matter, and energy flows) between aquatic and terrestrial ecosystems and their focus on conservation plans could make a difference with the rest of Chile. We propose to bring to the forefront the need for monitoring and recognition of the heritage value of the fraction of intact forests that protect the headwaters of the region's watersheds [4, 55].

Given the speed of tourism growth, there is an urgent need to link the management of PAs with regional development planning [22, 44]. This can occur in the processes of generating the Regional Land Use Plans and the Coastline Uses Zoning among other instruments. The growth of the tourism sector could generate adverse effects on the environment and some forms of biodiversity, by increasing the consumption of resources, production of waste, construction of roads that accelerate the introduction and propagation of species, and the introduction of new exotic species and increased probability of forest fires in remote areas [10]. It is therefore important to advance in planning the management of human activities in and around PAs, identifying areas with diverse opportunities to promote human development through the conservation of biodiversity and ecosystem services [28]. The National Forestry Corporation (administrator of the SNASPE until today) and agencies that in the future will safeguard Chile's environmental heritage, such as the Biodiversity and Protected Areas

Service, must be invited to participate in the entities that regulate the use of coastal zones, including the Regional Coastal Uses Commissions.

- **Financial consolidation of the protected areas system.** Considering that terrestrial, marine-terrestrial, and marine PAs are one of the fundamental instruments for the conservation of biodiversity and ecosystem services, and that Chilean Patagonia has a high coverage of PAs with low levels of management and public investment, a public and private investment plan is recommended to establish the foundations for coordinated management among all Patagonian conservation units. It is a priority to move towards stable, long-term financing of the PA system in Chilean Patagonia that creates incentives for its evaluation and continuous improvement [65, 68]. Annual budgets should ensure a minimum floor for the protection of all areas, thus reducing pressure for local tourism revenues channeled to conservation. Funding mechanisms for Patagonian conservation should consider the large gap that exists between the extensive protected area and the magnitude of the investment made by the country, and the corresponding payment for the services that protected nature provides to the inhabitants and to global sustainability. The explicit incorporation of regional academic centers, Indigenous communities, and citizen science groups in management and conservation tasks is essential. It is also necessary to design public and public–private financing and technical support mechanisms for all marine protection categories, including conventional marine protected areas those within the SNASPE, the Indigenous Peoples' Coastal Marine Spaces, and the Benthic Resource Management and Exploitation Areas (ECMPO and AMERB respectively in Spanish).

- **Consolidation of an integrated network of effectively protected marine conservation areas in Chilean Patagonia.** The unique archipelagic character of Chilean Patagonia represents a challenge for the protection and conservation of marine territory and land-marine-freshwater interfaces, which differs from the management formulas in most of Chile's PAs. Marine conservation still has serious deficits in Chilean Patagonia in terms of management plans, monitoring, follow-up, oversight, financing, and communication with the public. Currently, the vast majority of these marine areas are only on paper and some of them still coexist with aquaculture activities in their interior or are open to artisanal and medium-scale fisheries. There is an urgent need to move to proactive conservation action, considering the global environmental changes facing Chilean Patagonia [36]. New methodological approaches are also required, such as the development of high-resolution maps (marine and terrestrial) of carbon storage, together with maps of biodiversity and ecosystem services to identify and protect areas with the greatest co-benefits [60]. It is urgent to study and propose a network of marine conservation areas that is more comprehensive and representative than the current one, which on one hand differentiates and on the other unites northern, central, and southern Patagonia. The future network of Patagonian marine conservation areas should cover and adequately represent the different ecosystems, and also be compatible with current and future productive activities, considering the aspirations and rights of Indigenous peoples.

It is a priority to incorporate into management plans the coastal-marine portion of each of the units of the SNASPE that contain legally recognized areas within their perimeters [68]. In addition, it is proposed to develop a protocol and legal procedures to recognize the ECMPOs s and AMERBS as marine protected areas when their owners so request, as well as to generate a system of state support for such management, including the preparation and implementation of management and administration plans.

- **Design and implementation of standardized systems for monitoring biodiversity and the conservation status of ecosystems and their ecosystem services.** In conjunction with PA managers, we propose to design and implement a long-term, low-cost, monitoring system, with a minimum network of 40 monitoring sites distributed throughout the different terrestrial ecosystems, freshwater environments, the marine environment, and the cryosphere. The diverse users of biodiversity and ecosystems should be incorporated into these environmental monitoring activities, especially those related to aquaculture, fishing, tourism, transportation, resource extraction, and mining [44]. Long-term monitoring systems for the state of the most fragile or valuable ecosystems (e.g. peatlands, intact forests) are essential because projects such as the construction of the Chacao Bridge, the paving of the Austral Highway, the road that will connect the mainland with Yendegaia Bay, as well as mining exploration and exploitation permits, entail serious uncertainties for the future conservation of these ecosystems.
- **Conservation policies inspired by capacity building.** It is essential to generate policies and mechanisms for integrating local communities and visitors in Chilean Patagonia through a program of information, training, integration, and co-responsibility for effective and sustained conservation [38, 45]. There is an urgent need to increase the incorporation of local communities in the planning, management, implementation, and care of the PAs. Due to the scarcity of resources and the training needs, safeguarding natural heritage with personnel from parks and other PAs alone is unlikely in the short- and medium-term. Management must be reformulated, with training programs, and with funding for a horizon of a decade, encouraging coordinated citizen participation (citizen science, see [25]. Capacity-building should be based in particular on a deep knowledge of the value of the territory, natural and cultural heritages, and how these contribute to the sustainable growth of local economies, human well-being and the sustainability of the biosphere. These policies should enhance knowledge by integrating citizens from all backgrounds and promote the unification of values and behavior, under an ethic of collective and responsible socio-environmental stewardship of ecosystems and their resources [12, 45, 46].
- **Incentives for Chile-Argentina binational collaboration in the conservation of Patagonia.** Patagonia as a whole, with its eastern and western slopes, stands out globally for its numerous remote environments subject to reduced anthropogenic impacts [33]. This land and marine territory is currently exposed to different forces of accelerated global change (climate, oceanographic, fisheries, aquaculture, invasions of exotic species, tourism, overfishing), which may

affect its eastern and western sides differently. Collaboration between Chilean and Argentine academic and governmental entities, Indigenous peoples and NGOs in Patagonia is key to generate and disseminate new knowledge, promote ecosystem monitoring, and motivate joint conservation actions. As an example, in 2018 the establishment of one of the largest oceanic conservation areas was decreed in the extreme south of Chile: the Islas Diego Ramirez-Paso Drake Marine Park, with $140,200$ km^2, which is complemented by the Cabo de Hornos Biosphere Reserve ($48,000$ km^2). Both areas are adjacent to the Yaganes Marine Park in Argentina, with $68,843$ km^2. In this particular case, coordinated Chile-Argentina conservation management is indispensable.

- **Creation and funding of an Interdisciplinary Center for Conservation in Chilean Patagonia.** One of the most important needs in Patagonian conservation research and planning is to increase interdisciplinary knowledge about the main ecosystems, their conservation needs, and their relationships with human well-being. To break with fragmented and competitive research and conservation schemes, we propose to support interdisciplinary research that integrates science, society, and traditional ancestral knowledge, thus establishing a bridge between state actors, regional academic centers, native peoples, private entities, and NGOs. We recommend the creation of an "Interdisciplinary Center for the Conservation of Chilean Patagonia" with its own staff and infrastructure, complemented by the collaboration of institutions and other regional research centers in the natural, social, and humanistic sciences. This center should generate its own lines of research and strengthen the links between the sciences developed by different entities located in Chilean Patagonia. One of its important lines of action would be to implement the vision of Patagonian sea-land-society conservation developed in this chapter. The objectives of the center should be oriented to basic and applied research with high standards, publications in national and international journals, review of management plans, systematic monitoring in and outside PAs, professional training of park rangers, the educational system, and the implementation of citizen science programs [25].

- **Support Indigenous leadership in the conservation of Chilean Patagonia and encourage intercultural dialogue.** The conservation of Chilean Patagonia requires more inclusive and participatory forms of governance in PAs, as well as mechanisms for the recognition of the collective rights of Indigenous people over ancestral territories and territories. It is recommended that new policies and management capacities be developed in public services to facilitate these ends. These should include the establishment of modalities for the use and governance of PAs to support the survival of Indigenous peoples' ways of life and cultures, including mechanisms for sharing the benefits of economic activities in PAs. Indigenous conservation territories and areas conserved by local communities under the governance of Indigenous peoples should also be identified and recognized [6]. Dialogue should be promoted in each PA, bearing in mind the IUCN guidelines and recommendations regarding types of governance and the rights of Indigenous peoples over their lands and territories. We recommend the creation of an official mechanism to provide public follow-up and decisive support to these processes, when appropriate, and to propose specific forms of restitution of rights and new forms of shared management for conservation. We also recommend that the processes of creating new PAs include a review of their possible overlap with Indigenous territories and rights, thus avoiding the violation of ancestral rights.

It is important that the competent public bodies and private entities that manage protected areas in Chilean Patagonia consider the potential of ECMPO as conservation initiatives for Indigenous peoples in their marine coastal spaces, often adjacent to public PAs. ECMPOs assign access and management rights over marine areas to Indigenous communities in order to maintain the traditions and use of natural resources by communities linked to the coast (Tecklin et al., 2020) [26]. Although they are not currently recognized asmarine protected areas, the law establishes that ECMPOs must ensure the conservation of the natural resources within them. Therefore, it is recommended to: (i) advance in the study of the potential role of ECMPOs in biocultural conservation, including the analysis of political and legal obstacles [26] to accelerate the processes of processing ECMPO request within the stipulations of the law, (ii) provide support from CONADI and the Undersecretariat of Fisheries and Aquaculture to the communities and the ECMPO application processes for purposes compatible with conservation; (iii) provide support and strengthen the capacities of the communities for the collective governance of ECMPOs and the conservation and sustainable use of their resources.

- **Strengthening public policies and governance of the system of protected areas.** Effective and long-lasting conservation in Chilean Patagonia will only be possible under a governance system that ensures a continuous link between decision-makers, local communities, and scientific entities, both for resolutions based on scientific evidence and to promote capacity building. This process must ensure efficient participatory processes, based on principles of justice and equity [38] in which citizens can demand accountability.

The cross-cutting recommendation for conservation in Chilean Patagonia is that the current governance system be modified to produce adaptive and flexible processes in the face of new legislation. The Law on Biodiversity and Protected Areas Service represents a major opportunity for biodiversity management to include lessons learned and incorporate local knowledge and that of ancestral communities and their descendants [6, 68]. At present, the SNASPE operates under a dispersed, disjointed, and incomplete institutional framework, where the protection of terrestrial and marine ecosystems is weakly integrated and that of freshwater systems does not exist. Meanwhile, the Biodiversity and Protected Areas Service Law has been under legislative debate for more than a decade. It is urgent that this bill be approved in order to strengthen large-scale conservation in regions with such extensive PAs as those of Patagonia.

6   Conclusions

This chapter presents the results of a systematic review of the state of knowledge of conservation in Chilean Patagonia [38], and identifies knowledge gaps and areas where policy needs to be strengthened to safeguard Chilean Patagonian biodiversity, as well as identifying opportunities, challenges, and needs for action for effective land-sea conservation and its effects on human well-being.

The systematic mapping approach identified the main calls to action. For example, greater emphasis is needed on the study of freshwater systems and social systems in Chilean Patagonia, greater effort in the knowledge about management, and control of established PAs, to make a quantum leap from conservation on paper to integrated action. A positive aspect is the temporal trend evidencing an increase of studies that connect social and ecological aspects. The synthesis of research in Chilean Patagonia shows that the direct drivers of change, on which the evidence is concentrated, document greater concern for climate change and invasive species, including aquaculture of non-native species such as salmonids. Research lines such as the effects of pollution and species habitat loss are underrepresented in Patagonian studies. Research priorities on direct drivers of change differ among ecosystems. For example, for invasive non-native species, there is a similar number of studies in terrestrial and marine systems; however, research on the impacts of overexploitation of natural resources is concentrated mainly in marine systems, to the detriment of the terrestrial environment.

The issue that has generated the most recent concern in Chilean Patagonia is the expansion of industrial aquaculture activities, particularly salmonids, whose impact has been increasing, especially in the fjord and channel systems, many of which are relatively intact and little known. Another current issue is the recurrence of harmful algal bloom events. Although studies of terrestrial ecosystems and PAs predominate in number, those related to the state of conservation of marine systems have increased, especially along the coast and at the sea-land interface. Among the recommendations necessary for Chilean Patagonian ecosystems are the need for a more in-depth evaluation of the knowledge of some groups of organisms (particularly freshwater organisms) and to evaluate the impacts of recent anthropogenic alterations.

The region presents a globally unique opportunity for integrated land-sea-society conservation at a scale consistent with the most ambitious goals being discussed in the newly adopted international biodiversity conservation agreements that would help mitigate the effects of global warming and other global changes. The conservation platform provided by the regional SNASPE, which covers a large terrestrial and marine area, represents a unique opportunity in Chile, and unusual worldwide, to integrate the conservation of large areas covering important marine, terrestrial, and freshwater ecosystems.

An innovative and proactive recommendation for the Patagonian region's marine territory is to allocate permanent financial resources (public and private) to manage and conserve the existing and future system of public PAs and its complementation with auxiliary management-conservation areas managed by local communities, such as ECMPOs and AMERBs. The public system of Patagonian marine-terrestrial conservation should also be increased and integrated with private conservation initiatives.

Acknowledgements We are grateful to David Tecklin, who helped us update data and figures for this synthesis. We thank Aldo Farías for the preparation of the maps. JJA is grateful for the support of ANID (Chile) through project FB210006. JCC thanks the Faculty of Biological Sciences, the Interdisciplinary Center for Global Change and the Chair in Environmental Ethics, Catholic University of Chile. MJMH gratefully acknowledges the support of ANID (Chile) through Project No. 11201053.

References

1. Álvarez-Romero, J. G., Pressey, R. L., Ban, N. C., Vance-Borland, K., Willer, C., Klein, C. J., & Gaines, S. D. (2011). Integrated land-sea conservation planning: The missing links. *Annual Review of Ecology, Evolution, and Systematics, 42*, 381–409.
2. Armesto, J. J., Rozzi, R., Smith-Ramírez, C., & Arroyo, M. T. K. (1998). Conservation targets in South American temperate forests. *Science, 282*, 1271–1272.
3. Arroyo, M. T. K., Riveros, M., Peñaloza, A., Cavieres, L., & Faggi, A. M. (1996). Phytogeographic relationships and regional richness patterns of the cool temperate rainforest flora of southern South America. In: R. G., Lawford, E., Fuentes, & y P. B., Alaback (Eds.), *High-latitude rainforests and associated ecosystems of the west coast of the Americas* (pp. 134–172). Springer.
4. Astorga, A., Moreno, P., Rojas, P. and Reid, B. (2023). *Conserving the origin of rivers: Intact forested watersheds in Western Patagonia*. Springer.
5. Astorga, A., Moreno, P., & Reid, B. (2018). Watersheds and trees fall together: An analysis of intact forested watersheds in southern Patagonia (41–56° S). *Forests, 9*, 385.
6. Aylwin, J., Arce, L., Guerra, F., Núñez, D., Álvarez, R., Mansilla P., Alday, D., Caro, L., Chiguay, C., & y Huenucoy, C. (2023). *Conservation and indigenous people in Chilean Patagonia*. Springer.
7. Bachmann-Vargas, P., & van Koppen, C. S. A. (2020). Disentangling environmental and development discourses in a peripheral spatial context: The case of the Aysén Region, Patagonia, Chile. *The Journal of Environment & Development, 29*(3), 366–390.
8. Ban, N. C., Mills, M., Tam, J., Hicks, C. C., Klain, S., Stoeckl, N., Bottrill, M. C., Levine, J., Pressey, R. L., Satterfield, T., & Chan, K. M. (2013). A social-ecological approach to conservation planning: Embedding social considerations. *Frontiers in Ecology and the Environment, 11*, 194–202.
9. Barbier, E. B., Burgess, J. C., & Dean, T. J. (2018). How to pay for saving biodiversity. *Science, 360*(6388), 486–488.
10. Belsoy, J., Korir, J., & Yego, J. (2012). Environmental impacts of tourism in protected areas. *Journal of Environment and Earth Science, 2*, 64–73.
11. Bennett, N. J., Roth, R., Klain, S. C., Chan, K., Christie, P., Clark, D. A., Cullman, G., Curran, D., Durbin, T. J., Epstein, G., Greenberg, A., Nelson, M. P., Sandlos, J., Stedman, R., Teel, T. L., Thomas, R., Veríssimo, D., & Wyborn, C. (2017). Conservation social science: Understanding and integrating human dimensions to improve conservation. *Biological Conservation, 205*, 93–108.
12. Bennett, N. J., Whitty, T. S., Finkbeiner, E., Pittman, J., Bassett, H., Gelcich, S., & Allison, E. H. (2018). Environmental stewardship: A conceptual review and analytical framework. *Environmental Management, 61*, 597–614.
13. Buschmann, A. H., Niklitschek, E. J., & Pereda, S. (2023). *Aquaculture and its impacts on the conservation of Chilean Patagonia*. Springer.

14. D'agata, S., Mouillot, D., Wantiez, L., Friedlander, A. M., Kulbicki, M., & Vigliola, L. (2016). Marine reserves lag behind wilderness in the conservation of key functional roles. *Nature Communications, 7*, 12000.

15. Díaz, S., Pascual, U., Stenseke, M., Martín-López, B., Watson, R. T., Molnár, Z., Hill, R., Chan, K. M., Baste, I. A., Brauman, K. A., & Polasky, S. (2018). Assessing nature's contributions to people. *Science, 359*(6373), 270–272.

16. Durán, A. P., Casalegno, S., Marquet, P. A., & Gaston, K. J. (2013). Representation of ecosystem services by terrestrial protected areas: Chile as a case study. *PLoS One, 8*, e82643.

17. Ellis, E. C. (2015). Ecology in an anthropogenic biosphere. *Ecological Monographs, 85*(3), 287–331.

18. Glavovic, B. C., Limburg, K., Liu, K. K., Emeis, K. C., Thomas, H., Kremer, H., Avril, B., Zhang, J., Mulholland, M. R., Glaser, M., & Swaney, D. P. (2015). Living on the margin in the Anthropocene: Engagement arenas for sustainability research and action at the ocean-land interface. *Current Opinion in Environmental Sustainability, 14*, 232–238

19. González, H. E., Calderón, M. J., Castro, L., Clement, A., Cuevas, L. A., Daneri, G., Iriarte, J. L., Lizárraga, L., Martínez, R., Menschel, E., Silva, N., Carrasco, C., Valenzuela, C., Vargas, C. A., & Molinet, C. (2010). Primary production and plankton dynamics in the Reloncaví fjord and the interior sea of Chiloé, northern Patagonia, Chile. *Marine Ecology Progress Series, 402*, 13–30.

20. González, H. E., Castro, L., Daneri, G., Iriarte, J. L., Silva, N., Vargas, C. A., Giesecke, R., & Sánchez, N. (2011). Seasonal plankton variability in Chilean Patagonia fjords: Carbon flow through the pelagic food web of Aysén fjord and plankton dynamics in the Moraleda Channel basin. *Continental Shelf Research, 31*, 225–243.

21. Grantham, H. S., Duncan, A., Evans, T. D., Jones, K. R., Beyer, H. L., Schuster, R., Walston, J., Ray, J. C., Robinson, J. G., Callow, M., & Clements, T. (2020). Anthropogenic modification of forests means only 40% of remaining forests have high ecosystem integrity. *Nature Communications, 11*(1), 1–10.

22. Guala, C., Veloso, K., Farías, A., & Sariego, F. (2023). *Analysis of tourism development linked to protected areas in Chilean Patagonia*. Springer.

23. Haddaway, N. R., Macura, B., Whaley, P., & Pullin, A. S. (2018). ROSES reporting standards for systematic evidence syntheses: Proforma, flow-diagram and descriptive summary of the plan and conduct of environmental systematic reviews and systematic maps. *Environmental Evidence, 7*, 7.

24. Häussermann, V., Försterra, G., & Laudien, J. (2023). *Hard bottom macrobenthos of Chilean Patagonia: Emphasis on conservation of subtitoral invertebrate and algal forests*. Springer.

25. Hermoso, M. I., Martin, V. Y., Gelcich, S., Stotz, W., & Thiel, M. (2020). Exploring diversity and engagement of divers in citizen science: Insights for marine management and conservation. *Marine Policy, 124*, 104316.

26. Hiriart-Bertrand, L., Silva, J. A., & Gelcich, S. (2020). Challenges and opportunities of implementing the marine and coastal areas for indigenous people's policy in Chile. *Ocean & Coastal Management, 193*, 105233.

27. Hucke-Gaete, R., Viddi, F. A., & Simeone, A. (2023). *Marine mammals and seabirds of Chilean Patagonia: Focal species for the conservation of Marine ecosystems*. Springer.

28. Hull, V., Xu, W., Liu, W., Zhou, S., Viña, A., Zhang, J., Tuanmu, M. N., Huang, J., Linderman, M., & Chen, X. (2011). Evaluating the efficacy of zoning designations for protected area management. *Biological Conservation, 144*, 3028–3037.

29. Intergovernmental Science-Policy Platform on Biodiversity and Ecosystem Services (IPBES). (2019). Summary for policymakers of the regional assessment report on biodiversity and ecosystem services for the Americas of the Intergovernmental Science-Policy Platform on Biodiversity and Ecosystem Services. In Rice, J., Seixas, C. S., Zaccagnini, M. E., Bedoya-Gaitán, M., Valderrama, N., Anderson, C. B., Arroyo, M. T. K., Bustamante, M., Cavender-Bares, J., Diaz-de-Leon, A., Fennessy, S., García Marquez, J. R., García, K., Helmer, E. H., Herrera, B., Klatt, B., Ometo, J. P., Rodríguez Osuna, V., Scarano, F. R., Schill, S., & Farinaci, J. S. (Eds.). IPBES Secretariat. https://doi.org/10.5281/zenodo.3553579

30. Iriarte, J. L., González, H. E., & Nahuelhual, L. (2010). Patagonian fjord ecosystems in southern Chile as a highly vulnerable region: Problems and needs. *Ambio, 39*(7), 463–466.
31. Iriarte, J. L., Pantoja, S., & Daneri, G. (2014). Oceanographic processes in Chilean fjords of Patagonia: From small to large-scale studies. *Progress in Oceanography, 129*, 1–170.
32. James, K. L., Randall, N. P., & Haddaway, N. R. (2016). A methodology for systematic mapping in environmental sciences. *Environmental Evidence, 5*, 7.
33. Jones, K. R., Klein, C. J., Halpern, B. S., Venter, O., Grantham, H., Kuempel, C. D., Shumway, N., Friedlander, A. M., Possingham, H. P., & Watson, J. E. (2018). The location and protection status of Earth's diminishing marine wilderness. *Current Biology, 28*(15), 2506–2512.
34. Luebert, F., & Pliscoff, P. (2017). *Sinopsis bioclimática y vegetacional de Chile, Segunda edición*. Editorial Universitaria.
35. Mansilla, C. A., Domínguez, E., Mackenzie, R., Hoyos-Santillán, J., Henríquez, J. M., Aravena, J. C., & Villa-Martínez, R. (2023). *Peatlands in Chilean Patagonia: Distribution, biodiversity, ecosystem services, and conservation*. Springer.
36. Marquet, P. A., Buschmann, A. H., Corcoran, D., Díaz, P. A., Fuentes-Castillo, T., Garreaud, R., Pliscoff, P., & Salazar, A. (2023). *Global change and acceleration of anthropic pressures on patagonian ecosystems*. Springer.
37. Martínez-Harms, M. J., & Gajardo, R. (2008). Ecosystem value in the western Patagonia protected areas. *Journal for Nature Conservation, 16*, 72–87.
38. Martínez-Harms, M. J., Armesto, J. J., Castilla, J. C., Astorga, A., Aylwin, J., Buschmann, A. H., Castro, V., Daneri, G., Fernández, M., Fuentes-Castillo, T., Gelcich, S., González, H. E., Hucke-Gaete, R., Marquet, P. A., Morello, F., Nahuelhual, L., Pliscoff, P., Reid, B., Rozzi, R., & Tecklin, D. (2022). A systematic evidence map of conservation knowledge in Chilean Patagonia. *Conservation Science and Practice, e575*. https://doi.org/10.1111/csp2.575
39. Martínez-Tilleria, K., Núñez-Ávila, M., León, C. A., Pliscoff, P., Squeo, F. A., & Armesto, J. J. (2017). A framework for the classification Chilean terrestrial ecosystems as a tool for achieving global conservation targets. *Biodiversity and Conservation, 26*(12), 2857–2876.
40. Mazor, T., Doropoulos, C., Schwarzmueller, F., Gladish, D. W., Kumaran, N., Merkel, K., Di Marco, M., & Gagic, V. (2018). Global mismatch of policy and research on drivers of biodiversity loss. *Nature Ecology & Evolution, 2*, 1071–1074.
41. Millennium Ecosystem Assessment (2005). *Ecosystems and human well-being: current state and trends*. Island Press. Retrieved https://www.millenniumassessment.org/documents/document.766.aspx.pdf
42. Mittermeier, R. A., Mittermeier, C. G., Brooks, T. M., Pilgrim, J. D., Konstant, W. R., da Fonseca, G. A., & Kormos, C. (2003). Wilderness and biodiversity conservation. *Proceedings of the National Academy of Sciences, 100*, 10309–10313.
43. Molinet, C., * Niklischek, E. J. (2023). *Fisheries and marine conservation in Chilean Patagonia*. Springer.
44. Nahuelhual, L., & Carmona, A. (2023). *Drivers of change in ecosystems of Chilean Patagonia: Current and projected trends*. Springer.
45. National Research Council (2008). Increasing capacity for stewardship of oceans and coasts: A priority for the 21st century. The National Academies Press. Obtenido de: https://doi.org/10.17226/12043
46. Noble, I., Castilla, J. C., Choo, P. S., de Groot, R., Mooney, H. A., Naeem, S., Reed, W. V., Turpie, J. K., Williams, M., Shindong, Z., & Zhiyun, Z. (2003). Chapter II: Ecosystems and their services. In M. E. A. Board (Ed.), *Ecosystems and human well-being: A framework for assessment* (pp. 49–70). Island Press.
47. Núñez-Ávila, M., Corcuera, E., Farías, A., Pliscoff, P., Palma, J., Barrientos, M., & Sepúlveda, C. (2013). Diagnóstico y caracterización de iniciativas de conservación privada en Chile. Fundación Senda Darwin.
48. Pantoja, S., Iriarte, J. L., & Daneri, G. (2011). Oceanography of the Chilean Patagonia. *Continental Shelf Research, 31*, 149–153.
49. Pickard, G. L. (1971). Some physical oceanographic features of inlets of Chile. *Fisheries Research Board of Canada, 28*, 1077–1106.

50. Pickard, G. L., & Stanton, B. R. (1980). Pacific fjords—a review of their water characteristics. In H. J. Freeland, D. M. Farmer, & C. D. Levings (Eds.), *Fjord oceanography* (pp. 1–51). Springer.
51. Pisano, E. (1977). Fitogeografía de Fuego-Patagonia chilena. I.-Comunidades vegetales entre las latitudes 52 y 56° S. *Anales del Instituto de la Patagonia Chile, 8*, 121–250.
52. Pliscoff, P., & Fuentes-Castillo, T. (2011). Representativeness of terrestrial ecosystems in Chile's protected area system. *Environmental Conservation, 38*, 303–311.
53. Pliscoff, P., Martínez-Harms, M. J., & Fuentes-Castillo, T. (2023). *Representativeness assessment and identification of priorities for the protection of terrestrial ecosystems in Chilean Patagonia.* Springer.
54. Radic-Schilling, S., Corti, P., Muñoz, R., Butorovic, N., & Sánchez, L. (2023). *Steppe ecosystems in Chilean Patagonia: Distribution, climate, biodiversity, and threats to their sustainable management.* Springer.
55. Reid, B., Astorga, A., Madriz, I., Correa, C., & Contador, T. (2023). *A conservation assessment of freshwater ecosystems in Southwestern Patagonia.* Springer.
56. Rivera, A., Aravena, J. C., Urra, A., & Reid, B. (2023). *Chilean Patagonian glaciers and environmental change.* Springer.
57. Rovira, J., & Herreros, J. (2016). Clasificación de ecosistemas marinos chilenos de la zona económica exclusiva. Ministerio Medio Ambiente.
58. Rozzi, R., Rosenfeld S., Armesto, J. J., Mansilla, A., Núñez-Avila, M., & Massardo, F. (2023). *Ecological connections across the marine-terrestrial interface in Chilean Patagonia.* Springer.
59. Rozzi, R., Armesto, J. J., Gutiérrez, J. R., Massardo, F., Likens, G. E., Anderson, C. B., Poole, A., Moses, K. P., Hargrove, E., Mansilla, A. O., Kennedy, J. H., & Arroyo, M. T. (2012). Integrating ecology and environmental ethics: Earth stewardship in the southern end of the Americas. *BioScience, 62*(3), 226–236.
60. Soto-Navarro, C., Ravilious, C., Arnell, A., de Lamo, X., Harfoot, M., Hill, S. L. L., Wearn, O. R., Santoro, M., Bouvet, A., Mermoz, S., Le Toan, T., Xia, J., Liu, S., Yuan, W., Spawn, S. A., Gibbs, H. K., Ferrier, S., Harwood, T., Alkemade, R., … Kapos, V. (2020). Mapping co-benefits for carbon storage and biodiversity to inform conservation policy and action. *Philosophical Transactions of the Royal Society B: Biological Sciences, 375*, 20190128.
61. Spalding, M. D., Fox, H. E., Allen, G. R., Davidson, N., Ferdaña, Z. A., Finlayson, M., Halpern, B. S., Jorge, M. A., Lombana, A., Lourie, S. A., Martin, K. D., McManus, E., Molnar, J., Recchia, C. A., & Robertson, J. (2007). Marine ecoregions of the world: A bioregionalization of coastal and shelf areas. *BioScience, 57*(7), 573–583.
62. Stoms, D. M., Davis, F. W., Andelman, S. J., Carr, M. H., Gaines, S. D., Halpern, B. S., Hoenicke, R., Leibowitz, S. G., Leydecker, A., Madin, E. M., Tallis, H., & Warner, R. R. (2005). Integrated coastal reserve planning: Making the land-sea connection. *Frontiers in Ecology and the Environment, 3*, 429–436.
63. Strub, P. T., Corinne, J., Montecino, V., Rutllant, J. A., & Blanco, J. L. (2019). Ocean circulation along the southern Chile transition region (38°–46° S): Mean, seasonal and interannual variability, with a focus on 2014–2016. *Progress in Oceanography, 172*, 159–198.
64. Sullivan Sealey, K., & Bustamante, G. (1999). Setting geographic priorities for marine conservation in Latin America and the Caribbean. The Nature Conservancy. https://pdf.usaid.gov/pdf_docs/PNACH523.pdf
65. Tacón, A., Tecklin, D., Farías, A., Peña, M. P., & García, M. (2023). *Terrestrial protected areas in Chilean Patagonia: Characterization, historical evolution, and management.* Springer.
66. Tallis, H., & Lubchenco, J. (2014). Working together: A call for inclusive conservation. *Nature News, 515*, 27.
67. Tecklin, D. R., & Sepúlveda, C. (2014). The diverse properties of private land conservation in Chile: Growth and barriers to private protected areas in a market-friendly context. *Conservation and Society, 12*(2), 203–217.
68. Tecklin, D., Farías, A., Peña, M. P., Gelvez, X. Castilla, J. C., Sepúlveda, M., Viddi, F. A., & Hucke-Gaete, R. (2023). *Coastal-marine protection in Chilean Patagonia: Historical progress, current situation, and challenges.* Springer.

69. Veblen, T. T., Schlegel, F. M., & Oltremari, J. V. (1983). Temperate broad-leaved evergreen forests of South America. In: J. D. Ovington (Ed.), *Temperate broad-leaved evergreen forests. Ecosystems of the world* (Vol. 10, pp. 5–32). Elsevier.
70. Villagrán, C. (2018). Biogeografía de los bosques subtropical-templados del sur de Sudamérica. Hipótesis históricas. *Magallania (Chile), 46*, 27–48.
71. Watson, J. E., Dudley, N., Segan, D. B., & Hockings, M. (2014). The performance and potential of protected areas. *Nature, 515*, 67–73.
72. Watson, J. E., Venter, O., Lee, J., Jones, K. R., Robinson, J. G., Possingham, H. P., & Allan, J. R. (2018a). Protect the last of the wild. *Nature, 563*, 20–30.
73. Watson, J. E., Evans, T., Venter, O., Williams, B., Tulloch, A., Stewart, C., Thompson, I., Ray, J. C., Murray, K., Salazar, A., & McAlpine, C. (2018b). The exceptional value of intact forest ecosystems. *Nature Ecology & Evolution, 2*, 599–610.
74. Calvete, C., & Sobarzo, M. (2011). Quantification of the surface brackish water layer and frontal zones in southern Chilean fjords between Boca del Guafo (43°30′S) and Estero Elefantes (46°30′S). *Continental Shelf Research, 31*(3–4), 162–171. https://doi.org/10.1016/j.csr.2010.09.013.
75. Pliscoff, P. (2022). Actualización de las áreas protegidas de Chile: análisis de representatividad y riesgo climático. Centro de Estudios Públicos. Chile. https://policycommons.net/artifacts/3867163/actualizacion-de-las-areas-protegidas-de-chile/4673118/ on 30 Oct 2023. CID: 20.500.12592/c4043q.
76. Bachmann-Vargas, P. (2013). Ecosystem services modeling as a tool for ecosystem assessment and support for the decision-making process in Aysén region, Chile (Northern Patagonia) (Master Thesis. Faculty of Agriculture and Nutritional Sciences, Christian-Albrechts-Universität, Kiel–Germany).
77. Schröter, M., Berbés-Blázquez, M., Albert, C., Hill, R., Krause, T., Loos, J., Mannetti, L. M., Martín-López, B., Neelakantan, A., Parrotta, J. A., Quintas-Soriano, C., & van Oudenhoven, A. (2023). Science on ecosystems and people to support the Kunming-Montreal Global Biodiversity Framework. *Ecosystems and People, 19*(1), 2220913.

Part II

Global Changes

Global Change and Acceleration of Anthropic Pressures on Patagonian Ecosystems

2

Pablo A. Marquet, Alejandro H. Buschmann, Derek Corcoran, Patricio Andrés Díaz, Taryn Fuentes-Castillo, René Garreaud, Patricio Pliscoff, and Alejandro Salazar

Abstract

This chapter analyzes the available information regarding the main drivers of global change operating in Patagonia, including climate change and its impact on biodiversity, the introduction of exotic species, change in land use and cover, and some emerging drivers of global change such as harmful algal blooms (HABs) and the increase in connectivity of human populations associated with the construction and expansion of the Austral Highway, and the bridge over the Chacao Channel in Chiloé. We emphasize the complexities associated with global change due to the synergies of the different global change drivers in

P. A. Marquet (✉) · D. Corcoran
Departmento de Ecología, Facultad de Ciencias Biológicas, Pontificia Universidad Católica de Chile, Alameda Bernardo O'Higgins 340, Santiago, Chile
e-mail: pmarquet@bio.puc.cl

P. A. Marquet · D. Corcoran · P. Pliscoff
Centro de Modelamiento Matemático (CMM), Universidad de Chile, International Research Laboratory 2807, CNRS, Santiago, Chile

P. A. Marquet
Centro de Cambio Global UC, Pontificia Universidad Católica de Chile, Alameda Bernardo O'Higgins 340, Santiago, Chile

A. H. Buschmann · P. A. Díaz
i-Mar Center, Universidad de Los Lagos, Camino a Chinquihue Km 6, 5480000 Puerto Montt, Chile

T. Fuentes-Castillo · P. Pliscoff · A. Salazar
Instituto de Geografía, Facultad de Historia, Geografía y Ciencia Política, Pontificia Universidad Católica de Chile, Santiago, Chile

R. Garreaud
Departamento de Geofísica, Facultad de Ciencias Físicas y Matemáticas, Universidad de Chile, Santiago, Chile

Centro de Ciencia del Clima y la Resiliencia (CR)2, Santiago, Chile

© Pontificia Universidad Católica de Chile 2023   33
J. C. Castilla et al. (eds.), *Conservation in Chilean Patagonia*, Integrated Science 19,
https://doi.org/10.1007/978-3-031-39408-9_2

Patagonia, such as the introduction of exotic species, climate, and the increased likelihood of fires, and between HABs, climate, and nutrient inputs. Global climate models for Patagonia project that by 2070, and assuming a scenario of moderate change in greenhouse gas concentrations (RCP 4.5), average temperature will increase from 0, 9 to 1.4 °C. Similarly, precipitation is projected to decrease between 5.5 and 116 mm on average; the highest precipitation reduction is 221 mm, with a modal reduction of 21 mm. In some areas, however, precipitation is projected to increase up to 77 mm. Although Patagonian ecosystems have been resilient and able to adapt to Holocene climate modifications, evidence suggests large and abrupt changes associated with European colonization in the twentieth century that go hand in hand with the increase in the incidence of fires, habitat loss and invasion of exotic species. In particular, an increase in exotic species plantations—together with a drier and warmer climate and the abundance of exotic herbivores that affect the regeneration of native species—can have far-reaching consequences for Patagonian ecosystems. This chapter concludes with a series of recommendations that address both the knowledge gaps identified and the impacts of global change in the area. Some of these include the regulation of productive activities (tourism and aquaculture), periodic diagnostics of the state of Patagonian ecosystems and the services they provide, and improved knowledge of ecosystem functioning, particularly climate change and the synergic action of different global change drivers on their resilience.

Keywords

Patagonia · Global change · Climate change · Biodiversity · Harmful algal blooms · Fires

1   Introduction

Global change refers to the multiple impacts of our species' way of life on the biosphere. In order to study these impacts—which alter the Earth system as a whole—researchers must identify different global change drivers or processes. The modifications experienced by these drivers and processes help us evaluate the impacts, including land use changes, alteration of biogeochemical cycles, overexploitation of biotic and abiotic resources, introduction and removal of species, and climate change [174, 175]. More recently, the scale of the impact and its acceleration in the biosphere [157] has led to the concept of "planetary boundaries," referring to the space where human action is safe in that it does not compromise the sustainability of the biosphere,if this space is transgressed, however, abrupt changes associated with biosphere thresholds could be foreseen (e.g., [18, 87]) Faced with what could be imminent changes in the functioning of natural systems and the services they provide to humans or "Nature Contributions to People" (NCP) [46], undisturbed ecosystems need to be protected, and to improve understanding of the socioecological dynamics that occur within them, since their modification could result in irreversible changes.

Patagonia is one of the most extensive and pristine biomes in the Americas, along with the Amazon and the Pacific Northwest [74]. It plays an important role in the planet due to the relevance of its NCPs in aspects associated with the provision of freshwater, food, and recreation, among others [46, 77, 98]. Fortunately, an important portion of Chilean Patagonia's ecosystems are protected in national parks and reserves that cover more than 50% of the area between latitudes 41° and 56° S. However, despite high protection and its remote location, the region is still subject to threats associated with several global change drivers, including climate change, sustained increase of anthropogenic pressures associated with tourism, farming, and livestock, expansion of invasive exotic species, increased connectivity through growth in transportation infrastructure [74, 95, 140], and especially aquaculture-related impacts [31, 32, 61]. Climate change may have severe impacts on the region, especially for water supply due to glacial melting [76, 133], with serious consequences for the distribution of ecosystems such as forests and wetlands. Considerable impacts are also expected on the hydrological cycle [129] and the ecosystems associated with fjords, canals, and coastal archipelagos. A recent study [134] indicates that the loss of terrestrial water storage is occurring at an alarming rate in Patagonia as a result of glacial melting.

The following sections provide an overview of the main global change drivers in Chilean Patagonia (see also [124] for the Argentinean case), including: (i) climate change scenarios for the region from now until the end of the century, and the impact on species and ecosystems, (ii) the impact of invasive alien species on the region's biodiversity; (iii) current status and projections of land use changes; (iv) ultraviolet (UV) radiation; and (v) emerging global changes such as harmful algal blooms (HABs) and anthropogenic pressures associated with human population growth and the impact of tourism.

2   Scope and Purpose

This chapter discusses the causes and current and future impacts of global change on Patagonia, based on an analysis of the state of knowledge regarding the main global change drivers in the area.

3   Methods

The available scientific literature was reviewed on the main global change drivers in Chilean Patagonia, defined as the region located between Reloncaví Sound and the Diego Ramirez Islands (41° 42' S, 73° 02' W; 56° 29' S, 68° 44' W). Global change drivers include climate change and its impacts on biodiversity, introduced exotic species, land use and land cover changes, and other emerging global change drivers, namely HABs and increased connectivity associated with the construction and expansion of the southern highway and the bridge over the Chacao Channel on Chiloé Island. The literature review also included other global change drivers prevailing in Patagonia, including the impact of high UV radiation on ecosystems and the overexploitation of natural resources such as the removal of *Sphagnum* from peatlands.

Modeling of present and future climate was conducted with climate, temperature, and precipitation data, both for current conditions and future climate change scenarios, according to the fourth IPCC Report [76]. The climate baseline for current conditions comes from the repository published by Pliscoff et al. [122], of bioclimatic surfaces with a 1 km × 1 km spatial resolution representing a 50-year period (1950–2000) for southern South America. Following the recommendations of Fajardo et al. [48], four general circulation models (GCMs) were considered that represent the variability in the predictions of future climate (to 2070), plus the average ensemble of 30 models available in the "GCM CompareR" application Fajardo et al. [48], assuming a scenario of moderate greenhouse gas concentration changes (RCP 4.5) at a resolution of 10 min.

Land use layers from 1917 to 2016, with projections to 2100 [70], were used to study the evolution of land use changes over time. The three most common land uses in Chilean Patagonia were identified using these layers, which cover 75.2% or more of the area studied through the historical period. These three land uses are livestock (managed pasture and livestock), pristine land cover (primary forests and other primary vegetation types), and regenerating ecosystems (regenerating forest and other vegetation types).

4   Climate Change in Chilean Patagonia

Current climatic and vegetation patterns help understand the consequences of global change, and climate change in Patagonia in particular. The average annual mean temperature is 5.9 °C, ranging from −4.5 to 12 °C. Temperatures are relatively low and show fairly stable average values in the region, except for higher altitude areas around the northern and southern ice fields (Fig. 1).

Precipitation presents a gradient from west to east, with extremes ranging from 6,288 mm per year to 214 mm per year. Chilean Patagonia has mostly a humid climate, as shown by annual precipitation rates (Fig. 1) [92, 93]. Rainfall is unevenly distributed in space, its distribution has a positive bias verified by the fact that the mean is 1,653 mm per year, higher than the median (1,495 mm per year), which in turn is higher than the mode (1,072 mm per year). The most abundant vegetation formations in the region are associated with humid environments; peatlands, with 48,167 (20%) [96], followed by evergreen forests with 45,336 (18%), and deciduous forests with 35,342 (14%), steppes and grasslands are fourth, with 24,425 (9.7%).

The future climate projection results with the four selected models indicate that annual mean temperatures could increase from 0.9 °C to 1.4 °C on average (Table 1). All but two models project a decrease in precipitation. Precipitation could decrease between 5.5 and 116 mm on average in the four models (Table 1). The maximum drop in precipitation points to a 221 mm decline, with a reduction

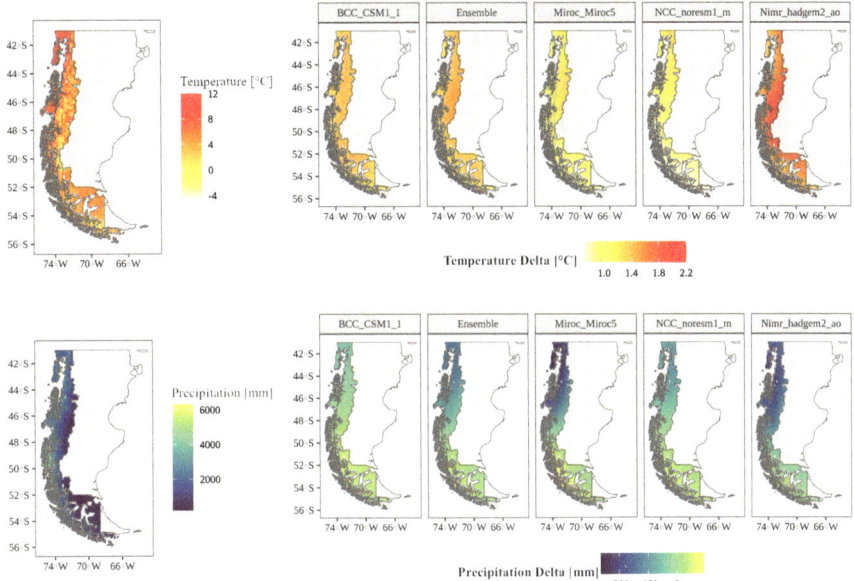

Fig. 1 Spatial distribution of mean annual temperature (top) and mean annual precipitation (bottom) in the Patagonian region. For each variable, to the left is the map with the current distribution of the variable, while to the right are the projections of the global climate models evaluated, and temperature (top) and precipitation (bottom) changes related to the current condition for the four models selected and the 30-model ensemble or average

mode of 21 mm and a maximum increase of 77 mm. Precipitation increases in the southern sector of Chilean Patagonia both in the models and in the ensemble, including the entire island of Tierra del Fuego, and decreases in the northern zone or temperate forests of the Chilean Patagonia region.

Table 1 Average annual mean temperature and precipitation for the entire region based on the four selected GCMs and the 30-model ensemble, including their differences with current conditions (baseline)

General circulation model	Average temperature (AT)	Annual precipitation (AP)	Delta (AT)	Delta AP
Baseline	5.97	1,667	0.0	0.0
ncc_noresm1_m	6.84	1,647	0.87	−20.7
miroc_miroc5	6.93	1,559	0.96	−108.3
Ensemble	7.11	1,603	1.15	−64.2
bcc_csm1_1	7.28	1,662	1.32	−5.6
nimr_hadgem2_ao	7.34	1,551	1.4	−116.5

The variability among models is not consistent in the different areas of Patagonia (Fig. 2). The fjords and channels area and the southern tip of Chilean Patagonia show less variation in their predictions compared to the central part and the region as a whole (Fig. 2). When changes are broken down based on vegetation formations, high altitude grassland and deciduous forest emerge as the two formations with the greatest variation in terms of temperature predictions, while peatlands and evergreen shrublands show the greatest consensus in terms of future changes in the four models.

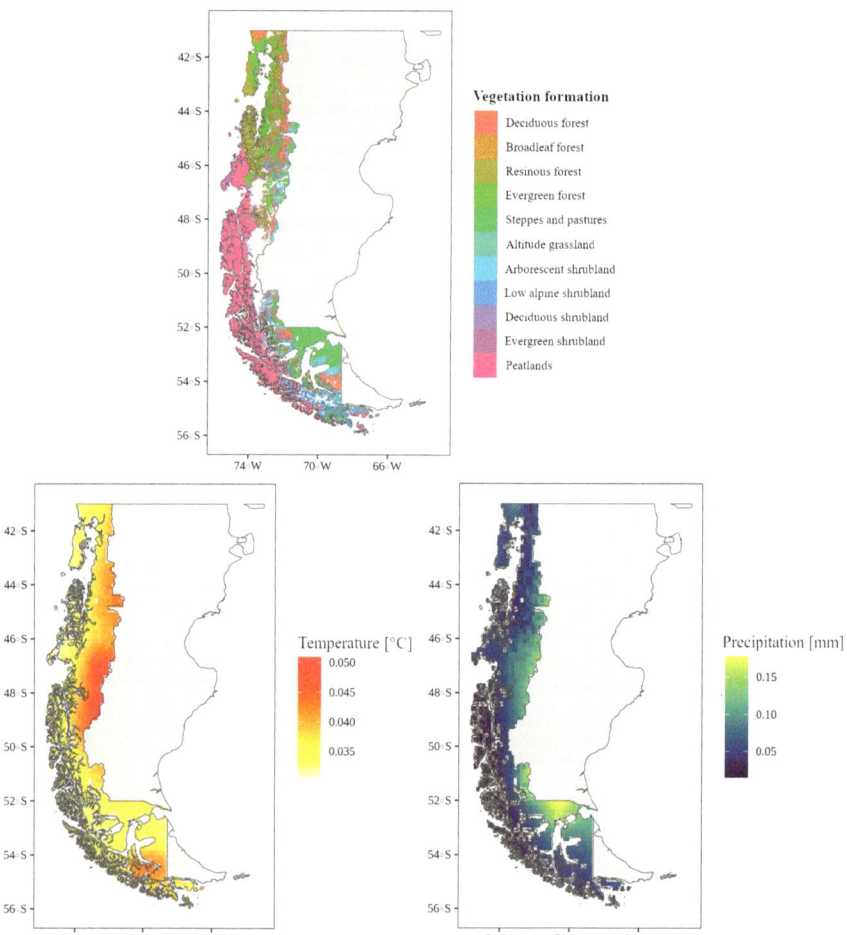

Fig. 2 Vegetation formations present in Chilean Patagonia according to Luebert and Pliscoff [93]. Standard deviation of the projected difference in temperature (top) and variation coefficient in projected rainfall for the 30 models (bottom) of the global circulation in Patagonia (bottom)

5 Impact of Exotic Species on Chilean Patagonia

Biological invasions are a major global change driver [175]. They may produce major changes in ecosystems that receive exotic species, and are often associated with biodiversity loss (e.g. [142, 176]), economic problems (e.g. [118]), and alterations in biogeochemical cycles [8, 15, 43]. However, the range of their impacts is so wide that they are generally uncertain, difficult to assess, occur with delayed effect, and are often sustained over time [151].

Chilean Patagonia is no exception in terms of species invasion. Terrestrial ecosystems have been invaded by species such as beaver and mink, with significant impacts. The same is true for marine ecosystems, specifically with the introduction of exotic salmonid species, as well as green crabs in Argentine Patagonia [37, 68, 101, 131, 173]. The main disturbance in river ecosystems has been the invasion of salmonids, with impacts on the aquatic biota both in Chile and Argentina. Additional impacts include alterations to the trophic webs, nutrient flow, and abundance of native vertebrate species [62]. In the past decade, Patagonian river ecosystems have been modified following the invasion of a diatom commonly known as Didymo (*Didymosphenia geminata*), with major ecosystem changes [130]. New exotic species are likely to continue invading in the short- and medium-term, while others will become invasive and increase their ranges in association with growing human population, tourism, commerce, land use, and climate changes, hence the relevance of assessing the current state of species invasions in Patagonian ecosystems. Table 2 summarizes invasive species most studied in the region, their ranges, and impacts.

Generally, knowledge of invasive species is concentrated in terrestrial vertebrates and plants. Little is known about terrestrial or marine invertebrates in the region. The synergistic effects of invasive species and other global change drivers (e.g. climate change) and between different exotic species constitute another knowledge gap. It is of great importance to measure the joint effects of all these species in river basins or landscapes, moving towards an integrative and ecosystem view.

An example of this synergic action is the way in which the North American beaver facilitates the spread of several invasive herbaceous species [100, 177]. There is also evidence that beaver habitat modifications make these habitats more prone to muskrats, which in turn can provide up to 50% of the minks' diet in environments far from marine coasts [39]. Minks have major impacts on native species and also encourage the dispersal of Didymo [91].

6 Land Use Changes

Climate change could act on its own to generate modification of Chile's Patagonian biomes. However, its impact on biodiversity depends on both climate forcing and human-generated biodiversity stress and degradation. Over 60% of the Chilean

Table 2 Most common exotic species in Patagonia: Date of introduction, status, and associated impacts

Species	Date of introduction	Current distribution	Status	Impact on biodiversity
North American Beaver (*Castor canadensis*)	1946 in Lake Fagnano, Argentina	Entire Tierra del Fuego Archipelago and adjacent Magallanes and Última Esperanza Provinces	Expanding	Ecosystem changes from forest to grassland favoring invasive species [5, 100]. Changes in the assemblage of river invertebrate communities [8]. Eutrophication of water bodies [8]
Exotic Bumblebee (*Bombus terrestris*)	1997 [105]	Throughout Patagonia, from Chiloé to Navarino Island [132, 144]	Expanding	Displacement of native giant bumblebee (*Bombus dahlbomii*), [107, 158]. Disease spread to native bumblebees [11], Exotic plant invasion [105]
Yelllow jacket (*Vespula germánica*)	2000 [152]	All Magallanes provinces [152]	Expanding, 37 km per year [171]	Competition for food with northwestern Patagonia ants [171]. Feeding on native arthropods [117]
American Mink (*Neovison vison*)	Punta Arenas 1934, Aysén 1967, and Los Lagos Regions 1972 [143]	Throughout Patagonia including Tierra del Fuego and Chonos archipelagos, and the entire continental zone [167]	Population expansion	Predation on native species [94, 115, 135, 167]. Potential competition with native carnivores [101, 166]. Disease vector [19, 20, 99, 148]
Didymo (*Didymosphenia geminata*)	2010 Futaleufú [34]	From Latitude 38° to 53° S [91]	Expanding	Changes to phosphorus cycle and pH modifications [130]

(continued)

Table 2 (continued)

Species	Date of introduction	Current distribution	Status	Impact on biodiversity
Salmonids: brown trout (*Salmo trutta*), Chinook salmon (*Oncorhynchus tshawytscha*), rainbow trout (*Oncorhynchus mykiss*)	Brown trout 1905, Chinook salmon, 1995 [155], Rainbow Trout, 1905	Throughout patagonia	Expanding	Competition with native species [153] causing local extinction and range reductions [63]. Impacts on nutrient cycles and ecosystem functioning (e.g. [41, 154]. Native species predators [180]
Pine (*Pinus contorta*)	Coyhaique 2010 [84]	Aysén region	Expanding [84, 112]	Competition with and exclusion of native species, such as *Araucaria* [112], promoting species associated with shade tolerance [28]. Highly flammable and potential impact on fire natural regime and dynamics [40, 159]
Gorse (*Ulex europaeus*)	Beginning of the nineteenth century	Between 33° and 43° S	Expanding	Negative impacts on exotic plantations (pine); invades livestock areas. Highly flammable with potential to alter fire dynamics [109]
Rabbit (*Oryctolagus cuniculus*)	Nineteenth century [57]	Throughout patagonia	Stable	Feeds on native plants [80]. Positive impact considering that hares or rabbits comprise 80% of native predators' diet [110]

(continued)

Table 2 (continued)

Species	Date of introduction	Current distribution	Status	Impact on biodiversity
European Hare (*Lepus europaeus*)	Nineteenth century [57]	Throughout patagonia	Stable	Feeds on native plants [80]. Positive impact considering that hares or rabbits comprise 80% of native predators' diet [110]
Dog (*Canis familiaris*)	Nineteenth century [116]	Throughout patagonia	Expanding	Spreads diseases to native carnivores [149]. Predator of native species [146]
Cat (*Felis catus*)	Nineteenth century	Throughout patagonia	Stable	Predator of native fauna [147]

Patagonian territory is currently considered natural land use (i.e. pristine or regenerating). Cattle ranching is the land use with the greatest impact, currently affecting 13% of the forest biome and 23% of the steppe biome.

Figure 3 shows how the area of pristine land cover has been shrinking over the years and is being replaced by regenerating land in the region's northeastern sector and by livestock land in Tierra del Fuego and Brunswick Peninsula. Future projections predict an increase in livestock land use to a maximum of 15.5% of the total area, decreasing to 8.3% by 2100. By 2100 (Fig. 3), 64.4% of the land is projected to have natural use (pristine or regenerating), similar to the current 63.1%. Despite this, most of this natural land cover will be in regeneration, whereas pristine land cover dominates at present.

The greatest conservation challenges in Patagonia will be in steppes and grasslands [123]. This is due to the following three factors: they are one of the least represented environments in the country's protected area system [121], their relevance for the movement of species in response to climate changes [65], and pressures to change land use from pristine conditions to cattle ranching. Given these soils' large carbon sequestration capacities, potential land use changes in steppes and grasslands can have serious consequences.

7   Ultraviolet Radiation

Since the discovery of a reduction in atmospheric ozone concentration over Antarctica there has been strong interest in the scientific community to measure the variability of ultraviolet radiation (UVR) and its impacts. Elevated UVR in Chilean Patagonia is a relevant global change driver for which there is abundant evidence in some taxonomic groups, for example fish [3]. However, little is known about

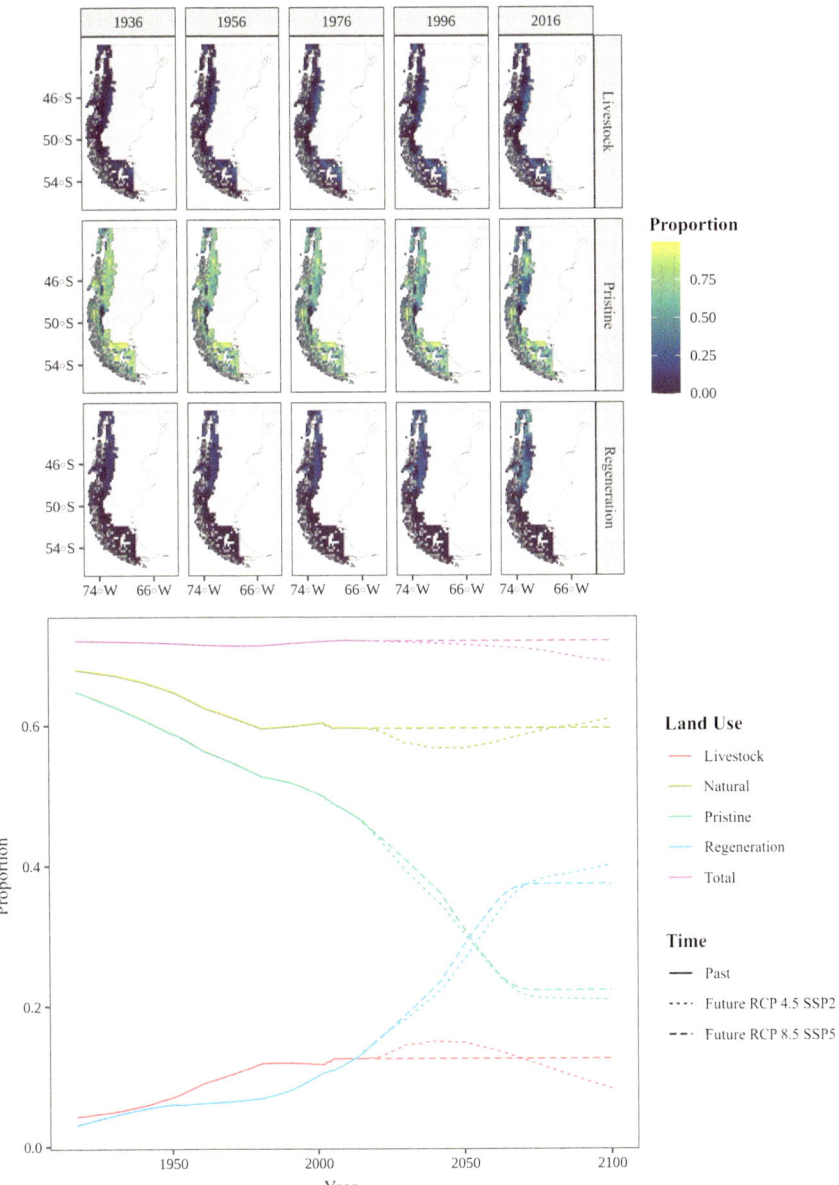

Fig. 3 Share of land grazed by livestock, pristine and regenerating land (top panel) in Chilean Patagonia. Projected land use change until 2100 for two emission scenarios. The natural land use category groups pristine and regenerating land uses (bottom panel)

its impacts on ecosystems, [17, 71, 172]. UV effects are known to have important impacts on microorganisms that form the basis of trophic webs in terrestrial and marine ecosystems. They lead to stress conditions resulting from the photooxidation of compounds associated with the generation of reactive oxygen species [51], increased persistence cost for zooplankton species such as *Daphnia* facing predation by native and introduced species (e.g. [41]), impacts on successional processes of intertidal algal species [35], and impacts on the microbial loop in the ocean with potential future effects on the biological or carbon pump [44]. Better understanding of UV impacts on terrestrial and marine ecosystems in Chilean Patagonia and their interaction with other global change drivers such as acidification and predation by introduced species are important knowledge gaps that need to be addressed in the near future.

8   Emerging Global Changes

Several impacts, such as Harmful Algal Blooms (or HABs), have emerged in the region as a consequence of synergistic effects among the different global change drivers. This is also the case of human landscape modifications, such as the extension of the Carretera Austral or Austral Highway and the bridge over the Chacao Channel to Chiloé Island—that can trigger major socioenvironmental transformations. All of this is associated with the introduction of new stakeholders into the territory's social configuration. These changes result from the pursuit of new territorial imaginaries that are transforming the environment, associated with a model of occupation linked to real estate development and tourist centers [69].

However, the most serious environmental issues have been associated with the latent threat of large-scale hydroelectric development, which has sparked opposition to projects to dam rivers, and conflict between different discourses about the Chilean Patagonia [136, 137, 163, 164]. Aquaculture has also driven important transformations. While cattle ranching and associated land clearing have been historically the main sources of territorial reconfiguration, the expansion of aquaculture and other developments in coastal waters has meant the expansion of territorial impacts on other ecosystems [24, 170], with effects on both natural and social environments, as it disrupts the ecological dynamics of natural ecosystems and social practices in Patagonia's coastal areas [141].

Ice flow of the San Rafael Glacier (Northern Ice Fields), Aysén Region. Photograph by María Paz Acuña

8.1 Austral Highway and Chiloé Bridge

The Austral Highway was planned in the 1960s and built in the 1980s. Its aim was to connect Chile's Patagonian territory with a longitudinal road that would overcome the limits of maritime transport and reduce the geographic isolation for a portion of the country's far southern area [12]. Its impacts were apparent from the moment of construction, as it traversed areas lacking adequate infrastructure. Therefore, the layout and irruption of a new logic of land use, since in many cases the road crosses areas without settlements or simply creates them, for example the case of Villa Lucía and of Hualaihué [111]. The project's main goal was to connect Patagonia with the rest of the country as well as with urban centers in the region, justified by higher population growth in towns and cities such as Coyhaique [16]. The old maritime traffic through Chiloé's inland sea and the cabotage through the channels was replaced by an amphibious route that meant adapting to new means of transportation, as well as new ways of understanding and experiencing Patagonian space [139].

The Austral Highway brought capital flow and people that had not existed in the area, and thus an increase in the intensity of occupation. This had an impact on human and non-human communities (for instance, the dissemination of exotic species), especially due to the emergence or increase in extractive activities including aquaculture and forestry, and services such as tourism. The flow of labor increased the fragmentation of the territory and intensified changes in land owner-ship structures [58]. It is significant that these two main activities produced tension that still exists in the area after the construction of the highway.

On one side there is a developmentalist view that promotes the connection of this "terrestrial island" with the rest of the country, and the potential exploitation of its resources and economic development. On the other side is the idea of con-servation and enhancement of its condition as a pristine territory, whose greatest value is in its landscapes and unique natural qualities. All of this has resulted in pressure on its resources, with higher pollution levels and spatial fragmentation due to population growth and economic activities. Finally, tourism and conserva-tion development have increased land values, leading to real estate speculation and the expulsion of traditional groups.

The construction of the bridge over the Chacao Channel will also increase traffic in Chilean Patagonia, especially in the northern zone, as it will become a complementary route to the Austral Highway and provide an overland route to the southernmost areas. Its construction, currently underway, will not only attract greater vehicle flow, but will also support the functional expansion of the Puerto Montt-Puerto Varas axis, with its well-known dysfunctional consequences regard-ing the peri-urban zones of metropolitan areas [2, 126]. It will also further intensify growth of the salmon industry by reducing logistical costs, increasing land values (agriculture-tourism) and thus deepening ongoing trends in the island in terms of land ownership structure with a consequent impact on the social reproduc-tion of traditional local groups in rural areas. Territorial transformations in urban spaces are also apparent in the monoculture of marine resources which regulate human-nature relations on the island of Chiloé [21, 138].

8.2 Harmful Algal Blooms

Harmful Algal Blooms (HABs) is the term coined by UNESCO's Intergovernmen-tal Oceanographic Commission to describe any microalga bloom, regardless of its concentration, perceived as harmful due to its socioeconomic impacts (damage to public health, coastal goods, and services). This socioeconomic definition includes blooms of various microalga species, including: (i) toxin-producing microalgae that accumulate through food webs (including emerging toxin producers); (ii) fish-killing microalgae (fish-killers); (iii) high-biomass bloom-forming microalgae (high-biomass HAB, HB-HAB), which although non-toxic, alter the environ-ments' physicochemical conditions; (iv) cyanobacteria [56], (v) harmful benthic microalgal blooms (benthic HABs, [23]). The main natural threat to bivalve farms and public health in Chilean Patagonia are HABs of toxin-producing species. Some of these phytotoxins are among the most potent bioactive compounds [47]. Filter-feeding bivalves accumulate toxins from plankton and when levels unfit for consumption (regulatory level) are reached, health and fishery authorities establish management measures (i.e. extraction bans) with considerable negative impacts on aquaculture and the exploitation of natural bivalve banks. In extreme cases, unregulated consumption of toxic bivalves has caused numerous human deaths [60].

An increase in toxic events has been observed globally over the past three decades. These have been partly associated with a progressive increase in the exploitation of coastal resources (aquaculture and tourism) and the exponential growth of monitoring programs (Hallegraeff, 1993) [64]. In addition to increased monitoring, there is growing evidence pointing to higher growth and dispersal of microalgae due to anthropogenic factors [9, 10]. among which nutrient enrichment of the water column (eutrophication) and sustained alterations in the temperature and precipitation regime (climate change) have become increasingly relevant [54, 55, 66].

Given the growing geographical extent, duration, and intensity of events [49], HABs have become one of the most important issues in the fisheries and aquaculture sectors worldwide, with an inauspicious prognosis [55]. HABs such as *Alexandrium catenella* can be complex given their biology, which involves resistant phases that depend on complex interactions with oceanographic and atmospheric factors and processes (Fig. 4).

Following the global trend, HABs in southern Chile's Patagonian fjords have caused recurrent problems in recent decades [45, 59, 60]. Over the past few years, an expansion of HAB events has been observed towards northern Chilean Patagonia, particularly outbreaks of paralytic shellfish poisoning caused by the dinoflagellate *Alexandrium catenella* [60, 102]. There is growing scientific evidence on the importance of atmospheric conditions in promoting HAB events [52] as well as the relationship with nutrient input from aquaculture practices [32], but

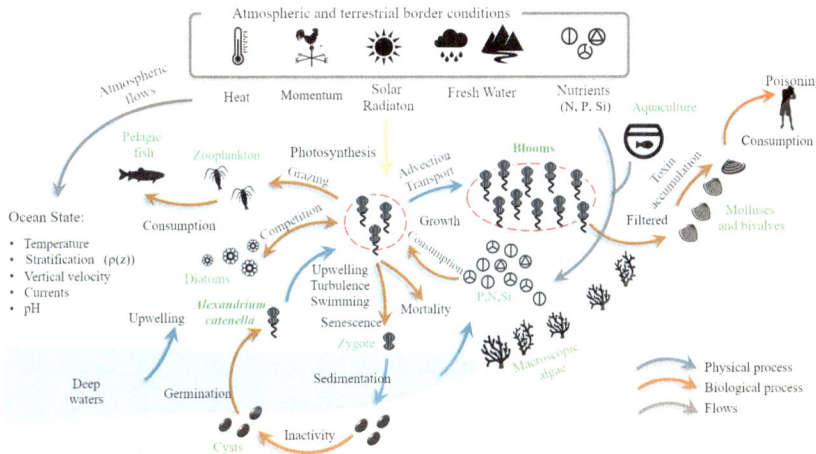

Fig. 4 Conceptual model of HAB events. The figure shows growth phases and associated biological processes in interaction with human activities and oceanographic and atmospheric processes

no comprehensive review of the evidence has been conducted. During the summer-autumn of 2016, toxic outbreaks reached the north of the Los Ríos Region (39° S) [67].

Very intense paralytic shellfish poisoning outbreaks are the main threat to public health and fisheries in the Patagonian fjords, particularly in the Aysén and Magallanes Regions [83]. In contrast, episodes of diarrheic shellfish poisoning caused by endemic species of the genus *Dinophysis*, mainly *D. acuta, D. acuminata*, producers of lipophilic toxins (okadaic acid and derivatives, pectenotoxins [128]), are the main threats in the Los Lagos Region [4, 86], which accounts for more than 95% of domestic mussel production (29×10^4 t yr^{-1}).

Proliferation of the dinoflagellate *Protoceratium reticulatum*, associated with the production of yessotoxins, has caused major problems in the Los Lagos Region in recent years. Other species, which can be called emerging species, such as those that produce ichthyotoxic toxins (*Chatonella, Pseudochattonella* and *Karenia*), have caused large salmon mortality in farms in the Aysén and Los Lagos regions. However, despite the serious economic disruptions caused by these microalgae, there are large knowledge gaps regarding their population dynamics, toxicology, triggering factors, and interannual variability, among others.

8.2.1 Causes of HABs in Chilean Patagonia

One of the greatest challenges to our understanding of HAB events is the diversity and multiplicity of biological and oceanographic processes [81, 119] that take place from the micro-scale (e.g. cyst germination) to the meso-scale (e.g. recirculation of water masses in a fjord) to the regional scale (e.g. freshwater discharge to the coast). All these processes are directly or indirectly modulated by atmospheric conditions, but the key variables depend on the spatiotemporal scale involved. For example, algal dispersal processes in the first few days or weeks of a HAB event are partially controlled by local ocean circulation, which is largely forced by surface wind strength [102]. Air temperature and solar energy on the surface determine to a large extent the temperature of the upper ocean layer, which in turn conditions the stability of the water column and may influence the cyst germination process (Fig. 4).

The link between atmospheric fluctuations and the occurrence of HAB events at the interannual scale is more elusive. Although the magnitude, spatial extent, and duration of HAB events in southern Chile vary substantially from year to year [60, 102, 103], the absence of a consolidated, long-term time series makes it difficult to connect them with climate variables. One hypothesis suggests that large-scale wind anomalies along Chilean Patagonia may be involved, due to their impact over coastal upwelling [102]. A long period of southerly wind may be able to increase the upwelling of nutrient-rich subsurface waters, favoring the occurrence of HAB events. However, the interannual fluctuations of the meridional wind off the coast of Patagonia are small, as the zonal (east–west) wind predominates in that area. Alternatively, León-Muñoz et al. [90] emphasize the role of freshwater discharge variations in the coastal zone of Chilean Patagonia. Their work discussed the worst HAB event recorded in the history of southern Chile, which occurred during the summer-autumn of 2016, with catastrophic environmental, social, and economic consequences [67]. This period had the most intense drought of the past 50 years [52], resulting from the superimposition of an intense El Niño event on a decreasing rainfall trend impacting Patagonia since the 1960s [26]. The scarce evidence available shows that the reduced freshwater input to Patagonian fjords and channels decreased the thermohaline stratification, allowing nutrient upwelling. Thus, the increased nutrients on the surface layer and increased solar radiation under drought conditions would have favored the explosive increase of *Pseudochattonella verruculosa* and *Alexandrium catenella* in early 2016 [90].

Although physically plausible, the hypothesis connecting interannual rainfall fluctuations and HAB events requires corroboration by considering other events (as well as years in which they were missing). If verified, a negative future scenario is then anticipated for harmful blooms in the Chilean Patagonia. Climate projections point to a constant trend of reduced precipitation [26], aggravated occasionally by El Niño years, as was the case in the summer of 2016 [52].

8.2.2 Global Change and Possible Responses of HAB Species

Hallegraeff [64] suggested some responses that can be expected in a global change scenario, (i) range expansion of warm-water species at the expense of colder water species, which would be displaced towards the poles; (ii) species-specific changes in HAB abundance and seasonality; (iii) changes in the phenology of some phytoplankton species (e.g. early onset, longer occurrence periods); (iv) secondary effects on the marine food web, mainly when zooplankton species and planktivorous fish are affected differently. The general hypothesis put forward by this author is that some harmful algal species may become more competitive, while others may decline considerably in areas where they are generally recurrent, i.e. there will be "winners" and "losers".

The predicted increase in sea surface temperature *(ca.* 2 °C) for the Patagonian fjord and channel system is expected to favor microalgae that cause the most problems in this area. Wells et al. [178] pointed out that higher temperatures, stratification, ocean acidification, and eutrophication of the water column will result in positive effects for taxonomic groups such as *Alexandrium* and *Heterosigma*. Moore et al. [106] suggested that temperature increases will widen the window of opportunity for *Alexandrium catenella*, reflected in the extension of the bloom period. Fu et al. [50] suggested that higher ocean acidification will increase cell toxicity in *A. catenella*, as well as in diatoms of the genus *Pseudo-nitzschia*.

Global change scenarios predicted for the coming decades in Chilean Patagonia suggest that HAB events could intensify in duration, toxicity, and even geographic range. Some of these issues are already apparent, as is the case of *A. catenella* and *Dinophysis acuminata*. Therefore, a thorough understanding of the dynamics and factors that govern these events is critical to put in place adequate management and mitigation measures.

8.3   Mining, Aquaculture, and Sphagnum Moss Extraction from Peatlands

Mining is undoubtedly an emerging threat to Patagonia. Although it has seen little development to date in Chilean Patagonia, it is still a relevant factor due to its potential consequences. In contrast, mining activities are widespread in Argentine Patagonia [22] where they caused major impacts during the mid-twentieth century, particularly in the area of Lake General Carrera [33]. According to Inostroza [73], there were 644 mining concessions in the Magallanes Region as of March 2010. Of these, 461 were for exploration and 183 for exploitation purposes (National Geology and Mining Service, SERNAGEOMIN), covering a total of 1,897 km^2, 1.4% of the region's total surface area. The author considers this to be a regional mining boom focused on coal, concentrated in five areas, located in the Magallanes coal basin: Natales, Skyring, Riesco, Brunswyck, and Tierra del Fuego, with Riesco being the area with the most mining activity. This coal boom is explained by the increase in coal prices and domestic demand associated with the energy sector. While the threat represented by this boom could be curbed under expected national energy decarbonization policies, the emergence of international markets is always a possibility. Aquaculture has also increased during the last 10 years, reflected in increasing production and in the number of aquaculture concession applications, which reached 979 in 2009 [73]. Unlike mining, however, this is a growing activity.

The extraction of *Sphagnum* moss, a key species in peatland ecosystems, is an emergent activity whose relevance is increasing on Chiloé Island and Patagonia. *Sphagnum* fibers are the second most important non-timber forest product in Chile [75, 89]. According to León et al. [89], its extraction rate has grown by more than 150% between 2007 and 2017, reaching annual exports of more than 3,500 tons, mainly to Taiwan, China, and the United States. This species provides important ecosystem services associated with carbon fixation and sequestration, as well as the production of fibers for horticulture. Its extraction negatively affects the diversity and composition of these plant communities, as well as the water and carbon cycles [88, 181]. Therefore, regulating the activity and enhancing knowledge for restoration and sustainable use is critical, especially in the context of the entry into force of Decree 25 of the Ministry of Agriculture, which governs extraction and requires harvesting plans.

9   Discussion

This chapter focused mainly on reviewing the available information on the primary global change drivers that operate, with varying intensity, in Chilean Patagonia. In addition, new information was presented on the impacts of climate change and its variability in the area, as well as biodiversity-related effects. However, our analyses are very preliminary and highlight the need to better address the impact of climate change on coastal zones, ecosystem processes, and protected areas in the region.

Many invasive species show high potential to lead to changes in ecosystem functioning, especially given positive feedback with other global change drivers, including climate change. Particularly worth noting is the potential of *Pinus contorta* and *Ulex europeus* to alter fire dynamics in northern Chilean Patagonia, contributing carbon dioxide into the atmosphere, increasing impacts (less precipitation, higher temperatures), which in turn would make both species more flammable [109, 159]. Positive feedbacks favor the presence of other exotic species and disturbances such as fire. This suggests that certain areas of Patagonia are vulnerable to an "invasion meltdown" [150], where facilitation phenomena among invasive species could increase their presence, distribution, and impact. This is particularly relevant in Tierra del Fuego, where the number of exotic mammals and freshwater fish exceeds that of native species [6, 165]. In addition to ecosystem monitoring programs, it is urgent to establish barriers to the introduction of more species and to prevent the expansion of those already introduced [145].

Other change drivers, tourism and land use change associated with cattle ranching, are also increasing. Tourism-related use of the landscape, accounting for 15.8% of the territory [73], is projected to increase in the coming years as a result of the southern highway expansion and the Chacao bridge. The importance of livestock, currently in about 24% of the Magallanes Region [73], will tend to increase in the long term, to then decrease (Fig. 3), thus basic knowledge to restore these ecosystems is needed.

The lack of knowledge about ecosystem functioning—particularly nutrient cycles—is a major gap in assessing and anticipating the impacts of the various global change drivers. This knowledge is essential in the context of climate change, where temperatures will increase, and precipitation events will become more extreme. Available information in Argentine Patagonia suggests a strong interaction between climate change and livestock, with impacts on the carbon cycle and particularly on soil organic carbon [114]. The authors suggest that livestock stocking management is essential for the maintenance of soil productivity. However, more long-term research is needed on key ecosystem processes associated with nutrient decomposition and cycling, as well as on soil microbiota.

Near-surface air temperature changes projected by global models for Patagonia are smaller in magnitude than those expected for other southern cone sectors. This is due in part to the thermal amelioration effect resulting from the reduced land mass relative to the surrounding ocean with its huge thermal inertia. There is consensus, however, that Chilean Patagonia will experience increased temperatures, with spatial variations from 1.1 °C to 1.7 °C by late in the century (2070), and under moderate greenhouse gas emission scenarios (RCP4.5). These values are comparable to the current interannual variability ranges for this region but may have important consequences on terrestrial and coastal ecosystems. Maximum rainfall reduction is between 5.5 mm and 116 mm, consistent with the reduction trends noted by other authors [26]. Although Patagonia has a hyper-humid climate condition that will remain even under the projected differences towards the end of the century, such changes could nonetheless have considerable impacts on terrestrial and marine systems [156]. In the latter case this is due to the drop in freshwater transport to the coastal zone, altering the area's complex hydrobiological balance. Changes predicted in the models also indicate that mean conditions will be altered.

Interannual variations (such as those resulting from the ENSO phenomenon) are also superimposed on this altered condition and can lead to an increase in the occurrence of extreme droughts. Take as an example the summer of 2016 [1, 52], which had severe socioenvironmental consequences due to the large HAB event in the autumn-summer of that same year [90], and is consistent with the exceedance analyses of global climate models, which predict an increase in the probability of minimum and maximum extreme temperature events, and longer and more intense droughts [42].

Changes in biodiversity and ecosystem functioning resulting from climate change are difficult to predict. Available literature for Chilean Patagonia points to a decreasing trend in the distribution of evergreen forests and peatlands [120], as well as major impacts on species [97], in interaction with other global change drivers such as fire [14, 169]. While there is evidence of Patagonian ecosystems' resilience and capacity to adapt to Holocene climate modifications, large and abrupt changes associated with the European colonization in the twentieth century are reported. The latter go hand in hand with an increase in fires, habitat loss, and invasion of exotic species [108, 168], making the interactions between global change drivers particularly relevant. As Iglesias and Whitlock [72] point out, "The weak relation between fire and prehistoric humans is in contrast to the influence that European settlement had on fire regimes. By altering the probability of ignition through accidental and deliberate burning, and converting large areas of native forest to fire-prone communities (e.g. pine and eucalyptus plantations), Europeans have gradually increased the risk of fire in Patagonia. This trend is likely to continue into the future with a drier climate, threatening the regeneration of fire-sensitive keystone species like *A. chilensis*". An increase in exotic plantations, together with a drier and warmer climate and an increase in the abundance of exotic herbivores that impact the regeneration of native species, may have profound consequences on the dynamics of Chilean and Argentinean Patagonian ecosystems (e.g. [125, 168]). Finally, the fjords and channels of Chilean Patagonia have been highlighted as an area relatively exposed to flooding and sea level rise as a result of climate change [36, 179]. This is explained by its large coastal area below 10 m and the predicted intensification of extreme weather events such as storm surges and floods [179], which are expected to have important negative effects on the flow of ecosystem services in the region [78].

This chapter suggests that one of the greatest threats to coastal ecosystems is associated with HABs. These have a major impact on biodiversity and the functioning of the area's socio-ecosystems, and result from the synergic action of different global change drivers associated with the climate regime and anthropogenic activities that discharge nutrients into rivers and cause coastal eutrophication events, in addition to climate change and salmon farming. Undoubtedly, HABs should be one of the main research priorities in Chilean Patagonia, particularly in their connections with exotic species. This requires basic research focused on public policies to regulate coastal productive activities.

Salmon farming in Chilean Patagonia has diverse environmental consequences [29, 31], that undoubtedly became intimately linked following the red tide event of the summer of 2016 [30]. The general perception is that aquaculture is related to HABs. Nutrient enrichment processes and HAB events are apparent in many coastal regions [53]. In Chile there is also evidence that algae can capture inorganic nitrogen produced by salmonids and intensify their growth at distances of up to at least 1 km from a farm. The amount of inorganic nitrogen that salmon farming introduces annually to the environment is very high and cannot be ignored [31]. This situation deserves more attention, as well as the development of technologies to control nutrient input to this extensive coastal zone. However, the environmental situation in Chilean Patagonia is even more complex. In addition to aquaculture, climate change and factors such as vessel traffic, coastal and seabed pollution, and overfishing are also present (Hucke et al., 2018) [104]. Emerging impacts such as microplastics have also been reported in Patagonian species [79], and require further monitoring (see [85]).

Finally, we must stress the importance of inland water ecosystems for the dynamics of terrestrial and coastal ecosystems and the scarce knowledge available. This is particularly important given the threats related to exotic species with relevant ecosystem effects, that interact with other global change drivers such as UV radiation, land use and land cover alterations, and climate change. This is compounded by weak environmental governance regarding ecosystem impact assessments [82]. Increasing our knowledge regarding these ecosystems and how to strengthen their resilience is certainly a priority.

10 Conclusions and Recommendations

There is an important tension between the intense threats associated with different global change drivers and the unique characteristics of Patagonia's ecosystems, their levels of protection and its pristine condition. Direct threats associated with reduced precipitation, increased temperatures, and exotic species are identified as important global change drivers in terrestrial and inland water ecosystems in Patagonia. While the recurrence of HAB events and wildfires are the main negative expressions of global systemic change in the area, major knowledge gaps on the functioning of Patagonian ecosystems persist. This is particularly true concerning the interactions between terrestrial, inland water, and marine ecosystems and the global change drivers that affect them. The former applies mainly to the synergies of different global change drivers, for example, the introduction of species, salmon farming, UVR, land use and land cover modifications, climate change, and the alteration of biogeochemical cycles. Tourism, salmon farming, cattle ranching, and *Sphagnum* extraction appear to be high-impact activities that require improved regulations to make the region's socioeconomic development goals consistent with conservation. While mining has a potential impact, it is still relatively minor. It could, however, become a major problem, depending on the behavior of internal and external markets.

Finally, one of the greatest conservation challenges is found in steppes and grasslands. This is due to the low predictability of climate change in areas with major variation—especially in precipitation—and to pressures associated with a shift to anthropogenic uses [123].

Considering the results of this chapter, we recommend the following:

- An assessment is required on the state of the Patagonian ecosystems; the state of the services they provide to people, and the impact of the different global change drivers. Specific focuses should include tourism and associated negative externalities, the introduction of exotic species, salmon farming, mining, and the potential impacts of these activities in the recurrence of HAB, fires and biodiversity loss. The creation of a long-term monitoring network is suggested, with a series of plots following the Ecology and Biodiversity Plots in Natural Environments in Southern Patagonia model [113], through a consortium of local universities, research centers, private organizations, and NGOs to promote scientific cooperation links with Argentinean researchers and research centers.
- In the short- to medium-term, monitoring programs are needed for exotic species and for the possibility of an "invasive meltdown", in which facilitation phenomena among invasive species could increase their establishment, range, and impacts. Research is also required for introduced pathogens that could be an important factor in the decline of natural populations, especially native fish species and native pollinators, where the co-introduction of pathogens has already been reported [11].
- In the short term, an evaluation is needed of the system of marine and terrestrial protected areas and their role in allowing for species' climate change adaptation. In particular, basic information on groups such as fish and invertebrates is crucial.
- Finally, an integrated and ecosystem perspective should guide land use, including a sustainable development logic that minimizes negative impacts on social ecosystems. In particular, those activities that require urgent attention and regulations are those associated with aquaculture and use of the coastline; livestock (promoting reduced impacts on soil carbon and land use changes), and tourism, promoting best practices among guides and tourists (e.g. [127]).

Acknowledgements The authors would like to thank an anonymous reviewer and the editors of this book for their support during the publication process. Pablo Marquet gratefully acknowledges funding from Grant AFB170008 and FB210005.

References

1. Aguayo, R., León-Muñoz, J., Vargas-Baecheler, J., Montecinos, A., Garreaud, R., Urbina, M., Soto, D., & Iriarte, J. L. (2019). The glass half-empty: Climate change drives lower freshwater input in the coastal system of the Chilean northern Patagonia. *Climatic Change, 155*(3), 417–435.
2. Allen, A. (2003). La interfase periurbana como escenario de cambio y acción hacia la sustentabilidad del desarrollo. *Cuadernos Del CENDES, 53*, 1–15.
3. Alves, R., & Agusti, S. (2020). Effect of ultraviolet radiation (UVR) on the life stages of fish. *Reviews in Fish Biology and Fisheries, 30*(2), 335–372.
4. Alves de Souza, C., Varela, D., Contreras, C., de La Iglesia, P., Fernández, P., Hipp, B., Hernández, C., Riobó, P., Reguera, B., Franco, J. M., Diogène, J., García, C., & Lagos, N. (2014). Seasonal variability of *Dinophysis* spp. and *Protoceratium reticulatum* associated to lipophilic shellfish toxins in a strongly stratified Chilean fjord. *Deep Sea Research Part II: Topical Studies in Oceanography, 101*, 152–162.
5. Anderson, C. B., Griffith, C. R., Rosemond, A. D., Rozzi, R., & Dollenz, O. (2006a). The effects of invasive North American beavers on riparian plant communities in cape Horn, Chile: Do exotic beavers engineer differently in sub-Antarctic ecosystems? *Biological Conservation, 128*(4), 467–474.
6. Anderson, C. B., Rozzi, R., Torres-Mura, J. C., Mcgehee, S. M., Sherriffs, M. F., Schüttler, E., & Rosemond, A. D. (2006b). Exotic vertebrate fauna in the remote and pristine sub-Antarctic cape Horn archipelago, Chile. *Biodiversity & Conservation, 15*(10), 3295–3313.
7. Anderson, C. B., Pastur, G. M., Lencinas, M. V., Wallem, P. K., Moorman, M. C., & Rosemond, A. D. (2009). Do introduced North American beavers *Castor canadensis* engineer differently in southern South America? An overview with implications for restoration. *Mammal Review, 39*(1), 33–52.
8. Anderson, C. B., & Rosemond, A. D. (2007). Ecosystem engineering by invasive exotic beavers reduces in-stream diversity and enhances ecosystem function in Cape Horn, Chile. *Oecologia, 154*(1), 141–153.
9. Anderson, D. (2014). HABs in a changing world: a perspective on harmful algal blooms, their impacts, and research and management in a dynamic era of climactic and environmental change. In H. Gyoon Kim, B. Reguera, G. M. Hallegraeff & C. K. Lee (Eds.), *Harmful algae 2012: Proceedings of the 15th international conference on harmful algae* (Vol. 2012, p. 3). CECO; NIH Public Access.
10. Anderson, D. M., Alpermann, T. J., Cembella, A. D., Collos, Y., Masseret, E., & Montresor, M. (2012). The globally distributed genus Alexandrium: Multifaceted roles in marine ecosystems and impacts on human health. *Harmful Algae, 14*, 10–35.
11. Arbetman, M. P., Meeus, I., Morales, C. L., Aizen, M. A., & Smagghe, G. (2013). Alien parasite hitchhikes to Patagonia on invasive bumblebee. *Biological Invasions, 15*(3), 489–494.
12. Arenas, F., Salazar, A., & Núñez, A. (Eds.). (2011). *El aislamiento geográfico:¿problema u oportunidad?: Experiencias, interpretaciones y políticas públicas*. Instituto de Geografía, Pontificia Universidad Católica de Chile.
13. Arismendi, I., Soto, D., Penaluna, B., Jara, C., Leal, C., & León-Muñoz, J. (2009). Aquaculture, non-native salmonid invasions and associated declines of native fishes in Northern Patagonian lakes. *Freshwater Biology, 54*(5), 1135–1147.
14. Armesto, J. J., Manuschevich, D., Mora, A., Smith-Ramírez, C., Rozzi, R., Abarzúa, A. M., & Marquet, P. A. (2010). From the Holocene to the Anthropocene: A historical framework for land cover change in southwestern South America in the past 15,000 years. *Land Use Policy, 27*(2), 148–160.
15. Ashton, I. W., Hyatt, L. A., Howe, K. M., Gurevitch, J., & Lerdau, M. T. (2005). Invasive species accelerate decomposition and litter nitrogen loss in a mixed deciduous forest. *Ecological Applications, 15*(4), 1263–1272.

16. Azócar García, G., Aguayo Arias, M., Henríquez Ruiz, C., Vega Montero, C., & Sanhueza-Contreras, R. (2010). Patrones de crecimiento urbano en la Patagonia chilena: El caso de la ciudad de Coyhaique. *Revista De Geografía Norte Grande, 46,* 85–104.
17. Barbieri, E. S., Marcoval, M. A., Hernández-Moresino, R. D., Spinelli, M. L., & Gonçalves, R. J. (2018). Global change and plankton ecology in the southwestern Atlantic. In M. Hoffmeyer, M. E. Sabatini, F. P. Brandini, D. L. Calliari & N. H. Santinelli (Eds.), *Plankton ecology of the Southwestern Atlantic* (pp. 565–574). Springer.
18. Barnosky, A., Hadly, E., Bascompte, J., Berlow, E., Brown, J., Fortelius, M., Getz, W. M., Harte, J., Hastings, A., Marquet, P. A., Martinez, N., Mooers, A., Roopnarine, P., Vermeij, G., Williams, J. W., Gillespie, R., Kitzes, J., Marshall, C., Matzke, N., ... Smith, A. B. (2012). Approaching a state shift in Earth's biosphere. *Nature, 486*(7401), 52–58.
19. Barros, M., Sáenz, L., Lapierre, L., Nuñez, C., & Medina-Vogel, G. (2014). High prevalence of pathogenic *Leptospira* in alien American mink (*Neovison vison*) in Patagonia. *Revista Chilena De Historia Natural, 87*(1), 19.
20. Barros, M., Cabezón, O., Dubey, J. P., Almería, S., Ribas, M. P., Escobar, L. E., Ramos, B., & Medina-Vogel, G. (2018). *Toxoplasma gondii* infection in wild mustelids and cats across an urban-rural gradient. *PLoS One, 13*(6).
21. Barton, J., Pozo, R., Román, Á., & Salazar, A. (2013). Reestructuración urbana de un territorio globalizado: Una caracterización del crecimiento orgánico en las ciudades de Chiloé, 1979–2008. *Revista De Geografía Norte Grande, 56,* 121–142.
22. Bechtum, A. (2018). La minería a gran escala en la Patagonia sur: el caso de cerro Vanguardia y la localidad de puerto San Julián, Santa Cruz. *Identidades, 8*(14).
23. Berdalet, E. P., Tester, A., & Zin-gone, A. (Eds.). (2012). Global ecology and oceanography of harmful algal blooms. In *GEOHAB Core research project: HABs in benthic systems.* IOC of UNESCO and SCOR.
24. Blanco, G., Arce, A., & Fisher, E. (2015). Becoming a region, becoming global, becoming imperceptible: Territorialising salmon in Chilean Patagonia. *Journal of Rural Studies, 42,* 179–190.
25. Blanco Wells, G., Arce, A., & Fisher, E. (2016). Intersubjetividad y domesticación en el devenir de una región global: Territorialización del salmón en la Patagonia chilena. *Revista De Ciencias Sociales, 54,* 125–145.
26. Boisier, J. P., Alvarez-Garretón, C., Cordero, R. R., Damiani, A., Gallardo, L., Garreaud, R. D., Lambert, F., Ramallo, C., Rojas, M., & Rondanelli, R. (2018). Anthropogenic drying in central-southern Chile evidenced by long-term observations and climate model simulations. *Elementa: Science of the Anthropocene, 6*(1).
27. Borla, M. L., Kizman, S., & Rey, A. R. (2010). *Compromiso Onashaga. Una experiencia sobre turismo y conservación. La fauna del canal Beagle como recurso vulnerable.* Retrieved from http://www.cadic-conicet.gob.ar/wp-content/uploads/2016/11/Onashaga-Borla-et-al.-2010.pdf
28. Bravo-Monasterio, P., Pauchard, A., & Fajardo, A. (2016). *Pinus contorta* invasion into treeless steppe reduces species richness and alters species traits of the local community. *Biological Invasions, 18*(7), 1883–1894.
29. Buschmann, A. H., Cabello, F., Young, K., Carvajal, J., Varela, D. A., & Henríquez, L. (2009). Salmon aquaculture and coastal ecosystem health in Chile: Analysis of regulations, environmental impacts and bioremediation systems. *Ocean y Coastal Management, 52*(5), 243–249.
30. Buschmann, A. H., Farías, L., Tapia, F., Varela, D., & Vásquez, M. (2016). *Scientific report on the 2016 southern Chile red tide.* Ministry of Economy. Retrieved from www.academiad eficiencias.cl/wp-contentuploads201704infofinal_comisionmarearoja_21nov2016-pdf
31. Buschmann, A. H., Niklitschek, E. J., & Pereda, S. (2023). *Aquaculture and its impacts on the conservation of Chilean Patagonia.* Springer.
32. Buschmann, A. H., Riquelme, V. A., Hernández-González, M. C., Varela, D., Jiménez, J. E., Henríquez, L. A., Vergara, P. A., Guiñez, R., & Filún, L. (2006). A review of the impacts of salmonid farming on marine coastal ecosystems in the southeast Pacific. *ICES Journal of Marine Science, 63*(7), 1338–1345.

33. Bustamante, L. P., Muñoz, M. D., & Contreras, R. S. (2010). Poblados mineros patagónicos: Paisajes culturales y estructura territorial. Registros. *Revista De Investigación Histórica, 7*, 49–61.
34. CIEP. (2010). Evaluación preliminar sobre la diatomea invasora exótica Didymosphenia geminata en cuencas de Futaleufú y Provincia de Palena, Región de Los Lagos, Chile. Preliminary Report, Centro de Investigación en Ecosistemas de la Patagonia, Coyhaique, Chile. 20 pp.
35. Campana, G. L., Zacher, K., Momo, F. R., Deregibus, D., Debandi, J. I., Ferreyra, G. A., & Quartino, M. L. (2020). Successional processes in Antarctic benthic algae. In I. Gómez, & P. Huovinen (Eds.), *Antarctic seaweeds* (pp. 241–264). Springer.
36. Camus, P., Losada, I. J., Izaguirre, C., Espejo, A., Menéndez, M., & Pérez, J. (2017). Statistical wave climate projections for coastal impact assessments. *Earth's Future, 5*(9), 918–933.
37. Castilla, J. C. & Neill, P. E. (2009). Marine bioinvasions in the southeastern Pacific: status, ecology, economic impacts, conservation and management. In G. Rilov, & J. A. Crooks (Eds.), *Biological invasions in marine ecosystems: ecological, management, and geographic perspectives* (pp. 439–457). Springer.
38. Center for International Earth Science Information Network, Columbia University and Internacional Centre for Tropical Agriculture (2005). *Gridded population of the world, version 3* (GPWv3): Population density grid.
39. Crego, R. D., Jiménez, J. E., & Rozzi, R. (2016). A synergistic trio of invasive mammals? Facilitative interactions among beavers, muskrats, and mink at the southern end of the Americas. *Biological Invasions, 18*(7), 1923–1938.
40. Cóbar-Carranza, A. J., García, R. A., Pauchard, A., & Pena, E. (2014). Effect of *Pinus contorta* invasion on forest fuel properties and its potential implications on the fire regime of *Araucaria araucana* and *Nothofagus antarctica* forests. *Biological Invasions, 16*(11), 2273–2291.
41. De los Ríos Escalante, P., Kies, F., & Correa-Araneda, F. (2017). An update on the global stressors and constraints affecting *Daphnia* populations in large Chilean Patagonian lakes (39–51° S). *Crustaceana, 90*(11–12), 1501–1516
42. Diffenbaugh, N. S., Singh, D., & Mankin, J. S. (2018). Unprecedented climate events: historical changes, aspirational targets, and national commitments. *Science Advances, 4*(2), eaao3354.
43. Dukes, J. S., & Mooney, H. A. (2004). Disruption of ecosystem processes in western North America by invasive species. *Revista Chilena De Historia Natural, 77*(3), 411–437.
44. Durán-Romero, C., Villafane, V. E., Valinas, M. S., Gonçalves, R. J., & Helbling, E. W. (2017). Solar UVR sensitivity of phyto-and bacterioplankton communities from Patagonian coastal waters under increased nutrients and acidification. *ICES Journal of Marine Science, 74*(4), 1062–1073.
45. Díaz, P.A., Molinet, C., Seguel, M., Díaz, M., Labra, G., & Figueroa, R. (2014). Coupling planktonic and benthic shifts during a bloom of Alexandrium catenella in southern Chile: Implications for bloom dynamics and recurrence. *Harmful Algae, 40*, 9–22.
46. Díaz, S., Pascual, U., Stenseke, M., Martín-López, B., Watson, R. T., Molnár, Z., Hill, R., Chan, K. M. A., Baste, I. A., Brauman, K. A., Polasky, S., Church, A., Lonsdale, M., Lariauderie, A., Leadley, P. W., van Oudehoven, A. P. E., van der Plaat, F., Schroter, M., Lavorel, S., & Shirayama, Y. (2018). Assessing nature's contributions to people. *Science, 359*(6373), 270–272.
47. van Egmond, H. P. (2004). Natural toxins: Risks, regulations and the analytical situation in Europe. *Analytical and Bioanalytical Chemistry, 378*(5), 1152–1160.
48. Fajardo, J., Corcoran, D., Roehrdanz, P. R., Hannah, L., & Marquet, P. A. (2020). GCM compareR: A web application to assess differences and assist in the selection of General Circulation Models for climate change research. *Methods in Ecology and Evolution, 11*, 656–663.
49. Food and Agriculture Organization (2004). *Marine biotoxins* (pp 278 +vi). Retrieved from http://www.fao.org/3/a-y5486e.pdf

50. Fu, F. X., Tatters, A. O., & Hutchins, D. A. (2012). Global change and the future of harmful algal blooms in the ocean. *Marine Ecology Progress Series, 470*, 207–233.
51. García, P. E., Queimaliños, C., & Diéguez, M. C. (2019). Natural levels and photo-production rates of hydrogen peroxide (H_2O_2) in Andean Patagonian aquatic systems: Influence of the dissolved organic matter pool. *Chemosphere, 217*, 550–557.
52. Garreaud, R. D. (2018). Record-breaking climate anomalies lead to severe drought and environ-mental disruption in western Patagonia in 2016. *Climate Research, 74*(3), 217–229.
53. Glibert, P. M., & Burford, M. A. (2017). Globally changing nutrient loads and harmful algal blooms: Recent advances, new paradigms, and continuing challenges. *Oceanography, 30*(1), 58–69.
54. Glibert, P. M., Icarus Allen, J., Artioli, Y., Beusen, A., Bouwman, L., Harle, J., Holmes, R., & Holt, J. (2014). Vulnerability of coastal ecosystems to changes in harmful algal bloom distribution in response to climate change: Projections based on model analysis. *Global Change Biology, 20*(12), 3845–3858.
55. Gobler, C. J., Doherty, O. M., Hattenrath-Lehmann, T. K., Griffith, A. W., Kang, Y., & Litaker, R. W. (2017). Ocean warming since 1982 has expanded the niche of toxic algal blooms in the North Atlantic and North Pacific oceans. *Proceedings of the National Academy of Sciences, 114*(19), 4975–4980.
56. Graneli, E., Codd, G. A., Dale, B., Lipiatou, E., Maestrini, S. Y., & Rosenthal, H. (1998). EUROHAB science initiative: Harmful algal blooms in European marine and brackish waters. *European Communities, 93*.
57. Grigera, D. E., & Rapoport, E. H. (1983). Status and distribution of the European hare in South America. *Journal of Mammalogy, 64*(1), 163–166.
58. Guala, C., Veloso, K., Farías, A., & Sariego, F. (2023). *Analysis of tourism development linked to protected areas in Chilean Patagonia*. Springer.
59. Guzmán, L., & Campodónico, I. (1975). *Marea roja en la Región de Magallanes*. Instituto de la Patagonia.
60. Guzmán, L., Pacheco, H., Pizarro, G., & Alarcón, C. (2002). *Alexandrium catenella* y veneno pa-ralizante de los mariscos en Chile. In E. Sar, M. Ferrario, & B. Reguera (Eds.). *Floracionnes algales nocivas en el cono sur americano* (pp. 235–255). Instituto Español de Oceanografía.
61. Gómez-Uchida, D., Sepúlveda, M., Ernst, B., Contador, T. A., Neira, S., & Harrod, C. (2018). Chile's salmon escape demands action. *Science, 361*(6405), 857–858.
62. Habit, E., Górski, K., Alò. D., Ascencio, E., Astorga, A., Colin, N., Contador, T., de los Ríos, P., Delgado, V., Dorador, C., Fierro, P., García, K., Parra, O., Quezada-Romegialli, C., Ried, B., Rivera, P., Soto-Azat, C., Valdovinos, C., Vera-Escalona, I. & Woelfl, S. (2019). Biodiversidad de ecosistemas de agua dulce. In: Marquet, P. A., Altamirano, A., Arroyo, M. T. K., Fernández, M., Gelcich, S., Górski, K., Habit, E., Lara, A., Maass, A., Pauchard, A., Pliscoff, P., Samaniego, H., & Smith-Ramírez, C. (Eds.) (2019). *Biodiversidad y cambio climático en Chile: Evidencia científica para la toma de decisiones. Informe de la mesa de biodiversidad*. Comité Científico COP25; Ministerio de Ciencia, Tecnología, Conocimiento e Innovación. Retrieved from http://www.minciencia.gob.cl/comitecientifico/
63. Habit, E., Piedra, P., Ruzzante, D. E., Walde, S. J., Belk, M. C., Cussac, V. E., González, J., & Colin, N. (2010). Changes in the distribution of native fishes in response to introduced species and other anthropogenic effects. *Global Ecology and Biogeography, 19*(5), 697–710.
64. Hallegraeff, G. M. (2010). Ocean climate change, phytoplankton community responses, and har- mful algal blooms: A formidable predictive challenge 1. *Journal of Phycology, 46*(2), 220–235.
65. Hannah, L., Roehrdanz, P. R., Marquet, P. A., Enquist, B. J., Midgley, G., Foden, W., Lovett, J. C., Corlett, R. T., Corcoran, D., Butchart, S. H. M., Boyle, B., Feng, X., Maitner, B., Fajardo, J., McGill, B. J., Merow, C., Morueta-Holme, N., Newman, E. A., Park, D. S., … Svenning, J. C. (2020). 30% land conservation and climate action reduce tropical extinction risk by more than 50%. *Ecography, 43*, 943–953.

66. Heisler, J., Glibert, P. M., Burkholder, J. M., Anderson, D. M., Cochlan, W., Dennison, W. C., Dortch, Q., Gobler, C. J., Heil, C. A., Humphries, E., & Lewitus, A. (2008). Eutrophication and harmful algal blooms: A scientific consensus. *Harmful Algae, 8*(1), 3–13.

67. Hernández, C., Díaz, P. A., Molinet, C., & Seguel, M. (2016). Exceptional climate anomalies and northwards expansion of paralytic shellfish poisoning outbreaks in Southern Chile. *Harmful Algae News, 54*, 1–2.

68. Hidalgo, F. J., Baron, P. J., & Orensanz, J. M. L. (2005). A prediction come true: The green crab invades the Patagonian coast. *Biological Invasions, 7*(3), 547–552.

69. Hidalgo, R., & Zunino, H. (2011). Negocios inmobiliarios en centros turísticos de montaña y nuevos modos de vida. El papel de los migrantes de amenidad existenciales en la comuna de Pucón-Chile. *Estudios y Perspectivas en Turismo, 20*(2), 307–326.

70. Hurtt, G. C., Chini, L. P., Frolking, S., Betts, R. A., Feddema, J., Fischer, G., Fisk, J. P., Hibbard, K., Houghton, R. A., Janetos, A., & Jones, C. D. (2011). Harmonization of land-use scenarios for the period 1500–2100: 600 years of global gridded annual land-use transitions, wood harvest, and resulting secondary lands. *Climatic Change, 109*(1–2), 117.

71. Häder, D. P., Helbling, E. W., Williamson, C. E., & Worrest, R. C. (2011). Effects of UV radiation on aquatic ecosystems and interactions with climate change. *Photochemical and Photobiological Sciences, 10*(2), 242–260.

72. Iglesias, V., & Whitlock, C. (2014). Fire responses to postglacial climate change and human impact in northern Patagonia (41–43 S). *Proceedings of the National Academy of Sciences, 111*(51), E5545–E5554.

73. Inostroza, L. (2015). El mito de pristinidad y los usos efectivos del territorio de la Región de Magallanes, Patagonia chilena: Forestal, minería y acuicultura. *Estudios Geográficos, 76*(278), 141–175.

74. Inostroza, L., Zasada, I., & König, H. J. (2016). Last of the wild revisited: Assessing spatial patterns of human impact on landscapes in southern Patagonia, Chile. *Regional Environmental Change, 16*(7), 2071–2085.

75. Instituto Forestal de Chile (2018). *Productos forestales no madereros.* Boletín N°31 marzo 2018. Retrieved from https://bibliotecadigital.infor.cl/handle/20.500.12220/27286

76. Intergovernmental Panel on Climate Change (2014). *Climate change 2014: Impacts, adaptation, and vulnerability. Part A: Global and sectoral aspects. Contribution of working group ii to the fifth assessment report of the intergovernmental panel on climate change. Climate change 2014: impacts, adaptation, and vulnerability.* Retrieved from https://www.ipcc.ch/report/ar5/syr/.

77. Intergovernmental Science-Policy Platform on Biodiversity and Ecosystems Services (2018). *The IPBES regional assessment report on biodiversity and ecosystem services for the Americas.* In J., Rice, C. S., Seixas, M. E., Zaccagnini, M., Bedoya-Gaitán, & N. Valderrama (Eds.), *Secretariat of the Intergovernmental Science-Policy Platform on Biodiversity and Ecosystem Services* (p. 656). IPBES. Retrieved from https://ipbes.net/assessment-reports/americas.

78. Iriarte, J. L., González, H. E., & Nahuelhual, L. (2010). Patagonian fjord ecosystems in southern Chile as a highly vulnerable region: Problems and needs. *Ambio, 39*(7), 463–466.

79. Jackson, G. D., Buxton, N. G., & George, M. J. (2000). Diet of the southern opah *Lampris immaculatus* on the Patagonian shelf; the significance of the squid *Moroteuthis ingens* and anthropogenic plastic. *Marine Ecology Progress Series, 206*, 261–271.

80. Jaksic, F. M. (1998). Vertebrate invaders and their ecological impacts in Chile. *Biodiversity and Conservation, 7*(11), 1427–1445.

81. Kudela, R. M., Berdalet, E., Enevoldsen, H., Pitcher, G., Raine, R., & Urban, E. (2017). GEOHAB: The global ecology and oceanography of harmful algal blooms program motivation, goals, and legacy. *Oceanography, 30*(1), 12–21.

82. Lacy, S. N., Meza, F. J., & Marquet, P. A. (2017). Can environmental impact assessments alone conserve freshwater fish biota? Review of the Chilean experience. *Environmental Impact Assessment Review, 63*, 87–94.

83. Lagos, N. (2003). Paralytic shellfish poisoning phycotoxins: Occurrence in South America. *Comments on Toxicology, 9*(2), 175–193.

84. Langdon, B., Pauchard, A., & Aguayo, M. (2010). Pinus contorta invasion in the Chilean Patagonia: Local patterns in a global context. *Biological Invasions, 12*, 3961–3971.
85. Law, K. L. (2017). Plastics in the marine environment. *Annual Review of Marine Science, 9*, 205–229.
86. Lembeye, G., Yasumoto, T., Zhao, J., & Fernández, R., (1993). DSP outbreak in Chilean fiords. In T. J., Smayda & Y., Shimizu (Eds.), *Toxic phytoplankton blooms in the sea* (pp. 525–529). Elsevier.
87. Lenton, T. M., Held, H., Kriegler, E., Hall, J. W., Lucht, W., Rahmstorf, S., & Schellnhuber, H. J. (2008). Tipping elements in the Earth's climate system. *Proceedings of the National Academy of Sciences, 105*(6), 1786–1793.
88. León, C. A., Martínez, G. O., & Gaxiola, A. (2018). Environmental controls of cryptogam composition and diversity in anthropogenic and natural peatland ecosystems of Chilean Patagonia. *Ecosystems, 21*(2), 203–215.
89. León, C. A., Neila-Pivet, M., Benítez-Mora, A., & Lara, L. (2019). Effect of phosphorus and nitrogen on *Sphagnum* regeneration and growth: An experience from Patagonia. *Wetlands Ecology and Management, 27*(2–3), 257–266.
90. León-Muñoz, J., Urbina, M. A., Garreaud, R., & Iriarte, J. L. (2018). Hydroclimatic conditions trigger record harmful algal bloom in western Patagonia (summer 2016). *Scientific Reports, 8*(1), 1–10.
91. Leone, P. B., Cerda, J., Sala, S., & Reid, B. (2014). Mink (*Neovison vison*) as a natural vector in the dispersal of the diatom *Didymosphenia geminata*. *Diatom Research, 29*(3), 259–266.
92. Luebert, F., & Pliscoff, P. (2006). *Sinopsis bioclimática y vegetacional de Chile*. Editorial Universitaria.
93. Luebert, F., & Pliscoff, P. (2009). Depuración y estandarización de la cartografía de pisos de vegetación de Chile. *Chloris Chilensis, 12*(1).
94. Maley, B. M., Anderson, C. B., Stodola, K., & Rosemond, A. D. (2011). Identifying native and exotic predators of ground-nesting songbirds in subantartic forests in southern Chile. *Anales Del Instituto De La Patagonia, 39*(1), 51–57.
95. Mansilla, A., Ojeda, J., & Rozzi, R. (2012). Cambio climático global en el contexto de la ecorregión subantártica de Magallanes y la Reserva de Biósfera Cabo de Hornos. *Anales Del Instituto De La Patagonia, 40*(1), 69–76.
96. Mansilla, C. A., Domínguez, E., Mackenzie, R., Hoyos-Santillán, J., Henríquez, J. M., Aravena, J. C., & Villa-Martínez, R. (2023). *Peatlands in Chilean Patagonia: Distribution, biodiversity, ecosystem services, and conservation*. Springer.
97. Marquet, P. A., Abades, S., Armesto, J. J., Barría, I., Arroyo, M. T. K., Cavieres, L. A., Gajardo, R., Garín, C., Labra, F., Meza, F., Pliscoff, P., Prado, C., Ramírez de Arellano, P. I., and Vicuña, S. (2010). *Estudio de vulnerabilidad de la biodiversidad terrestre en la eco-región mediterránea, a nivel de ecosistemas y especies, y medidas de adaptación frente a escenarios de cambio climático*. Centro de Cambio Global UC, CASEB, IEB. Retrieved from https://cambioglobal.uc.cl/images/proyectos/Documento_10_Impactos-CC-biodiversidad-pdf
98. Marquet P. A., Altamirano, A., Arroyo, M. T. K., Fernández, M., Gelcich, S., Górski, K. Habit, E., Lara, A., Maass, A., Pauchard, A., Pliscoff, P., Samaniego, H., & Smith-Ramírez, C. (Eds.) (2019). *Biodiversidad y cambio climático en Chile: Evidencia científica para la toma de decisiones*. Informe de la Mesa de Biodiversidad. Comité Científico COP25; Ministerio de Ciencia, Tecnología, Conocimiento e Innovación. Retrieved from http://www.mincie ncia. gob.cl/comitecientifico/
99. Martino, P. E., Samartino, L. E., Stanchi, N. O., Radman, N. E., & Parrado, E. J. (2017). Serology and protein electrophoresis for evidence of exposure to 12 mink pathogens in free-ranging American mink (*Neovison vison*) in Argentina. *Veterinary Quarterly, 37*(1), 207–211.
100. Martínez-Pastur, G. M., Lencinas, M. V., Escobar, J., Quiroga, P., Malmierca, L., & Lizarralde, M. (2006). Understorey succession in *Nothofagus* forests in Tierra del Fuego (Argentina) affected by *Castor canadensis*. *Applied Vegetation Science, 9*(1), 143–154.

101. Medina-Vogel, G., Barros, M., Organ, J. F., & Bonesi, L. (2013). Coexistence between the southern river otter and the alien invasive North American mink in marine habitats of southern Chile. *Journal of Zoology, 290*(1), 27–34.
102. Molinet, C., Lafon, A., Lembeye, G., & Moreno, C. A. (2003). Patrones de distribución espacial y temporal de floraciones de *Alexandrium catenella* (Whedon y Kofoid) Balech 1985, en aguas interiores de la Patagonia noroccidental de Chile. *Revista Chilena De Historia Natural, 76*(4), 681–698.
103. Molinet, C., Niklitschek, E., Seguel, M., & Díaz, P. (2010). Trends of natural accumulation and detoxification of paralytic shellfish poison in two bivalves from the Northwest Patagonian inland sea. *Revista De Biología Marina y Oceanografía, 45*(2), 195–204.
104. Molinet, C., & Niklitschek, E. J. (2023). *Fisheries and marine conservation in Chilean Patagonia.* Springer.
105. Montalva, J., Arroyo, M. T. K., & Ruz, L. (2008). *Bombus terrestris* Linnaeus (Hymenoptera: Apidae: Bombini) en Chile: Causas y consecuencias de su introducción. *Revista Del Jardín Botánico Chagual, 6*(6), 13–20.
106. Moore, S. K., Trainer, V. L., Mantua, N. J., Parker, M. S., Laws, E. A., Backer, L. C., & Fleming, L. E. (2008). Impacts of climate variability and future climate change on harmful algal blooms and human health. *Environmental Health, 7*(S2), S4.
107. Morales, C. L., Arbetman, M. P., Cameron, S. A., & Aizen, M. A. (2013). Rapid ecological replacement of a native bumble bee by invasive species. *Frontiers in Ecology and the Environment, 11*(10), 529–534.
108. Moreno, P. I., Simi, E., Villa-Martínez, R. P., & Vilanova, I. (2019). Early arboreal colonization, postglacial resilience of deciduous *Nothofagus* forests, and the southern westerly wind influence in central-east Andean Patagonia. *Quaternary Science Reviews, 218*, 61–74.
109. Norambuena, H., Escobar, S., & Rodríguez, F. (2000). The biocontrol of gorse, *Ulex europaeus*, in Chile: A progress report. In: N. R. Spencer (Ed.), *Proceedings of the international symposium on biological control of weeds* (pp. 955–961). Montana State University.
110. Novaro, A. J., Funes, M. C., & Walker, R. S. (2000). Ecological extinction of native prey of a carnivore assemblage in Argentine Patagonia. *Biological Conservation, 92*(1), 25–33.
111. Olea, J., & Román, J. (2017). Ordenamiento territorial y modernización en la Patagonia norte chilena. El caso de la comuna de Hualaihué: Borde costero, salmoneras y comunidades indígenas. *Revista Planeo, 70*, 2–11.
112. Pena, E., Hidalgo, M., Langdon, B., & Pauchard, A. (2008). Patterns of spread of *Pinus contorta* Dougl. Ex Loud. Invasion in a natural reserve in southern South America. *Forest Ecology and Management, 256*(5), 1049–1054.
113. Peri, P. L., Lencinas, M. V., Bousson, J., Lasagno, R., Soler, R., Bahamonde, H., & Martinez-Pastur, G. M. (2016). Biodiversity and ecological long-term plots in southern Patagonia to sup- port sustainable land management: The case of PEBANPA network. *Journal for Nature Conservation, 34*, 51–64.
114. Peri, P. L., Rosas, Y. M., Ladd, B., Toledo, S., Lasagno, R. G., & Martínez-Pastur, G. (2018). Modelling soil carbon content in south Patagonia and evaluating changes according to climate, vege- tation, desertification and grazing. *Sustainability, 10*(2), 438.
115. Peris, S. J., Sanguinetti, J., & Pescador, M. (2009). Have Patagonian waterfowl been affected by the introduction of the American mink *Mustela vison*? *Oryx, 43*(4), 648–654.
116. Philippi, R. A. (1885). Zoolojía: Sobre los animales introducidos en Chile desde su conquista por los españoles. *Anales de la Universidad de Chile, 319–335.*
117. Pietrantuono, A. L., Moreyra, S., & Lozada, M. (2018). Foraging behaviour of the exotic wasp *Vespula germanica* (Hymenoptera: Vespidae) on a native caterpillar defoliator. *Bulletin of Entomological Research, 108*(3), 406–412.
118. Pimentel, D., Lach, L., Zuniga, R., & Morrison, D. (2000). Environmental and economic costs of nonindigenous species in the United States. *BioScience, 50*(1), 53–65.
119. Pitcher, G., Moita, T., Trainer, V., Kudela, R., Figueiras, F., & Probyn, T. (Eds.). (2005). *Global ecology and oceanography of harmful algal blooms. GEOHAB core research project: HABs in upwelling systems.* IOC of UNESCO and SCOR.

120. Pliscoff, P., Arroyo, M. T. K., & Cavieres, L. (2012). Changes in the main vegetation types of Chile predicted under climate change based on a preliminary study: Models, uncertainties and adapting research to a dynamic biodiversity world. *Anales Del Instituto De La Patagonia, 40*(1), 81–86.

121. Pliscoff, P., & Fuentes-Castillo, T. (2011). Representativeness of terrestrial ecosystems in Chile's protected area system. *Environmental Conservation, 38*(3), 303–311.

122. Pliscoff, P., Luebert, F., Hilger, H. H., & Guisan, A. (2014). Effects of alternative sets of climatic predictors on species distribution models and associated estimates of extinction risk: A test with plants in an arid environment. *Ecological Modelling, 288*, 166–177.

123. Pliscoff, P., Martínez-Harms, M. J., & Fuentes-Castillo, T. (2023). *Representativeness assessment and identification of priorities for the protection of terrestrial ecosystems in Chilean Patagonia.* Springer.

124. Raffaele, E., de Torres Curth, M., Morales, C. L., & Kitzberger, T. (Eds.). (2014). *Ecología e Historia Natural de la Patagonia Andina. Un cuarto de siglo de investigación en biogeografía, ecología y conservación.* Fundación de Historia Natural Féliz de Azara.

125. Raffaele, E., Veblen, T. T., Blackhall, M., & Tercero-Bucardo, N. (2011). Synergistic influences of introduced herbivores and fire on vegetation change in northern Patagonia, Argentina. *Journal of Vegetation Science, 22*(1), 59–71.

126. Ravetz, J., Fertner, C., & Nielsen, T. (2013). The Dynamis of peri-urbanization. In K. Neilsson, S. Pauleit, S. Bell, C. Aalbers, & T. S. Nielson (Eds.), *Peri-urban futures: Scenarios and models for land use change in Europe* (pp. 13–44). Springer.

127. Raya Rey, A. N., Pizarro Pinochet, J. C., Anderson, C. B., & Huettmann, F. (2017). Even at the uttermost ends of the Earth: How seabirds telecouple the Beagle Channel with regional and global processes that affect environmental conservation and socio-ecological sustainability. *Ecology and Society, 22*(4), 31.

128. Reguera, B., Riobó, P., Rodríguez, F., Díaz, P. A., Pizarro, G., Paz, B., Franco, J. M., & Blanco, J. (2014). *Dinophysis* toxins: Causative organisms, distribution and fate in shellfish. *Marine Drugs, 12*(1), 394–461.

129. Reid, B., Astorga, A., Madriz, I., Correa, C., & Contador, T. (2023). *A conservation assessment of freshwater ecosystems in Southwestern Patagonia.* Springer.

130. Reid, B., & Torres, R. (2014). *Didymosphenia geminata* invasion in South America: Ecosystem impacts and potential biogeochemical state change in Patagonian rivers. *Acta Oecologica, 54*, 101–109.

131. Relva, M. A., Damascos, M. A., Macchi, P., Mathiasen, P., Premoli, A. C., Quiroga, M. P., Radovani, N. I., Raffaele, E., Sackmann, P. Speziale, K., Svriz, M., & Vigliano, P. H. (2014). Impactos humanos en la Patagonia. In E. Raffaele, M. de Torres Curth, C.L. Morales & T. Kitzberger, (Eds.), *Ecología e historia natural de la Patagonia andina: Un cuarto de siglo de investigación en biogeografía, ecología y conservación*, pp. 157–182. Fundación de Historia Natural Féliz de Azara.

132. Rendoll-Carcamo, J. A., Contador, T. A., Saavedra, L., & Montalva, J. (2017). First record of the invasive bumblebee Bombus terrestris (Hymenoptera: Apidae) on Navarino Island, southern Chile (55 S). *Journal of Melittology, 71*, 1–5.

133. Rivera, A., Aravena, J. C., Urra, A., & Reid, B. (2023). *Chilean Patagonian glaciers and environmental change.* Springer.

134. Rodell, M., Famiglietti, J. S., Wiese, D. N., Reager, J. T., Beaudoing, H. K., Landerer, F. W., & Lo, M. H. (2018). Emerging trends in global freshwater availability. *Nature, 557*(7707), 651–659.

135. Roesler, I., Imberti, S., Casanas, H., & Volpe, N. (2012). A new threat for the globally endangered hooded grebe *Podiceps gallardoi*: The American mink *Neovison vison*. *Bird Conservation International, 22*, 383–388.

136. Romero Toledo, H. (2014). Ecología política y represas: Elementos para el análisis del proyecto HidroAysén en la Patagonia chilena. *Revista De Geografía Norte Grande, 57*, 161–175.

137. Romero Toledo, H., & Romero Aravena, H. (2015). Ecología política de los desastres: Vulnerabilidad, exclusión socio-territorial y erupciones volcánicas en la Patagonia chilena. *Magallania (punta Arenas), 43*(3), 7–26.

138. Román, Á., Barton, J. R., Bustos, B., & Salazar, A. (Eds.). (2015). *Revolución salmonera: paradojas y transformaciones territoriales en Chiloé.* RIL Editores. P. Universidad Católica de Chile, Instituto de Estudios Urbanos, Colección Estudios Urbanos UC.

139. Rossetti, F. (2018). De infraestructura a paisaje. La carretera austral como motor de resignificación. *ARQ (santiago), 99,* 86–95.

140. Rozzi, R., Armesto, J. J., Gutiérrez, J. R., Massardo, F., Likens, G. E., Anderson, C. B., Poole, A., Moses, K. P., Hargrove, E., Mansilla, A. O., & Kennedy, J. H. (2012). Integrating ecology and environmental ethics: Earth stewardship in the southern end of the Americas. *BioScience, 62*(3), 226–236.

141. Saavedra Gallo, G. (2015). Los futuros imaginados de la pesca artesanal y la expansión de la salmonicultura en el sur austral de Chile. *Chungará (arica), 47*(3), 521–539.

142. Sala, O. E., Chapin, F. S., Armesto, J. J., Berlow, E., Bloomfield, J., Dirzo, R., Huber-Sanwald, E., Huenneke, L. F., Jackson, R. B., Kinzig, A., & Leemans, R. (2000). Global biodiversity scenarios for the year 2100. *Science, 287*(5459), 1770–1774.

143. Sandoval, R. J. (1994). Estudio ecológico del visón asilvestrado (Mustela vison, Schreber) en la XI Región. DVM Thesis, Universidad Austral de Chile.

144. Schmid-Hempel, R., Eckhardt, M., Goulson, D., Heinzmann, D., Lange, C., Plischuk, S., … & Schmid-Hempel, P. (2014). The invasion of southern South America by imported bumblebees and associated parasites. *Journal of Animal Ecology, 83*(4), 823–837.

145. Schüttler, E., Crego, R. D., Saavedra-Aracena, L., Silva-Rodríguez, E. A., Rozzi, R., Soto, N., & Jiménez, J. E. (2019). New records of invasive mammals from the sub-Antarctic cape Horn archipelago. *Polar Biology, 42*(6), 1093–1105.

146. Schüttler, E., Klenke, R., McGehee, S., Rozzi, R., & Jax, K. (2009). Vulnerability of ground-nesting waterbirds to predation by invasive American mink in the Cape Horn Biosphere Reserve Chile. *Biological Conservation, 142*(7), 1450–1460.

147. Schüttler, E., Saavedra-Aracena, L., & Jiménez, J. E. (2018). Domestic carnivore interactions with wildlife in the Cape Horn Biosphere Reserve, Chile: Husbandry and perceptions of impact from a community perspective. *PeerJ, 6,* e4124.

148. Sepúlveda, M. A., Muñóz-Zanzi, C., Rosenfeld, C., Jara, R., Pelican, K. M., & Hill, D. (2011). Toxoplasma gondii in feral American minks at the Maullín river, Chile. *Veterinary Parasitology, 175,* 60–65

149. Sepúlveda, M. A., Singer, R. S., Silva-Rodríguez, E. A., Eguren, A., Stowhas, P., & Pelican, K. (2014). Invasive American mink: Linking pathogen risk between domestic and endangered carnivores. *EcoHealth, 11*(3), 409–419.

150. Simberloff, D., & Von Holle, B. (1999). Positive interactions of nonindigenous species: Invasional meltdown? *Biological Invasions, 1*(1), 21–32.

151. Simberloff, D., Martín, J. L., Genovesi, P., Maris, V., Wardle, D. A., Aronson, J., Courchamp, F., Galil, B., García-Berthou, E., Pascal, M., & Pyšek, P. (2013). Impacts of biological invasions: What's what and the way forward. *Trends in Ecology y Evolution, 28*(1), 58–66.

152. Sola, F. J., J Valenzuela, A. E., Anderson, C. B., Martínez Pastur, G., & Lencinas, M. V. (2015). Reciente invasión del Archipiélago de Tierra del Fuego por la avispa Vespula germanica (Hymenoptera: Vespidae). *Revista de la Sociedad Entomológica Argentina, 74*(3–4), 197–202.

153. Soto, D., Arismendi, I., González, J., Sanzana, J., Jara, F., Jara, C., … & Lara, A. (2006). Southern Chile, trout and salmon country: Invasion patterns and threats for native species. *Revista chilena de Historia Natural, 79*(1), 97–117.

154. Soto, D., Arismendi, I., Di Prinzio, C., & Jara, F. (2007). Establishment of chinook salmon (*Oncorhynchus tshawytscha*) in Pacific basins of southern South America and its potential ecosystem implications. *Revista Chilena De Historia Natural, 80,* 81–98.

155. Soto, D., Jara, F., & Moreno, C. (2001). Escaped salmon in the inner seas, southern Chile: Facing ecological and social conflicts. *Ecological Applications, 11*(6), 1750–1762.

156. Soto, D., León-Muñoz, J., Dresdner, J., Luengo, C., Tapia, F. J., & Garreaud, R. (2019). Salmon farming vulnerability to climate change in southern Chile: Understanding the biophysical, socioeconomic and governance links. *Reviews in Aquaculture, 11*(2), 354–374.

157. Steffen, W., Broadgate, W., Deutsch, L., Gaffney, O., & Ludwig, C. (2015). The trajectory of the Anthropocene: The great acceleration. *The Anthropocene Review, 2*, 81–98.

158. Tavie, J. D., Vieli, L., & Montalva, J. (2015). Nuevos antecedentes acerca de la presencia *de Bombus dahlbomii* Guérin-Méneville (Hymenoptera: *Apidae*) en la isla Grande de Tierra del Fuego. *Anales Del Instituto De La Patagonia, 43*(1), 171–174.

159. Taylor, K. T., Maxwell, B. D., McWethy, D. B., Pauchard, A., Nuñez, M. A., & Whitlock, C. (2017). *Pinus contorta* invasions increase wildfire fuel loads and may create a positive feedback with fire. *Ecology, 98*(3), 678–687.

160. Thomas, W. H., & Gibson, C. H. (1990a). Effects of small-scale turbulence on microalgae. *Journal of Applied Phycology, 2*(1), 71–77.

161. Thomas, W., Vernet, M., & Gibson, C. (1995). Effects of small-scale turbulence on photosynthesis, pigmentation, cell division, and cell size in the marine dinoflagellate *Gonyaulax polyedra* (Dinophyceae). *Journal of Phycology, 31*, 50–59.

162. Thomas, W. H., & Gibson, C. H. (1990b). Quantified small-scale turbulence inhibits a red tide dinoflagellate, *Gonyaulax polyedra* Stein. *Deep Sea Research Part A. Oceanographic Research Papers, 37*(10), 1583–1593.

163. Torres-Salinas, R., García-Carmona, A., & Rojas-Hernández, J. (2017). Privatizando el agua, produciendo sujetos hídricos: Análisis de las políticas de escala en la movilización socio-hídrica contra Pascua Lama e HidroAysén en Chile. *Agua y Territorio, 10*, 149–166.

164. Torres-Salinas, R., & Carmona, A. G. (2009). Conflictos por el agua en Chile: El gran capital contra las comunidades locales. Análisis comparativo de las cuencas de los ríos Huasco (desierto de Atacama) y Baker (Patagonia austral). *Espacio Abierto, 18*(4), 695–708.

165. Valenzuela, A. E., Anderson, C. B., Fasola, L., & Cabello, J. L. (2014). Linking invasive exotic vertebrates and their ecosystem impacts in Tierra del Fuego to test theory and determine action. *Acta Oecologica, 54*, 110–118.

166. Valenzuela, A. E., Rey, A. R., Fasola, L., & Schiavini, A. (2013). Understanding the interspecific dynamics of two co-existing predators in the Tierra del Fuego archipelago: The native southern river otter and the exotic *American mink*. *Biological Invasions, 15*(3), 645–656.

167. Valenzuela, A. E., Sepúlveda, M. A., Cabello, J. L., & Anderson, C. B. (2016). El visón ameri- cano en Patagonia: Un análisis histórico y socioecológico de la investigación y el manejo. *Mastozoología Neotropical, 23*(2), 289–304.

168. Veblen, T. T., Holz, A., Paritsis, J., Raffaele, E., Kitzberger, T., & Blackhall, M. (2011). Adapting to global environmental change in Patagonia: What role for disturbance ecology? *Austral Ecology, 36*(8), 891–903.

169. Veblen, T. T., & Markgraf, V. (1988). Steppe expansion in Patagonia? *Quaternary Research, 30*(3), 331–338.

170. Vila, A. R., Falabella, V., Gálvez, M., Farías, A., Droguett, D., & Saavedra, B. (2016). Identifying high-value areas to strengthen marine conservation in the channels and fjords of the southern Chile ecoregion. *Oryx, 50*(2), 308–316.

171. Villacide, J. M., Masciocchi, M., & Corley, J. C. (2014). Avispas exóticas en la Patagonia: La impor tancia de la ecología de invasiones en el manejo de plagas. *Ecología Austral, 24*(2), 154–161.

172. Villafaña, V. E., Helbling, E. W., & Zagarese, H. E. (2001). Solar ultraviolet radiation and its impact on aquatic systems of Patagonia, South America. *AMBIO: A Journal of the Human Environment, 30*(2), 112–117.

173. Villaseñor-Parada, C., Pauchard, A., Ramírez, M. E., & Macaya, E. C. (2018). Macroalgas exóti- cas en la costa de Chile: Patrones espaciales y temporales en el proceso de invasión. *Latin American Journal of Aquatic Research, 46*(1), 147–165.

174. Vitousek, P. M. (1994). Beyond global warming: Ecology and global change. *Ecology, 75*(7), 1861–1876.

175. Vitousek, P. M., Mooney, H. A., Lubchenco, J., & Melillo, J. M. (1997). Human domination of Earth's ecosystems. *Science, 277*(5325), 494–499.
176. Vázquez, D. P. (2002). Multiple effects of introduced mammalian herbivores in a temperate forest. *Biological Invasions, 4*(1–2), 175–191.
177. Wallem, P. K., Jones, C. G., Marquet, P. A., & Jaksic, F. M. (2007). Identifying the mechanisms underlying the invasion of *Castor canadensis* (Rodentia) into Tierra del Fuego Archipiélago Chile. *Revista Chilena De Historia Natural, 80*(3), 309–325.
178. Wells, M. L., Trainer, V. L., Smayda, T. J., Karlson, B. S., Trick, C. G., Kudela, R. M., Ishikawa, A., Bernard, S., Wulff, A., Anderson, D. M., & Cochlan, W. P. (2015). Harmful algal blooms and climate change: Learning from the past and present to forecast the future. *Harmful Algae, 49*, 68–93.
179. Winckler, P., Contreras-López, M., Vicuña, S., Larraguibel, C., Mora, J., Esparza, C., Salcedo, J., Gelcich, S., Fariña, J. M., Martínez, C., Agredano, R., Melo, O., Bambach, N., Morales, D., Marinkovic, C., and Pica, A. (2019). *Determinación del riesgo de los impactos del cambio climático en las costas de Chile.* Ministerio del Medio Ambiente. Retrieved from: https://cambioclimatico.mma.gob.cl/wp-content/uploads/2020/04/2019-10-22-Informe-V02-CCCostas-Exposicio%CC%81n-Rev1.pdf
180. Young, K. A., Dunham, J. B., Stephenson, J. F., Terreau, A., Thailly, A. F., Gajardo, G., & García de Leaniz, C. (2010). A trial of two trouts: Comparing the impacts of rainbow and brown trout on a native galaxiid. *Animal Conservation, 13*(4), 399–410.
181. Zegers, G., Larraín, J., Díaz, M. F., & Armesto, J. (2006). Impacto ecológico y social de la explotación de pomponales y turberas de *Sphagnum* en la isla grande de Chiloé. *Revista Ambiente y Desarrollo, 22*(1), 28–34.

Part III

Terrestrial Ecosystems

Representativeness Assessment and Identification of Priorities for the Protection of Terrestrial Ecosystems in Chilean Patagonia

Patricio Pliscoff, María José Martínez-Harms, and Taryn Fuentes-Castillo

Abstract

Protected area systems are the primary tool to guarantee the conservation of biodiversity and the multiple ecosystem services vital for human well-being. The protected area system is even more relevant in Chilean Patagonia, since it is one of the most pristine areas of the planet, with a great diversity of ecosystems, species richness and diversity. The identification of conservation gaps and priorities is a first step in the evaluation of a protection system. Chilean Patagonia has been described as a zone with a large amount of protected area, but some of its ecosystems have been identified as under-represented in protected areas. This chapter analyzes the representativeness of the system of protected areas in Chilean Patagonia, including an assessment of priorities for in situ protection of the terrestrial system. The results show underrepresentation of ecosystem and faunal species diversity in Chilean Patagonia. The current network of protected areas represents only 20% of terrestrial ecosystems. A bias in representation is identified towards higher altitude zones, glaciers-ice fields and areas of lower opportunity cost in the region. Protection gaps indicate a representation of less than 17% (Aichi Target 11) in steppe and deciduous forest ecosystems. The representation of faunal diversity is not adequately considered by the current protection network, including areas where less than 30%

P. Pliscoff (✉) · M. J. Martínez-Harms
Department of Ecology, Faculty of Biological Sciences, Pontificia Universidad Católica de Chile, Santiago, Chile
e-mail: pliscoff@uc.cl

P. Pliscoff · T. Fuentes-Castillo
Faculty of History, Geography and Political Science, Institute of Geography, Pontificia Universidad Católica de Chile, Santiago, Chile

P. Pliscoff · M. J. Martínez-Harms
Center for Applied Ecology and Sustainability (CAPES), Universidad Católica de Chile, Santiago, Chile

© Pontificia Universidad Católica de Chile 2023
J. C. Castilla et al. (eds.), *Conservation in Chilean Patagonia*, Integrated Science 19, https://doi.org/10.1007/978-3-031-39408-9_3

of the total diversity is concentrated. The priority areas identified are concentrated in the northern zone of Patagonia (Chiloé and Palena province), steppes and in the steppe-deciduous forest-steppe transition zone, both in the Aysén and Magallanes Regions. The gaps in representation prevent adequate adaptation to the conservation challenges that arise with the impact of climate change. Effects of this include the loss of biodiversity components and the redistribution of species and ecosystems. Recommendations are proposed to improve this type of representativeness assessment at different time horizons. Participatory instances should be sought for the definition of conservation targets and goals which analyze terrestrial and marine environments as an integrated study area. The current deficits in the representation of these ecosystems show the urgent need to address the gaps in representation of the current network of Patagonian protected areas and the need to improve the representation of under-represented ecosystems.

Keywords

Patagonia • Chile • Terrestrial ecosystems • Species • Representativeness • Gap analysis • Species distribution models • Spatial prioritization

1   Introduction

Chilean Patagonia is considered one of the most pristine areas of the planet due to its low human footprint, which is the product of a reduced and relatively recent history of occupation [14, 36]. In addition, its unique position in the global context, as the continental land mass closest to the South Pole and the Antarctic, confers unique characteristics to its terrestrial and marine biota [9]. This particularity is also expressed in the presence of climatic factors that define its current characteristics, including the proximity to the South Pole ice mass and the presence of the westerly wind belt [10]. This landscape has the particularity of being a territory of recent conformation, where traces of the retreat of the last glacial maximum (*ca.* 12 thousand years ago) can be seen along its entire length [33].

The large ecosystems present in Chilean Patagonia decrease latitudinally in plant diversity, except for bryophytes and lichens, which dominate in the extreme south of the area [28, 35]. There are three types of forests in the northern sector of Patagonia (41°–47° S): (i) the evergreen forest that extends towards the north of Patagonia; (ii) the conifer-dominated forests, where alerce (*Fitzroya cupressoides*) and Guaitecas cypress (*Pilgerodendron uviferum*) are the dominant species, and finally a deciduous forest that marks the transition with the Patagonian steppe. This last ecosystem is exclusive to this area of the world; in Chile it is located in the eastern part of the Andes. In the coastal sector, the evergreen forest is intertwined with moorlands-dominated soils in the lower elevation areas with less slope [42]. A longitudinal pattern can be recognized in the distribution of the main ecosystems, evergreen and coniferous forest with moorlands in the coastal and inland zone, deciduous forest that marks the forest-steppe transition, and the Patagonian steppe, which dominates the entire inland zone and the border with Argentina throughout the area [16]. Fifty-four percent of the land area of Chilean Patagonia (from Reloncaví sound to the Diego Ramírez islands) is protected, which represents the majority of land area under protection in the country [40] and 86.4% of the National System of Protected Areas (in Spanish SNASPE) in Chile. Chilean Patagonia is also home to the three largest protected areas in the country: Bernardo O'Higgins National Park, Alberto de Agostini National Park and Kawésqar National Reserve.

The science of biological conservation offers tools to analyze the different actions and options that can be carried out in the territory, in response to the protection policies and international commitments subscribed to by countries. One of these approaches is systematic conservation planning, which facilitates the step-by-step determination of conservation goals and objectives through optimal solutions in a transparent and replicable process [17, 32]. Systematic planning seeks to represent previously defined conservation targets (e.g. biodiversity, ecosystem services) through the best possible solution that fulfills the established goals. The definition of conservation targets can be through "surrogates" that represent different dimensions of biodiversity, which due to lack of data or the impossibility of collecting information, cannot be considered in the planning process. The resulting solutions then permit the most strategic approach to incorporating new protected areas to the protected area system so as to minimize the associated costs (e.g. land suitability and value) and surface area involved.

The natural environment in Chilean Patagonia is a unique landscape in the world, which due to its pristine characteristics must be effectively and efficiently conserved and protected. The evaluation of the current representativeness and the identification of priorities for the protection of terrestrial ecosystems is a fundamental exercise for the definition of conservation priorities for the natural biota present in Chilean Patagonia.

2 Scope and Objectives

The general objectives of this chapter are: (i) to determine if there are gaps in the representation of terrestrial ecosystems in Chilean Patagonia; (ii) to identify priorities for the protection of terrestrial ecosystems and species of flora and fauna; and (iii) to identify opportunities to improve the representation of the diversity of terrestrial ecosystems in the SNASPE across Chilean Patagonia, which is defined as the area between Reloncaví Sound and the Diego Ramírez islands (41° 42′ S 73° 02′ W; 56° 29′ S 68° 44′ W).

The first stage included a review of the existing literature on representativeness analysis (terrestrial ecosystems and species) and spatial prioritization, both nationally and for Chilean Patagonia. This review included analysis of a database of scientific articles [2], as well as articles, reports and theses of national coverage. The second stage consisted of the application of a spatial prioritization method, including species and ecosystems as conservation targets. Subsequently, an analysis of the representativeness of terrestrial ecosystems and species of flora and fauna present in Chilean Patagonia was carried out. Finally, recommendations are presented for the adequate conservation of the ecosystem processes and services within the SNASPE. These also indicate actions necessary for the adjustment and improvement of the system of protected areas in Chilean Patagonia.

3   Methods

3.1   Literature Review

Three search strategies were carried out with the objective of compiling the available literature related to the current representativeness and conservation gaps in the protected areas of Chilean Patagonia. The first strategy involved searching for references using the search terms "conservation", "gaps", "priorities", "representativeness", "protected areas", "ecosystems" and "biogeographic regions". As a second search strategy, Google Scholar was used to repeat the previous search arguments, adding the words: Chile and Patagonia and Aysén and Magallanes. Finally, the topics used in the previous search were consulted in the "ISI web of knowledge" search engine, which was extended to include reports, theses and books.

3.2   Assessment of Protection Gaps and Priorities

3.2.1   Spatial Prioritization Analysis

A spatial prioritization analysis was performed [15] considering two types of conservation targets: terrestrial ecosystems (vegetation belts) and species (flora and fauna). The objective was to determine a set of priority areas to evaluate the gaps in representativeness of the diversity of terrestrial ecosystems in Chilean Patagonia, and to identify opportunities to strengthen the representation of terrestrial biodiversity of the SNASPE in Patagonia. Ecosystems and species were selected as conservation targets because they represent the two levels of biodiversity with the greatest amount of available information. Other types of targets, such as those related to ecosystem processes, were not considered in this chapter. Spatial prioritization was carried out using the Zonation software [26]. This software applies a prioritization meta-algorithm according to different cell removal rules that minimize the marginal loss of the total landscape [26].

The vegetation belts of Luebert and Pliscoff [16] were used as descriptors of terrestrial ecosystems in Chilean Patagonia. This proposal has been defined by the Ministry of the Environment (in Spanish Ministerio de Medio Ambiente, MMA) as the official classification of terrestrial ecosystems in Chile. Vegetation classification systems have been widely used as surrogates for ecosystems, because vegetation is an integral element of the biophysical attributes of an area [3]. Species Distribution Models (SDMs) were used to analyze the set of flora and fauna species present in

Chilean Patagonia. The SDMs are empirical models that relate records of occurrences with predictive environmental variables, with the purpose of modeling the distribution of species or set of species in geographic space [12].

Choique post-Roballos border crossing, Patagonia Park, Aysén Region. Photograph by Jorge López

Occurrence records of vascular flora species in Chilean Patagonia were compiled, based on information from herbaria in Chile (Universidad de Concepción, Museo Nacional de Historia Natural, Herbarium de la Facultad de Ciencias Forestales y de la Conservación de la Naturaleza, Universidad de Chile). Additionally, the vascular flora database compiled by Scherson et al. [37] was considered. If the locality of the herbarium record did not have geographic coordinates, these were assigned manually. Occurrence records were considered for four groups of fauna: mammals, reptiles, amphibians and birds. This information was obtained from the report by Marquet et al. [18] and from the specimen database of the Chilean MMA (Global Environmental Inforamtion Facility-Chile). Records with geographic coordinates for all taxonomic groups were corrected for locality inconsistencies, duplicates and synonymies.

Temperature and precipitation data used as environmental predictor variables to develop SDMs. The climatic basis for current conditions was obtained from the repository published by Pliscoff et al. [31], from which bioclimatic surfaces with a spatial resolution of 1 km × 1 km were used, representing a 50-year period (1950–2000) for southern South America. The "Maxent" software was used to model the species distributions of all taxonomic groups [7, 29]. To avoid spatial correlation of occurrences, records that were less than 4 km apart were removed, and

Table 1 Number of species, relationship to national diversity by taxonomic group [23] and occurrence records used for SDM

Taxonomic group	Number of species included in the modeling	Percentage of species with respect to the national total (%)	Number of records included in the modeling
Vascular flora	783	13.8	18.511
Amphibians	7	11.3	112
Birds	40	8.6	392
Mammals	10	6.2	136
Reptiles	5	4.1	50
Total	844	–	–

those species with at least 10 records were selected. The final database for Chilean Patagonia is composed of 18,511 records for flora and fauna. The breakdown by taxonomic group is presented in Table 1 and its spatial distribution in Fig. 1.

The less correlated bioclimatic variables were selected as predictor variables for the SDM. The selected variables were: mean annual temperature, seasonality of temperature, maximum temperature of the warmest month, annual precipitation and seasonality of precipitation. The records of occurrence of each species were partitioned into percentages of 80% to train the model and 20% to generate the tests of the final models by species. The SDMs were estimated according to the probability of occurrence of each species, and binary SDMs were estimated (presence-absence of a species), applying the sensitivity–specificity threshold. The modeling was carried out considering the limits of Chile, and then restricting the results to the study area defined for Chilean Patagonia.

Species richness of flora and fauna was estimated by summing the distributions of each taxonomic group modeled with SDM. These were grouped into the total species richness per taxonomic group (plants, mammals, amphibians, reptiles and birds), and the richness of threatened species per taxonomic group. The classification of species of the Ministry of the Environment of the Chile [24] was used for the definition of threatened species, up to process n°16.

For the spatial prioritization in Chilean Patagonia, we considered as input data the vegetation belts present in the area (36) and the distribution models of the 844 species of flora and fauna, from which four scenarios were generated (see Table 2). The aim for total species richness is to identify the areas with the greatest diversity of the five selected taxonomic groups, using the core-area (CA) prioritization method. By considering the diversity of threatened species separately, we seek to prioritize the areas in which these species are represented, in this case using the additive benefit function (ABF) prioritization method. Finally, scenarios were analyzed to identify the contribution of SNASPE areas to fill gaps in representation. A prioritization was first carried out for the entire area of Chilean Patagonia, and then the exercise was repeated considering only the areas outside the SNASPE (Table 2).

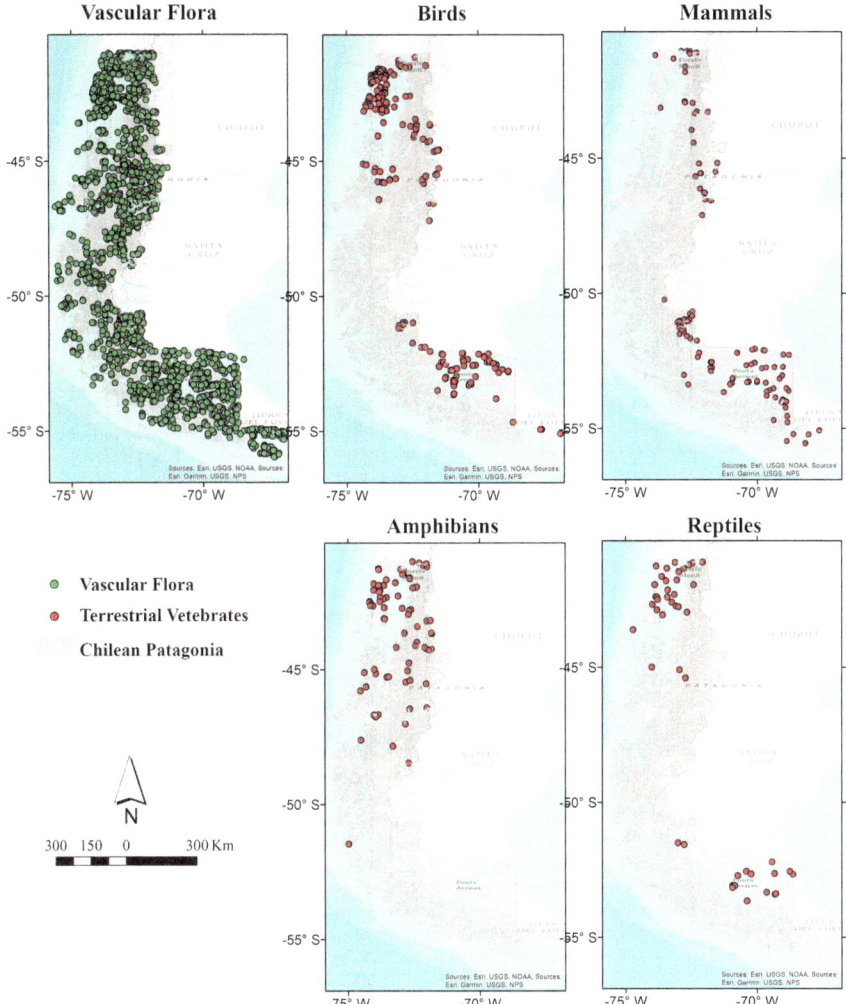

Fig. 1 Database by taxonomic groups present in Chilean Patagonia

The final result of the prioritization is a percentage ranking of the total area of Chilean Patagonia, from which the 17% with the highest priority was selected, following the protection target defined by the Convention on Biological Diversity (CBD) for terrestrial ecosystems [5]. Finally, the spatial correspondence between the highest priority areas and those belonging to the SNASPE was analyzed using the ArcGis 10.6 Geographic Information System [8].

Table 2 Prioritization scenarios for terrestrial species and ecosystems in Chilean Patagonia

Scenario	Conservation objects		Removal rule	Spatial dimension
	Ecosystems	Species		
1 (41)	36	Richness of taxonomic groups	ABF	All of Chilean Patagonia
2 (37)	36	Threatened species richness	CA	All of Chilean Patagonia
3 (41)	36	Richness of taxonomic groups	ABF	Areas outside SNASPE
4 (37)	36	Threatened species richness	CA	Areas outside SNASPE

The number of conservation features used for each scenario is indicated in brackets
* SNASPE (National System of State Protected Areas)

3.2.2 Representativeness Analysis of Terrestrial Ecosystems and Species of Flora and Fauna in Chilean Patagonia

The percentage of surface area protected by the SNASPE in each of the vegetation belts of Luebert and Pliscoff [16] present in Chilean Patagonia was estimated by superimposition mapping. The SNASPE boundaries were obtained from the MMA's national registry of protected areas [25]. The total area of each ecosystem was estimated by defining the area of the vegetation belt and the remaining area of vegetation. This remaining area is obtained from the anthropic land use categories (agricultural, urban and forest plantation areas) defined in the Cadastre of Native Vegetation Resources of the National Forestry Corporation [4], in which regional updates were used for Los Lagos (2013), Aysén (2013) and Magallanes (2005). The estimate of the percentage of protected area was calculated as the ratio between the total area of the ecosystem and the remaining area.

The Aichi target, in addition to setting a target of 17% protection of terrestrial ecosystems [5], considers all levels of biodiversity organization, so it is also relevant to analyze the representation of flora and fauna species present within the network of protected areas. The SDMs were analyzed in two groups: flora (vascular plants) and fauna (mammals, birds, reptiles and amphibians). Finally, the species richness of the two groups was grouped into deciles and the surface areas within the SNASPE areas were calculated.

4 Results

4.1 Bibliographic Review

Twelve documents referring to the terrestrial and marine environments of Chilean Patagonia were identified and reviewed (Table 3). These documents presented either representation analyses or gaps in the conservation of terrestrial environments, both for Chilean Patagonia and for the whole country. Ten of the documents analyzed address the terrestrial system and two address the marine system. The documents that analyze the terrestrial system are primarily national technical reports [34], Geobiota [11], only one scientific article focuses on Chilean Patagonia [19]. The rest of the studies analyzed the identification of conservation priorities and the analysis of the representativeness of the SNASPE, also considering other categories of protection (e.g. private conservation initiatives, nature sanctuaries, protected national assets).

Two approaches were to define terrestrial ecosystems. The first uses vegetation as a proxy of terrestrial ecosystems; this is the case of Luebert and Pliscoff's vegetation belts [27, 30, 39]. The second uses the classification of ecoregions to assess national protection gaps [21, 38, 39]. Two studies considered other conservation targets to evaluate their representation in the SNASPE; Tognelli et al. [41] established conservation priorities based on the distribution of terrestrial vertebrates and Durán et al. [6] used the ecosystem services approach.

4.2 Assessment of Protection Gaps and Priorities

4.2.1 Spatial Prioritization Analysis

The first prioritization scenario considered terrestrial ecosystems and the taxonomic groups modeled (plants, amphibians, mammals, reptiles and birds). The results of this scenario suggest that the greatest concentration of areas of importance is in the northern sector of Patagonia (between 41° and 47° S); specifically, the areas of highest priority are concentrated in the area of the Palena Province and south of Chiloé Island. Other priority zones are identified in the interior of the Aysén Region and in the forest-steppe transition zone in the Magallanes Region.

The results for the second scenario, which considers the distribution of threatened species, presented differences with respect to the first scenario due to the fact that the priority area within the Aysén Region is increased, including the western zone of the archipelagos in the Southern Ice Fields and expanding the priority areas in the Magallanes Region. Excluding the current areas of the SNASPE (third and fourth scenarios) did not modify the results substantially, and the same priority areas shown in the two previous scenarios were maintained (Fig. 2).

4.2.2 Analysis of the Representativeness of Terrestrial Ecosystems and Species of Flora and Fauna

The analysis of the representativeness of terrestrial ecosystems allows us to identify the current gaps in the SNASPE (Fig. 3) and the representation of the different percentages of diversity of flora and fauna species (Fig. 4). The results show an imbalance of current protection between the ecosystems present in the archipelago zone (which are over-represented) versus the interior-south zone of Chilean Patagonia (which are under-represented), and are located on the border with Argentina (Fig. 3B). The archipelago zone has moorlands and evergreen forest ecosystems and the interior-south zone has steppe and deciduous forest ecosystems (Fig. 3B).

Table 3 Documents identified in the literature review

Year	Authors	Title	Font type
2018	Núñez-Ávila, M., E. Corcuera, A. Farias, P. Pliscoff, J. Palma, M. Barrientos, and C. Sepúlveda	Diagnosis and characterization of private conservation initiatives	Technical report
2018	Schutz, J.	Creating an integrated protected area network in Chile: a GIS assessment of ecoregion representation and the role of private protected areas	Scientific article
2017	Luebert, F., and Pliscoff, P.	Bioclimatic and vegetational synopsis of Chile	Book
2015	Martínez-Tilleria, K.	Optimizing a marine-terrestrial conservation portfolio for Chile: effects and consequences of integration	Thesis
2015	Pliscoff, P.	Application of the International Union for Conservation of Nature (IUCN) criteria for risk assessment of Chile's terrestrial ecosystems	Technical report
2013	Durán, A. P., Casalegno, S., Marquet, P., and Gaston, K. J.	Representation of ecosystem services by terrestrial protected areas: Chile as a case study	Scientific article
2012	Squeo, F., Estévez, R., Stoll, A., Gaymer, C., Letelier, L., and Sierralta, L.	Towards the creation of an integrated system of protected areas in Chile: achievements and challenges	Scientific article
2011	Pliscoff, P., and Fuentes-Castillo, T.	Representativeness of terrestrial ecosystems in Chile's protected area system	Scientific article
2011	Geobiota Consultants	Systematization and proposal of national conservation objectives, criteria for re-presentation and prioritization, and qualification and management at national, regional and local levels of priority sites for biodiversity conservation. Ministry of the Environment. Chile	Technical report
2008	Tognelli, M. F., Ramírez de Arellano, P., and Marquet, P.	How well do the existing and proposed reserve networks represent vertebrate species in Chile?	Scientific article
2008	Martínez-Harms, M. J., and Gajardo, R.	Ecosystem value in the western Patagonia protected areas	Scientific article
2007	Ramírez de Arellano, P.	Systematic conservation planning in Chile: sensitivity of reserve selection procedures to target choices, cost surface, and spatial scale	Thesis

See full citations in the references section

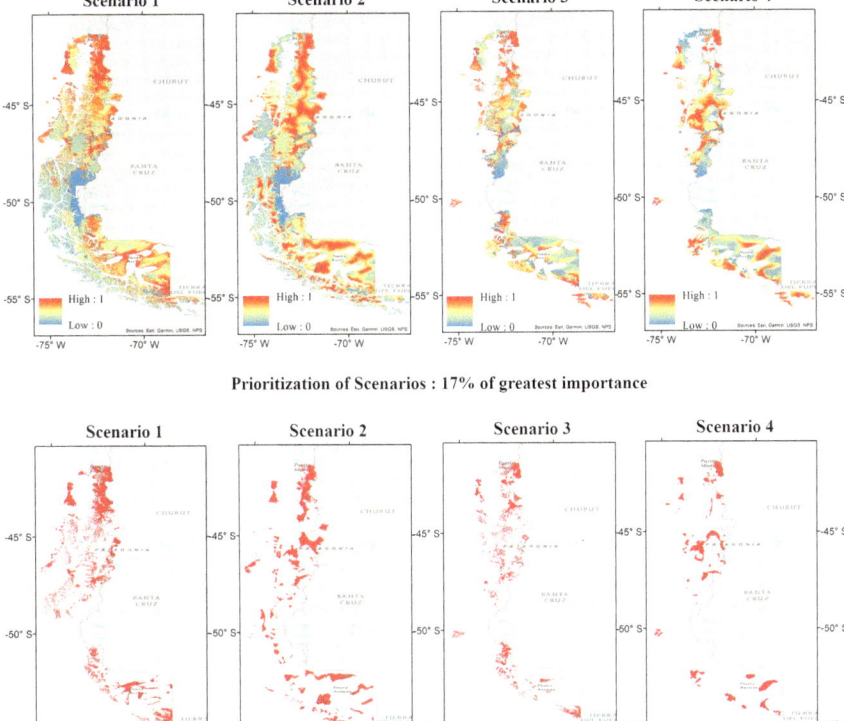

Fig. 2 Prioritization of scenarios analyzed in Chilean Patagonia. Above: scenarios prioritized by ranking. Below: indicates the 17% of greatest importance (protection goal)

The analysis of species representativeness (Fig. 4A, B) shows that the highest concentrations of richness for both flora and fauna are found in the northern zone of Chilean Patagonia between 41° and 44° S. These areas of greatest richness are located in Chiloé in the case of flora and in the province of Palena for fauna. The SNASPE adequately represents the areas of greatest floral richness, however, for fauna, only areas with richness less than 30% are represented (Fig. 4C), indicating an under-representation of the areas where fauna is concentrated in Chilean Patagonia.

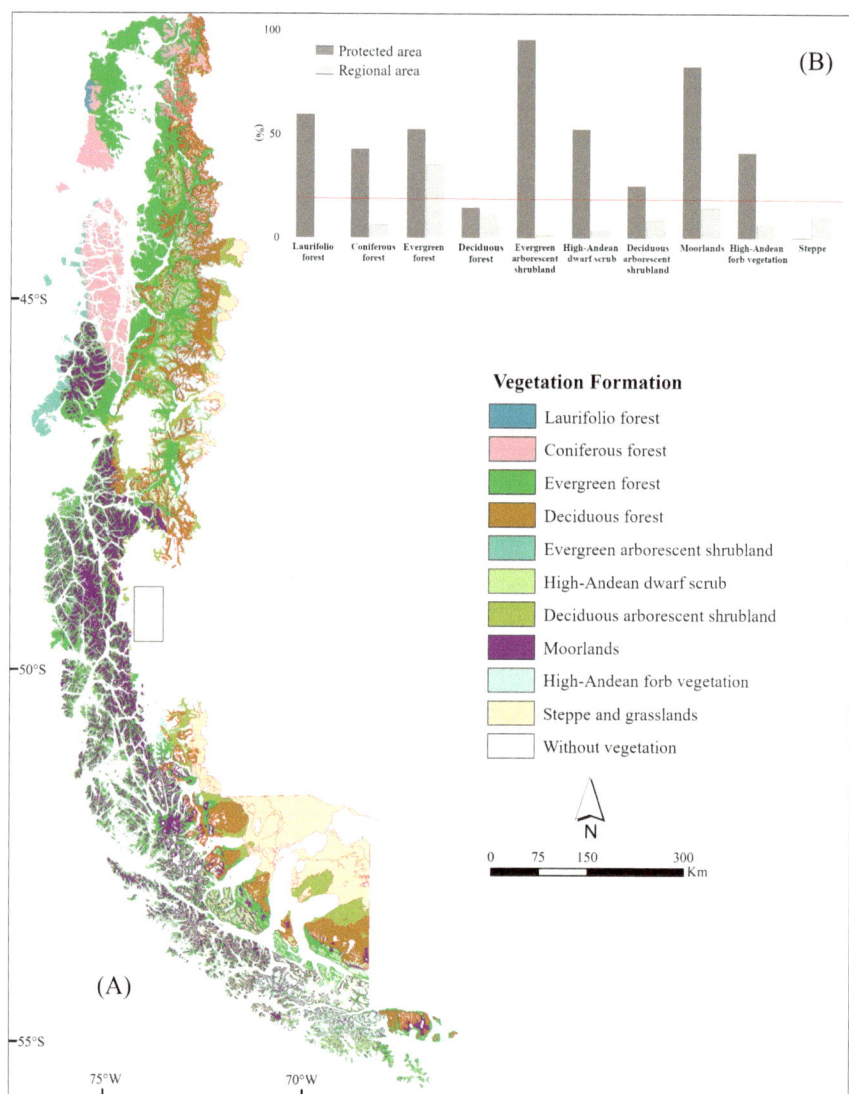

Fig. 3 Representativeness of terrestrial ecosystems in Chilean Patagonia. **A** Shows the distribution of vegetation formations (VF); red border indicates ecosystems below the 17% protection target. **B** Bar graph showing the protected area versus the total regional area of each VF. Ecosystems below the 17% target are indicated by the red line

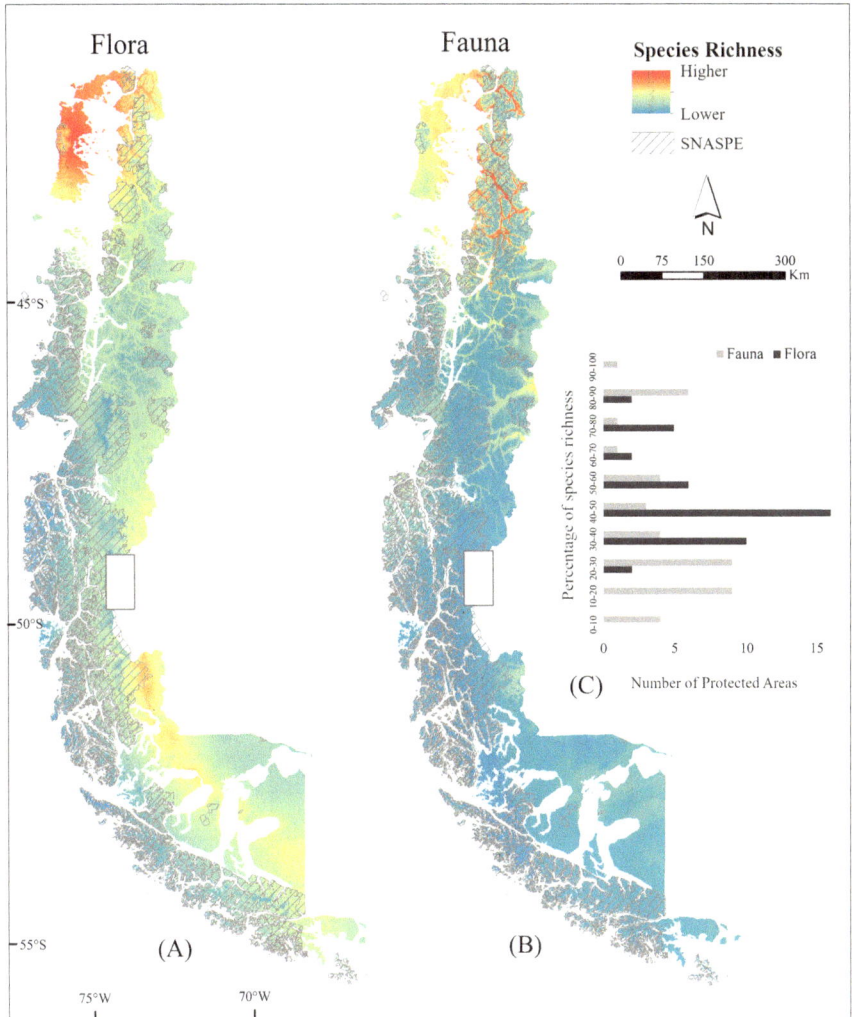

Fig. 4 Species representativeness in Chilean Patagonia. **A** Flora species richness. **B** Fauna species richness. **C** Bar graph shows the representation of percentages of flora and fauna richness versus the number of SNASPE units

5 Discussion

The results obtained regarding the representativeness of terrestrial ecosystems in Chilean Patagonia allow us to determine some key elements for discussion. Although the identification of protection gaps in deciduous forest-steppe and steppe transition ecosystems had already been reported previously in representativeness analyses [16, 30], one of the contributions of this chapter is to demonstrate that it is possible to use both the distribution of species and terrestrial ecosystems to prioritize areas for conservation, including or excluding the SNASPE.

The distribution of flora and fauna species had not been evaluated for Chilean Patagonia using a prioritization approach, but there was a history that the current protected areas did not consider the representation of the variety of existing ecosystems, concentrating only on some types of ecosystems (e.g. moorlands, evergreen forest) [39]. For terrestrial ecosystems, we worked with the most detailed spatial definition currently existing for Chile (vegetation belts), but due to the focus of this work, certain types of ecosystems have been excluded (e.g. azonal ecosystems).

The main information gaps detected in this chapter are the lack of data to analyze ecological processes and threats to biodiversity resulting from human activities. To improve the approach developed in this chapter, it is necessary to consider different cost categories (socioeconomic, biodiversity) in an integrated manner, in order to meet conservation goals [20]. Also, land-use cover needs to be included to model the current distribution of ecosystems.

The analyses in this chapter considered the potential distribution of each ecosystem, which does not affect the results presented since the areas with the greatest human intervention in Chilean Patagonia are small (in surface area) compared to other intervened areas in Chile. If the analysis had considered anthropically intervened areas, the results would probably have been different for the transition zone between the deciduous forest and the steppe, which is one of the most disturbed areas in Chilean Patagonia [14]. New databases of flora and fauna occurrences are also needed to obtain more robust SDMs, as well as to include new species that were not contemplated in this study. This aspect is especially relevant for fauna, where the available databases do not have representative information for some taxonomic groups, for example, birds and mammals in Chilean Patagonia.

In the literature there are national and global methodological approaches that could complement the approach of this work. For example, Durán et al. [6], developed a mapping of ecosystem services at the national scale using a prioritization approach similar to the one proposed in this chapter. At the global scale there is greater availability of mapped information on ecosystem services (e.g. carbon sequestration, tourism and recreation, among others) relevant to develop future regional prioritization exercises [13].

6   Conclusions and Recommendations

The definition of the coastal-terrestrial interface and the generation of adequate data that allow both systems to be analyzed in an integrated manner should be the focus of research in the short term. The literature review conducted for this chapter found only one study that prioritized both marine and terrestrial systems at the national level, and this defined separate conservation targets for each [21]. With regard to the representativeness of terrestrial ecosystems, the greatest gaps in representation were identified in steppe ecosystems and in the deciduous forest-steppe transition, both of which are located in the Aysén Region and in Magallanes.

The diversity of fauna species is not adequately represented in the SNASPE, which is currently concentrated in areas of lower richness. This situation is different for flora, where the areas of highest species richness have greater representation in the SNASPE.

The prioritization scenarios allowed us to analyze the patterns of distribution and richness of flora and fauna in Chilean Patagonia. The results indicate that the SNASPE does not represent the 17% of areas with the greatest importance (Fig. 2). Priority areas were identified in the insular Chiloé and Palena province (Los Lagos Region), and in the inland area of the Coyhaique and General Carrera provinces (Aysén Region), Última Esperanza province, Magallanes and Tierra del Fuego (Magallanes Region).

The following is a set of recommendations aimed at different stakeholder groups, such as decision-makers, scientists and others.

- **Short-term recommendations (1–2 years)**. (i) Conduct a new terrestrial prioritiztion analysis using updated information generated by publications and repositories of global biodiversity and ecosystem services information. The study should consider other protected area categories (e.g., private conservation initiatives) in the representativeness analysis. The level of administrative management of protected areas should also be considered, as it is possible that areas identified as priorities in future analyses may not have effective protection in conservation category areas. It is also important to analyze the costs of conservation in the prioritization analysis, in case priority areas are not represented in protected areas. The new analysis should include researchers from Chilean Patagonia. (ii) Develop a participatory and open process for the definition of conservation objectives and goals, in order to be included in the next prioritization exercises. There are studies carried out in the marine system [22, 43] which could be a methodological guide to be replicated in the terrestrial system.
- **Medium-term recommendations (up to 5 years)**. (i) Develop a prioritization analysis of the marine-terrestrial complex that accounts for the coastal interface, for example, considering runoff or glacial dynamics models; (ii) develop species distribution models (SDMs) with marine-terrestrial climate data in coastal zones in an integrated manner. This would be relevant to analyze when considering future scenarios and establishing conservation priorities [1]; (iii) re-evaluate the gaps in ecosystem representation (terrestrial and marine) by the incorporation or removal of protected areas and reserve systems.
- **Long-term recommendations (>5 years)**. Develop a system for monitoring the system of terrestrial and marine protected areas at different levels: ecosystems, species, processes and ecosystem services.

Acknowledgements The authors thank The Pew Charitable Trusts for financial support of this work. The suggestions and contributions of the editors of this book and two anonymous referees who helped improve the chapter are gratefully acknowledged. TFC gratefully acknowledges the support of ANID through Project No. 3190433.

References

1. Álvarez-Romero, J. G., Mills, M., Adams, V. M., Gurney, G. G., Pressey, R. L., Weeks, R., & Storlie, C. J. (2018). Research advances and gaps in marine planning: Towards a global database in systematic conservation planning. *Biology and Conservation, 227*, 369–382.
2. Armesto, J. J., Martínez, M. J., Castilla, J. C., & Fuentes-Castillo, T. (2023). *An integrated conservation vision for Chilean Patagonia.* Springer.
3. Austin, M. P. (1991). Vegetation: Data collection and analysis. In C. R. Margules & M. P. Austin (Eds.), *Nature conservation: Cost effective biological surveys and data analysis* (pp. 34–37). Australia CSIRO.
4. CONAF-CONAMA y BIRF. (1999). *Catastro y evaluación de los recursos vegetacionales nativos de Chile.* Informe nacional con variables ambientales. Corporación Nacional Forestal (CONAF). http://sit.conaf.cl/
5. Convention on Biological Diversity. (2010). *United Nations convention on biological diversity conference of the parties 10 Decision X/2. Strategic Plan for Biodiversity 2011–2020.* https://www.cbd.int/decision/cop/?id=12268
6. Durán, A. P., Casalegno, S., Marquet, P., & Gaston, K. J. (2013). Representation of ecosystem services by terrestrial protected areas: Chile as a case study. *PLoS ONE, 8*(12), e82643.
7. Elith, J., Phillips, S. J., Hastie, T., Dudík, M., Chee, Y. E., & Yates, C. J. (2011). A statistical explanation of "Maxent" for ecologists. *Diversity and Distributions, 17*(1), 43–57.
8. Environmental Systems Research Institute. (2017). *ArcGIS desktop: Release 10.6.* Environmental Systems Research Institute.
9. Fraser, C. I., Nikula, R., Ruzzante, D. E., & Waters, J. M. (2012). Poleward bound: Biological impacts of Southern Hemisphere glaciation. *Trends in Ecology and Evolution, 27*, 462–471.
10. Garreaud, R., López, P., Minvielle, M., & Rojas, M. (2013). Large-scale control on the Patagonian climate. *Journal of Climate, 26*, 215–230.
11. Geobiota Consultores. (2011). *Proyecto PNUD 125/2010. Sistematización y proposición de objetivos nacionales de conservación, criterios de representatividad y priorización, y calificación y gestión a nivel nacional, regional y local de sitios prioritarios para la conservación de la biodiversidad.* Ministerio del Medio Ambiente. http://bdrnap.mma.gob.cl/recursos/privados/Recursos/CNAP/GEF-SNAP/Geobiota_2012.pdf
12. Guisan, A., & Thuiller, W. (2005). Predicting species distribution: Offering more than simple habitat models. *Ecology Letters, 8*(9), 993–1009.
13. Hansen, M. C., Potapov, P. V., Moore, R., Hancher, M., Turubanova, S. A., Tyukavina, A., Thau, D., Stehman, S. V., Goetz, S. J., Loveland, T. R., Kommareddy, A., Egorov, A., Chini, L., Justice, C. O., & Townshend, J. R. G. (2013). High-resolution global maps of 21st-century forest cover change. *Science, 342*, 850–853.
14. Inostroza, L., Zasada, I., & König, H. J. (2016). Last of the wild revisited: Assessing spatial patterns of human impact on landscapes in southern Patagonia, Chile. *Regional Environmental Change, 16*, 2071–2085.
15. Kukkala, A., & Moilanen, A. (2013). The core concepts of spatial prioritization in systematic conservation planning. *Biological Review, 88*, 443–464.
16. Luebert, F., & Pliscoff, P. (2017). *Sinopsis bioclimática y vegetacional de Chile. Segunda edición.* Editorial Universitaria.
17. Margules, C. R., & Pressey, R. L. (2000). Systematic conservation planning. *Nature, 405*(6783), 243–253.
18. Marquet, P. A., Abades, S., Armesto, J. J., Barría, I., Arroyo, M. T. K., Cavieres, L. A., Gajardo, R., Garín, C., Labra, F., Meza, F., Pliscoff, P., Prado, C., Ramírez de Arellano, P. I., & Vicuña, S. (2010). *Estudio de vulnerabilidad de la biodiversidad terrestre en la eco-región mediterránea, a nivel de ecosistemas y especies, y medidas de adaptación frente a escenarios de cambio climático.* Centro de Cambio Global UC, CASEB, IEB. https://cambio-global.uc.cl/images/proyectos/Documento_10_Impactos-CC-biodiversidad-.pdf
19. Martínez-Harms, M. J., & Gajardo, R. (2008). Ecosystem value in the western Patagonia protected areas. *Journal for Nature Conservation, 16*(2), 72–87.

20. Martínez-Harms, M. J., Bryan, B. A., Wood, S. A., Fisher, D. M., Law, E., Rhodes, J. R., Dobbs, C., Biggs, D., & Wilson, K. A. (2018). Inequality in access to cultural ecosystem services from protected areas in the Chilean biodiversity hotspot. *Science of the Total Environment, 636,* 1128–1138.
21. Martínez-Tilleria, K. (2015). *Optimización de un portafolio de conservación marino-terrestre para Chile: efectos y consecuencias de la integración.* Tesis doctoral, Universidad de La Serena, Facultad de Ciencias. http://www.biouls.cl/public_php/docencia/dr-bea/conservacion/Tesis_Karina_Martinez.pdf
22. Miethke, S., & Gálvez, M. (2009). *Marine and coastal high conservation value areas in southern Chile.* International Workshop Report. WWF Chile. https://www.yumpu.com/en/document/view/22200774/marine-and-coastal-high-conservation-value-areas-in-southern-chile
23. Ministerio del Medio Ambiente. (2019). *Sexto informe nacional de biodiversidad de Chile ante el Convenio sobre la Diversidad Biológica (CDB).* Ministerio del Medio Ambiente. https://mma.gob.cl/wp-content/uploads/2020/01/6NR_FINAL_ALTA-web.pdf
24. Ministerio del Medio Ambiente. (2019). *Reglamento de clasificación de especies. Decimosexto proceso de clasificación de especies silvestres según categoría de conservación.* Decreto Supremo N°79/2018. https://clasificacionespecies.mma.gob.cl/
25. Ministerio del Medio Ambiente. (2019). *Registro nacional de áreas protegidas.* http://areasprotegidas.mma.gob.cl/
26. Moilanen, A., Meller, L., Leppänen, J., Montesino Pouzols, F., Arponen, A., & Kujala, H. (2012). *"Zonation": Spatial conservation planning framework and software version 3.1 user manual.* University of Helsinki.
27. Núñez-Ávila M., Corcuera, E., Farías, A., Pliscoff, P., Palma, J., Barrientos, M., & Sepúlveda, C. (2013). *Diagnóstico y caracterización de iniciativas de conservación privada.* Informe final. Fundación Senda Darwin en colaboración con ASI Conserva Chile A.G. para el Proyecto MMA/GEF-PNUD "Creación de un Sistema Nacional Integral de Áreas Protegidas para Chile: Estructura financiera y operacional". MMA, GEF-PNUD. http://bdrnap.mma.gob.cl/recursos/privados/Recursos/CNAP/GEF-SNAP/FundSendaDarwin_2013.pdf
28. Patiño, J., & Vanderpoorten, A. (2018). Bryophyte biogeography. *Critical Reviews in Plant Sciences, 37*(2–3), 175–209.
29. Phillips, S. J., Anderson, R. P., & Schapire, R. E. (2006). Maximum entropy modeling of species geographic distributions. *Ecological Modelling, 190*(3–4), 231–259.
30. Pliscoff, P., & Fuentes-Castillo, T. (2011). Representativeness of terrestrial ecosystems in Chile's protected area system. *Enviromental Conservation, 38*(3), 303–311.
31. Pliscoff, P., Luebert, F., Hilger, H. H., & Guisan, A. (2014). Effects of alternative sets of climatic predictors on species distribution models and associated estimates of extinction risk: A test with plants in an arid environment. *Ecological Modelling, 288,* 166–177.
32. Pressey, R. L., Cabeza, M., Watts, M. E., Cowling, R. W., & Wilson, K. A. (2007). Conservation planning in a changing world. *Trends in Ecology and Evolution, 22*(11), 583–592.
33. Rabassa, J., Coronato, A., & Martínez, O. (2011). Late Cenozoic glaciations in Patagonia and Tierra del Fuego: An updated review. *Biological Journal of Linnean Society, 103,* 316–355.
34. Ramírez de Arellano, P. A. (2007). *Systematic conservation planning in Chile: Sensitivity of reserve selection procedures to target choices, cost surface, and spatial scale.* Ph.D. thesis, State University of New York. https://pqdtopen.proquest.com/doc/304826098.html?FMT=AI
35. Rozzi, R., Rosenfeld, S., Armesto, J. J., Mansilla, A., Núñez-Ávila, M., & Massardo, F. (2023). *Ecological connections across the marine-terrestrial interface in Chilean Patagonia.* Springer.
36. Sanderson, E. W., Jaiteh, M., Levy, M. A., Redford, K. H., Wannebo, A. V., & Woolmer, G. (2002). The human footprint and the last of the wild. *BioScience, 52,* 891–904.
37. Scherson, R. A., Thornhill, A. H., Urbina-Casanova, R., Freyman, W. A., Pliscoff, P., & Mishler, B. D. (2017). Spatial phylogenetics of the vascular flora of Chile. *Molecular Phylogenetics and Evolution, 112,* 88–95.
38. Schutz, J. (2018). Creating an integrated protected area network in Chile: A GIS assessment of ecoregion representation and the role of private protected areas. *Environmental Conservation, 45*(3), 269–277.

39. Squeo, F. A., Estévez, R. A., Stoll, A., Gaymer, C. F., Letelier, L., & Sierralta, L. (2012). Towards the creation of an integrated system of protected areas in Chile: Achievements and challenges. *Plant Ecology and Diversity, 5*(2), 233–243.
40. Tacón, A., Tecklin, D., Farías, A., Peña, M. P., & García, M. (2023). *Terrestrial protected areas in Chilean Patagonia: Characterization, historical evolution, and management*. Springer.
41. Tognelli, M. F., Ramírez De Arellano, P. A., & Marquet, P. A. (2008). How well do the existing and proposed reserve networks represent vertebrate species in Chile? *Diversity and Distribution, 14*, 148–158.
42. Veblen, T. (2007). Temperate forests of the southern Andean region. In T. Veblen, K. Young, & A. O. Orme (Eds.), *The physical geography of South America* (pp. 217–231). Oxford University Press.
43. Wildlife Conservation Society. (2019). *Una red de áreas marinas protegidas para la Patagonia chilena. Fortaleciendo la conservación del mar en el largo plazo*. Wildlife Conservation Society. https://chile.wcs.org/Portals/134/INFORME%20WCS.spreadpdf_fn.pdf

Patricio Pliscoff Geographer and Associate Professor, Institute of Geography and Department of Ecology, Pontificia Universidad Católica de Chile. M.Sc. Biological Sciences, Universidad de Chile. Ph.D. Ecology, Université de Lausanne, Lausanne, Switzerland.

María José Martínez-Harms Ph.D. Biodiversity and Conservation Science, University of Queensland, Australia. Researcher at the Centre for Research and Innovation in Climate Change, Universidad Santo Tomas, Chile. Principal investigator at the Institute of Ecology and Biodiversity. Adjunct researcher at the Coastal Socio-Ecological Millenium Institute.

Taryn Fuentes-Castillo Engineer in renewable natural resources, Universidad de Chile. Master in Wild Areas and Nature Conservation. Ph.D. in Forestry and Veterinary Sciences, mention in Biological Conservation.

Terrestrial Protected Areas in Chilean Patagonia: Characterization, Historical Evolution, and Management

Alberto Tacón, David Tecklin, Aldo Farías, María Paz Peña, and Magdalena García

Abstract

Chile's Patagonian region houses globally unique ecosystems whose conservation has been addressed principally through the National Protected Areas System (in Spanish SNASPE). In order to improve understanding of the region's current level of protection, we analyze the history, coverage, and management status of legally protected areas. Patagonia's SNASPE accounts for a high percentage of the total land under protection in Chile, and includes archipelagos, fjords, channels, glaciers, icefields, and large areas of globally unique and highly intact forests. Management of the National System of State Wild Protected areas by the National Forestry Corporation has advanced substantially over the last century. Nonetheless, Areas our evaluation, which was carried out using official data, indicates the persistence of important limitations in almost all protected areas evaluated. There is a need to strengthen institutional capacities in order to overcome historic problems and raise levels of management. We present recommendations that highlight the importance of strengthening the legal framework, as well as the need to bring planning up to date, and improve management inputs through public policies that address gaps in funding.

A. Tacón (✉) · D. Tecklin · A. Farías · M. P. Peña
Programa Austral Patagonia, Faculty of Economic and Administrative Sciences, Universidad Austral de Chile, Valdivia, Chile
e-mail: alberto@calahuala.cl

A. Tacón
Cooperativa Calahuala, Servicios para la Conservación, García Reyes 455 of. 9, Valdivia, Chile

M. García
Department of Geography, University of Montreal, 1375 Avenue Thérèse Lavoie Roux, Montréal, Québec, Canada

© Pontificia Universidad Católica de Chile 2023
J. C. Castilla et al. (eds.), *Conservation in Chilean Patagonia*, Integrated Science 19, https://doi.org/10.1007/978-3-031-39408-9_4

Keywords

Patagonia • Chile • Protected areas • National parks management • National System of State Wild Protected Areas • National Forestry Corporation

1   Introduction

Protected areas (PAs) are the most standard conservation tool worldwide, defined by the International Union for Conservation of Nature (IUCN) as "A clearly defined geographical space, recognized, dedicated and managed, through legal or other effective means, to achieve the long-term conservation of nature, with associated ecosystem services and cultural values" [12]. An extensive literature has been generated over the last decades on evaluating the effectiveness of PAs in preventing habitat loss [22, 23], and ensuring the provision of ecosystem goods and services [70]. The literature has also documented that many of these PAs only have legal protection on paper and lack effective management on the ground [52]. According to Aichi Target No. 11 agreed at the tenth Conference of the Parties to the Convention on Biological Diversity (CBD), signatory countries have the commitment to protect by 2020, "at least 17% of terrestrial and inland water areas and 10% of marine and coastal areas, (…) through effectively and equitably managed systems of protected areas (…)" [7]. Although this commitment was incorporated in the Chilean National Biodiversity Strategy 2017–2030 (Ministry of the Environment [39], in Spanish Ministerio del Medio Ambiente, MMA), assessments to date indicate that protected areas in our country are far from being effectively and equitably managed, reaching only 50% of their optimal level [19, 20, 62].

In a context of increasing environmental vulnerability and anthropic pressures resulting from historical processes of colonization and displacement of native peoples, Patagonian ecosystems have been profoundly transformed and invasive species have been introduced. The establishment of intensive productive activities such as mining and aquaculture in the fjords and channels has generated emblematic socio-environmental conflicts in Chilean Patagonia [11]. The weakness of the Environmental Impact Assessment System and other land use planning and regulation instruments reinforces the importance of PAs in Chile as a tool for biodiversity conservation (Organization for Economic Cooperation and Development [46]).

Chilean Patagonia, between Reloncavi Sound and the Diego Ramirez Islands (41° 42′ S 73° 02′ W; 56° 29′ S 68° 44′ W), concentrates more than 70% of the total surface of terrestrial and coastal areas protected by the State, and about 28% of the PAs legally recognized nationally. Increasing their management effectiveness is of key importance for the achievement of international species and ecosystem conservation commitments, including the Washington Convention and the Convention on Biological Diversity. This chapter offers a historical analysis of the establishment and current management status of the primarily terrestrial PAs administered by the National Forestry Corporation (in Spanish Corporación

Nacional Forestal, CONAF) as a contribution to the design of management standards for the National Protected Area System (in Spanish Sistema Nacional de Áreas Protegidas del Estado, SNASPE).

2   Scope and Objectives

In order to contribute to knowledge and public debate regarding PAs in Chile, and especially in Chilean Patagonia, we present an analysis of the establishment and historical evolution of PAs in the region stretching from the Reloncaví sound to Cape Horn, complemented by a description of their main biogeographical characteristics and an analysis of their current level of management. Finally, four typical situations are proposed that reflect different levels of management, along with a general discussion of the needs and opportunities to advance in strengthening the management of the NPWAs. This chapter does not address other key issues contained in other chapters of this volume, e.g. coastal-marine protection within the NPWAs and marine protected areas [24, 28, 65].

Neither does the chapter address ecological representativeness and relations with Indigenous peoples, which are discussed in other chapters of this volume [2, 54]. The Chiloé archipelago is not included here, since the chapter is based on an earlier analysis carried out by the Austral Patagonia team that did not address that subregion.[1] While our historical and biogeographic analysis covers all official PA categories, the analysis of management status is limited to the NPWAs.

3   Methods

Given its wide coverage in Chilean Patagonia and the greater availability of information, the analysis concentrates on the legal categories that make up the NPWAs, which include National Parks (NPs), Natural Reserves (NRs) and Natural Monuments (NMs), together these are referred to as National Protected Wild Areas (NPWAs; in Spanish Areas Silvestres Protegidas del Estado). Other protected areas, such as Protected National Assets (PNAs) and Nature Sanctuaries (NSs) will be considered in a separate section. These categories are not included in the analysis of the level of management, due to lack of publicly available information. Biosphere Reserves (BRs) are included in the discussion but not in the PA statistics, since they are not yet a legally recognized form of protection in Chile. The review includes a synthesis of information from various secondary sources summarized in Table 1. The compilation of cartographic information presented for the NPWAs comes from the Ministry of National Assets (in Spanish Ministerio

---

[1] This compilation of information is based primarily on a series of studies carried out by the Austral Patagonia Program of the Austral University aimed at improving the coverage and management of PAs in Chilean Patagonia. For the purposes of this program, the area of interest was limited to the Palena Province in the Los Lagos Region and the Aysén and Magallanes Regions.

de Bienes Nacionales, MBN) and CONAF and for the other PA categories the Ministry of Environment's National Registry of PAs was used. The geographic coverage calculations include only the terrestrial portions of the areas analyzed, since marine coverage is dealt with separately in [65]. The updates of the Native Forest Cadastre available in CONAF's Territorial Information System were used for the geographic and resource characterization associated with the PAs. ArcGIS 10.5 software [14] was used for the analysis and processing of geographic information. The historical evolution of the establishment of PAs was generated based on a review of available historical and contemporary literature, including CONAF archives, and other sources, including laws and decrees.

CONAF and the Austral Patagonia Program previously evaluated the management of the terrestrial NPWAs in Chilean Patagonia [63]. A management model was developed that represents the main activities, results, inputs, processes, and outputs that form the NPWAs management cycle in Chile (Fig. 1). Subsequently, the main indicators were identified to analyze the level of management achieved for each stage of the management cycle, following international recommendations for this type of protected areas management evaluation (PAME) instrument [35]. Information from the following was collected and systematized from different sources in a single database, using a binary coding system (complies/does not comply) for each of the 38 indicators in the 34 NPWAs units in force as of December, 2017. From a first initial analysis of results, explicit rating ranges (1–4) were established based on an adaptation of internationally accepted methodologies [61]. The indicators were validated through a management effectiveness evaluation workshop with 30 park rangers and NPWAs managers. The information provided by CONAF's regional protected area administration in the Los Lagos, Aysén, and Magallanes regions for the period 2014–2017 was reviewed, to complement the information on the provision of management inputs and the type of management activities carried out in each unit.

It is worth mentioning that with the land donation and park expansion agreement signed in 2018 between the collection of NGOs working under the umbrella of Tompkins Conservation and the government, the NPWAs in Patagonia increased from 34 to 36 units, and several NR were reclassified as NP. This analysis addresses the units as they existed prior to that change. Subsequently, a partial update of the data corresponding to the 18 national parks was carried out. These data were not included in the final quantitative analysis, but the main qualitative changes in the management situation of the NPWAs are described in the chapter.

Table 1 Summary of variables analyzed and sources of information consulted

Variables	Indicators/Information layers	Sources of information
Surface area of study area	Terrestrial areas	INE map service: http://www.censo2 017.cl/servicio-de-mapas/
PAs at national level, and for the study area	NPWAs categories	CONAF's Territorial Information System. https://sit.conaf.cl/
	Nature sanctuaries	National Register of Protected Areas Ministry of Environment (https://areasp rotegidas.mma.gob.cl/areas-proteg idas/) MinBBNN
	Biosphere reserves	
	National protected assets	
Types of land use at the national level and for the study area	Land use and vegetation	CONAF's Territorial Information System https://sit.conaf.cl/
Geographic characteristics	Slopes	CONAF's Territorial Information System https://sit.conaf.cl/
	Island and coastal geography	INE map service: http://www.censo2 017.cl/servicio-de-mapas/
PA establishment	Dates of establishment	National PA Registry Ministry of the Environment. http://areasprotegidas. mma.gob.cl/
	Conservation objectives	
	Establishment processes	Historical and contemporary files CONAF reports, laws, and decrees
Key elements for context, NPWAs management level	Contents of the creation decree	[44] National Environmental Information System Official documentation provided by CONAF regional protected area administration
	Technical rationale	
	Socio-environmental baseline information	
Planning tools for NPWAs management	Management plan	
	Operational plans	
	Specific plans	

(continued)

Table 1 (continued)

Variables	Indicators/Information layers	Sources of information
Provision of NPWAs management inputs	Operating budget	
	Staffing, coverage, and training	
	Infrastructure and equipment for handling	
Handling procedures for NPWAs management	Administrative management	
	Biodiversity conservation	
	Public use management	
	Community outreach	
	Conservation monitoring	

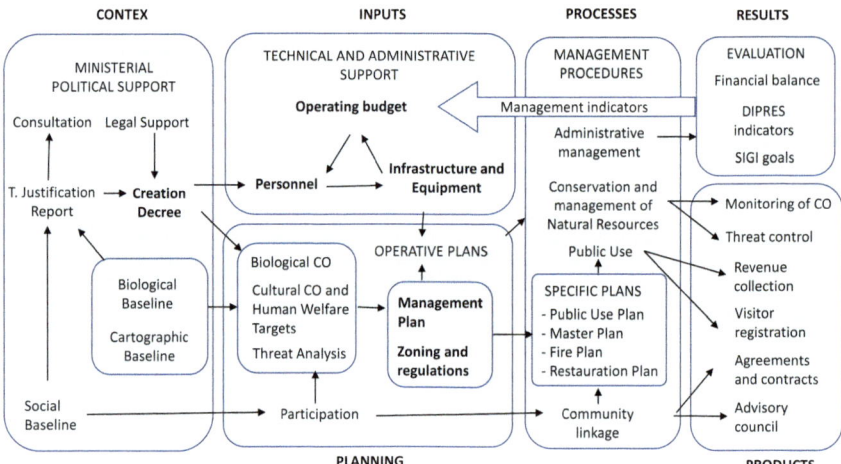

OC: conservation objects.

Fig. 1 Flowchart of the NPWA management cycle in Chile indicating the different processes, inputs, activities, and critical products involved in management. OC: conservation objects

Visitors to Bernardo O'Higgins National Park, Magallanes and Chilean Antarctica Region. Photo courtesy of Conaf

4   Results

4.1 Establishment and Historical Evolution of NPWAs in Chilean Patagonia

The history of NPWA creation can be seen as progressing through four main stages: (i) oriented to the protection of State forests during the colonization era; (ii) for scientific, tourist, or sovereignty reasons; (iii) as part of an administrative and territorial reorganization of the system; (iv) as part of an expansion driven by non-State actors. Over the last century many of the current NPWAs have changed their name, surface areas, and objectives. In this chapter the units are identified according to their current names and surface areas.

4.1.1 Stage 1: Forest Reserves and Forest Exploitation (1913–1939)
The oldest NPWA in Chile dates to 1907 and the earliest in Chilean Patagonia to 1913, when the Llanquihue Forest Reserve was created for an area now encompassed in the Alerce Andino NP. This first stage in the creation of NPWA emerged in response to the "ecological disaster" caused by the extensive burning of forests to make way for agricultural and grazing lands [4, 50]. At this time, however, NPWAs established in areas with better access were generally later fully or partially converted and settled as part of the colonization of the region [21]. With the first Forestry Law of 1925, the categories of Forest Reserves and National Tourism

Parks were created, with new areas located in territories unsuitable for colonization and with little value for forestry exploitation. In 1931 the 1925 Forestry Law was modified and the former Forest Reserves were reclassified as National Parks, and surplus land was allocated to colonists. The Magallanes Forest Reserve was established in Patagonia in 1932 and the Las Guaitecas Forest Reserve in 1938 (Fig. 3), and given their difficult access, they were not exploited.

4.1.2 Stage 2: Scientific Explorers, Tourism and Sovereignty in Chilean Patagonia (1940–1971)

In the next phase of NPWAs development, scientists and explorers pushed for the protection of various territories in Patagonia for scientific and tourism purposes [21, 69]. The creation of the first national parks in border territories of Chile and Argentina was carried out as a strategy to establish sovereignty in remote and border areas [21, 32, 45, 58]. During the 1940s and 1950s, Carlos Muñoz Pizarro, scientist and Director of Forests within the Ministry of Lands and Colonization, promoted the establishment of a Network of National Parks and Forest Reserves in Chile and the expansion of NPWAs [42]. Patagonia's oldest and southernmost NP, Cabo de Hornos, was established in 1945 as a National Tourism Park and Virgin Region Reserve (Fig. 2). Scientists and travellers promoted the creation of Lago Grey National Tourism Park in 1959. Three years later, Torres del Paine was declared a National Tourism Park, thanks to land donations and the incorporation of public lands. Laguna San Rafael National Tourism Park was created in the same year (Fig. 2). Andean clubs took possession of parts of NPWAs in this period, as in the case of the Magallanes Forest Reserve.

The most significant period of growth in terms of number and surface area of NPWAs occurred during the government of Eduardo Frei Montalva (1964–1970). Important land policies were implemented during this administration including the agrarian reform[2] and the ratification of Washington Convention (1967).[3] Twenty-six NPWAs were established in Patagonia, of which 10 were National Tourism Parks; some of which were very extensive, covering >1 million hectares (ha) (Fig. 2). The Administration of National Parks and Forest Reserves (in Spanish Administración de Parques y Reservas Forestales, APARFO) was created in 1964 within the Ministry of Agriculture, which at that time managed *ca.* 3 million ha of terrestrial NPWAs nationally [60]. Subsequently, its functions were transferred to the Agriculture and Livestock Service (in Spanish Servicio Agrícola y Ganadero, SAG), created in 1967.

There was no proposed technical management system for NPWAs until 1965. At this time, many NPWAs were reclassified, merged, degazetted, or reconfigured

---

[2] Within the framework of the Agrarian Reform, a new process of review of the fiscal property took place in order to transfer land to the Agrarian Reform Corporation, which implied the creation and reclassification of parks and reserves created to date [16].

[3] Decreto Supremo N° 531 de 1967. Convención para la Protección de la Flora, Fauna y las Bellezas Escénicas Naturales de América (https://bcn.cl/2jzac).

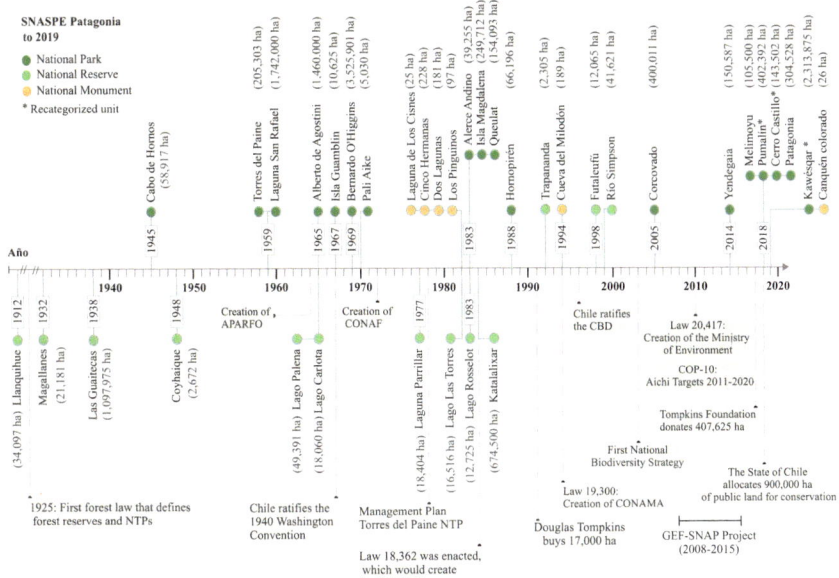

Fig. 2 Timeline of the establishment of terrestrial protected areas in Chilean Patagonia and significant milestones in the process, as for Decembre 2019

as part of a significant reorganization of the system. A process of title reorganization and area delimitation began and intensified a decade later [13]. However, the occupation by settlers and burning to clear land for agriculture continued in areas under protection. This created difficulties in expanding or creating new PAs, given that the settlers had interests different from those of the State.

In Aysén, for example, colonists' resisted the expansion of the Lago Carlota, Cochrane, and Jeinimeni Forest Reserves. They managed to prevent the expansion of the last of these, given that they wanted the land for cattle raising [4]. The State also required extensive NPWA land areas in Magallanes to install police checkpoints and other public offices (Ministry of Lands and Colonization [36, 37]; in Spanish Ministerio de Tierras y Colonizacion, MTC). There was no management for any NPWA in Patagonia until the end of 1960 [59], with the exception of the Llanquihue Forest Reserve, which had an administrator since 1925.

4.1.3 Stage 3: Creation of CONAF and Re-categorization of the NPWAs (1972–1999)

By the end of the 1970s, 41 NPWAs had been declared in Patagonia (15 National Tourist Parks and 26 Forest Reserves).[4] The vast majority of the NPWAs had no administration or management practices [13, 50], there were only seven administrators and 14 park rangers. Torres del Paine NP was a pioneer in Patagonia with the development of the first management plan and establishment of minimum impact infrastructure (1978), with the support of the US Peace Corps [15]. The first administrator started his duties in 1981, along with seven park rangers.

In 1972 CONAF was created as a private non-profit corporation under the Ministry of Agriculture and it incorporated other units of the ministry such as APARFO, which until then had jurisdiction over protected areas. At that date there were already close to 10 million ha under some form of protection nationally [60]. With the support of CONAF, the Food and Agriculture Organization (FAO) generated a first planning document for NPWAs between 1974 and 1975.[5] The first technical policies for NPs were drafted in 1975, and in 1988 the first technical policies were established for NR [48].

CONAF undertook an extensive process of reclassification, redefinition of boundaries, and management categories in all the Patagonian NPWAs. The reorganization was based on the categories defined by the Washington Convention and the 1978 IUCN categories. In this process, the NPWAs that did not meet the standards of these categories, or that had been colonized, were abolished [49]. In both Aysén and Magallanes, given the lack of control of the protected territories or the demands for other uses, 11 Forest Reserves were disaffected [21], including four in the Cisnes and Palena areas of Aysén, for a total of 8,606 ha [38]. This process of reclassification and disaffection of NPWAs led to the degazetting of around 1 million ha nationally by the end of 1980 [21].

The designation of new areas began to be oriented towards the protection of ecological values in this period. As a result of the process of national deforestation and reorganization, according to CONAF in 1989 only 5% of protected areas nationally had conflicts with private properties [48]. However, given the lack of administration, this reality was different in Patagonia, where such conflicts were accentuated. In Aysén there were property conflicts with settlers in the Río Simpson, Cochrane, and Cerro Castillo Forest Reserves; and in Magallanes, in the Magallanes, Pali Aike, and Torres del Paine National Tourism Parks. Since 1984 the NPWAs have been administered by CONAF as part of the NPWAs.[6] During the 1980s a more

---

[4] The FAO-APARFO project (1970–1976) "Strengthening of the National Forestry Program DP/CHI/66/526" promoted the beginning of NPWA planning, following the method proposed by K. Miller in the USA. The project collaborated in the development of the institutional framework and procedures for NPWAs in CONAF [17].

[5] Wilderness Systems Planning [66].

[6] Law No. 18,362 which created the NPWA was passed In 1984. However, its entry into force was subject to the promulgation of Law No. 18,348 which was to convert CONAF into a decentralized public service. Its article 19 stipulated that the law would enter into force "on the day on which

systematic process of area planning began, with the design of the technical policies for the management of the NPWAs [8].

4.1.4 Stage 4: Creation of Philanthropic, State-Driven PAs and Development of Private Conservation Initiatives (2000–2018)

The last two decades have seen an expansion of protected areas in Patagonia, largely driven by Non-Governmental Organizations (NGOs) and various private conservation initiatives (PCIs). These PCIs involve companies, individuals, or Indigenous communities who wish to preserve all or part of their properties, or philanthropists who buy land to protect nature for non-profit purposes. As of 2014, 47 PCIs have been identified between the Los Lagos and Magallanes regions covering an area of 964,000 ha, equivalent to *ca.* 57% of the total area nationwide [43]. Three of the five largest PCIs were established in Patagonia: Pumalín Park in Palena (now an NP), Tantauco Park in insular Chiloé, and Karukinka Natural Reserve in Tierra del Fuego (Fig. 3). Despite the fact that PCIs have been consolidating their role as a necessary and alternative conservation mechanism nationally, they remain in an institutionally precarious position [64], and generally demonstrate low management effectiveness [43].[7]

The most emblematic case are those initiatives undertaken by the foundations linked to Douglas and Kristine Tompkins, later consolidated under the name Tompkins Conservation, which purchased a series of properties in Patagonia for conservation. The first was Pumalín, which opened the debate on private conservation policy in Chile [11, 27, 64] and generated a new management model with infrastructure of high aesthetic quality, free public use, and significant presence of management staff. An initiative of donations from Tompkins Conservation to the State of Chile began in 2005 with the creation of Corcovado NP. This concluded in 2018 with a donation of 407,625 ha of land by Tompkins Conservation, and a series of measures by the Chilean government aimed at consolidating a set of NPs denominated as the Patagonian Parks Network. These measures included the creation of Pumalín Douglas Tompkins NP from the private land donation and adjacent public lands, the creation of Patagonia NP from the donation and incorporation of the Lago Cochrane Forest Reserve and Lago Jenimeni NR, the creation of Melimoyu NP, and the reclassification of the Alacalufes Forest Reserve and its expansion with adjacent public lands to create Kawésqar NP. The Hornopirén,

---

the decree by virtue of which the President of the Republic dissolves the private law corporation known as the National Forestry Corporation is published in the Official Gazette", a decree which has not been issued. Consequently, the NPWA is not legally in force either [60].

[7] As part of this consolidation, the first organization bringing together small and medium-sized owners of private PAs and Indigenous peoples was created in 2010 and denominated Así Conserva Chile A.G. Private PAs are recognized in Article 35 of the General Environmental Framework Law (No. 19,300 of 1994), however, the country lacks basic operational definitions, standards, and administrative procedures that establish what criteria and conditions these initiatives must meet in order to be officially recognized.

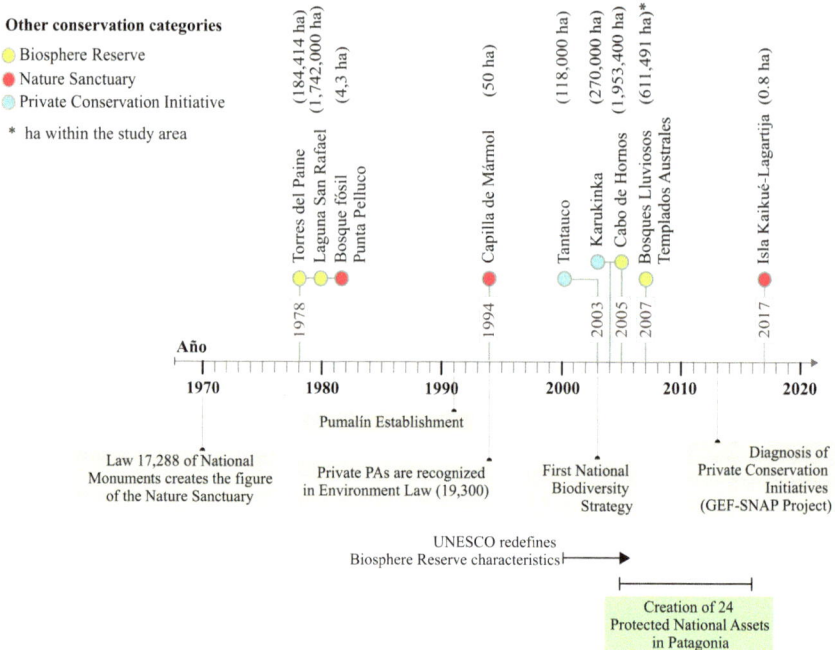

Fig. 3 Timeline of the creation and significant milestones for other land protection figures in Chilean Patagonia, as for December 2019

Corcovado, and Isla Magdalenad NPs were also expanded with adjacent public lands (Fig. 3).[8]

4.2 Establishment and Evolution of Other Figures of Terrestrial Protection

Other categories of protected areas, which are not part of the NPWAs, are the Nature Sanctuaries (NSs) established under the National Monuments Law No. 17,288 of 1970 and Protected National Assets (PNAs) established by MBN policy and self-designated by decree of this ministry. The first NS to be established in Patagonia was the Punta Pelluco Fossil Forest in 1978 (Fig. 3). Four others have been declared after 1990: Capilla de Mármol and Estero Quitralco in Aysén; Pumalín in Palena, (later reclassified to NP) and Isla Kaikué-Lagartija in the Los Lagos Region. Twenty-four PNAs were decreed from 2003 to 2016 in Patagonia

[8] See 'Protocolo de acuerdo: Proyecto Red de Parques Nacionales en la Patagonia chilena', signed by foundations linked to Tompkins Conservation and various Chilean State departments on March 15, 2017.

Table 2 Number and surface area of PAs in Chilean Patagonia and nationally. In addition, for a more detailed analysis of ecological representativeness, see Pliscoff et al. [54]

Protected area category	IUCN category	Chile		Chilean Patagonia			
		No. of areas	Surface area (ha)	No. of areas	Surface area (ha)	Percent of category nationally (%)	Percent of Chilean Patagonia (%)
National parks	II	41	12,676,295	18	11,337,427	89.4	42.1
National reserves	IV, VI	46	5,380,188	12	1,967,424	36.6	7.3
Natural monument	III	18	34,466	6	747	2.2	0.003
Subtotal NPWAs		105	18,090,949	36	13,305,571	73.5	49.4
Nature sanctuary	III, IV	56	511,555	3	55	0.01	0.0002
National protected assets		58	616,524	24	293,957	47.7	1.1
Total		219	19,219,028	63	13,599,583	70.8	50.5

(Fig. 3). Their administration is the responsibility of the MBN, but their management is granted in concession to third parties. Finally, a series of Biosphere Reserves (BRs) have been established that are not legally recognized as PAs in the country, but which can generate a conservation framework for the designated territories.[9] Four BRs have been established (Fig. 3): Torres del Paine and Laguna San Rafael in 1978, whose surfaces are equivalent to the NPs of the same name [41]; and Cabo de Hornos in 2005, which was the first to integrate marine and terrestrial environments[10] [55] and Bosques Templados Lluviosos in 2007, which incorporates nine NPWAs units, four of which are located in Patagonia [41].

4.2.1 General Description of NPWAs of Chilean Patagonia

The Chilean Patagonian region is eminently a conservation territory, with almost 51% of its surface area under protection. There are currently 63 PAs, distributed in five protection categories, covering 13.6 million ha of land in the study area (Table 2, Fig. 4). The results show Patagonia's gravitational weight in terms of terrestrial ecosystem protection, covering 71% of the national total of the categories analyzed (19.2 million ha).

---

[9] Biosphere Reserve is an official recognition by UNESCO applicable to terrestrial, coastal and/or marine ecosystem areas of international importance, with the objective of promoting the harmonious integration of people and nature.

[10] Includes Alberto de Agostini NP and Cabo de Hornos NP.

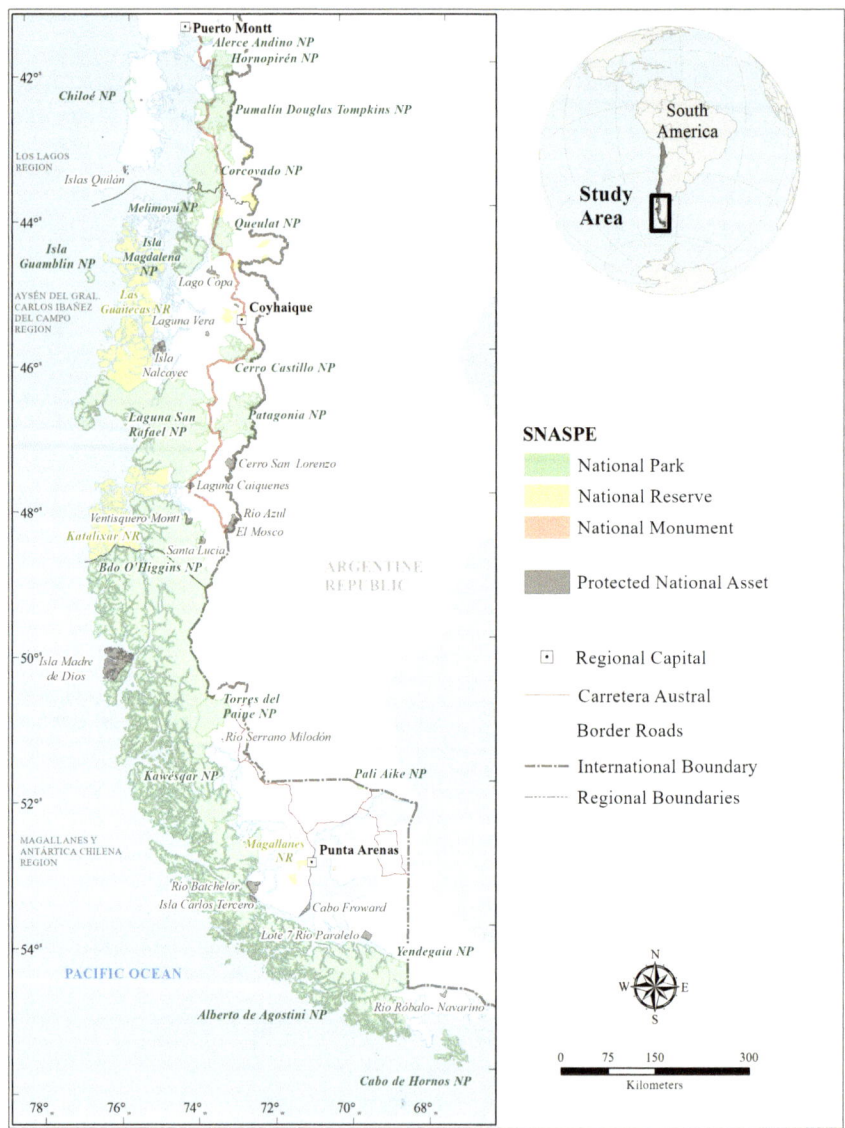

Fig. 4 Terrestrial protection by legal category in Chilean Patagonia

The NPWAs contributes the largest proportion of this protection, totaling 13.3 million ha in 36 units, primarily under the category of NP and NR, and to a much lesser extent MN (Table 2). Of note are the following large NPs and NRs (>1 million ha): Laguna San Rafael, Bernardo O'Higgins, Kawésqar, and Alberto de Agostini (Fig. 4). Las Guaitecas and Katalalixar in Aysén are among the NR with areas greater than 500,000 ha (Fig. 4). The other PAs contribute a relatively low area in ha despite their significant number (Table 2).

The NPAs in Patagonia represent 48% of the area in NPAs nationally, and only 5 of the 24 existing units contribute 70% of the area covered by this category: Nalcayec (22,934 ha) and Cerro San Lorenzo (19,400 ha) in Aysén, and Isla Madre de Dios (123,668 ha), Río Serrano Milodón (24,124 ha) and Lote 7 Río Paralelo (15,347 ha) in Magallanes. The NS present in Patagonia, which contribute a smaller area (55 ha), are Punta Pelluco Fossil Forest and Isla Kaikué Lagartija in the Los Lagos Region, and Capilla de Mármol in the Aysén Region (Fig. 4).

Many of the aforementioned PAs and their areas of influence have been recognized by UNESCO as Biosphere Reserves, although this category is not yet considered an official protection category in Chile. The four BRs present in Patagonia cover a total of 4,491,305 ha, that is, 41% of the total national surface area of the BRs (10.9 million ha).

According to the Native Forest Cadastre, the NPWAs represents an important and diverse portion of land use types in Patagonia (Fig. 5). The NPWAs concentrate a large area of snow and glaciers (83% of the total present), mainly in the regions of Aysén and Magallanes, as well as peatlands (68%) in Magallanes. Shrublands and native forest have a similar proportion within the NPWAs (*ca.* 40%), with greater coverage in the Aysén Region. The greater representation of native forest, snow, glaciers, and areas without vegetation is consistent with the large proportion of steep slopes within the NPWAs (>45%), and therefore soils with little potential for agricultural use. This is the situation for 50% of the surface of the NPWAs in Los Lagos, 32% in Magallanes and 23% in Aysén. These data are relevant in the face of discussions regarding the impact of NPWAs coverage on the development of silvicultural-agricultural industry in the regions. There are also still natural uses with less representation such as the steppes (9%, 59,126 ha), which are primarily represented within the NPWA of the Magallanes Region (with 25,000 ha). Finally, it is important to note that the coastal geography of Chilean Patagonia is predominantly archipelagic and is composed of more than 40,000 islands, islets, and rock outcroppings [65]. Thus it is important to recognize that the majority of the NPWA's land area is archipelagic and therefore presents particular management challenges.

4.3 Current Management Situation of the NPWAs

4.3.1 Development of Tools for Assessing the Effectiveness of NPWA Management

Interest in assessing the effectiveness of PAs in adequately representing different ecosystems and providing effective protection on the ground prompted the IUCN World Commission on Protected Areas to propose a conceptual framework that evaluates PA management based on an analysis of the management cycle [25, 26]. Different tools related to PAME have been developed based on this conceptual framework over the last decades, and are validated by Convention on Biological

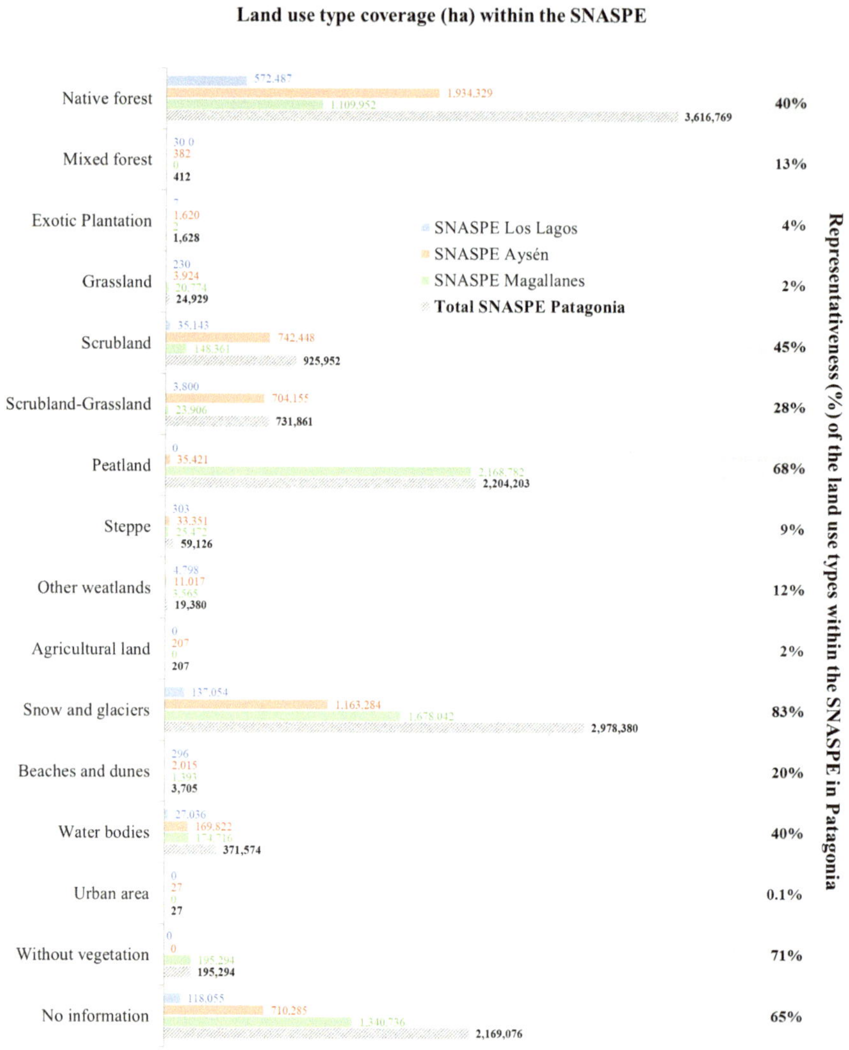

Land use type coverage (ha) within the SNASPE

Native forest	572,487 / 1,934,329 / 1,109,952 / 3,616,769 — **40%**
Mixed forest	30.0 / 382 / 0 / 412 — **13%**
Exotic Plantation	1,620 / 5 / 1,628 — **4%**
Grassland	230 / 3,924 / 20,774 / 24,929 — **2%**
Scrubland	35,143 / 742,448 / 148,361 / 925,952 — **45%**
Scrubland-Grassland	3,800 / 704,155 / 23,906 / 731,861 — **28%**
Peatland	0 / 35,421 / 2,168,782 / 2,204,203 — **68%**
Steppe	303 / 33,351 / 25,472 / 59,126 — **9%**
Other weatlands	4,798 / 11,017 / 3,565 / 19,380 — **12%**
Agricultural land	0 / 207 / 0 / 207 — **2%**
Snow and glaciers	137,054 / 1,163,284 / 1,678,042 / 2,978,380 — **83%**
Beaches and dunes	296 / 2,015 / 1,393 / 3,705 — **20%**
Water bodies	27,036 / 169,822 / 174,717 / 371,574 — **40%**
Urban area	0 / 27 / 0 / 27 — **0.1%**
Without vegetation	0 / 0 / 195,294 / 195,294 — **71%**
No information	118,055 / 710,285 / 1,340,736 / 2,169,076 — **65%**

Legend:
- SNASPE Los Lagos
- SNASPE Aysén
- SNASPE Magallanes
- **Total SNASPE Patagonia**

Representativeness (%) of the land use types within the SNASPE in Patagonia

Fig. 5 Land use coverage in the NPWAs of Chilean Patagonia

Diversity as indicators to verify compliance with the commitments of the Parties [6]. Specific tools have also been developed to assess social participation [18], quality of governance [3], equity in management [40, 71] and generation of social benefits [33], among other aspects.

However, use of these self-assessment tools implies a high degree of subjectivity [5], which is why external evaluation processes have recently been implemented with verifiable indicators, including PA certification processes that allow accreditation of compliance with sufficient levels of management effectiveness.[11] The most important initiative is the IUCN Green List programme, which already has a procedures manual [30] and a global standard based on four principles, 17 criteria and 70 indicators [31]. The Green List standard has already been applied in different Latin American countries, including Colombia, Peru, and Mexico [68].

Different tools have also been applied in Chile to evaluate the effectiveness of NPWAs management. The first manual of operations and technical policies was published in the 1980s, which made it possible to standardize the NPWA management cycle [8]. Subsequently, the first experiences of effectiveness evaluation were implemented in the 1990s [9] and the Management Planning and Control System was implemented in the 2000s; an Institutional Management Information System (SIGI) was adopted based on indicators that represent relevant processes or strategic products related to the objective of the program [10]. Implementation of SIGI No. 14 [44] was initiated in 2018; this is a tool for evaluating compliance with the legal objectives of the NPWAs based on three principles, 33 criteria and 65 PA management indicators that assess the achievement of certain regional and national outputs/outcomes. However, this tool was not designed to evaluate the performance of the units, providing only results grouped by administrative region. CONAF does not currently have a standardized procedure to evaluate the effectiveness of individual NPWA management.

4.3.2 Evaluation of the Level of Management in NPWA

Based on the conceptual model that represents the NPWA management cycle and the selection of 37 indicators grouped into 5 areas and 12 sub-areas of management (Table 3), an evaluation was carried out using official information provided by CONAF's department of NPWA administration. The verifiers available for each of the indicators were identified through a detailed review of procedure manuals, planning instruments, management reports, resolutions, and other background information available at CONAF [63]. A summary of the results obtained for each of the areas evaluated is provided below.

4.4 Context Area

Contextual information makes it possible to determine whether management efforts are consistent with the importance or degree of pressure on the units. Only 12 of the 34 units evaluated (35%) have detailed baseline information and digital cartography that is less than 10 years old, while 11 others have information that is more than 10 years old, and 13 have no biological baseline information.

---

[11] The coordinated audit of PAs developed by the Special Technical Commission on the Environment of the Latin American and Caribbean Organization of Supreme Audit Institutions was applied to 1,120 PAs from 10 countries in the region. The Office of the Comptroller General of the Republic of Chile did not participate in this process [47].

Table 3 Indicators evaluated for effectiveness of NPWA management, by scope and criteria

Areas	Criteria	Indicators
Context	Legal	Decree for designation
		Optimization processes (legal limits)
	Information	Official cartography
		Biological baseline
		Cultural heritage baseline
		Social baseline
Planning	Strategic plans	Technical justification report
		Management plan
		Zoning of uses
		Area of influence
		Threat analysis and prioritization
	Operative plans	Annual or multi-annual Operational Plan (AOP)
		Public use plan
		Other specific plans
Inputs	Staff	Park rangers
		Sectors served
		Professionalization
	Equipment	Larger vehicles
		Minor vehicles
		Driving equipment
	Infrastructure	Infrastructure for public use
		Administrative infrastructure
	Budget	Operating budget
		Investment
		Collection
Handling processes	Community engagement	Advisory Council
		Community information
		Agreements or collaboration agreements
		Indigenous peoples
	Conservation	Threat control
		Species conservation
		Restoration
		Cultural resources
	Public use	Visitor management
		Concession control

(continued)

Table 3 (continued)

Areas	Criteria	Indicators
Results	Evaluation	Conservation monitoring
		Results monitoring
		Visitation monitoring

Terrestrial and marine baseline studies are being developed for the planning of only three units at present, so there is no biodiversity database in the NPWAs units based on primary information. Although the NPWAs have been integrated into planning instruments and regional development strategies, especially in the Aysén Region, there is only one unit with a valuation study of the environmental goods and services of the NPWAs. However, in 2017 the first systematic survey of threats to ecosystems was applied to units, and there are also systematized data on the numbers and types of visitors (Table 4), which allows estimates of the degree of public use pressure that each unit receives.

4.5 Planning Scope

Of the 34 units evaluated, 26 (75%) have some kind of management instrument. Of these, 22 have a management plan while three have management guides, and only one has a resource register. However, only eight management instruments (35%) are less than five years old (Fig. 6). The drafting of Management Guides for Pumalín Douglas Tompkins NP and Patagonia NP, along with three Management Plans under development, have improved this situation. The elaboration of Public Use Plans is much less widespread, with the exception of Aysén, which has completed this instrument for 80% of the units with current management. Only 50% of the units with a planning instrument develop Annual Operating Plans, and less than 25% of the units report having monitoring programs designed or being implemented.

4.6 Management Inputs

The information provided by NPWA administrators for the period 2014–2017 allows a historical evaluation of the operating budget, staffing, and staff training for each of the units. However, it has not been possible to obtain systematized information on the provision of infrastructure and equipment.

NPWA expenditures in Patagonia were around CLP$ 5 billion annually (about US$ 7 million) for the 2014–2017 period (Table 5).[12] This figure is very close to the operational income from entrance fees and concession payments (*ca.* CLP$ 4.5

---

[12] Average US$ value in 2019 http://www.sii.cl/valores_y_fechas/dolar/dolar2019.htm.

Table 4 Basic data of the NPWAs units evaluated in 2018

	Management category	Unit name	Surface area (ha)	Region	Budget (average 2014–2017) CLP	Visitors (average 2014–2017) CLP	Income (average 2014–2017) CLP
1	NP	Alerce Andino	39,053	Los Lagos	27,858,500	22,851	28,117,871
	NP	Hornopirén	34,775	Los Lagos	7,552,500	1,510	0
	NP	Corcovado	295,938	Los Lagos	0	0	0
	NR	Futaleufú	13,096	Los Lagos	7,825,750	1,806	0
5	NR	Lago Palena	38,565	Los Lagos	0	0	0
	NR	Lago Rosselot	12,472	Aysén	0	0	0
	NP	Queulat	158,747	Aysén	30,735,250	32,396	82,178,626
	NR	Lago Carlota	26,990	Aysén	6,115,000	0	0
	NR	Lago Las Torres	16,334	Aysén	0	0	0
	NP	Isla Magdalena	150,315	Aysén	0	0	0
	NP	Isla Guamblin	14,482	Aysén	0	0	0
	RN	Las Guaitecas	991,991	Aysén	0	0	0
	NM	Cinco Hermanas	212	Aysén	0	0	0
	NR	Rio Simpson	40,740	Aysén	21,520,000	10,369	25,252,800
	NR	Coyhaique	2,672	Aysén	13,407,500	21,369	24,998,885
	NR	Dos Lagunas	208	Aysén	5,817,587.5	2,153	2,314.250
	NR	Trapananda	2,303	Aysén	5,762,600		0
	NR	Cerro Castillo	135,008	Aysén	20,162,525	3,250	5,800,286
	NR	Laguna San Rafael	1,698,092	Aysén	66,172,500	5,647	30,147,706
	NR	Lago Jeinimeni	170,381	Aysén	26,225,000	2,651	4,008,540
	NR	Lago Cochrane	8,406	Aysén	20,600,000	3,770	8,048,676
	NR	Katalalixar	665,432	Aysén	0	0	0

(continued)

Table 4 (continued)

Management category	Unit name	Surface area (ha)	Region	Budget (average 2014–2017) CLP	Visitors (average 2014–2017) CLP	Income (average 2014–2017) CLP
NP	Bernardo O'Higgins	3,906,127	Aysén and Magallanes	61,999,960	32,346	0
NR	Alacalufes	2,032,614	Magallanes	7,750,000	1,675	800,461
NP	Torres del Paine	205,303	Magallanes	588,517,500	231,659	3,305,432,607
MN	Cueva del Milodon	180	Magallanes	45,719,500	122,966	279,353,830
NR	Pali Aike	5,090	Magallanes	22,688,250	2,473	2,930,000
MN	Los Pinguinos	73	Magallanes	26,994,750	29,559	6,651,437
NR	Magallanes	21,181	Magallanes	56,263,000	13,266	13,334,284
NR	Laguna Parrillar	22,156	Magallanes	28,408,250	5,566	7,107,975
MN	Laguna Cisnes	1,736	Magallanes	0	0	0
NP	Yendegaia	150,695	Magallanes	0	0	0
NP	Alberto De Agostini	1,190,458	Magallanes	0	0	0
NP	Cabo de Hornos	58,917	Magallanes	0	7,824	61,280,388

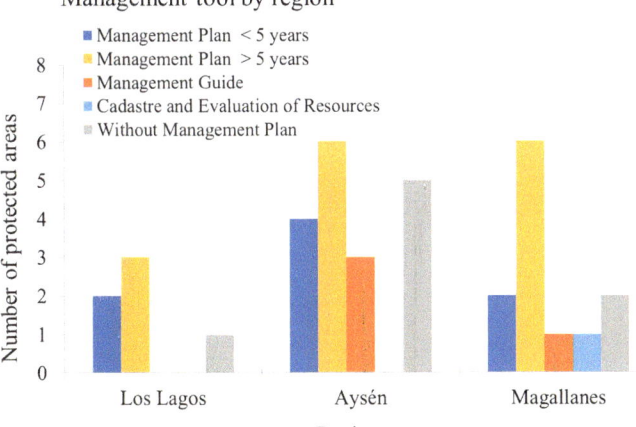

Management tool by region

Fig. 6 NPWA planning instruments in Chilean Patagonia by type and region

billion annually), so the net fiscal contribution was very marginal (*ca.* CLP\$ 500 million annually). Although there was a sustained increase in self-generated revenues (Fig. 7), the year-on-year budget increase was very low, following a pattern similar to the national one, although much more evident in Patagonia. There is also a marked difference in the budget between regions, with Magallanes registering the highest ratio of executed expenditures to revenues received.

Spending was concentrated on personnel in all three regions, which represents about 80% of the allocated budget. On average, only 25% of the regional budget was allocated to cover the operating budgets in the NPWA units (Table 5). However, the main source of operational income was visitor entry fees (>80%, Table 5), and this item was proportionally more important for Aysén (97.4%).

In December, 2017, 22 of the 34 NPWAs had their own operating budget (Table 4). The data reviewed shows a huge disparity in this aspect, with an average of CLP\$ 36 million per year, and a range from 13 units with no operating

Table 5 Regional budget of CONAF's Department of Protected Wild Areas (2014–2017)*

Budget item	Los Lagos	Aysén	Magallanes	Patagonia
Employees	1,141,988	1,165,234	1,703,149	4,010,371
Goods and services	255,991	185,890	612,233	1,054,114
Total expenses	**1,397,979**	**1,351,124**	**2,315,382**	**5,064,485**
Revenue from entrance fees	520,257	178,075	3,273,231	3,971,562
Collection of concessions	113,274	4,787	448,684	566,745
Total public use collection	**633,531**	**182,862**	**3,721,915**	**4,538,307**

Values in thousands of Chilean pesos (CLP\$)

* Average annual expenditures and collections for each region in the years 2014–2017. Expenditure figures are from the regional ASP departments, including both regional office expenditures and those of all units in that region

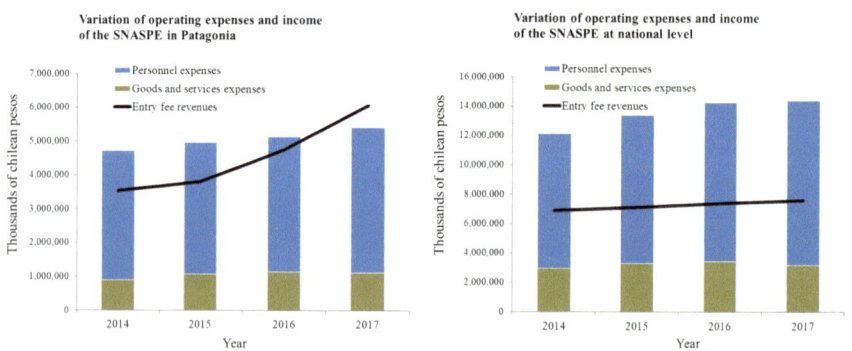

Fig. 7 Distribution of the NPWAs budget in Chilean Patagonia in 2014–2017. *Employee expenses: considers per diems and overtime; permanent and temporary wages. **Data obtained from Toledo [67]

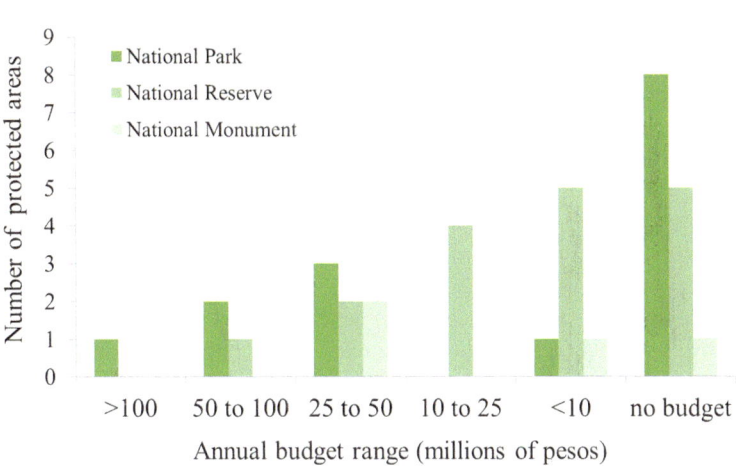

Fig. 8 Distribution of NPWAs by operating budget range

budget (38%) to a single unit with more than CLP$ 100 million per year (3%) (Fig. 8). It is worth mentioning that updated budget data per unit could not be obtained following the expansion of the NPWAs in 2018. The percentage of units with enabling infrastructure reached 90% in Los Lagos in December, 2017, 66% in Aysén and only 25% in Magallanes. There are 10 units with universal access facilities, although it has not been possible to access an inventory of the infrastructure and administrative equipment available in each unit.

As of December, 2017 there were 113 people working in various functions (administrators, purchasing assistants, park rangers) in the 24 NPWA for which there are personnel assigned; Torres del Paine NP had the largest number of personnel, with 28 park rangers (Table 4). All these numbers changed considerably during 2019, with the expansion of the system through the incorporation of Pumalín Douglas Tompkins NP and Patagonia NP, which add park guards to the system. The area protected by these 24 units is estimated to average 63,000 ha per ranger, varying from less than 50 ha/ranger in Los Pingüinos or Cueva del Milodón NM to more than 2.5 million ha/ranger in Kawésqar NP. Only 12% or 12 of the 113 rangers in 2017 were women. Most of the staff in these units have completed formal education, although a very low percentage have a technical/professional academic degree. Finally, 55% have more than 15 years experience in the system; 59% of these are over 45 years old.

4.7 Scope of Management Processes

The main sources of information for the management processes developed in Patagonia's NPWAs are the Annual Operating Plans and the NPWAs regional indicators of effectiveness [44]. In relation to conservation actions, although most of the units report having carried out patrols, the percentage of units addressing threats is very low, or even nil in the case of Aysén. The scope of the conservation actions for fauna species listed in the National Conservation Plan is variable; from 100% of the units in the Los Lagos Region to 37% in Aysén and 66% in Magallanes. Surprisingly, none of the units report conservation actions for threatened flora species. However, the system used by the Annual Operating Plans for recording activities makes it difficult to classify them by type of action. The number of actions taken to monitor impacts of projects licensed through the Environmental Impact Assessment System is very high in Magallanes and Los Lagos; for Aysén it is 36% of the projects licensed.

As of December 2017, a total of nine units (25%) reported having constituted consultative councils for community outreach, especially concentrated in Aysén, which has six consultative councils, compared to two in Los Lagos and one in Magallanes. Although all the regions report other community outreach actions, very few units have established agreements or usufructs with local communities. The Magallanes Region leads the development of six institutional agreements with other public services, compared to one in Aysén, and none in Los Lagos. It has not been possible to obtain information on other participation and consultation procedures, and there is an absence of surveys of perceptions of the NPWAs among the local population.

In relation to public use, a total of 21 units report visitor control and registration procedures, although only 16 charge entrance fees. A total of 28 tourism service concessions have been registered in 11 units, concentrated in Aysén and Magallanes, with 13 and 11, respectively. The number of tour operators is still very low, with Magallanes standing out with 16.2% of the operators making use of the NPWAs. Satisfaction levels among users are around 80% for visitors surveyed. The expansion of the system in 2018 significantly improved this situation, with the incorporation of Douglas Tompkins Pumalín NP and Patagonia NP, although 50% of the new parks lack infrastructure and visitor management actions. The total number of units with administration processes as of December, 2017 is 28 out of 34 NPWAs (80%), with less development in Aysén (11 out of 18) and Magallanes (8 out of 12). Only 5 of the 8 NPs created or expanded during 2018 have administrative processes in place.

Despite the fact that the 34 NPWAs analyzed as of December, 2017 exist in legal terms, at least 11 lack permanent staff in the field, eight do not have planning instruments and 14 have no budget allocation, leaving only 20 units with some level of effective management. Only one new unit with effective management was incorporated with the reconfiguration of the system during 2018, while the other newly created park lacks staff, budget, and management plan. Four type-situations are evident as of December, 2017, representing different management levels. There are 13 units at the initial management level which do not have a management plan, staff, or budget, representing around 2,400,000 ha (20% of the surface area); eight units are at an intermediate level, with a management plan in force, a team of park rangers, and sufficient infrastructure or equipment to carry out basic management activities, representing about 5,800,000 ha (48% of the surface area). The other 12 units are at a basic level, with some outdated management instruments, a minimum number of park rangers and a budget that severely limits management, which represent about 3,500,000 ha (29% of the surface area). Only one NPWA is at the consolidated level of management, with a team of specialized park rangers, specific planning instruments, and sufficient infrastructure or equipment to carry out advanced management activities (Figs. 9 and 10).

5   Discussion

As a result of different historical phases of PA establishment, the areas protected and corresponding management capacities are very uneven in Patagonia. The processes of PA establishment in the region show a sustained growth with defined peaks of expansion, the latest being the period 2016–2018 with the expansion and creation of NPs (Fig. 2). The substantial increase in the area protected in PAs in Patagonia during the twentieth century is a consequence of changes in the approach to land use and occupation, new valuations of natural resources (particularly biodiversity), and the creation of public policies associated with pAs. This process has been strengthened in recent decades with the new environmental institutional framework, international commitments assumed under agreements such as the Convention on Biological Diversity, and national biodiversity conservation goals.

PAs represent an important proportion of the Patagonian region (50.5%) which in turn represent a large majority of the national system (87%). This percentage exceeds the minimum targets established in international agreements, and the coverage and representativeness of PAs is optimal in relation to that of other regions in Chile such as coastal Mediterranean deciduous forests, coastal Mediterranean sclerophyllous forests, or coastal desert cactus scrub, among others [53].

Most of this area is protected under the national parks and reserves recognized under the Forest Law. Despite their smaller contribution in coverage, the other forms of protection offer an important complement in protecting specific conservation values such as vegetation formations, water bodies, and habitat for emblematic species. The Protected National Asset is a recent legal category that has allowed the protection of 24 areas widely distributed in the region (Fig. 3). Biosphere Reserves also present an opportunity to complement the protection of terrestrial and marine ecosystems and socio-cultural aspects regionally and nationally [41], but they still lack legal recognition and integration into the policy framework. Private conservation initiatives of great importance in other regions of the country represent a minor contribution in Patagonia, and their surface area is concentrated in a few areas.

PAs harbor resources of great value globally, with a uniqueness that provides a basis for a conservation system without parallels globally [29]. The results show that Patagonia's NPWAs provide a significant representation of the main

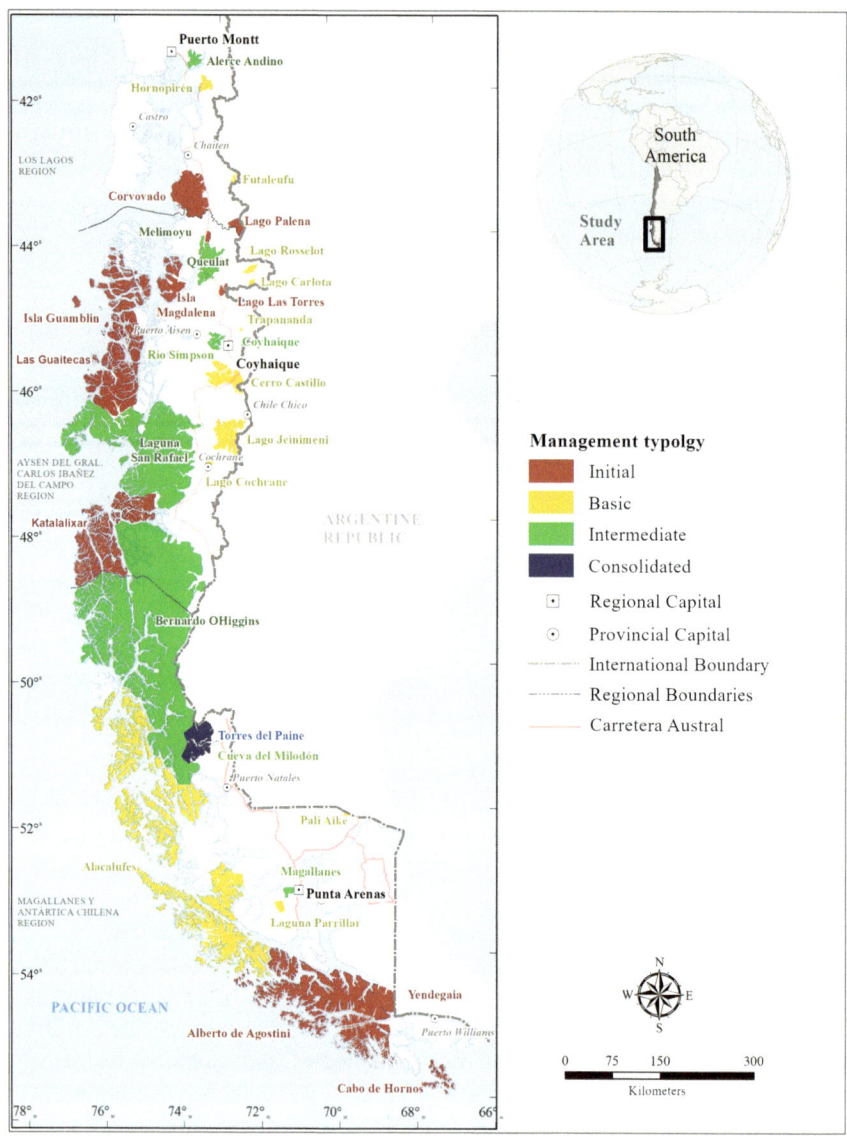

Fig. 9 Distribution of terrestrial NPWAs in Chilean Patagonia in 2018, indicating their levels of management effectiveness

ecosystems in the region, including primary native forests and peatlands which are important sources of regulating ecosystem services such as carbon sequestration, as well as snow and glaciers that provide cultural services such as recreation and leisure opportunities (Fig. 5). However, there is still a need to protect the steppes, an ecosystem with a high degree of anthropogenic pressure [56]. Finally,

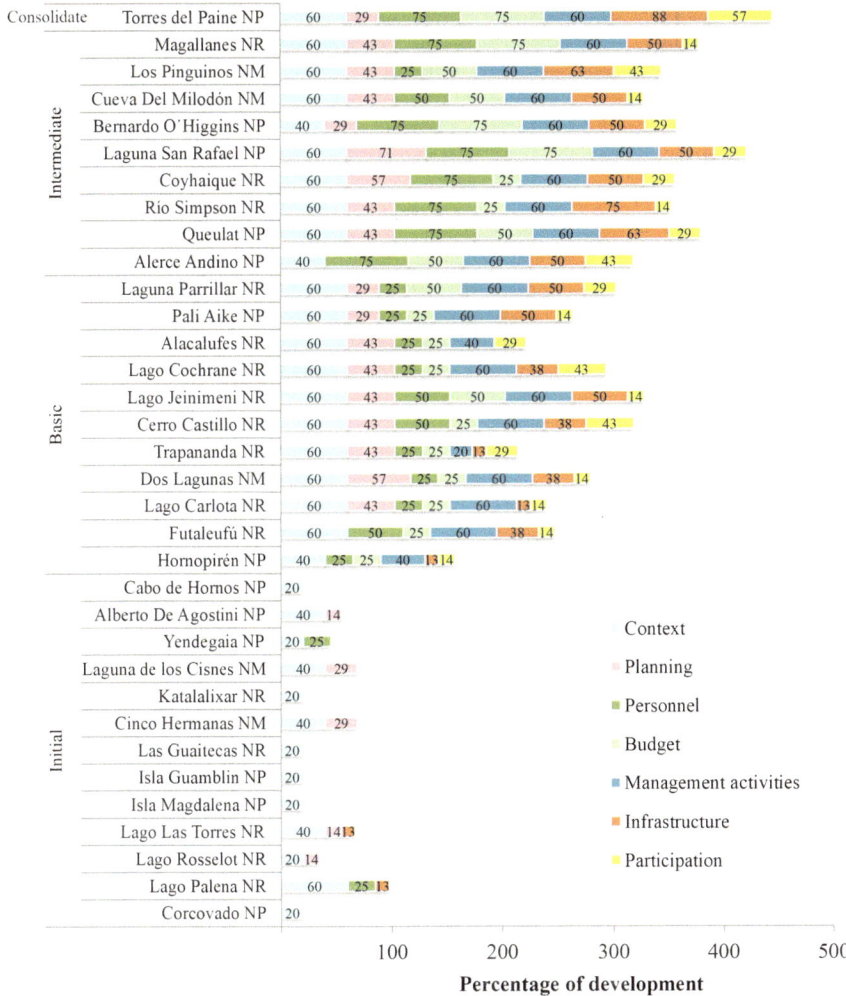

Fig. 10 Management levels of terrestrial NPWAs in Chilean Patagonia according to the percentage of the optimum achieved in different management subfields (AV: advanced)

it is interesting to note that the vast majority of the protected area affects areas with restrictive soils and slopes (>45%), which indicates a very low potential for extractive and silvo-agro-livestock activities, which are often promoted despite the essential conservation vocation of these lands.

One of the main geographic characteristics of Patagonia's NPWAs is their archipelagic condition, including more than 40,000 islands. A discussion of marine protection is not included here as it is included in another contribution in this volume [65], but it is noteworthy that NPs and NRs include important coastal-marine ecosystems in the fjords and channels located within their boundaries. This highlights the importance of generating integrated terrestrial-marine planning and management formulas, since terrestrial and marine systems are interconnected [1].

The situation of the NPWAs in Patagonia is not unrelated to what has been observed nationally. Although considerable progress has been made in many aspects of management over the 20 years since the first NPWAs evaluations, the results described here are consistent with previous diagnoses of the system's major challenges and limitations [9, 19, 51–53, 62].

Differentiating the NPWAs into four levels of management according to the indicators evaluated by CONAF, 13 units stand out as having made no progress in management since their establishment, and 11 have remained for years or even decades at a basic level of management that limits effective conservation actions (Fig. 10). There is no evident correlation between a higher level of management and variables such as the age, management category, accessibility, or proximity of the unit to population centers. Therefore, it appears that the main barrier affecting the effectiveness of the management of the NPWAs in Patagonia is of a systemic order, and in particular an insufficient budget that limits units from developing adequate personnel, infrastructure, and equipment to achieve the planned activities. The current budgetary and personnel deficiency of the NPWAs requires medium and long-term planning to ensure a minimum and stable fiscal contribution for management and administration [67]. This situation, in addition to keeping a significant percentage of the units at an initial level of development–without effective management due to lack of staff or operational budget– generates a marked disconnect between planned activities and actions developed in each unit, as well as a shortage of research and monitoring mechanisms to evaluate and provide feedback on conservation and community outreach strategies.

Other factors must be considered along with budget gaps in order to achieve management effectiveness. This is reflected in those NPWA at intermediate or advanced levels of management, which, despite having a larger budget allocation, still present deficiencies in various aspects of their management (Fig. 10). Among the least developed indicators are those related to social connections, that is, those that evaluate effective governance processes with social participation and that involve communities and other local actors.

The lack of updated management plans and monitoring programs for conservation objectives are also an important gap for the adequate management of NPWAs. These require field information that is still deficient for many Patagonian units, and which must be collected and analyzed in collaboration with different stakeholders such as NGOs, scientists, partners, and agencies working in the area. Such efforts will allow for development of shared visions to optimize procedures and consequently achieve results [52].

Based on the results presented here, there is an evident need for a systematic evaluation of management effectiveness for the NPWA of Patagonia that allows us to measure progress. Many of the indicators proposed through the PAME tools are not considered in the current evaluations carried out by CONAF or are absent in several NPWAs. Thus the scenario facing the NPWAs remains far from meeting the Aichi 11 targets and achieving international management certifications such as the IUCN Green List.

6   Conclusions and Recommendations

The synthesis presented in this chapter is a comprehensive assessment of the processes of PA establishment and management in Patagonia. The current protection system for terrestrial environments shows multiple gaps that require new regulatory frameworks and a new PA system. The current conditions of PAs in Chile and Patagonia place Chile in a scenario that is still far from fulfilling its international commitments to biodiversity conservation and the objectives of contributing to the collective functioning of the global PA network. We present the following recommendations:

- **Public policies**: In order to guide the management of Patagonia's NPWAs towards the achievement of international standards, it is necessary in the first place to overcome the barriers and limitations associated with the gaps in and fragmentation of the national legal framework. In particular, this requires the development of modern legislation that defines an institutional and regulatory framework appropriate to the management needs of the different existing PA categories and in accordance with international recommendations [34]. The bill to create the Biodiversity and Protected Areas Service, currently in Parliament, could be a relevant step forward in this regard, but greater consensus among the different institutions and political actors is required to finalize this process and thus ensure budgets in line with conservation challenges.

  There is an urgent priority to increase staffing levels and to generate a stable funding system that meets minimum needs, is transparent, and creates appropriate incentives for continuous improvement. Along with improving annual budgets to ensure stability for all areas, new systems for generating revenues are needed that include incentives for autonomous and decentralized management. One alternative to this end is an improvement in the collection of entry fees, and an in-depth analysis of the multiple concession systems that currently operate across different units in order to improve their coordination.

  Given the NPWA limitations in coverage and management, specific policies and regulations should be developed to promote the implementation and management of complementary conservation categories and auxiliary conservation measures, such as the private protected areas recognized in article 35 of Law 19,300 and Nature Sanctuaries.

- **Administration and management**: There are two clear priorities for future investment in the NPWAs. One is to establish management plans, minimum monitoring, and budgets for parks and reserves which are still at the initial level of management. Only with the installation of standardized planning will it be possible to evaluate the pressures and needs of the units and thus test different management hypotheses. The second is the need for clear planning, park ranger staffing, and infrastructure in the units that are experiencing a boom in visitation due to favorable access provided by the Austral Highway and links to tourist destinations. As part of the process aimed at approaching international standards, attention should be focused on the human dimension of PA management, generating standardized procedures for information management, consultation, and participation of the different stakeholders that coexist in the territory where each of the units is located.

- **Research and knowledge management**: One of the first measures recommended is to focus on standardizing the evaluations of management effectiveness and homologating these to international standards, in order to generate an accurate understanding of the current management situation, both at systemic and individual unit levels, and to determine the priorities for State investment. It is recommended that the State invest in adapting and applying methodologies to quantify the contribution of PA visitation to local economies, in order to document the return on State investment in each PA. A joint investment by State agencies and universities is required to generate a cost-effective methodology that can be replicated periodically for Patagonia.

Acknowledgements The authors are grateful to CONAF's Department of Protected Wild Areas for providing the information analyzed and to The Pew Charitable Trusts for its financial support for this work. In addition, the suggestions and contributions of an anonymous referee who helped to improve the chapter are gratefully acknowledged.

References

1. Álvarez-Romero, J. G., Pressey, R. L., Ban, N. C., Vance-Borland, K., Willer, C., Klein, C. J., & Gaines, S. D. (2011). Integrated land-sea conservation planning: The missing links. *Annual Review of Ecology, Evolution and Systematics, 42*, 381–409.
2. Aylwin, J., Arce, L., Guerra, F., Núñez, D., Álvarez, R., Mansilla P., Alday, D., Caro, L., Chiguay, C., & Huenucoy, C. (2023). *Conservation and indigenous people in Chilean Patagonia*. Springer.
3. Borrini-Feyerabend, G., Dudley, N., Jaeger, T., Lassen, B., Broome, N. P., Philips, A., & Sandwith, T. (2014). *Protected area governance: From understanding to action. No. 20 in the Protected Areas Good Practice Guidelines Series*. IUCN.
4. Camus, P. (2006). *Ambiente, bosques y gestión forestal en Chile*. Ediciones LOM. https://lom. cl/ea63570f-8c98-497d-801c-8e5da9965763/Ambiente-bosques-y-gestión-forestal-en-Chile. aspx
5. Carranza, T., Manica, A., Kapos, V., & Balmford, A. (2014). Mismatches between conservation outcomes and management evaluation in protected areas: A case study in the Brazilian Cerrado. *Biological Conservation, 173*, 10–16.
6. Coad, L., Leverington, F., Burgess, N., Cuadros, I., Geldmann, J., Marthews, T., Mee, J., Nolte, C., Stoll-Kleemann, S., Vansteelant, N., Zamora, C., Zimsky, M., & Hockings, M. (2013). Progress towards the CBD protected area management effectiveness targets. *Parks, 19*(1), 13–23.
7. Convention on Biological Diversity. (2010). Strategic plan for biodiversity 2011–2020. *Provisional technical basis, possible indicators and suggested milestones for the Aichi targets*. https://www.cbd.int/doc/meetings/cop/cop-10/official/cop-10-27-add1-eS.pdf

8. Corporación Nacional Forestal. (1989). *Políticas técnicas para el manejo de los Parques Nacionales y Monumentos Naturales* (75 pp.). Manual técnico N° 12, Ediciones CONAF.
9. Corporación Nacional Forestal. (1997). *Evaluación de la situación del manejo en el Sistema Nacional de Áreas Silvestres Protegidas del Estado* (36 pp.). Ediciones CONAF.
10. Corporación Nacional Forestal. (2013). *Manual de procedimientos, requisitos y obligaciones para proyectos de investigación científica en el Sistema Nacional de Áreas Silvestres Protegidas del Estado* - SNASPE. https://www.conaf.cl/wp-content/uploads/2013/03/Reglamento-de-Investigaciones-en-el-SNASPE-2013-11.pdf
11. Cuevas, C. (2015). Protected areas in Chilean Patagonia. In G. Wuerthner, E. Crist, & T. Butler (Eds.), *Protecting the wild: Parks and wilderness, the foundation for conservation* (pp. 226–241). Island Press.
12. Dudley, N. (2008). *Guidelines for the application of protected area management categories.* IUCN. https://portals.iucn.org/library/efiles/documents/PAPS-016-Es.pdf
13. Elizalde, R. (1970). *The survival of Chile. La conservación de sus recursos naturales renovables* (2nd ed.). Servicio Agricola y Ganadero, Ministerio de Agricultura. http://www.memoriachilena.cl/archivos2/pdfs/mc0027346.pdf
14. Environmental Systems Research Institute. (2016). *ArcGIS Desktop: Release 10.5.* Environmental Systems Research Institute.
15. Fernández, M., & Recabarren, N. (2018). *Memoria histórica y cultural del Parque Nacional Torres del Paine. Informe final proyecto FONDART Memoria histórica y cultural del Parque Nacional Torres del Paine. Informe final proyecto FONDART, Folio 415853.* http://www.cequa.cl/cequa/documentos/Documento_MemoriaHistóricaCulturaldelParqueNacionalTorresdelPaine.pdf
16. Fischman, E. I. (2007). *Areas protegidas: tierra de nadie.* Universidad de Chile. http://repositorio.uchile.cl/handle/2250/112631
17. Food and Agriculture Organization. (2006). *FAO in Chile: 60 years of collaboration 1945–2005.* FAO. www.fao.org/3/a-a0816s.pdf
18. Franks, P., & Small, R. (2016). *Social assessment for protected areas (SAPA). Methodology manual for SAPA facilitators.* IIED. http://pubs.iied.org/14659IIED
19. Fuentes, E., & Domínguez, R. (2011). *Aplicación y resultados de la encuesta sobre efectividad de manejo de las principales áreas protegidas de Chile.* Ministerio del Medio Ambiente, Global Environment Facility (GEF), Programa de las Naciones Unidas para el Desarrollo (PNUD). http://bdrnap.mma.gob.cl/recursos/privados/Recursos/CNAP/GEF-SNAP/Fuentes_Dominguez_2011.pdf
20. Fuentes, E., Domínguez, R., & Gómez, N. (2015). *Consultoría para la Aplicación y Análisis de Resultados del Management Effectiveness Tracking Tool (METT) a las Principales Áreas Protegidas en Chile: Informe 2015.* Ministerio del Medio Ambiente, Global Environment Facility (GEF), Programa de las Naciones Unidas para el Desarrollo (PNUD). https://areasprotegidas.mma.gob.cl/wp-content/recursos/privados/CNAP/GEF SNAP/Fuentes_Dominguez_Gomez_2015.pdf
21. García, M., & Mulrennan, M. (2020). Tracking the history of protected areas in Chile: Territorialization, strategies and shifting state rationalities. *Journal of Latin American Geography, 19*, 199–234.
22. Geldmann, J., Barnes, M., Coad, L., Craigie, I. D., Hockings, M., & Burgess, N. D. (2013). Effectiveness of terrestrial protected areas in reducing habitat loss and population declines. *Biological Conservation, 161*, 230–238.
23. Geldmann, J., Joppa, L., & Burgess, N. (2014). Mapping change in human pressure globally on land and within protected areas. *Conservation Biology, 28*(6), 1604–1616.
24. Häussermann, V., Försterra, G., and Laudien, J. (2023). *Hard bottom macrobenthos of Chilean Patagonia: Emphasis on conservation of sublitoral invertebrate and algal forests.* Springer.
25. Hockings, M., Stolton, S., & Dudley, N. (2000). *Evaluating effectiveness: A framework for assessing the management of protected areas.* IUCN. https://portals.iucn.org/library/sites/library/files/documents/PAG-006.pdf

26. Hockings, M., Stolton, S., Leverington, F., Dudley, N., & Courrau, J. (2006). *Evaluating effectiveness a framework for assessing management effectiveness of protected areas* (2 ed.). IUCN. https://portals.iucn.org/library/efiles/documents/PAG-014.pdf

27. Holmes, G. (2014). What is a land grab? Exploring green grabs, conservation, and private protected areas in southern Chile. *The Journal of Peasant Studies, 41*(4), 547–567.

28. Hucke-Gaete, R., Viddi, F. A., & Simeone, A. (2023). *Marine mammals and seabirds of Chilean Patagonia: Focal species for the conservation of marine ecosystems.* Springer.

29. Inostroza, L., Zasada, I., & König, H. J. (2016). Last of the wild revisited: Assessing spatial patterns of human impact on landscapes in Southern Patagonia, Chile. *Regional Environmental Change, 16*(7), 2071–2085.

30. International Union for Conservation of Nature and World Commission on Protected Areas. (2016). *IUCN green list of protected and conserved areas: Standard, version 1.1.* IUCN-WCPA. https://www.iucn.org/sites/dev/files/content/documents/iucn_green_list_standard_version_1.0_september_2016_0117.pdf

31. International Union for Conservation of Nature and World Commission on Protected Areas. (2017). *IUCN green list of protected and conserved areas: Standard, version 1.1.* IUCN. https://www.iucn.org/sites/dev/files/content/documents/iucn_2017_standard_green_list_v1.1.pdf

32. Keller, P. (2007). Transboundary protected area proposals along the southern Andes of Chile and Argentina: Status of current efforts introduction: Chilean and Argentine area efforts in Patagonia. In A. Watson, J. Sproull, L Dean (Eds.), *Science and stewardship to protect and sustain wilderness values: Eigth World Wilderness Congress symposium* (pp. 244–248). Department of Agriculture, Forest Service, Rocky Mountain Research Station. https://www.fs.usda.gov/treesearch/pubs/31036

33. Kettunen, M., & Brink, P. (2013). *Social and economic benefits of protected areas: An assessment guide.* Routledge. https://www.routledge.com/Social-and-Economic-Benefits-of-Protected-Areas-An-Assessment-Guide/Kettunen-ten-Brink/p/book/9780415632843

34. Lausche, B. (2012). *Guidelines for protected area legislation.* IUCN. https://portals.iucn.org/library/sites/library/files/documents/EPLP-081-Es.pdf

35. Leverington, F., Costa, K. L., Pavese, H., Lisle, A., & Hockings, M. (2010). A global analysis of protected area management effectiveness. *Environmental Management, 46*(5), 685–698.

36. Ministerio de Tierras y Colonización. (1945). *Decreto 782.* A través del cual se destina territorios de las Reservas Forestales Springhill, Bahía Felipe y Estancia Nueva para los Servicios de Gendarmería, de Ganadería y Sanidad Ambiental y Dirección General de Carabineros (1945). Archivo Nacional. República de Chile.

37. Ministerio de Tierras y Colonización. (1953). *Decreto 2251.* Destina terrenos de la Reserva Forestal Bahía Felipe en Tierra del Fuego para instalación de la Dirección General de Carabineros de Chile (1953). Archivo Nacional. Republica de Chile.

38. Ministerio de Tierras y Colonización. (1980). *Desafecta de su calidad de tal las reservas forestales que indica (1980). Archivo Nacional.* República de Chile.

39. Ministerio del Medio Ambiente. (2017). *Estrategia nacional de biodiversidad 2017–2030.* Ministerio del Medio Ambiente, Gobierno de Chile. http://portal.mma.gob.cl/wp-content/uploads/2018/03/Estrategia_Nac_Biodiv_2017_30.pdf.

40. Moreaux, C., Zafra-Calvo, N., Vansteelant, N. G., Wicander, S., & Burgess, N. D. (2018). Can existing assessment tools be used to track equity in protected area management under Aichi Target 11? *Biological Conservation, 224*, 242–247.

41. Moreira-Muñoz, A., & Troncoso, J. (2014). *Representatividad biogeográfica de las Reservas de la Biósfera de Chile.* In A. Moreira-Muñoz & A. Borsdorf (Eds.), *Reservas de la Biósfera de Chile: laboratorios para la sustentabilidad* (pp. 24–61). Academia de Ciencias Austriaca, Pontificia Universidad Católica de Chile, Instituto de Geografía. Serie Geolibros. https://www.zobodat.at/pdf/Sonderbaende-Inst-Interdisz-Gebirgsforsch_1_0023-0061.pdf

42. Muñoz, C. (1947). Los parques nacionales. Charla bajo el auspicio de la Comisión Nacional de Protección a la Vida Silvestre. *Chloris Chilensis, 14*(1). http://www.chlorischile.cl/cmuñoz-PPNN/parquesnacionalesCM.htm

43. Núñez-Ávila, M., Corcuera, E., Farías, A., Pliscoff, P., Palma, J. M., Barrientos, M., & Sepúlveda, C. (2016). *Diagnóstico y caracterización de las iniciativas de conservación privada en Chile.* http://bdrnap.mma.gob.cl/recursos/privados/Recursos/CNAP/GEF-SNAP/DT_Diagnóstico_ICP_Web.pdf

44. Núñez, E. (2018). *Tasa de variación de efectividad del cumplimiento de los objetivos legales del SNASPE.* Anexo 1. Manual SIGI n°14. Corporación Nacional Forestal.

45. Núñez, P., & Guevara, T. (2015). La frontera argentino-chilena y la integración social. San Carlos de Bariloche, 1966–1983. *Revista Austral de Ciencias Sociales, 28,* 137–162.

46. Organization for Economic Cooperation and Development-UN-ECLAC. (2016). *ECD Environmental performance reviews: Chile 2016.* ecd.org/chile/oecd-environmental-performance-reviews-chile-2016-9789264252615-en.htm

47. Organización Latinoamericana y del Caribe de Entidades Fiscalizadoras Superiores-OLACEFS. (2015). *Auditoría coordinada de las Áreas Protegidas de América Latina, Primera Edición.* Comisión Técnica Especial de Medio Ambiente. https://www.olacefs.com/wp-content/uploads/2015/10/Resumen-ejecutivo-Auditoria-en-Areas-Protegidas-de-America-Latina-web.pdf

48. Ormazábal, C. (1991). *Landmarks and obstacles in the development of Chile's national system of protected wildlands.* Special project submitted to Professor Stephen R. Kellert, School of Forestry and Environmental Studies, Yale University.

49. Ormazábal, C. (1992). *Proposición para el fortalecimiento del Sistema Nacional de Áreas Silvestres Protegidas del Estado.* Working document. Plan de acción forestal. Consultancy for the Food and Agriculture Organization.

50. Otero, L. (2006). *La huella del fuego: historia de los bosques nativos. Poblamiento y cambios en el paisaje del sur de Chile* (1st ed.). Pehuén. https://tienda.pehuen.cl/products/la-huella-del-fuego

51. Pauchard, A., & Villarroel, P. (2002). Protected areas in Chile: History, current status, and challenges. *Natural Areas Journal, 22*(4), 318–330.

52. Petit, I. J., Campoy, A. N., Hevia, M. J., Gaymer, C. F., & Squeo, F. A. (2018). Protected areas in Chile: Are we managing them? *Revista Chilena de Historia Natural, 91*(1), 1–8.

53. Pliscoff, P. (2015). *Aplicación de los criterios de la Unión Internacional para la Conservación de la Naturaleza (IUCN) para la evaluación de riesgo de los ecosistemas terrestres de Chile.* http://portal.mma.gob.cl/wp-content/uploads/2016/01/Informe-final-Eval_ecosistemas_para_publicacion_16_12_15.pdf

54. Pliscoff, P., Martínez-Harms, M. J., & Fuentes-Castillo, T. (2023). *Representativeness assessment and identification of priorities for the protection of terrestrial ecosystems in Chilean Patagonia.* Springer.

55. Praus, S., Palma, M., & Domínguez, R. (2011). *La situación jurídica de las actuales áreas protegidas de Chile.* Andros Impresores. http://bdrnap.mma.gob.cl/recursos/privados/Recursos/CNAP/GEF-SNAP/Praus_Palma_Dominguez_2011.pdf

56. Radic-Schilling, S., Corti, P., Muñoz, R., Butorovic, N., and Sánchez, L. (2023). *Steppe ecosystems in Chilean Patagonia: Distribution, climate, biodiversity, and threats to their sustainable management.* Springer.

57. Rodríguez, J. C., Gissi, N., & Medina, P. (2015). Lo que queda de Chile: la Patagonia, el nuevo espacio sacrificable. *Andamios, 12*(27), 335–356.

58. Sepúlveda, B., & Guyot, S. (2016). A lo largo y a través de la frontera: áreas protegidas y gestión participativa en la norpatagonia (Chile-Argentina), Capítulo 9. In M. A. Nicoletti, P. Núñez, & A. Núñez (Eds.), *Araucanía norpatagónica* (p. 442). Editorial UNRN. https://books.openedition.org/eunrn/549

59. Servicio Agricola Ganadero. (1969). *Política técnica del patrimonio forestal.* Servicio Agrícola y Ganadero. División Forestal (Patrimonio Forestal).

60. Sierralta, L., Serrano, R., Rovira, J., & Cortés, C. (2011). *Las áreas protegidas de Chile. Antecedentes, institucionalidad, estadísticas y desafíos.* División de Recursos Naturales Renovables y Biodiversidad Ministerio del Medio Ambiente, Gobierno de Chile. http://bibliotecadigital.ciren.cl/handle/123456789/6990

61. Stolton, S., & Dudley, N. (2016). *METT handbook: A guide to using the management effectiveness tracking tool (METT)*. WWF. https://www.protectedplanet.net/system/comfy/cms/files/files/000/000/045/original/WWF_METT_Handbook_2016_FINAL.pdf.
62. Tacón, A., Fernández, U., Wolodarsky-Franke, A., & Núñez, E. (2006). *Evaluación rápida de la efectividad de manejo en las áreas silvestres protegidas de la ecorregión Valdiviana*. WWF Chile. http://www.wwf.cl/?145047/Evaluacion-Rapida-de-la-Efectividad-de-Manejo-en-las-reas-Silvestres-Protegidas-de-la-Ecorregion-Valdiviana-RAPPAM
63. Tacón, A., Gerding, J., & Almonacid, A. (2019). *Informe técnico de la consultoría "Desarrollo de estándares de calidad de gestión de lasÁreas Silvestres Protegidas del Estado (SNASPE) en la Patagonia Chilena"*. CONAF.
64. Tecklin, D., & Sepúlveda, C. (2014). The diverse properties of private land conservation in Chile: Growth and barriers to private protected areas in a market-friendly context. *Conservation and Society, 12*(2), 203.
65. Tecklin, D., Farías, A., Peña, M. P., Gélvez, X. Castilla, J. C., Sepúlveda, M., Viddi, F. A., & Hucke-Gaete, R. (2023). *Coastal-marine protection in Chilean Patagonia: Historical progress, current situation, and challenges*. Springer.
66. Thelen, K., & Miller, K. (1976). *Planificación de sistemas de áreas silvestres (con una aplicación a los parques nacionales de Chile)*. Documento Técnico de Trabajo No. 16. Proyecto FAO-RLAT TF-199. http://catalogo.corfo.cl/cgi-bin/koha/opac-detail.pl?biblionumber=3143
67. Toledo, C. (2017). *Análisis económico de los ingresos y egresos del Sistema Nacional de Áreas Silvestres Protegidas del Estado (SNASPE)*. Publicaciones Fundación TERRAM, n° 65 diciembre. https://www.terram.cl/descargar/naturaleza/biodiversidad/app_-_analisis_de_politicas_publicas/APP-65-Analisis-Economico-de-los-Ingresos-y-Egresos-del-SNASPE.pdf
68. United Nations Environmental Program-WCMC, IUCN and NGS. (2018). *Protected planet report 2018*.
69. Wakild, E. (2017). *Protecting Patagonia: Science, conservation and the pre-history of the nature state on a South American frontier, 1903–1934. The Nature State: Rethinking the history of conservation*. https://scholarworks.boisestate.edu/history_facpubs/113
70. Wells, M., Brandon, K., & Hannah, L. J. (1992). *People and parks: Linking protected area management with local communities*. World Bank. http://documents.worldbank.org/curated/en/171421468739524360/People-and-parks-linking-protected-area-management-with-local-communities
71. Zafra-Calvo, N., Garmendia, E., Pascual, U., Palomo, I., Gross-Camp, N., Brockington, D., Cortes-Vázquez, J. A., Coolsaet, B., & Burgess, N. (2019). Progress toward equitably managed protected areas in Aichi target 11: A global survey. *BioScience, 69*(3), 191–197.

Alberto Tacón BS in Biology, Universidad Autónoma de Madrid, Spain. Master in Rural Development, Universidad Austral de Chile. Consultant in protected areas and associate researcher of the Austral Patagonia Program and partner of the Calahuala Work Cooperative.

David Tecklin BA, Swarthmore College, MSc University of California, Berkeley and PhD in Geography, University of Arizona, USA. Research Associate, Austral Patagonia Program, Universidad Austral de Chile. Principal Officer, The Pew Charitable Trusts.

Aldo Farías Forestry Engineer, Universidad Austral de Chile. Executive Coordinator of the Austral Patagonia Program, Universidad Austral de Chile, Valdivia.

María Paz Peña Biologist in Natural Resources and Environment, Pontificia Universidad Católica de Chile. Ph.D. in Sciences, mention Systematics and Ecology, Universidad Austral de Chile. Researcher, Austral Patagonia Program, Universidad Austral de Chile.

Magdalena García M.Sc. in Geography and Environmental Studies, Concordia University, Canada. Ph.D. (c) in Geography, University of Montreal, Canada. Prior work at the Ministry of Environment and the Undersecretariat of Tourism, Chile.

Conserving the Origin of Rivers: Intact Forested Watersheds in Western Patagonia

5

Anna Astorga Roine, Paulo Moreno-Meynard, Paulina Rojas R., and Brian Reid

Abstract

Forests without significant human intervention are globally uncommon and rapidly declining, especially in the temperate biome. These ecosystems provide essential habitat for the conservation of biodiversity, mitigation of climate change and regulation of freshwater ecosystems. Efforts towards mapping intact forests globally have advanced via forest patches as the basic conservation unit. We propose an alternative ecosystem-based perspective that takes into consideration the connection between forests and riverine processes, using small watersheds as the unit of analysis. This chapter assesses the distribution of intact forested watersheds in Patagonia, as a more conservative approach to highlight conservation priorities for both terrestrial and aquatic systems. The study consisted of mapping all catchments in Western Patagonia (41–55°S), followed by a two-stage validation based on satellite images and field observations. We summarize the patterns in their distribution based on categories of land use, bioclimatic gradients, forest types and inclusion within protected areas. Large areas of temperate forest watersheds without major human impacts were documented across much western drainage in Patagonia, potentially representing a globally significant biodiversity refuge. However, observations on terrestrial and aquatic biodiversity in these ecosystems is limited, and since nearly half of these systems are located outside of protected areas, there is an urgent need for general baseline observations on the biodiversity of intact watersheds. The chapter concludes with a summary of the challenges and recommendations for the conservation of intact forested watersheds in western Patagonia.

A. A. Roine (✉) · P. Moreno-Meynard · B. Reid
Centro de Investigaciones en Ecosistemas de la Patagonia, Coyhaique, Chile
e-mail: anna.astorga@ciep.cl

P. Rojas R.
University of California, Davis, CA, USA

© Pontificia Universidad Católica de Chile 2023
J. C. Castilla et al. (eds.), *Conservation in Chilean Patagonia*, Integrated Science 19,
https://doi.org/10.1007/978-3-031-39408-9_5

123

Keywords

Patagonia • Western • Chile • Intact watershed • Temperate forests •
Head-water streams • Old growth forests • Conservation

1   Introduction

Temperate forest ecosystems without major human intervention in western Patagonia are a natural treasure and global heritage [7, 79, 115]. Numerous rivers and winding crystalline streams are born in these forests. A closer look reveals that the stream-forest interface is a complex and dynamic zone that is teeming with life. River ecosystems play essential roles in the transport of sediment, rocks, leaves, nutrients and dead wood, demonstrating a close relationship between what happens in the entire watershed, the banks and the watercourse [43, 53]. Considering the importance of water for life, this connection between forest and water is essential for our survival as a society.

Ancient forest ecosystems, scarcely modified by human action, are important due to their capacity to sequester and store carbon [63, 93, 92] and water [36], their role in soil protection and water quality, regulation of hydrological regimes, nutrient cycling and retention [47, 100] and the geomorphology of associated river systems [16, 28, 113]. Due to the global decline in forest area, an increasing number of publications and reports address issues such as the rate of forest cover loss [42, 77], remaining biomass in intact forest landscapes [91], habitat fragmentation levels [29, 30, 41], the human footprint on the landscape [101, 111] and priority areas for conservation and restoration [117]. However, comparable regional-scale studies are generally scarce in Patagonia [90].

As in other regions, major forest losses in western Patagonia are due to human actions such as large fires, timber extraction, cattle ranching, and to a lesser extent, the replacement of forests with plantations of exotic trees and agriculture [15, 40, 77]. These anthropogenic impacts continue to expand, resulting in the decrease of intact forest area [40]. After 100 years of intensive landscape use, Chile's forest area has suffered significant reductions [6]. It is estimated that most native forests in western Patagonia were still pristine only 100–200 years ago, despite some localized impacts from management by Indigenous people [8, 13, 57]. Despite the losses, Patagonia is home to potentially the largest area worldwide of temperate forests and wetlands without major human intervention and has become a refuge for the biota of these ecosystems [9, 10, 65, 99].

Global-scale forest mapping efforts have advanced in recent years by defining areas of remaining intact forests or landscapes as critical conservation units, and analyzing their size and connectivity [42, 91, 115]. However, the watershed seems to be the most relevant unit for conservation from an ecosystem perspective [45, 61]. Considering the close relationship between forest, water, and fluvial ecosystem in watersheds, it is especially relevant to study the forests of western Patagonia that are intact or with very little human impact, because they represent a global ecological reference condition. The identification and mapping of watersheds with forests displaying very little intervention requires a set of restrictive conditions: a forest ecosystem unaltered by human activity, adjusted to the contour of a watershed whose watershed boundary lines are identifiable up to a downstream drainage point, and without significant human impact along the stream drainage network.

2   Scope and Objectives

The objective of this chapter is to determine the distribution of forest watersheds without any major anthropogenic intervention (Table 1) in the area of western Patagonia stretching from Petrohué River to Tierra del Fuego including Chiloé Island (42–56°S).

We also highlight the conservation value of these terrestrial and freshwater ecosystems, considering their global importance and as a local conservation unit (see Fig. 1 for diagrams that include some typical ecosystem transitions at the terrestrial-aquatic interface in valleys of mountain and coastal basins of western Patagonia).

The results include a map of intact forested watersheds for western Patagonian region (41°42′S, 73°02′W–56°29′S, 68°44°W), followed by an analysis of the distribution of these intact forest watersheds according to land use categories, bioclimatic gradient, forest types and their representation within protected areas (e.g. National System of Protected Wild Areas (in Spanish SNASPE) and private conservation initiatives). Finally, some of the main threats currently facing these forest ecosystem watersheds in the Patagonian region are discussed.

Box 1

What is considered to be an intact/pristine forested watershed in this study? In this study the intactness or alteration of a forest or watercourse is based on the fact that all environments exhibit varying degrees of human intervention and cannot be treated in absolute values. One view (International Conference: Intact forests of the twenty-first century, University of Oxford, June 2018) that considers landscapes along a gradient of degrees of intervention is:

No intervention ⟵―――――――――――――⟶ Totally degraded

The concept of intact forested watershed in our study is the watersheds that are closest to presenting a state of no intervention by human beings. In order to arrive at a definition of systems that are "intact" or display "little intervention," it is necessary to start with a summary of the disturbances that

Table 1 Some flagship species of freshwater and terrestrial fauna using mapped intact forest and river ecosystems

Species (common name)	Habitat	Historical distribution	Current distribution	Conservation category	References
Lontra provocax (huillín or southern river otter)	River banks, streams, lakes, estuaries, channels and rocky coastline. Frequently in places of evergreen vegetation and dense forest Feeds primarily on crustaceans	Historically it was distributed from central Chile (Cachapoal river Valley) to Magallanes	Only inhabits inland waters with very little anthropogenic intervention in southern Chile and Argentina, from around 38°S to 56°S From the Taitao peninsula to the south, and can also be found on the fjord coast	*Endangered* A >50% reduction in the size of the population is projected over the next 30 years, for those populations that primarily use rivers and lakes	Medina [71] Cassini and Sepúlveda [19] Sepulveda et al. [103] MMA [76]
Aplochiton taeniatus (peladilla)	Rivers and lakes. It inhabits rivers such as those inhabited by introduced salmonids Currently found in IFW streams that drain into lakes	No information	From the IX Region of Araucanía to the Magallanes Region	*Endangered* Reduction in the observed population greater or equal to 50% over the last 10 years	McDowall [70] Campos et al. [18] MMA [76]
Rhinoderma darwini (Darwin's frog)	Associated with the austral temperate forest, lives on the forest floor among the vegetation close to areas of flowing water	No information	In Chile, from sea level up to 1800 m (from Concepción (Bío Bío Region) to Aysén Region	Endangered	Díaz-Páez and Ortiz [27] Crump [25] MMA [76]

(continued)

Table 1 (continued)

Species (common name)	Habitat	Historical distribution	Current distribution	Conservation category	References
Hippocamelus bisulcus (huemul)	Forests of Nothofagus with sparsely dense understory, periglacial areas and borders around continental ice. In areas with more anthropogenic impacts, it is restricted to mountain areas with high elevation and slope	Historically abundant from the Cachapoal River (34°S) to the Strait of Magellan. Uses a wide range of habitats, including valley bottoms during winter	Estimated population total <1500 (<1% of its historic population), distributed mainly in the Patagonian Andes and in some of the coastal areas of the inland archipelago of southern Chile	*Endangered*	Vila et al. [114] Iriarte et al. [50] MMA [76]
Lama guanicoe (guanaco)	Although it is not a forest species, the historical use of habitats in Chile was the bottoms of the valleys of large rivers, the highlands of the basins, and transition to the pampas	It is estimated that the historic population reached up to between 30 and 50 million at the end of the 1500s and was widely distributed throughout the Southern Cone, Chile and Argentina	The majority (95%) are located in the Argentine pampa. It is estimated that the historical distribution in Chile and Peru decreased by 75%. Currently, the largest populations in Chilean Patagonia are in the region of Cochrane (Patagonia Park) and Torres del Paine	*Vulnerable* Loss of habi-tat, construction of fences and competing with cattle	Franklin and Johnson [34] MMA [76]

(continued)

Table 1 (continued)

Species (common name)	Habitat	Historical distribution	Current distribution	Conservation category	References
Puma concolor (puma)	Due to human impact, it has been displaced towards more mountainous sectors of the coast and the Andes, although also, occasionally in the Central Valley and even close to human installations	No information	The puma is the most widely distributed feline in America, inhabiting everywhere from the south of Alaska and northwest Canada to the Strait of Magellan. It lives throughout continental Chile, in many diverse environments, excluding Chiloé, the Chonos and Guaitecas islands, and south of the Strait of Magellan	*Near threatened vulnerable*	Cofré and Marquet [20] Franklin et al. [35] MMA [76]
Leopardus guigna (güiña)	Only carnivore endemic to the Andean. Patagonian forest of Chile and Argentina. Mixed temperate rainforests of sectors of the Andes and the coast of southern Chile	No information	Chile and Argentina. In Chile it is distributed from Coquimbo to about 70 km south of Cochrane (Aysén), mainly associated with environments with forest and dense scrubland	*Vulnerable*	Cofré and Marquet [20] Acosta and Lucherini [2] MMA [76]

(continued)

Table 1 (continued)

Species (common name)	Habitat	Historical distribution	Current distribution	Conservation category	References
Leopardus geoffroyi (Geoffroy's cat)	Lives in the forest, scrub and steppe. In Chile it has been found in environments of transition between forest and Patagonian steppe, up to 1000 meters above sea level	No information	Bolivia, northern and southern Brazil, the Paraguayan Chaco, Uruguay and a large part of Argentina, from sea level to 3300 m. Its distribution is rather marginal in Chile; it is present in a variety of environments: steppe, scrubland and Patagonian forest in Magallanes and Aysén	*Near threatened*	Cofré and Marquet [20] MMA [76]
Pudu pudu (pudu)	Species confined to the temperate—Valdivian forests, from sea level to 1700 m above sea level in altitude. It is associated preferably to shady and humid places of the understory	No information	The species only lives in Argentina and Chile. In Chile, it lives both in mountain areas and the foothills of the Andes and the coast, from Curicó (Maule Region) to Aysén (in the sector of the Pascua river, about 70 km to the south of Cochrane), from the coast up to 1700 m	*Vulnerable*	Muñoz-Pedreros [81] Cofré and Marquet [20] MMA [76]

(continued)

Table 1 (continued)

Species (common name)	Habitat	Historical distribution	Current distribution	Conservation category	References
Campephilus magellanicus (Magellan woodpecker)	Mature forests of Nothofagus; needs large native trees	No information	From the Bio Bio Region to Magallanes Region	*Endangered/vulnerable*	Jaramillo et al. [52] MMA [76]
Macrocyclis peruvianus (black snail)	The only species of the genus Macrocyclis. It inhabits temperate rainforests	No information	Mainly in the south of Chile, but also, in border areas with Argentina. In Patagonia it has been recorded in Queulat by National Forestry Corporation (in Spanish Corporación Nacional Forestal, CONAF), but it has also been seen by our group in Murta and Río Huiña (southernmost recordings)	Although it is not endemic to Chile, its populational distribution Is limited and vulnerable to fragmentation. It does not fall under any category of conservation, although within its range in the country, it is frequently classified as "rare"	Stuardo and Vega [104]

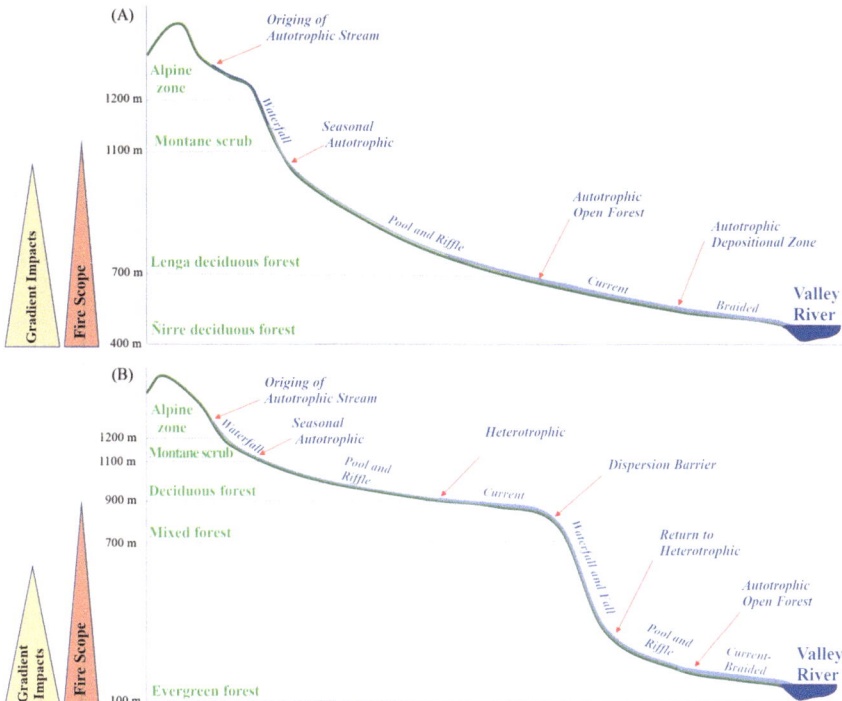

Fig. 1 Conceptual scheme of the altitudinal gradient of terrestrial ecosystem-water interface of the study area in Chilean Patagonia. **a** Zonation of coupled terrestrial-aquatic ecosystems in the deciduous forest zone. Typical scheme of mountain range river valleys. The streams that originate in the high mountains pass through areas of steeper slopes and stunted forests. In some cases, they pass through primary forests with little intervention before reaching the zone affected by anthropogenic effects (in areas where the slope is less steep, accessed by trails and secondary roads). This zone is usually the altitudinal limit of historical fires (sometimes affecting only the understory). The geomorphological gradients of the valley, as well as anthropogenic impacts, are changes in the physical habitat of the streams and rivers in this zone. **b** Zonation of coupled terrestrial-aquatic ecosystems in the evergreen forest zone. Shown is a U-shaped valley formed by glaciation, often at low elevations near the coast. The forest and stream usually have little intervention after the change of slope to higher elevations (1200–1100 m), because they are hard to reach. Below this zone, the transition to the impacts of fires from the last century begins (often mild at this altitude, affecting only the understory). These geomorphological gradients in the valley, as well as the anthropogenic impacts, are characteristic changes in the physical habitat of the streams and rivers: (i) waterfall: high slope zone where the water falls almost vertically; (ii) pool and riffle: sequence of rapids and turbulent flow zones with low velocity and deeper pools; (iii) stream: laminar flow zone; (iv) braided: braided channel, network of small channels separated by small, temporary islands, which are produced by the decrease in slope of the channel

these ecosystems may experience, considering anthropogenic impacts and natural disturbances.

Sometimes natural or anthropogenic disturbances of watersheds (e.g. fires, landslides, geological processes and ecological succession) can even be decoupled from disturbances in the aquatic system within the same watershed (hydrological cycles and events, a river that has a series of disturbances annually or with greater frequency). In light of this, it is evident that the geomorphology, vegetation cover and hydrology of watersheds can be very diverse along the morphological and bioclimatic gradients of the region. Streams and rivers are also diverse in their morphology at various scales, from a few meters (habitat) to hundreds of meters (stretch of the river) to kilometers. This variability sometimes complicates the interpretation of images from mapping. We propose a basic classification system for intact watersheds located in Patagonia, with the following criteria:

1. Absence of evidence of large-scale fires resulting in massive tree mortality and soil transformation, although the effect of low intensity fire on the understory may be less evident in areas near these large fires.
2. No evidence of large-scale logging or harvesting, although selective logging from many years ago is sometimes difficult to identify.
3. No evidence of road construction or of impacts associated with the entry of heavy vehicles, or of streams crossing over these roads, or the generation of areas of impact on soils due to compaction or erosion.
4. No evidence of the effect of intensive livestock pressure such as trail networks, usually shown by the presence of exotic species associated with livestock (clover meadows, etc.), or artificial wetlands and riparian wetlands being altered by cattle hooves trampling them.

In summary, our definition of an intact forest watershed does not mean that there is no human presence. Rather, the fundamental criterion is that there is no evidence of significant impacts on soils (such as erosion, changes in composition), the water network (drainage, dams, channel modification) or vegetation (changes in horizontal and vertical vegetation structure).

There are unavoidable anthropogenic impacts occurring globally, such as the deposition of atmospheric pollutants and climate change. But these effects are most likely moderate to mild in western Patagonia compared to other parts of the world (e.g. areas of lower atmospheric pollution: [44, 88, 87], and an anomaly in terms of climate change [32, 37].

Typical ecosystem of a mountain stream of Nothofagus pumilio (lenga) forests, Lago Atravesado sector, Aysén Region. Photograph by Rubén Isaí Madriz

3 Pristine Forest Watersheds in Western Patagonia

3.1 The Watershed as a Conservation Unit

A fundamental issue in conservation ecology is the planning unit: an endangered species, populations with a disjunct distribution (along with their genetic components), uniform, species-rich forest patches, diverse plant communities and ecosystems are frequently used units. The territorial unit used here, the watershed, is another variant that has few examples to date of being implemented in biodiversity conservation plans [1, 68]. One of the most notable advantages of the watershed as a conservation planning unit is that its topographically defined environment also integrates several conservation elements which are usually considered separately. It includes both the forest and its transition to the zone above the vegetation boundary known as the alpine zone, the riparian zone ecotone, and the aquatic systems as a whole. Not just the river or stream, but the overall water network that feeds the main river and the system that regulates the effects of precipitation events in the river. It involves species, and also a landscape that better encapsulates the possibilities of movement and the set of interactions between species; this is unlike a patch of forest or the arbitrary boundaries of a park or forest reserve. For example, in the watershed there is an altitudinal gradient that includes the ecotones between forest, water and wetland, which may be relevant for the distribution of species sensitive to environmental changes [21, 23, 96] and are equally important for generalist species with wider dispersal ranges. Finally, the above applies both to conservation within a defined space and the downstream effects, connecting diverse ecosystems and human beings.

Given these general criteria as a starting point, watersheds with intact forests (Table 1) are likely to be a much narrower subset, considering only forest area, both in terms of their distribution and area. Assuming that intact watersheds are globally rare and are usually linked to headwater streams that are of great importance in the conservation of biodiversity and downstream ecosystems [1], this intact watershed approach is of enormous value for the global conservation of freshwater systems [62, 102].

3.2   Western Patagonia as a Study Area Based on Its Watersheds

We define western Patagonia as the bioregion of watersheds that originate in the Andes and drain into the Patagonian fjords on the Chilean coast. Our analysis begins in the north in the Petrohué River basin (which discharges at 41.5°S, Los Lagos administrative region), westward to include the island of Chiloé and southward to Cape Horn, 55.91°S. Towards the south it excludes some contiguous forests of Atlantic basins of the Chubut River (north) and Río Grande of Tierra del Fuego to the east (Fig. 2). The area includes the political jurisdictions of the regions of Los Lagos (Palena and Chiloé provinces), Aysén and Magallanes in Chile, and the provinces of Chubut, Santa Cruz and Tierra del Fuego in Argentina. The maximum altitude in the entire region is Mount San Valentín, at 3910 m. The linear distances of river courses from the source to the sea are relatively short, with altitudinal gradients and steep river slopes. The study area includes some of the largest ice fields in the temperate zone: the Northern Ice Field, Southern Ice Field and Darwin Range [97], as well as some of the largest lakes in South America and deepest in the world. Trans-Andean basins are an important feature of the rivers of southern Patagonia, with westward drainage of the Puelo, Yelcho, Palena, Aysén and Baker rivers, while the drainage of the Rio Grande in Tierra del Fuego is eastward through Argentina to the Atlantic Ocean. Small coastal rivers abound in a continental region dominated by extensive fjords and islands (approximately 60% of the Chilean area, Fig. 2, with significant runoff despite the small size of their watersheds. A main characteristic of the region is the strong climatic gradient from west to east, with hyper-humid coastal systems and extreme annual precipitation (>6000 mm/year, to dry and cold steppe systems (<250 mm/year; [94]). The hydrographs of Patagonian rivers vary according to origin, from cold steppes with peak flows from September to October, mountain snowmelt with peak flows from November to December, glacial melt streams with peak flows from January to February, to coastal temperate forest systems with no consistent hydrographs and pulses occurring at any time of the year. The occurrence of this range of hydrographic characteristics is typical for an area the size of western Patagonia.

Fig. 2 Map of the large coastal basins, or hydrological and mountain units of western Patagonia that drain into the Patagonian fjords and channels between 41° and 56°S. The different hydrological units are highlighted with numbers (1–23) in the geographic range, and binational basins are highlighted in darker shades

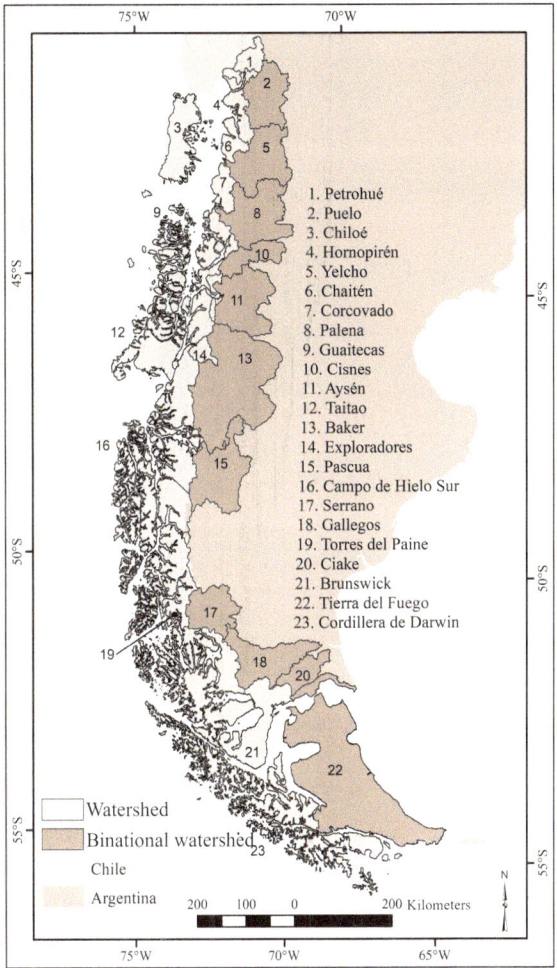

1. Petrohué
2. Puelo
3. Chiloé
4. Hornopirén
5. Yelcho
6. Chaitén
7. Corcovado
8. Palena
9. Guaitecas
10. Cisnes
11. Aysén
12. Taitao
13. Baker
14. Exploradores
15. Pascua
16. Campo de Hielo Sur
17. Serrano
18. Gallegos
19. Torres del Paine
20. Ciake
21. Brunswick
22. Tierra del Fuego
23. Cordillera de Darwin

Watershed
Binational watershed
Chile
Argentina

200 100   0   200 Kilometers

3.3 Distribution of Intact Forested Watersheds

Temperate forests in South America are distributed along a long, narrow strip on both sides of the Andes in the western part of the Southern Cone [9]. Of the total area of watersheds with mature forests in Chilean Patagonia (*ca.*, 134,000 km^2), 66,000 km^2 were identified as intact forested watersheds (IFWs) with no large-scale logging or harvesting interventions, fires or roads (Fig. 3a).[1]

The Gualas Glacier watershed (365 km^2) in the Aysén Region is considered to be one of the largest intact watersheds; it has low forest cover, but large periglacial

---

[1] See methods in Astorga et al. [11], data at: https://ide.goreaysen.cl/index.php/documentos-menu/category/4-documents-aysen.

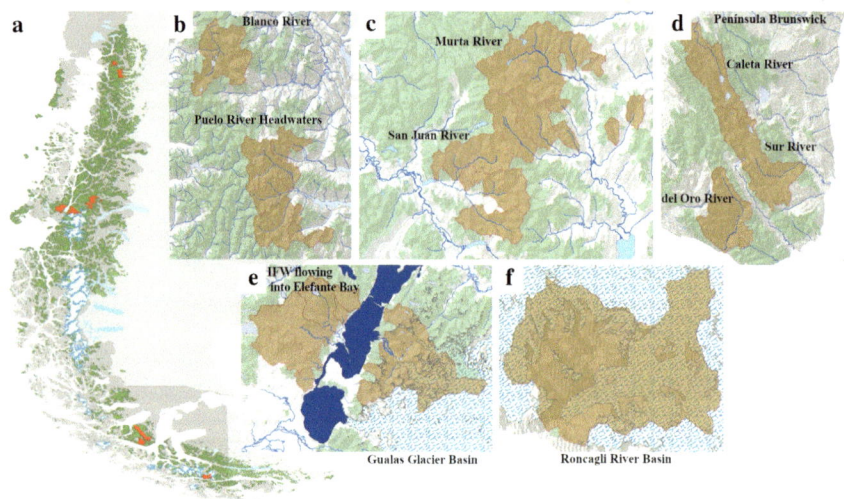

Fig. 3 Distribution of the largest undisturbed forested watersheds in western Patagonia (41–56°S). **a** Areas in green are watersheds mapped as intact (following the methods of Astorga et al. [11]), in red are watersheds or concentrations of larger intact forest watersheds (IFWs). Details in panels **b–f** from north to south; **b** Blanco River in Los Lagos Region and headwaters of the Puelo River in Argentina; **c** Murta River and San Juan River in the Aysén Region; **d** IFWs in Brunswick peninsula; **e** Gualas Glacier watershed and several agglomerated IFWs flowing into Elephant Bay, within Laguna San Rafael National Park in the Aysén Region; **f** Roncagli River basin in the Darwin Range, Magallanes Region

areas (Fig. 3e). This is also the case of the Roncagli River basin (234 km^2) in the Cordillera Darwin Range, with a forest cover that is mainly limited to valley bottoms and coastal areas (Fig. 3f).

There are also intact watersheds with large forest cover whose waters originate in the Andes, including the Blanco River (190 km^2) and the headwaters of the Puelo River (several intact watersheds totaling 389 km^2), all in areas that are not included in a formal conservation category (Figs. 3b and 4). Despite its proximity to Punta Arenas, one of the largest cities in Chilean Patagonia, the Brunswick Peninsula conserves large contiguous intact watersheds, with the Caleta River (235 km^2), South River (281 km^2) and Gold River (256 km^2), which flow into the Strait of Magellan (Fig. 3d). Finally, the Murta/San Juan River area (452 and 258 km^2, respectively) and Elefante Bay (several contiguous IWCs totaling 479 km^2) are very large coastal watersheds with 100% evergreen forest cover that have been conserved without major impacts, due to the characteristics of their forests and because they are difficult to access (Fig. 3c).

The pie chart in the upper right corner indicates the proportion of the area in each conservation category: public protected areas (SNASPE), private protected areas, and unprotected or without any conservation category, which are mainly public lands (MiBN, 2018)).

Fig. 4 Distribution and overlap of intact forested watersheds with private (Private Protected Area) and public, National System of Protected Areas (in Spanish SNASPE), conservation categories in the mapping area. *Note* This map was prepared prior to the declaration of Pumalín and Patagonia Park as SNASPE public areas, and prior to the Rewilding Foundation project in Brunswick Peninsula in 2021

The pie chart in the upper right corner indicates the proportion of the area in each conservation category: public protected areas (SNASPE), private protected areas, and unprotected or without any conservation category, which are mainly public lands [75].

According to updates from the National Forestry Corporation's Native Forest Inventory [24], land use in these intact watersheds was classified with 44.5% as forest, 41.6% as alpine zone (above the tree line) and smaller percentages of wetlands, grasslands and scrubland (Fig. 5). The forest cover within these intact watersheds totals *ca.* 30,000 km^2, with the highest coverage in the Aysén Region, then in the Los Lagos Region and finally in Magallanes, where there is also naturally less forest cover (Fig. 6b).

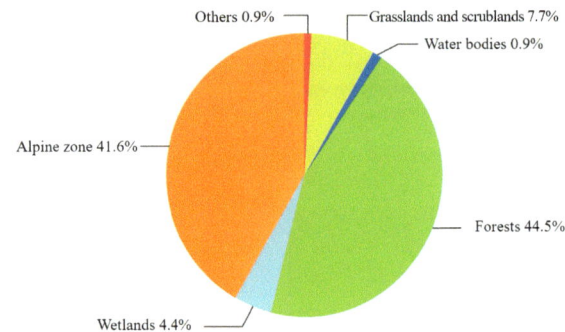

Fig. 5 Distribution of land cover types in intact watersheds in Chilean Patagonia, according to the National Forestry Corporation's Native Forest Inventory [49]

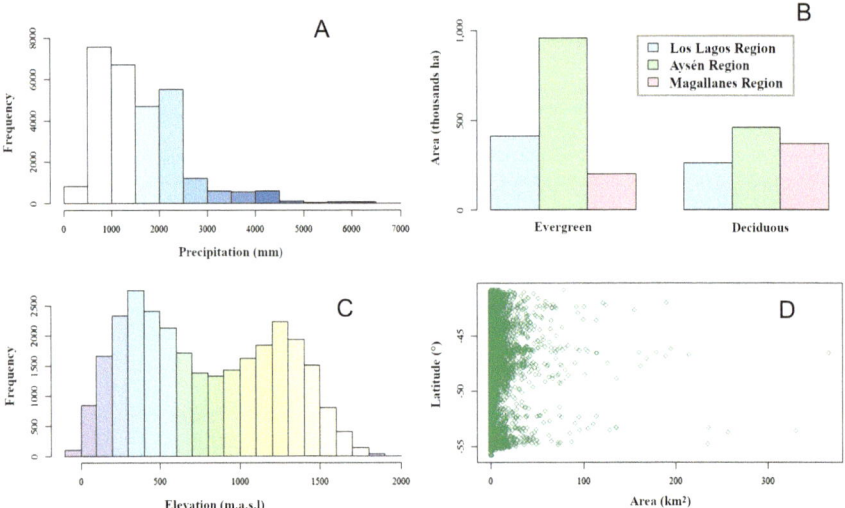

(a)Distribution along precipitation gradient. (B) Distribution among forest types, evergreen and deciduous, by administrative region. (C) Distribution along elevation range. (D) Latitudinal distribution of IFW area

Fig. 6 Attributes of intact watersheds and their frequency distribution along the bioclimatic gradient. **a** Distribution along precipitation gradient. **b** Distribution among forest types, evergreen and deciduous, by administrative region. **c** Distribution along elevation range. **d** Latitudinal distribution of IFW area

3.4 Biodiversity

The conservation status of freshwater biodiversity is broadly discussed in Reid et al. [95], while in this chapter we address biodiversity on two different levels:

(i) one related to ecosystem diversity in intact forested micro-watersheds, analyzing how these intact watersheds are distributed along the bioclimatic gradient of western Patagonia (Fig. 6), assuming that the bioclimatic gradient per se creates patterns of terrestrial and freshwater ecosystem biodiversity; (sii) a review of the state of knowledge on biodiversity associated with the ecotone of the riparian and aquatic zones present in primary forest watersheds in western Patagonia.

3.4.1 First Level: Representation of Ecosystems in Intact Watersheds

The pronounced variation in precipitation across Patagonia represents one of the steepest bioclimatic gradients in the world [38, 110], and is one of the factors that explains the large changes in biodiversity and the composition of vegetation species. The results of the mapping and analysis of the distribution of intact micro-watersheds in the different vegetation types by elevation and bioclimatic zones provide a first look at the diversity of ecosystems present in intact watersheds (Fig. 6). Intact watersheds are mostly distributed in precipitation ranges between 500 and 2500 mm per year (Fig. 6a). The high frequency of watersheds with precipitation between 500 and 1000 mm may be related to the high frequency of micro-watersheds in the *lenga* (*Nothofagus pumilio*) forests in the headwaters of western Patagonia's central valley, which are frequent, but not large (Fig. 6d). Intact micro-watersheds occur mostly in evergreen forests in the Aysén and Los Lagos Regions, while in Magallanes these watersheds occur mainly in deciduous forests (Fig. 6b).

The histogram of elevation frequencies of the intact watersheds is bimodal, with a first peak between 200 and 400 m. and another above 1000 m. in the upper parts or headwaters of the larger continental basins (Fig. 6c). The first peak is related to coastal areas such as Elefante Bay, the watershed of the Gualas Glacier and the San Juan River (Exploradores Valley) whose access is limited, or forest exploitation has been restricted because they are currently included within the Laguna San Rafael National Park conservation unit (Fig. 6c). The second peak corresponds to headwater basins that frequently contain strips of primary forests located above the forest boundary in the alpine zone dominated by sparse high-altitude vegetation, rocks, ice, lagoons and glaciers (Fig. 6c). Wetlands are also present in small headwater basins, especially in mountain rivers (Fig. 1). This tendency is typical of watersheds with forests of mainly *lenga* (*Nothofagus pumilio*) and *ñirre* (*Nothofagus antarctica*), but not necessarily of the zone of channels and fjords where many streams originate from the forest and have a much less pronounced zonation (Fig. 1).

3.4.2 Second Level: Conservation in Terrestrial and Aquatic Ecosystems

The global biodiversity crisis is the product of anthropogenic threats to forests [115] and inland waters [26, 69]. This section is divided into three subsections: (i) threatened flagship species that are generally the most well-known by the greater public, (ii) preliminary review of the list of species under a conservation category (Species Sheets of the Ministry of Environment, in Spansih Ministerio

del Medio Ambiente, MMA) listed for Los Lagos and Magallanes regions, and
(iii) preliminary observations on aquatic invertebrate and bryophyte biodiversity
in Patagonia.

(i) Flagship species

The temperate forest ecosystems of Chile and Argentina are the habitat of several
Patagonian flagship species, such as the huemul (*Hippocamelus bisulcus*), puma
(*Puma concolor*), huillín (southern river otter) (*Lontra provocax*) and Darwin's
frog (*Rhinoderma darwinii*), among others. They are defined as flagship species
because they are well-known by the public and often promote a general interest
in conservation (e.g. Macdonald et al. [64]). A preliminary list of these species is
presented in Table 1, together with a summary of the knowledge about them based
mainly on a literature review, using keywords (species names) and the species
sheets in the MMA database. Some of these species belong to the category of keystone species, defined as generalist species that have wide distribution ranges and
require contiguous blocks of habitat with buffer zones and connectivity between
landscapes for their conservation (according to Forbes and Chase [30], Echeverría
et al. [33]).

The mapping considered micro-watersheds of approximately 1 km^2 on average as units of analysis. However, neighboring intact watersheds add up to form
extensive landscapes of watersheds with habitats without intervention, reaching
surfaces of contiguous areas of approximately 500 km^2. The fact that Chilean
Patagonia is a naturally fragmented landscape should also be considered. Many
animal species show a significant biodiversity turnover in the gradient from the
evergreen Valdivian forests to the Patagonian steppe, and they have high degrees
of endemism in the different zones [55, 54, 86]. We can consider these aggregated
micro-watersheds as larger watersheds, analogous to the concept of "habitat core"
[17].

Expeditions by naturalists to Chilean Patagonia indicate that many of the mammal species present today once occupied a greater diversity of habitats, including
river valleys, forests, and mountainous areas [39, 84]. The geographic range of
several flagship Patagonian species in Chile and Argentina has been described
as much more extensive latitudinally than longitudinally. Impacts generated from
the twentieth century onwards by fires, erosion, cattle ranching, and urban centers
resulted in losses of these large continuous areas of intact landscapes and forests in
western Patagonia [57, 77]. The distribution of several flagship species of Chilean
Patagonia has receded to the last inaccessible habitats with little intervention in
the upper parts of the watersheds, and in some cases to environments unfavorable
for their populations. This may be the case of the huemul, considered a species of
high mountains and extreme zones [51, 97], but during the early stages of European colonization, its presence was greater in the productive valleys in the vicinity
of Coyhaique such as Simpson River [67].

(ii) Species with conservation problems

The list of species in the precarious conservation category [76] between Los Lagos and Magallanes Regions of Chile includes 367 taxa, of which more than 100 are associated with aquatic ecosystems: rivers, streams, lakes, wetlands, large rivers and transition from freshwater to coastal systems. Most of these species are discussed in detail in Reid et al. [95]. Below, we refer in particular to terrestrial species listed in some conservation category in the MMA Species Sheets, which are often associated with riparian zones. The list includes 30 species of fungi (22 genera), including the two species of *Cyttaria*, known as "Indian bread," which parasitize living trees and probably face the same threats that affect forests. However, the diversity of fungi in Patagonian forest systems is still poorly studied. There are 52 species of ferns, many of which are epiphytes (e.g. *Grammitis* and *Hymenophyllum* spp.). This diversity of pteridophytes is also mainly associated with intact forest systems [89]. The only vascular plant species associated with streams in forests is *Hebe salicifolia*, although its relationship with intact systems is not clear. One noteworthy point is that ferns outnumber vascular plants (35 species) and that plants, such as Juncaceae and the graminoids require more effort for taxonomic identification; Poaceae and Cyperaceae are virtually absent (B. Reid, personal observation). This is probably an indication of the lack of more complete studies and/or assessments of the conservation status of this flora, at least in remote locations (with the exception of Tierra del Fuego, Moore [78]. An alternative explanation may be that ferns are better represented in the Patagonian flora, as is the case for lichens and bryophytes [82, 83, 99].

We should not only consider flagship species, or species already recognized as species with conservation problems, but also the diversity of lesser-known groups whose potential species richness indicate that they are possibly the ones that contribute most to regional biodiversity. We refer to aquatic insects, mosses, ferns, lichens and freshwater algae [58, 80, 85]. Unfortunately, knowledge of these species in the mapped ecosystems is still very poor.

(iii) Freshwater species richness

Aquatic species richness is concentrated in insects (Plecoptera, Ephemeroptera, Trichoptera), crustaceans (*Aegla* and *Hyalella*), gastropods and bryophytes [95, 108–108]. The diversity of these groups in Chilean Patagonia has been described as slightly lower than in the Valdivian region [109]. Many taxonomic groups undoubtedly reach the southern limit of their geographic range in the latitudinal gradient of Chilean Patagonia, although there has been less sampling effort than in regions farther north [80, 85]. Freshwater diversity along east–west bioclimatic gradients in less accessible watercourses such as headwater streams and in the island and fjord region is represented with very few examples [95]. A prominent exception is research conducted in Magallanes, primarily on Navarino island, where macroinvertebrate biodiversity has been described in aquatic systems in steep elevation gradients with very little intervention [22, 21, 96].

Preliminary results on macroinvertebrate diversity in headwater streams with watersheds dominated by primary forests in Aysén indicate a total of 89 species, of which almost 40% had not been recorded before in the region [85, 108]. Preliminary results indicate that macroinvertebrate diversity in 102 micro-watersheds studied to date is strongly related to natural climatic and hydrological variation among basins [12]. Among the climatic variables that best explain macroinvertebrate diversity are regional precipitation and temperature gradients [12]. Forest cover and hydrological properties of the channel are closely related to the composition of stream invertebrates [12]. The latter is probably related to an altitudinal gradient and the origin of the stream (e.g. alpine zone or forest).

Biodiversity studies of bryophytes carried out in the riverbeds and riparian zone of the micro-watersheds of the headwaters of the Aysén rivers and lake General Carrera identified 258 taxa: 3 anthocerotes, 101 liverworts (100 species and one variety) and 154 mosses (153 species and one variety). Two of the species are new records for Chile: the liverworts *Austrololophozia andina* R.M. Schust. and *Riccardia theliophora* Hässel.); 41 are new records for the Aysén Region; 105 are new records for the General Carrera province; 77 are new records for the Coyhaique province and 22 are new records for the Aysén province [48, 58, 82, 83], in addition to one new record for science (*Syntrichia lamellaris*, Gallego et al., [106]). These results show us that the distribution of bryophytes in the Aysén Region, as well as in other sectors of Chilean Patagonia, is still poorly known, as is true for other taxonomic groups as well.

The analysis of Reid et al. [95] complements this biodiversity section on two important topics: (i) it discusses the literature review of other freshwater taxonomic groups associated with headwater streams, such as amphibians and native fishes, which also contribute to watershed conservation values, (ii) it refers to the state of knowledge of invasive exotic species, which are currently affecting freshwater systems in western Patagonia, such as salmonids, mink and beaver among vertebrates, and among plants, species such as lupine (*Lupinus polyphyllus*) and willow (*Salix fragilis*), which are invading riparian zones and larger riverbeds [60, 72].

3.5   Threats

The forested watersheds of western Patagonia with little intervention provide valuable ecosystem services such as drinking water supply and irrigation, flood control, provision of habitat for biodiversity, carbon sequestration and cultural services, which collectively can provide economic and non-economic values for local communities [14]. These services in Chilean Patagonia are threatened by the impacts of human activities, including the introduction of invasive species, fires, cattle ranching, road construction and the unsustainable use of timber resources, added to the effects of global climate change [66]. These processes have gradually led to habitat degradation and fragmentation, soil erosion, the spread of invasive species, pollution and overexploitation of forests [40]. The most serious threats currently facing these ecosystems are detailed below.

3.5.1 Climate Change

Climate projections based on global and regional models are generally rather uncertain for the complex terrain of the region [59, 112]. Temperature increase is expected to be less, in part due to regional cooling of ocean temperatures [32, 38, 37]. Effects on the headwater stream temperature may be weak due to this oceanic climate, but water temperature also depends on rainwater input and snowmelt, and this equilibrium (snow line or zero isotherm) is changing across the region, based on local observations [46]. Historical data in the region exhibit a decreasing trend in precipitation in recent decades [5]. Unfortunately, current regional climate models do not clarify the uncertainties of the projections, since a significant portion of the Chilean Patagonian region is located in a transition between areas of decreasing (Los Lagos Region) and increasing precipitation (Magallanes Region) that also vary between the east and west [31]. Specific changes in precipitation and temperature in Chilean Patagonia will affect forest regeneration, hydrology, erosion, soil quality and alpine zone boundaries, among others [66].

3.5.2 Timber and Firewood Extraction

One of the most intense current anthropogenic pressures on the forested watersheds of Chilean Patagonia is firewood extraction. The Los Lagos and Aysén regions are home to the greatest expanses of native forests, and the extraction of firewood and timber is also the most intense, mainly large volumes of lenga (67%) and ñirre (31%). Coihue (*Nothofagus dombeyi*), pine, and tepa occupy the remaining 2% (Ministry of Agriculture [73, 74]; in Spanish Ministerio de Agricultura). Timber extraction is generally carried out on roads that frequently cross or run parallel to bodies of water, leading to soil erosion and sedimentation in streams during rain events. This sediment load in streams affects primary production, habitat availability and aquatic biodiversity [3]. Such effects could be exacerbated by changes in precipitation and seasonality [32, 31, 38], as many of the slopes and streams without vegetation cover could be eroded.

Energy needs are met in the Aysén Region by the use of firewood (approximately 687,000 cubic meters of firewood are consumed annually), but only 40% of the harvest has a management plan. The largest volumes of firewood and timber are harvested near population centers [105]. The forests that supply the city of Coyhaique, the capital of the Aysén Region, with firewood, for example, include the areas of Lake Pollux and Frío, Cerro La Virgen, Cerro Galera, and Villa Ortega, among others [98, 105]. As timber resources are depleted, the distance to the harvest centers increases and the area moves towards the high-altitude forests.

Atmospheric pollution due to the intense domestic use of firewood, plus the thermal inversion effect that occurs during fall and winter months in Coyhaique (April–August) have made the city the most polluted in Latin America in some periods, according to data analyzed by the World Health Organization between 2013 and 2016 (*The Guardian*, May 12, 2016). The government enacted the first Environmental Decontamination Plan (PDA in Spanish) for the area in 2015. This plan involves several measures aimed at diversifying the energy grid, improving the thermal insulation of homes, replacing heaters and environmental education, to improve air conditions in Coyhaique and other urban centers in the region. The long-term plan could also theoretically reduce threats to Aysén's intact forested watersheds.

3.5.3 Loss of Connectivity and Fragmentation

One of the greatest threats to the large tracts of intact landscapes in western Patagonia is the expansion of the road network. The growth of the road network allows access to primary forests that for cost reasons were excluded from timber production, facilitating the entry of livestock and other exotic animals, and generating impacts that accelerate their exploitation and degradation [56, 91]. The network of logging roads has also been shown to have negative impacts on river channels, changing channel geomorphology and sediment loads in watersheds [116]. In Chilean Patagonia these problems are due to poor territorial planning, which should consider not only the need for connectivity, but also the conservation status of terrestrial ecosystems and watersheds crossed by roads. An example of this urgent need for planning and zoning is the Special Development Plan for Extreme Zones, where the Agricultural Development Institute (in Spanish INDAP) has established a special road construction program with a budget of 3 billion Chilean pesos that proposes to build a total of 550 km of roads connecting properties within three years [17].

3.5.4 Summer Livestock Grazing in Headwater Watersheds and Alpine Areas of Public Lands

As has been said, headwater streams in forests with little intervention safeguard the quality and quantity of water and the integrity of aquatic ecosystems downstream, but their usefulness is undermined by the use of equally pristine, but much more fragile areas in the alpine zone. A clear example is the management and control of summer grazing. This activity takes place in pastures belonging to the State located in the high-altitude mountain areas, which are used only four months a year due to climatic characteristics (Ministry of National Asssets 2018; in Spanish Ministerio de Bienes Nacionales, MBN; and personal communication). The MBN currently grants permits for use; however, there is no regular oversight of uses and impacts. Summer livestock grazing is a culturally rooted practice in the region, on public and inaccessible lands where indiscriminate use is made of the resources of an area where the watershed's watercourses are born. The lack of management results in the degradation of vegetation resources, native fauna, and water sources.

4   Conclusions and Recommendations

Headwater basins could be likened to water towers; objects in the landscape that capture, store, and deliver water to downstream ecosystems, both those with little intervention and those with more intense productive use. Although freshwater ecosystems are among the most threatened in the world, their point of greatest vulnerability is their origin in the upper parts of the mountain range, where small, crystalline, meandering streams are born. When these streams pass through forests, the risk of losing their integrity increases due to the demand for wood products, such as firewood, as well as road construction, especially in the temperate biome. Therefore, intact forested watersheds should be among the primary targets for freshwater and intact landscape conservation (Fig. 4).

According to our analysis, the watersheds with forest cover that drain into the Patagonian fjords and the Pacific represent a great opportunity for global conservation, but a significant portion of them are outside existing protected areas. In this chapter, information on the specific biodiversity of these ecosystems was presented and discussed, highlighting the limited knowledge of the species present there and their natural history. The analysis of the spatial distribution and bioclimatic diversity of these systems is now available for planning [90] and redefining conservation priorities.

Although Chilean Patagonia has approximately 50% of its total area under some category of conservation, the unprotected areas contain approximately half of the intact forested watersheds mapped in this research (Fig. 4). These watersheds without major interventions also contain some of the ecosystems most vulnerable to climate change and land use change, the headwater watersheds together with their mature primary forest, stunted and alpine zone ecosystems (Figs. 4 and 5). Therefore, we present the following recommendations for conservation and sustainable use:

- Ensure that public policy and/or regional conservation and land use planning instruments recognizes the importance of the watershed concept which integrates water, forests, and soils. This should be accompanied by a system of incentives to landowners in order to effectively conserve the most pristine or valuable areas for their biodiversity, together with improved management and related public policies. More specific mechanisms for these purposes could be, for example.
- Forest management plans or recreational use plans in the SNASPE should include the micro-watershed as a planning and zoning unit, considering the connection between water, forests, and soil.
- Fodder subsidies are required as an alternative practice to high-altitude summer livestock grazing, along with improvements in the registry/mapping of active summer grazing areas and carrying capacities.
- Protection of wetlands associated with headwater streams is required, starting with a wetland registry for this region [65, 95].
- Implementation of integrated management pilot projects in small watersheds (10–30 km^2), which due to their size and lesser degree of conflicts between different uses have a greater chance of success. At the same time, valuable experience could be gained in carrying out planning and conservation projects on larger scales, for example, pilot programs in watersheds that provide drinking water in rural communities, where planning and/or zoning of uses has a

direct impact on the quantity and quality of drinking water. It is also necessary to develop local models for the provision of ecosystem services (including payment for these services) in different land use scenarios of watersheds [4].

- Chile-Argentina binational efforts. Conservation of watersheds has to respect hydrological boundaries and not just political ones, as in freshwater biodiversity conservation [95]. There is a large percentage of binational headwater watersheds that contribute to the large rivers that flow into the fjord zone of Chilean Patagonia (Fig. 2). Binational conservation management in Patagonia would be more effective in several aspects related to freshwater biodiversity, along with the idea of anticipating conflicts.

References

1. Abell, R., Allan, J. D., Lehner, B. (2007). Unlocking the potential of protected areas for freshwaters. *Biological Conservation, 134*, 48–63.
2. Acosta, G., & Lucherini, M. (2008). *Leupardus guigna* (on-line). *The IUCN Red list of threatened species.* Retrieved from: https://www.iucnredlist.org/details/15311
3. Allan, D. J., & Castillo, M. M. (2007). *Stream ecology: Structure and function of running waters.* Springer.
4. Álvarez-Garretón, C., Lara, A., Boisier, J. P., & Galleguillos, M. (2019). The impacts of native forests and plantation on water supply in Chile. *Forests, 10*, 473.
5. Aravena, J. C., & Luckman, B. (2009). Spatio-temporal rainfall patterns in southern South America. *International Journal of Climatology, 29*(14), 2106–2120.
6. Armesto, J. J., Manuschevich, D., Mora, A., Smith-Ramírez, C., Rozzi, R., Abarzua, A., & Marquet, P. (2010). From the Holocene to the Anthropocene: A historical framework for land cover change in southwestern South America in the past 15,000 years. *Land Use Policy, 27*, 148–160.
7. Armesto, J. J., Rozzi, R., Smith-Ramirez, C., & Arroyo, M. T. K. (1998). Effective conservation targets in South American temperate forests. *Science, 282*, 1271–1272.
8. Armesto, J. J., Villagrán C., & Donoso, C. (1994). La historia del bosque templado chileno. *Revista Ambiente y Desarrollo*, 67–72.
9. Armesto, J. J., Smith-Ramírez, C., Carmona, M., Celis-Diez, J., Díaz, I., Gaxiola, A., Gutiérrez, A., Núñez-Ávila, M., Pérez, C., & Rozzi, R. (2009). Old-growth temperate rainforests of South America: conservation, plant-animal interactions, and baseline biogeochemical processes. In: C. Wirth, G. Gleixner, M. Heimann (Eds.), *Old-growth forests. Ecological studies (Analysis and synthesis)* (Vol. 207, pp. 367–390). Springer.
10. Arroyo, M. T. K. (1998). Los bosques de lenga de Chile ¿qué debemos hacer para asegurar su adecuada preservación y uso sustentable? In: Defensores del Bosque Nativo (Eds). *La tragedia del bosque chileno* (pp. 110–116). Ocho Libros Editores.
11. Astorga, A., Moreno, P. C., & Reid, B. (2018). Watersheds and trees fall together: An analysis of intact forested watersheds in southern Patagonia (41–56°S). *Forests, 9*, 385.
12. Astorga, A., Reid, B., Uribe, L., Moreno-Meynard, P., Fierro, P., Madriz, I., & Death, R. G. (2022). Macroinvertebrate community composition and richness along extreme gradients: The role of local, catchment, and climatic variables in Patagonian headwater streams. *Freshwater Biology, 67*(3), 445–460.
13. Aylwin, J., Arce, L., Núñez, D., Guerra, F., Álvarez, R., Mansilla P., Alday, D., Caro, L., Chiguay, C., & Huenucoy, C. (2023). *Conservation and indigenous people in Chilean Patagonia.* Springer.

14. Bachmann-Vargas, P., De la Barrera, F., & Tironi, A. (2014). *Recopilación y sistematización de información relativa a estudios de evaluación, mapeo y valorización de servicios eco-sistémicos en Chile.* Cienciambiental Consultores S.A. Retrieved from: http://portal.mma.gob.cl/wp-content/uploads/2014/10/Informe-final.pdf

15. Bizama, G., Torrejón, F., Aguayo, M., Muñoz, M. D., Echeverría, C., & Urrutia, R. (2011). Pérdida y fragmentación del bosque nativo en la cuenca del río Aysén (Patagonia-Chile) durante el siglo XX. *Revista de Geografía Norte Grande, 49,* 125–138.

16. Brookhuis, B., & Hein, L. (2016). The value of the flood control service of tropical forests: A case study for Trinidad. *Forest Policy and Economics, 62,* 118–124.

17. Buchanan, J. B., Fredrickson, R. J., & Seaman, D. E. (1998). Mitigation of habitat "take" and the core area concept. *Conservation Biology, 12*(1), 238–240.

18. Campos, H., Dazarola, G., Dyer, B., Fuentes, L., Gavilán, J., Huaquín, L., Martínez, G., Meléndez, R., Pequeño, G., & Ponce, F. (1998). Categorías de conservación de peces nativos de aguas continentales de Chile. *Boletín del Museo Nacional de Historia Natural Chile, 47,* 101–122.

19. Cassini, M., & Sepúlveda, M. (2006). El huillín *Lontra provocax*: Investigaciones sobre una nutria patagónica en peligro de extinción. *Profauna, Serie Fauna Neotropical, 1,* 162.

20. Cofre, H., & Marquet, P. A. (1999). Conservation status, rarity, and geographic priorities for conservation of Chilean mammals: An assessment. *Biological Conservation, 88,* 53–68.

21. Contador, T., Kennedy, J., Ojeda, J., Feinsinger, P., & Rozzi, R. (2014). Ciclos de vida de insectos dulceacuícolas y cambio climático global en la ecorregión subantártica de Magal-lanes: Investigaciones ecológicas a largo plazo en el Parque Etnobotánico Omora, Reserva de Biósfera Cabo de Hornos (55°S). *Bosque, 35,* 429–437.

22. Contador, T. A., Kennedy, J. H., & Rozzi, R. (2012). The conservation status of southern South American aquatic insects in the literature. *Biodiversity and Conservation, 21,* 2095–2107.

23. Contador, T., Kennedy, J. H., Rozzi, R., & Ojeda, J. (2015). Sharp altitudinal gradients in Magellanic sub-antarctic streams: Patterns along a fluvial system in the Cape Horn Biosphere Reserve (55S). *Polar Biology, 38*(11), 18531866.

24. Corporación Nacional Forestal, Chile. (2020). *Catastro de Uso de Suelo y Vegetación.* Obtenido de: http://www.ide.cl/descarga/capas/item/catastros-de-uso-de-suelo-y-vegetacion.html

25. Crump, M. (2002). Natural history of Darwin's frog, *Rhinoderma darwinii. Herpetological Natural History, 9,* 21–30.

26. Dudgeon, D., Arthington, A., Gessner, M., Kawabata, Z., Knowler, D., Leveque, C., Naiman, R., Prieur-Richard, A., Soto, D., Stiassny, M., & Sullivan, C. (2006). Freshwater biodiversity: importance, threats, status and conservation challenges. *Biological Reviews, 81,* 163–178.

27. Díaz-Páez, H., & Ortiz, J. C. (2003). Evaluación del estado de conservación de los anfibios en Chile. *Revista Chilena de Historia Natural, 76,* 509–525.

28. D'Odorico, P., Laio, F., Porporato, A., Ridolfi, L., Rinaldo, A., & Rodríguez-Iturbe, I. (2010). Ecohydrology of terrestrial ecosystems. *BioScience, 60*(11), 898–907.

29. Echeverría, C., Coomes, D., Salas, J., Rey-Benayas, J. M., Lara, A., & Newton, A. (2006). Rapid deforestation and fragmentation of Chilean temperate forests. *Biological Conservation, 130,* 481–494.

30. Echeverría, C., Newton, A., Lara, A., Benayas, J., & Coomes, D. (2007). Impacts of for-est fragmentation on species composition and forest structure in the temperate landscape of southern Chile. *Global Ecology and Biogeography, 16,* 426–439.

31. Falvey, M. (2012). Simulación regional con el modelo PRECIS. *Anales del Instituto de la Patagonia (Chile), 40*(1), 45–50.

32. Falvey, M., & Garreaud, R. D. (2009). Regional cooling in a warming world: Recent temper-ature trends in the southeast Pacific and along the west coast of subtropical South America (1979–2006). *Journal of Geophysical Research: Atmospheres, 114,* D04102.

33. Forbes, A. E., & Chase, J. M. (2002). The role of habitat connectivity and landscape geometry in experimental zooplankton metacommunities. *Oikos, 96,* 433–440.

34. Franklin, W. L., & Johnson, W. E. (1994). Hand capture of newborn open-habitat ungulates: The South American huanaco. *Wildlife Society Bulletin, 22,* 253–259.
35. Franklin, W. L., Johnson, W. E., Sarno, R. J., & Iriarte, J. A. (1999). Ecology of the Patagonia puma *Felis concolor* in southern Chile. *Biological Conservation, 90,* 33–40.
36. Frêne, C., Dörner, J., Zúñiga, F., Cuevas, J., Alfaro, F., & Armesto, J. (2020). Eco-hydrological functions in forested catchments of southern Chile. *Ecosystems, 23*(2), 307–323.
37. Garreaud, R., Boisier, J., Rondanelli, R., Montecinos, A., Sepúlveda, H., & Veloso-Águila, D. (2019). The central Chile mega drought (2010–2018): A climate dynamics perspective. *International Journal of Climatology, 40,* 421–439.
38. Garreaud, R., Lopez, P., Minvielle, M., & Rojas, M. (2013). Large-scale control on the Patagonian climate. *Journal of Climate, 26*(1), 215–230.
39. Gay, C. (1854). *Atlas de la Historia Fisica y Política de Chile, Tomo I y II. 3a Edición.* LOM Ediciones 2010.
40. Gutiérrez, A., & Díaz-Hormazábal, I. (2019). *Monitoreo de la alteración del dosel en bosques nativos.* Universidad de Chile. Retrieved from: https://www.researchgate.net/publication/332083104_MONITOREO_DE_LA_ALTERACION_DEL_DOSEL_EN_BOSQUES_N ATIVOS
41. Haddad, N. M., Brudvig, L. A., Clobert, J., Davies, K. F., González, A., Holt, R. D., Lovejoy, T. E., Sexton, J. O., Austin, M. P., & Collins, C. D. (2015). Habitat fragmentation and its lasting impact on earth's ecosystems. *Science Advances, 1*(2), e1500052.
42. Hansen, M. C., Potapov, P. V., Moore, R., Hancher, M., Turubanova, S., Tyukavina, A., Thau, D., Stehman, S., Goetz, S., & Loveland, T. (2013). High-resolution global maps of 21st-century forest cover change. *Science, 342,* 850–853.
43. Harmon, M. E., Franklin, J. F., Swanson, F. J., Sollins, P., Gregory, S. V., Lattin, J. D., Anderson, N. H., Cline, S. P., Aumen, N. G., Sedell, J. R., Lienkaemper, G. W., Cromack, K. Jr., & Cummins, K. W. (1986). Ecology of coarse woody debris in temperate ecosystems. In: A. MacFadyen & E. D. Ford (Eds.), *Advances in ecological research* (pp. 133–302). Academic Press.
44. Hedin, L. O., Armesto, J. J., & Johnson, A. (1995). Patterns of nutrient loss from unpolluted, old-growth forests: Evaluation of biogeochemical theory. *Ecology, 76,* 493–509.
45. Hedin, L., & Campos, H. (1991). Importance of small streams in understanding and comparing watershed ecosystem processes. *Revista Chilena Historia Natural, 64,* 583–596.
46. Helman, M. I. (2015). *Perceptions of climate change and water governance vulnerability in the Aysén region of Chile.* [Thesis]. University of Montana, Missoula, MT. Retrieved from: https://www.cfc.umt.edu/grad/icd/files/Helman_Michal_Thesis.pdf
47. Huygens, D., Rütting, T., Boeckx, P., Van Cleemput, O., Godoy, R., & Müller, C. (2007). Soil nitrogen conservation mechanisms in a pristine south Chilean *Nothofagus* forest ecosystem. *Soil Biology and Biochemistry, 39*(10), 2448–2458.
48. Hässel de Menéndez, G. G., & Rubies, M. (2009). *Catalogue of Marchantiophyta and Anthocerotophyta of southern South America: Chile, Argentina and Uruguay, including Easter Is. (Pascua I.), Malvinas Is. (Falkland Is.), South Georgia Is., and the subantarctic South Shetland Is., South Sandwich Is., and South Orkney Is.* Nova Hedwigia Beiheft, Borntraeger Gebrueder.
49. Instituto de Desarrollo Agropecuario. (2015). Programa de caminos intraprediales. Obtenido de: http://www.indap.gob.cl/noticias/detalle/2015/10/01/caminos-intraprediales-acortan-dis tancias-para-la-agricultura-campesina-de-ays%C3%A9n
50. Iriarte, J. A., Feinsinger, P., & Jaksic, F. M. (1997). Trends in wildlife use and trade in Chile. *Biological Conservation, 81,* 9–20.
51. Iriarte, A., Donoso, D. S., Segura, B., & Tirado, M. (2017). *El huemul de Aysén y otros rincones.* Ediciones Secretaría Regional Ministerial de Agricultura de la Región de Aysén y Flora y Fauna Chile Ltda.
52. Jaramillo, A., Burke, P., and Beadle, D. (2005). *Aves de Chile.* Barcelona, España: Lynx Ediciones.

53. Keeton, W. S., Kraft, C. E., & Warren, D. (2007). Mature and old growth riparian forests: Structure, dynamics and effects on Adirondack stream habitats. *Ecological Applications, 17*(3), 852–868.
54. Kelt, D. A. (1994). The natural history of small mammals from Aysén region, southern Chile. *Revista Chilena de Historia Natural, 67*, 183–207.
55. Kelt, D. A. (1989). *Biogeography and assemblage structure of small mammals across a transition zone in southern Chile.* [Thesis]. Northern Illinois University, IL, USA.
56. Kleinschroth, F., & Healey, J. R. (2017). Impacts of logging roads on tropical forests. *Biotropica, 49*(5), 620–635.
57. Lara, A. M., Solari, M., Prieto, M., & Peña, M. P. (2012). Reconstrucción de la cobertura de la vegetación y uso del suelo hacia 1550 y sus cambios a 2007 en la ecorregión de los bosques valdivianos lluviosos de Chile (35°–43° 30´ S). *Bosque, 33*(1), 13–23.
58. Larraín, J. (2016). The mosses (Bryophyta) of Capitán Prat province, Aysén Region, southern Chile. *PhytoKeys, 68*, 91–116.
59. Lenaerts, J. T., Van Den Broeke, M. R., Van Wessem, J. M., Van De Berg, W. J., Van Meijgaard, E., Van Ulft, L. H., & Schaefer, M. (2014). Extreme precipitation and climate gradients in Patagonia revealed by high-resolution regional atmospheric climate modeling. *Journal of Climate, 27*(12), 4607–4621.
60. Lewerentz, A., Eggers, G., Householder, E., Reid, B., Garofano-Gómez, V., & Braun, A. (2019). Functional assessment of invasive *Salix fragilis* L. in north-western Patagonian flood plains: A comparative approach. *Acta Oecologica, 95*, 36–44.
61. Likens, G. (2013). *Biogeochemistry of a forested ecosystem* (3rd ed.). Springer Science+Business Media New York.
62. Lowe, W. H., & Likens, G. E. (2005). Moving headwater streams to the head of the class. *BioScience, 55*(3), 196–197.
63. Luyssaert, S., Schulze, E. D., Börner, A., Knohl, A., Hessenmöller, D., Law, B. E., Ciais, P., & Grace, J. (2008). Old-growth forests as global carbon sinks. *Nature, 455*, 213–215.
64. Macdonald, E., Hinks, A., Weiss, D. J., Dickman, A., Burnham, D., Sandom, C. J., Malhi, Y., & Macdonald, D. W. (2017). Identifying species for conservation marketing. *Global ecology and Conservation, 12*, 204–214.
65. Mansilla, C. A., Domínguez, E., Mackenzie, R., Hoyos-Santillán, J., Henríquez, J. M., Aravena, J. C., & Villa-Martínez, R. (2023). *Peatlands in Chilean Patagonia: Distribution, biodiversity, ecosystem services, and conservation.* Springer.
66. Marquet, P. A., Buschmann, A. H., Corcoran, D., Díaz, P. A., Fuentes-Castillo, T., Garreaud, R., Pliscoff, P., &Salazar, A. (2023). *Global change and acceleration of anthropic pressures on Patagonian ecosystems.*
67. Martinic, M. (2005). *De la trapananda al Aysén: una mirada reflexiva sobre el acontecer de la región de Aysén desde la prehistoria hasta nuestros días* (Vol. 38). Pehuén Editores Ltda.
68. Martínez-Tilleria, K., Núñez-Ávila, M., & León, C. A., et al. (2017). A framework for the classification Chilean terrestrial ecosystems as a tool for achieving global conservation targets. *Biodiversity and Conservation, 26*, 2857–2876. https://doi.org/10.1007/s10531-017-1393-x
69. Master, L., Flack, S. R., & Stein, B. A, (eds.). (1998). *Rivers of life: Critical watersheds for protecting freshwater biodiversity.* The Nature Conservancy.
70. McDowall, R. M. (1969). A juvenile of *Aplochiton taeniatus. Copeia*, 631–632.
71. Medina, G. (1996). Conservation and status of *Lutra provocax* in Chile. *Pacific Conservation Biology, 2*, 414–419.
72. Meier, C., Reid, B., & Sandoval, O. (2013). Effects of the invasive plant *Lupinus polyphyllus* on vertical accretion of fine sediment and nutrient availability in bars of the gravel-bed Paloma River. *Limnologica, 43*(5), 381–387.
73. Ministerio de Agricultura. (2015a). Caracterización de centros de acopio rurales y periurbanos de leña en las regiones de O'Higgins, Maule, Bío Bío, Araucanía, Los Ríos, Los Lagos y Aysén. Unidad Dendroenergía.

74. Ministerio de Agricultura. (2015b). *Medición del consumo nacional de leña y otros combustibles sólidos derivados de la madera.* Unidad Dendroenergía. MINAGRI.
75. Ministerio de Bienes Nacionales. (2018). *Bienes nacionales protegidos.* Obtenido de: http://www.ide.cl/descarga/capas/item/bienes-nacionales-protegidos.html
76. Ministerio del Medio Ambiente. (2019). *Fichas de especies.* Obtenido de: http://especies.mma.gob.cl/CNMWeb/Web/WebCiudadana/pagina.aspx?id=87&pagId=85
77. Miranda, A., Altamirano, A., Cayuela, L., Lara, A., & González, M. (2016). Native forest loss in the Chilean biodiversity hotspot: Revealing the evidence. *Regional Environmental Change, 17*(1), 285–297.
78. Moore, D. M. (1983). Flora of the Fuego-Patagonian cordilleras: Its origins and affinities. *Revista Chilena de Historia Natural, 56*, 123–136.
79. Morales-Hidalgo, D., Oswalt, S. N., & Somanathan, E. (2015). Status and trends in global primary forest, protected areas, and areas designated for conservation of biodiversity from the global forest resources assessment 2015. *Forest Ecology Management, 352*, 68–77.
80. Moya, C., Valdovinos, C., Moraga, A., Romero, F., Debels, P., & Oyanedel, A. (2009). Patrones de distribución espacial de ensambles de macroinvertebrados bentónicos de un sistema fluvial andino patagónico. *Revista Chilena de Historia Natural, 82*, 425–442.
81. Muñoz-Pedreros, A. (2000). Orden Rodentia. In A. Muñoz-Pedreros & J. Yáñez (Eds.), *Mamíferos de Chile* (pp. 201–203). Ediciones CEA.
82. Müller, F. (2009). An updated checklist of the mosses of Chile. *Archive for Bryology, 58*, 1–124.
83. Müller, F. (2009). New records and new synonyms for the Chilean moss flora. *Tropical Bryology, 30*, 77–84.
84. Osorio, M., Saavedra, G., & Velázquez, H. (2007). *Otras narrativas en Patagonia. Tres miradas antropológicas a la región de Aysén.* Editorial Ñirre Negro.
85. Oyanedel, A., Valdovinos, C., Azócar, M., Moya, C., Mancilla, G., Pedreros, P., & Figueroa, R. (2008). Patrones de distribución espacial de los macroinvertebrados bentónicos de la cuenca del río Aysén (Patagonia chilena). *Gayana Botanica, 72*, 241–257.
86. Patterson, G. (1992). The ecology of a New Zealand grassland lizard guild. *Journal of the Royal Society of New Zealand, 22*(2), 91–106.
87. Perakis, S., & Hedin, L. (2001). Fluxes and fates of nitrogen in soil of an unpolluted old-growth temperate forest, southern Chile. *Ecology, 82*(8), 2245–2260.
88. Perakis, S., & Hedin, L. O. (2002). Nitrogen loss in unpolluted South American forests mainly via dissolved organic compounds. *Nature, 415*, 416–419.
89. Pincheira-Ulbrich, J., Rau, J. R., & Smith-Ramírez, C. (2012). Diversidad de plantas trepadoras y epífitas vasculares en un paisaje agroforestal del sur de Chile: Una comparación entre fragmentos de bosque nativo. *Boletín de la Sociedad Argentina de Botánica, 47*(3–4), 411–426.
90. Pliscoff, P., Martínez-Harms, M. J., & Fuentes-Castillo, T. (2023). *Representativeness assessment and identification of priorities for the protection of terrestrial ecosystems in Chilean Patagonia.* Springer.
91. Potapov, P., Hansen, M. C., Laestadius, L., Turubanova, S., Yaroshenko, A., Thies, C., Smith, W., Zhuravleva, I., Komarova, A., & Minnemeyer, S. (2017). The last frontiers of wilderness: Tracking loss of intact forest landscapes from 2000 to 2013. *Science Advances, 3*(1), e1600821.
92. Pérez-Quezada, J., Célis, J. L., Brito, C., Gaxiola, A., Núñez-Ávila, M., Pugnaire, F., & Armesto, J. J. (2018). Carbon fluxes from a temperate rainforest site in southern South America reveal a very sensitive sink. *Ecosphere, 9*, e2193.
93. Pérez-Quezada, J. F., Olguín, S., Fuentes, J. P., & Galleguillos, M. (2015). Tree carbon stock in evergreen forests of Chiloe, Chile. *Bosque, 36*, 27–39.
94. Radic-Schilling, S., Corti, P., Muñoz, R., Butorovic, N., & Sánchez, L. (2023). *Steppe ecosystems in Chilean Patagonia: distribution, climate, biodiversity, and threats to their sustainable management.* Springer.

95. Reid, B., Astorga, A., Madriz, I., Correa, C., & Contador, T. (2023). *A conservation assessment of freshwater ecosystems in southwestern Patagonia.* Springer.
96. Rendoll, J., Contador, T., Gañán, M., Pérez, C., Maldonado, A., Convey, P., Kennedy, J., & Rozzi, R. (2020). Altitudinal gradients in Magellanic sub-Antarctic lagoons: The effect of elevation on freshwater macroinvertebrate diversity and distribution. *PeerJ, 7,* e7128.
97. Rivera, A., Aravena, J. C., Urra, A., & Reid, B. (2023). *Chilean Patagonian glaciers and environmental change.* Springer.
98. Rojas, P. (2020). *Assessment of ecosystem services in the Coyhaique watershed, Chilean Patagonia. Recommendations for sustainable development.* [Thesis]. University of California, Davis, CA.
99. Rozzi, R., Armesto, J. J, Goffinet, B., Buck, W., Massardo, F., Silander, J., Arroyo M. T. K., Russell, S., Anderson, C., Cavieres, Lohengrin, J., & Callicott, J. (2008). Changing biodiversity conservation lenses: Insights from the subantarctic non-vascular flora of southern South America. *Frontiers in Ecology and the Environment, 6,* 131–137.
100. Salmon, C., Walter, M., Hedin, L., & Brown, M. (2001). Hydrological controls on chemical export from an undisturbed old-growth Chilean forest. *Journal of Hydrology, 253,* 69–80.
101. Sanderson, E. W., Jaiteh, M., Levy, M. A., Redford, K. H., Wannebo, A. V., & Woolmer, G. (2002). The human footprint and the last of the wild: The human footprint is a global map of human influence on the land surface, which suggests that human beings are stewards of nature, whether we like it or not. *BioScience, 52*(10), 891–904.
102. Saunders, D., Meeuwig, J., & Vincent, A. (2002). Freshwater protected areas: Strategies for conservation. *Conservation Biology, 16*(1), 30–41.
103. Sepúlveda, M., Valenzuela, A., Pozzi, C., Medina-Vogel, G., & Chehébar, C. (2015). *Lontra provocax.* The IUCN red list of threatened species: E. T12305a21938042. Obtenido de: https://www.iucnredlist.org/species/12305/21938042
104. Stuardo, J., & Vega, R. (1985). Synopsis of the land mollusca of Chile: With remarks on distribution. *Studies on Neotropical Fauna and Environment, 20*(3), 125–146.
105. Sáez, N. (2009). Distribución y caracterización de lugares boscosos proveedores de leña nativa destinada a la ciudad de Coyhaique (45°30′s). *Tiempo y Espacio, 23,* 7–24.
106. Teresa Gallego, M., Cano, M. J., Larraín, J., & Guerra, J. (2020). Syntrichia lamellaris M.T.Gallego, M.J.Cano & Larraín (Pottiaceae), a new moss species from Chilean Patagonia. *Journal of Bryology, 42*(2), 128–132. https://doi.org/10.1080/03736687.2019.1706036
107. Valdovinos, C. (2006). Estado de conocimiento de los gastrópodos dulceacuícolas de Chile. *Gayana, 70*(1), 100–113.
108. Valdovinos, C., Kiessling, A., Mardones, M., Moya, C., Oyanedel, A., Salvo, J., Olmos, V., & Parra, O. (2010). Distribución de macroinvertebrados (Plecoptera y Aeglidae) en ecosistemas fluviales de la Patagonia chilena: ¿muestran señales biológicas de la evolución geomorfológica postglacial? *Revista Chilena de Historia Natural, 83,* 267–287.
109. Valdovinos, C. (2008). *Invertebrados dulceacuícolas. Biodiversidad de Chile 204-225. Patrimonio y Desafíos.* Ocho Libros Editores.
110. Veblen, T. T., & Lorenz, D. C. (1988). Recent vegetation changes along the forest/steppe ecotone of northern Patagonia. *Annals of the Association of American Geographers, 78,* 93–111.
111. Venter, O., Sanderson, E. W., Magrach, A., Allan, J. R., Beher, J., Jones, K. R., Possingham, H. P., Laurance, W. F., Wood, P., & Fekete, B. M. (2016). Global terrestrial human footprint maps for 1993 and 2009. *Science Data, 3,* 160067.
112. Vera, C., Silvestri, G., Liebmann, B., & González, P. (2006). Climate change scenarios for seasonal precipitation in South America from IPCC-AR4 models. *Geophysical Research Letters, 33*(13), L13707.
113. Vertessy, R. A., Watson, F. G., & Sharon, K. (2001). Factors determining relations between stand age and catchment water balance in mountain ash forests. *Forest Ecology Management, 143,* 13–26.
114. Vila, A., López, R., Pastore, H., Faúndez, R., & Serret, A. (2004). *Distribución actual del huemul en Argentina y Chile.* Publicación técnica de WCS, FVSA y CODEFF.

115. Watson, J. E., Evans, T., Venter, O., Williams, B., Tulloch, A., Stewart, C., Thompson, I., Ray, J. C., Murray, K., & Salazar, A. (2018). The exceptional value of intact forest ecosystems. *Nature Ecology & Evolution, 2*, 599–610.
116. Wemple, B. C., Swanson, F. J., & Jones, J. A. (2001). Forest roads and geomorphic process interactions, cascade range, Oregon. *Earth Surface Processes and Landforms, 26*, 191–204.
117. World Resources Institute. (2017). *Annual report*. Obtenido de: https://wriorg.s3.amazonaws.com/s3fs-public/uploads/wri-2017-annual-report.pdf

Anna Astorga Roine BS in Biology, Pontificia Universidad Católica de Chile. Ph.D. in Ecology, specializing in Stream Ecology, Oulu University, Finland. Researcher at Centro de Investigaciones en Ecosistemas de la Patagonia, Coyhaique, Región de Aysén, Chile.

Paulo Moreno-Meynard Forestry Engineer, University of Chile. M.Sc. Forest Resources and Conservation, University of Florida, USA. Ph.D. in Science of Marine Ecosystems, University of Genoa, Italy. Resident researcher at CIEP, Chile.

Paulina Rojas R. Forestry Engineer. M.Sc. International Agricultural Development. Ph.D. student, Ecology & Water Management Laboratory, Department of Land, Air and Water Resources, University of California, Davis, USA.

Brian Reid BS in Neurobiology, Cornell University, USA. Ph.D. Aquatic Ecology, Montana University, USA. Between 1994 and 2001 worked in conservation NGOs in the USA. Resident researcher at CIEP, Coyhaique, Aysén Region, Chile.

Peatlands in Chilean Patagonia: Distribution, Biodiversity, Ecosystem Services, and Conservation

Claudia A. Mansilla, Erwin Domínguez, Roy Mackenzie, Jorge Hoyos-Santillan, Juan Marcos Henríquez, Juan Carlos Aravena, and Rodrigo Villa-Martínez

Abstract

Peatlands are wetland ecosystems characterized by the accumulation of large amounts of organic matter over centuries and millennia; they are the most important long-term carbon sink in terrestrial ecosystems. Peatlands are also valuable ecosystems for biocultural and biodiversity conservation, as paleo-climatic archives, and as providers of ecosystem services to human society.

C. A. Mansilla (✉) · J. C. Aravena · R. Villa-Martínez
GAIA-Antarctic Research Center, Universidad de Magallanes, Punta Arenas, Chile
e-mail: claudiaandrea.mansilla@umag.cl

C. A. Mansilla · J. M. Henríquez
Instituto de la Patagonia, Universidad de Magallanes, Punta Arenas, Chile

C. A. Mansilla · J. Hoyos-Santillan · J. C. Aravena
Network for Extreme Environments Research (NEXER), Universidad de Magallanes, Punta Arenas, Chile

C. A. Mansilla · R. Mackenzie · J. C. Aravena · R. Villa-Martínez
Cape Horn International Center, Universidad de Magallanes, Puerto Williams, Chile

E. Domínguez
Instituto de Investigaciones Agropecuarias (INIA), Punta Arenas, Chile

R. Mackenzie
Programa de Conservación Biocultural Subantártica, Universidad de Magallanes, Puerto Williams, Chile

Instituto de Ecología y Biodiversidad, Parque Etnobotánico Omora, Puerto Williams, Chile

J. Hoyos-Santillan
School of Biosciences, University of Nottingham, Sutton Bonington, Loughborough, UK

Environmental Biogeochemistry in Extreme Ecosystems Laboratory, Universidad de Magallanes, Punta Arenas, Chile

Center for Climate and Resilience Research (CR)2, Santiago, Chile

© Pontificia Universidad Católica de Chile 2023
J. C. Castilla et al. (eds.), *Conservation in Chilean Patagonia*, Integrated Science 19,
https://doi.org/10.1007/978-3-031-39408-9_6

The primary current anthropogenically-driven threats to peatlands in Patagonia include changes in land use leading to land desiccation, the introduction of invasive species, and *Sphagnum magellanicum* moss and peat extraction for export. Inappropriate management of peatlands could have major environmental and social impacts. This chapter aims to provide updated and synthesized information to support decision-making for the management and conservation of peatlands, and the potential contribution of pristine peatland ecosystems in mitigating climate change. A national inventory, conservation of peatland ecosystems in their multiple levels of ecosystem functioning, improvement of restoration practices, and the prevention of the degradation are among the urgent priorities in order to reduce negative socio-ecological and economic impacts over the short, medium and long term.

Keywords

Patagonia • Chile • Peatlands • Biodiversity • Ecosystem services • Carbon storage • *Sphagnum* • Peat extraction

1   Introduction

Peatlands have gained prominence in the world literature, due to the importance of the ecosystem services they provide to society [4]. Along with harboring vast biological diversity, these ecosystems have played a key role in water, soil, and climate regulation over thousands of years [1]. Peatlands are wetlands vertically stratified in two layers, called the acrotelm and catotelm. The acrotelm is the layer from the surface to the water table (located at 30–50 cm, it is aerobic and hosts vegetation that fixes carbon through photosynthesis (primary productivity and maintains active recycling of organic matter below the surface. The catotelm is located from the water table to the basal mineral substrate (water saturation zone, which can be several meters deep; it is essentially anaerobic, with a large number of microorganisms capable of decomposing organic matter slowly, mainly by fermentation, which generates a net accumulation of organic matter and carbon in the form of peat [33] (Fig. 1).

Peatlands occupy approximately 3% of the world land area and contain 21% of soil organic carbon [60]. With an approximate storage of 644 Gt C (Gt C = 1×10^9 ton of carbon), the carbon pool in peatlands is equivalent to *ca.* 1.7 times the carbon stored in all forests on earth [35, 52]. The peatlands of Chilean (western) Patagonia account for *ca.* 1.1% (*ca.* 45,000 km^2) of the area of all peatlands worldwide (*ca.* 3.9 million km^2) [51, 65]. Although Patagonian peatlands represent a small proportion of the planet's peatlands, they are the largest terrestrial carbon pool in the temperate zones of the Southern Hemisphere (>30°S), with a small representation in Australia, New Zealand, some South Atlantic islands, and the Antarctic Peninsula [37]. Peatlands in Chilean Patagonia are mostly located far from urban centers, so some of them are in a quasi-pristine state [34]. However, within peatlands in the southernmost part of Patagonia (*ca.* 53°S) there are records of atmospheric deposition of chemical elements and pollution of anthropogenic

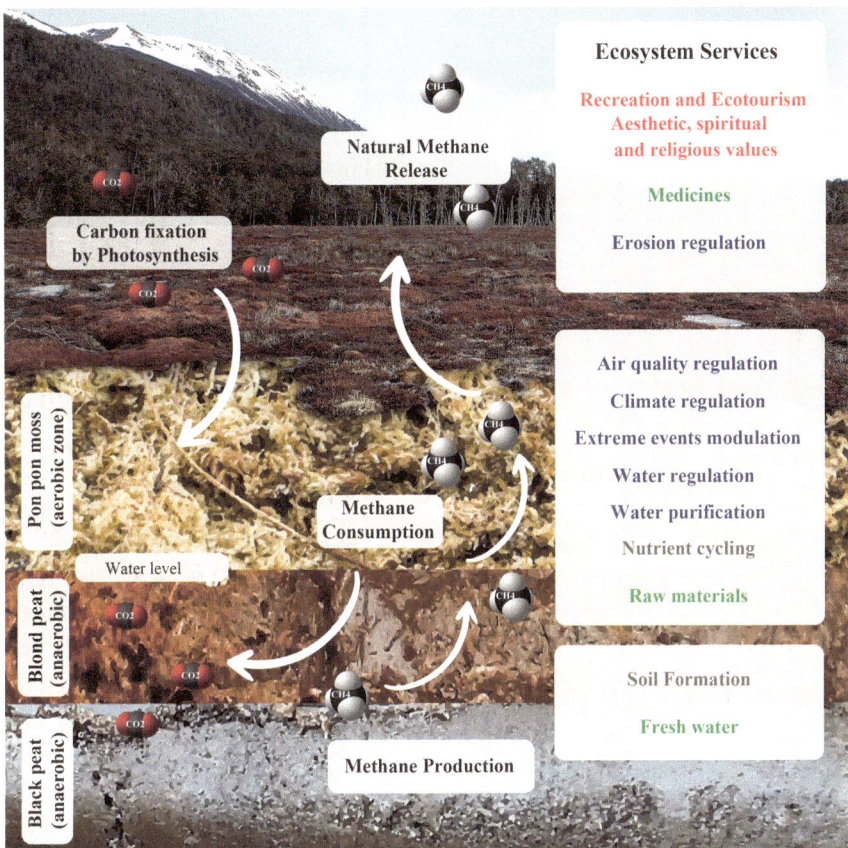

Fig. 1 Main ecosystem services and carbon cycling in an ombrotrophic peatland of *Sphagnum magellanicum* on Navarino Island, Magallanes Region. The different metabolic steps in the carbon cycle and their location in the vertical profile are represented. The strata of pompom moss, blond and black peat were obtained at different depths: black peat: 5–5.5 m, blond peat: 2–2.5 m, pompom: 0–0.5 m. On the right side of the figure, the most relevant ecosystem goods and services are shown in colors in the acrotelm and catotelm of the peatland: red: Cultural; blue: Regulating; gray: Sustaining; green: Provisioning (*Photos* R. Mackenzie)

origin (e.g. Cu, Sb, S and Hg), originating in other latitudes, whose concentrations in the substrate have fluctuated over the last few hundred years [3, 64].

2   Scope and Objectives

This chapter presents an overview of current knowledge on the peatland ecosystems of Chilean Patagonia. It highlights their wide distribution and geographic extension, their importance as a waterlogged habitat with a remarkable biodiversity of vascular plants, bryophytes, vertebrates, invertebrates, and microorganisms, as well as their value in the provision of ecosystem services associated with the well-being of society, and as regulators of water, soil, and climate cycles. The great longevity and importance of peatlands as millenary climatic-environmental records and their role as carbon sinks are also highlighted. The direct and indirect anthropogenic threats and pressures are discussed and the legal norms and other actions for the protection and regulation of the harvesting of peat moss, *Sphagnum magellanicum*, peat extraction, and the integral conservation of the components of peatland ecosystems are analyzed.

3   Peatlands of Chilean Patagonia

3.1   Distribution of Peatlands in Chilean Patagonia

The spatial distribution of peatland communities in the landscape of southern and Patagonian Chile spans about 14° latitude, from the Los Ríos Region (*ca.* 41°S) to Cape Horn (*ca.* 56°S) [38, 40] (Fig. 2). Their distribution is determined by a negative precipitation gradient from west to east, caused by the contact of westerly winds with the Andes Range, establishing the greatest extension of peatland communities in Chilean Patagonia and in the Tierra del Fuego archipelago [34]. Patagonian peatlands contrast with their counterparts in the Northern Hemisphere, because they are located in bioclimatic zones with a greater range of precipitation (600–4000 mm per year) and with more oceanic influence [37].

The most characteristic communities in the diverse mosaic of Patagonian peatland systems are: (i) graminiform peatlands, dominated mainly by species of cyperaceae of the genus *Schoenus: S. antarcticus* and *S. andinus.* sedges such as *Marsippospermum grandiflorum*, various poaceae, and other species such as *Carpha alpina* in conditions where precipitation is <1000 mm per year; (ii) natural sphagnum peatlands, dominated exclusively by the moss *Sphagnum magellanicum*, under a precipitation regime between 600 and 1500 mm per year; (iii) pulvinate peatlands, dominated by cushion-shaped vascular plant species such as *Donatia fascicularis*, *Astelia pumila* and *Drosera uniflora* that form compact folds and develop under hyper-oceanic conditions with rainfall that can exceed 4000 mm per year and is homogeneously distributed throughout the year; (iv) ecotonal peatlands, where sphagnum and pulvinate peatland communities intermingle [53]. These peatland communities can also form part of the basal stratum of forests of endemic conifers such as the Guaitecas cypress (*Pilgerodendron uviferum*) (Fig. 3), evergreen forests throughout Chilean Patagonia, and deciduous forests in the southern Patagonian region. There are anthropogenic peat bogs (in Spanish pomponales), mainly in areas of Llanquihue and Chiloé (*ca.* 42°S., mostly landscapes of anthropogenic origin) dominated by the moss *Sphagnum magellanicum*, showing different floristic composition, that accumulate less quantities of peat than natural *Sphagnum* peatlands [11].

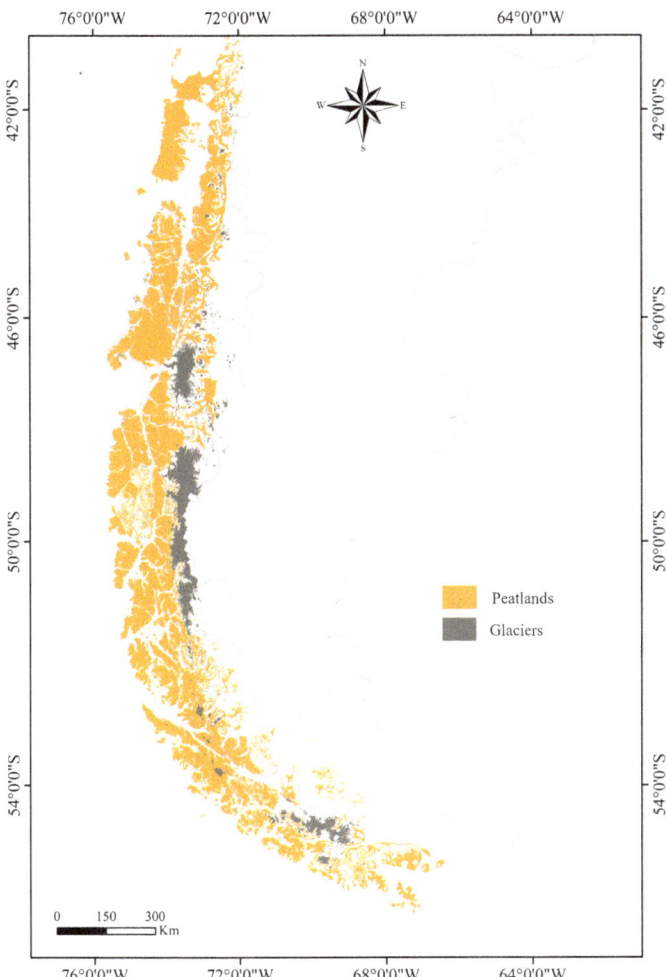

Fig. 2   Distribution of Chilean peatlands between *ca.* 41 and 56°S. Included are those areas where peatlands are part of the basal stratum of endemic conifer forests, as well as evergreen forests throughout Chilean Patagonia and the southernmost deciduous forests [38, 40]

3.2   Importance of Biological Diversity in Chilean Patagonian Peatlands

The peatlands of Chilean Patagonia are seasonally flooded ecosystems that harbor a wide biodiversity of organisms, some of which are highly specialized and capable of living in conditions that are adverse to other species, such as acidic, flooded soils and low nutrient availability [33]. Peatlands are the habitat of a rich flora of cryptogams, including mosses and plant species [14]. Brown mosses, mainly species of the genus *Polytrichum*, form extensive mats and column-like structures

Fig. 3 *Sphagnum* peat bog community with Guaitecas cypress (*Pilgerondendron uviferum*) in Tortel commune, Aysén Region (*Photo* R. Mackenzie)

associated with peatlands, and have a relevant role in the regeneration of *Sphagnum magellanicum* in peatlands that have been intervened or exploited [20]. *S. magellanicum* peatlands are also the habitat of endemic mosses such as *Tayloria dubyi*, a moss that grows exclusively on Caiquén or southern goose (*Chloephaga picta*) feces and uses volatile compounds to attract coprophilic insects that disperse their spores. *T. dubyi* belongs to the only family of cryptogams in the Southern Hemisphere that exhibits entomophily and is restricted to *S. magellanicum* formations in the Subantarctic forest of Magallanes [32].

Graminiform and pulvinate peatlands have higher species richness and vascular plant diversity (angiosperms and ferns) than *Sphagnum magellanicum* peatlands, with more than 70% of the species being endemic to southern South America. Vascular plants with highly specialized strategies able to live in soils with low nutrient availability (e.g. nitrogen) include the carnivorous plants *Pinguicula antarctica* and *D. uniflora* [14], which are associated with the endemic conifer *Lepidothamnus fonkii* (dwarf cypress) [5], considered to be the southernmost in the world together with *P. uviferum* [63].

The diverse microtopography and internal hydrology of a peatland generates a number of microhabitats (e.g. phorophytes, pools, soil). Environmental microgradients are formed in these that can even be occupied by other *Sphagnum* species, which are arranged in a microtopographic moisture gradient, with *S. falcatulum* being the most hygrophilic species, usually growing underwater where it takes on a feathery appearance [39]. These microhabitats also possess functions, processes and specific biodiversity. For example, in areas of an ecotonal peatland community dominated by *S. magellanicum* (*Sphagnum*–pulvinates), the effect of carbon sequestration is greater than in areas dominated by angiosperms [43]. Therefore, changes in the pattern and abundance of plant species inhabiting peatlands, either as a result of climate change or associated with direct anthropogenic factors, would alter microhabitats, food webs, and complex biogeochemical processes, including those that control peat accumulation rates, including carbon dioxide (CO_2) and methane (CH_4) emissions. In other words, a peatland can change from a carbon sink to a net emitter of greenhouse gases to the atmosphere, depending on its conservation status.

As in other ecosystems, the entomofauna of peatland communities plays a key role in the food chain and in the transfer of energy to higher trophic links, supporting vertebrate populations that include amphibians and insectivorous birds [31]. Although freshwater insects have been considered bioindicators of the ecological quality of freshwater ecosystems [8], they are not considered as part of the composition, structure and functioning of peatlands, nor are they used as an alternative to evaluate the environmental quality of these ecosystems when planning management and conservation strategies.

Information on the fauna associated with the peatlands of Chilean Patagonia is still scarce and fragmented. Although to date no birds have been described as exclusive to peatland communities, certain species are well adapted to terrestrial ecosystems of high humidity and low temperatures such as some Charadriformes, in particular species of *Gallinago* [57]. These habitats are crucial for certain bird communities for feeding, reproduction, and shelter [27]. Amphibians are highly dependent on humid environments and the presence of water for reproduction and larval stages. There are amphibian species exclusive to peatland environments such as *Nannophryne variegata, Chaltenobatrachus* aff *grandisonae,* and *Batrachyla antartandica* [49]. Other vertebrates such as mammals (e.g. huemul deer, foxes, and rodents) are relatively scarce, even more so in Chilean Patagonia due to its insular characteristics. The rodents *Oligoryzomys longicaudatus, Abrothrix olivaceus,* and *A. lanosus* visit peatlands; the first of these most often, due to its preference for humid environments [22]. Prokaryotes and eukaryote microorganisms play a preponderant role in the biogeochemical cycles of peatlands [18]. However, there is a profound lack of knowledge of their functions and structure in the peatlands of southern South America. These microbiota form consortia in the hyalocysts of *S. magellanicum* and develop as symbionts in the acrotelm, also known as the *Sphagnum* microbiome [6, 18]. The methanotrophic and diazotrophic symbionts of *S. magellanicum* are able to fix between 5 and 20% of CO_2 from methane and up to 35% of cellular nitrogen, respectively; microbial consortia located in the catotelm perform the latter, step in peat decomposition by methanogenesis, using CO_2 as an electron acceptor [18]. The balance between the microbial processes of methane production and consumption in the catotelm and acrotelm, respectively, attenuate the contribution of this greenhouse gas naturally made by peatlands towards the atmosphere [36].

The communities of amoeboid protists with a shell (test) have hardly been studied in Chilean Patagonia and can represent almost 50% of the total biomass of microorganisms in a peatland. Chilean Patagonian peatlands harbor a high specific diversity and a significant proportion of these species, which are endemic to southern South America [16].

In the implementation of plans for the restoration of peatland ecosystems, it is crucial to recover the diversity of microhabitats, as well as the processes that occur in them in order to maintain the trophic webs and biogeochemical processes that promote the accumulation of carbon prior to the exploitation of peatlands, among other reasons.

3.3   Ecosystem Services of Peatlands

Ecosystem goods and services correspond to the benefits, products, and services that human societies obtain from natural ecosystems and their biological diversity. Most of the ecosystem services that peatlands provide to society do not have a direct economic value in the classical sense, therefore, they usually do not form part of a country's indicators of economic activity, are thus undervalued, and are suffering a worrying rate of degradation. This deficiency in the economic system requires the attention of society as a whole, since the sustainable development of a country and the well-being of its inhabitants depend directly on the services and functions of ecosystems and their biological diversity [55].

Peat bog with presence of moss, *Sphagnum*, Río Mayer sector, Villa O'Higgins, Aysén Region. Photograph by Erwin Domínguez.

Therefore, the ecosystem services of environments are a powerful tool for the design of environmental policies and for decision-making aimed at sharing the benefits of natural areas throughout society. Ecosystem services can be classified according to the main benefits that the ecosystem provides, which in sphagnum peatland ecosystems are: (i) provisioning: goods and products obtained from ecosystems (e.g. products with economic value such as *Sphagnum* fibers and peat substrate); (ii) regulation: key ecosystem functions that help to reduce the impact of certain local and global events (e.g. regulation of the water cycle and storage of water and nutrients in the ecosystem); (iii) cultural: these are the intangible benefits of ecosystems (e.g. spiritual, scientific, educational, and artistic value); (iv) support/habitat: is the capacity to host biodiversity, natural processes and ecosystem biogeochemical fluxes (e.g. climate change mitigation by acting as carbon sinks) [4] (Fig. 1).

3.4   Value of Peatlands for Paleoenvironment and Climate Reconstruction

Peatlands are natural deposits of organic matter, sediments, and pollen that continuously record information about past ecological, environmental, and climatic conditions. These records include evidence of natural events such as volcanic eruptions, sea level changes, and glaciations.

The peatlands of Chilean Patagonia, due to their geographic location, are long-lived and deep, which allows access to high-resolution paleoenvironmental and climatic records. These peatlands are located in the southernmost area of South America and the Southern Hemisphere, directly influenced by the southern westerlies and the Antarctic Circumpolar Current, which makes these ecosystems unique terrestrial archives for investigating regional and global land–ocean-atmosphere interactions. The vast majority of the peatlands existing today in Chilean Patagonia have originated and developed since the last glacial maximum (LGM: about 20,000 years ago). Glacial and post-glacial processes modeled the topography of the landscape, originating shallow water bodies, and impermeable soils favorable for the onset of sediment and organic matter deposition over millennia [24]. Peatlands in Chilean Patagonia are mainly associated with areas of early deglaciation, near the maximum limits of ice extent during the LGM. For example, in the northern region of Chilean Patagonia (*ca.* 41°S) there are peatlands with ages close to 20,000 years [24] and in the central region (*ca.* 45°S) close to 18,000 years [41]. Peatlands appear to be associated with later deglaciations in the southern zone, *ca.* 17,000 years on Navarino island (*ca.* 54°S) [44]. Peatlands in Chilean Patagonia reach great depths, for example 14 and 11 m in the central and southern zones, respectively [41, 44]. The information obtained from paleoenvironmental and climatic records in Chilean Patagonian peatlands has high resolution and reliability, for example to analyze changes in carbon accumulation rates over time, they are important for understanding climate changes in the Southern Hemisphere and globally.

3.5 Chilean Patagonian Peatlands as Carbon Sinks

The plant biomass and humus that accumulate in peatlands contain carbon as part of their molecular structure; consequently, peat deposits represent important reservoirs of carbon [61]. For this reason, peatlands are a structural component of the global carbon cycle, actively participating in the regulation of the planet's climate [65]. Peatlands also are one of the most important sources methane emissions (CH_4) into the atmosphere [54]. Despite emitting carbonin the long term, carbon sequestration in peatlands has been greater than emissions, so they have consistently behaved as carbon sinks [17]. Peatlands cover an area of at least 31,000 km^2 in the Magallanes Region [9] (in Spanish CONAF), about 90% of the total area nationally [10], and store between 3.6 and 4.8 Gt C [25]. This reservoir is five times greater than the total amount of carbon present in the aboveground biomass of Chile's forests [25, 58], making peatlands the most important natural carbon reservoir in the country. It is currently not possible to establish whether the peatlands of Chilean Patagonia will behave as sinks or net carbon emitting sources to the atmosphere in the coming decades under anticipated global climate change scenarios [42]. However, during the last *ca.* 18,000 years [41], peatlands in Chilean Patagonia (Aysén) have accumulated peat at an average rate of 0.43 mm per year (n = 96), which translates into an average long-term carbon accumulation rate of *ca.* 12.25 gC $m^{-2}y^{-1}$ (Long-term Rate of Carbon Accumulation, LORCA) [26]. If the average rates of carbon accumulation in Chilean Patagonia are maintained over the next 30 years, which is the period projected by Chile to reach carbon neutrality, peatlands could represent one of the most important carbon sinks in the country, sequestering *ca.* 12 million t of carbon between 2020 and 2050 [26]. It is worth mentioning that peatlands which have been directly impacted by human activities, as well as those that will be affected by global climate change, could behave as carbon emitters and not as carbon sinks. Thus, both conservation and restoration of peatlands in Chile represent effective natural alternative mechanisms to achieve carbon neutrality, and thus contribute to mitigating the effects of climate change.

3.6 Main Anthropogenic Threats to Peatland Biodiversity

Land use change, introduction of invasive species, harvesting of *Sphagnum* moss, peat exploitation, and climate change are the main direct and indirect anthropogenic threats that affect the composition, diversity, and functioning of peatlands. Land use change for the purpose of increasing connectivity via new roads, especially in the Aysén and Magallanes regions, and the consequent installation of infrastructure for urban areas with limited planning represent a direct threat to peatlands [30]. These activities fragment peatland habitats and significantly affect ecosystem services, resulting in serious ecosystem degradation.

The beaver (*Castor canadensis*) is one of the invasive species that most severely impact peatlands and wetlands in Chilean Patagonia. The beaver is an ecosystem

engineer that drastically alters the ecosystems it inhabits. This species was introduced in the Argentine sector of Tierra del Fuego in 1946; its current population on
this island exceeds 100,000 individuals [2]. In 2016 it was estimated, by analysis
of satellite images, that at least 0.6% of the peatlands had been affected by flooding
caused by the construction of dams by this species [23]. However, an undetectable
form of impact by beavers is the construction of galleries in the peat, which generate drainage of water from the peatland and thereby degrade the ecosystem [21].
This disrupts peat accumulation and increases the decomposition rate, transforming peatlands into net emitters of CO_2 to the atmosphere. Exploited and abandoned
peatlands are invaded by exotic vascular plants such as *Holcus lanatus* and *Carex
canescens*, which generate significant changes in the cover and composition of
the original flora of a *Sphagnum* peatland [12]. This floristic replacement prevents
the recovery of degraded peatlands, and consequently they lose or diminish the
ecosystem services that benefit society.

S. magellanicum moss in the country is mainly harvested for use in vertical
gardening, biodegradable planters, kokedama, and as hydrocarbon adsorbents [14].
The export of dehydrated *S. magellanicum* for these purposes increased by approximately 200% between 2004 and 2014 [48] (in Spanish Oficina de Planificación
Agrícola, ODEPA); Domínguez et al. [15]; Fig. 4). However, the economic benefits of *S. magellanicum* export have not reached the local communities that extract
it manually, as the sale price remains between $2500 and $5000 Chilean pesos
(CLP) per 50 kg green bag (wet basis). This causes rural producers to increase the
volume of extraction to increase their income, leaving aside traditional techniques
or good practices which are intended to make the harvest sustainable. An example
of this situation was identified by Vaccarezza [62], who described how the unsustainable extraction of *S. magellanicum* in the Los Lagos Region has caused a major
problem of water availability for rural populations and associated ecosystems.

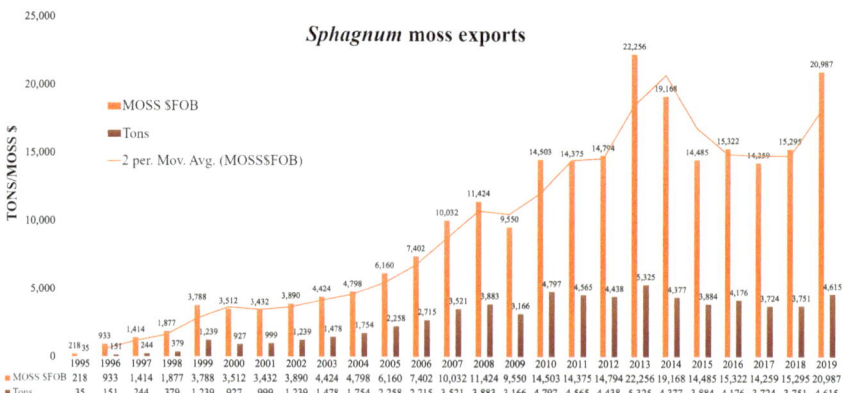

Fig. 4 National export of plant fiber from dehydrated *Sphagnum magellanicum* moss, in t and dollar FOB [28, 48] (*Source* ODEPA, Forestry Institute of Chile; in Spanish Instituto Forestal de Chile,
IF)

According to Domínguez et al. [12], in the Magallanes Region there have been eight peatland harvests in operation historically, with a total affected area of 444 hectares (ha), a low amount compared to the peat extraction carried out in the Argentinean area of Tierra del Fuego, where the extraction area reached an area of 4,600 ha [7] (Fig. 5). Only two projects for peat mining in the Magallanes Region have been approved by the Environmental Impact Assessment System (SEIA), equivalent to *ca.* 40% (178 ha) of the area under historical exploitation in the region [56]. This practice is very risky without an adequate environmental impact study, as the extraction causes the drainage of ecosystems, which are then left in the hands of the local communities [56].

This leaves them vulnerable to fires, radically modifying the hydrology, topography, and net carbon emissions to the atmosphere, and consequently may cause the degradation and total loss of the functions of this ecosystem in the biosphere.

Fig. 5 Peat extraction. The changes generated in the landscape, hydrology, topography and vegetation cover can be seen, the vegetation being completely eliminated, which collapses the functions and ecosystem services of the peatland. (Municipality of Punta Arenas, Magallanes Region) (*Photo* E. Domínguez)

3.7   Peatland Conservation in Patagonia: Regulations, Opportunities and Barriers

Chile has recently committed to update its Nationally Determined Contribution (NDC) under the Paris Agreement framework, incorporating peatlands as mitigators of the impacts of climate change, committing to carry out a national inventory that involves the identification of large peatland areas [46]. According to Yu et al. [65], peatlands in Patagonia (Chile and Argentina) cover an area of approximately 45,000 km^2. However, this figure could overestimate the real area occupied by these ecosystems, since they are located in regions that are remote and often little explored. Furthermore, this estimate considers only natural peatland communities located south of *ca.* 45°S and does not consider anthropogenic peatlands. It is remarkable that despite this huge area in southern South America, peatlands have not been explicitly included in the national wetland inventory (Ministry of the Environment and Center for Applied Ecology [45], Saavedra and Villarroel [59], in Spanish Ministerio del Medio Ambiente, MMA). Peatlands, being a type of wetland, are under the international area of action of the Ramsar Convention, adopted by Chile in DS No. 771 of the Ministry of Foreign Affairs (in Spanish Ministerio de Relaciones Exteriores, MRE).

The exploitation of peat and the harvesting of *Sphagnum magellanicum* generate productive activities that differ in their regulatory framework and type of impact. Peat extraction is considered a mining activity, regulated by the MMA through Law No. 19,300 and its Supreme Decree No. 40, which stipulate that all peat extraction, regardless of its magnitude, is considered industrial. For this reason, all peat mining initiatives must be submitted to the Environmental Evaluation Service through an Environmental Declaration or Environmental Impact Study, depending on the surface area of the project and its impact on hydrology, topography, and biodiversity. In addition, as peat is considered a type of fossil soil, it is a non-renewable natural resource that can be requested as a concession under the Mining Code, Law No. 18,248. As in the case of water, it is considered a "national asset of public use" (article 595 of the Civil Code), for which use rights can be obtained under the Water Code (D.F.L. no. 1.122/1981, Ministry of Justice (in Spanish Ministerio de Justicia, MJ).

The lack of national regulation between 1995 and 2017, together with poor harvesting practices of *S. magellanicum* fiber, drainage alterations, and land use change, resulted in many cases in the replacement of peatlands by *Eucalyptus* plantations, which has caused irreversible deterioration of peatland ecosystems [50]. Particularly on the island of Chiloé, *S. magellanicum* peatlands fulfill strategic environmental services for agricultural activity and livestock farming by providing filtered fresh water that supplies rural wells. Mosses also have a great capacity to retain water and inhibit the growth of fungi and bacteria, properties that make them attractive as substrates for orchid cultivation in Asian countries. These qualities have generated interest in harvesting and marketing this resource for export, mainly as raw material with no added value. The first exports began in 1995 and maintained continuous growth until 2013 (5325 t), the year when environmental problems caused by the absence of regulations began to be manifest in Chile (Fig. 4). However, recently the Ministry of Agriculture (in Spanish Ministerio de Agricultura, MA) recognized *S. magellanicum* as a silvoagricultural resource, creating in 2017 the first regulatory instrument that aims to protect this resource and mitigate environmental, social and economic impacts in the areas where it is harvested. Subsequently, in 2019 Supreme Decree No. 25 was enacted in Chile for the protection of *S. magellanicum*; it gave powers to the Agricultural and Livestock Service, as the governing body, to require and oversee harvesting plans, as well as to conduct training on sustainable moss collection and harvesting practices in Chile. DS No. 25 prohibits the use of large machinery in harvesting the resource and proposes to encourage a sustainable resource management plan. However, it does not specify how training will be carried out, and even more critically, no distinction is made as to whether the moss is harvested from pomponales or natural peat bogs.

Although much of the range of peatland ecosystems (>75%) in western Patagonia is protected in national parks and reserves, and new regulations have recently been created, they do not regulate all the components of peatlands, such as the conservation of biodiversity, and biogeochemical flows. The main problem lies in the separation of different biotic elements of an ecosystem into different institutional jurisdictions and regulations regarding their regulation and management, omitting the role and value of biodiversity as an integrator of the multiple levels of ecosystem functioning. The "Moss Layer Transfer Technique" has been developed for restoration of exploited peatlands in the country; it consists of the collection of live fibers of *S. magellanicum* and their subsequent implantation in intervened areas. This technique has several limiting factors, the main ones being the high cost and lack of explicitness of the objective to be achieved; for example: repair, recovery, rehabilitation, or ecological restoration of the previously intervened peatland [13, 12]. These situations keep peatlands in a state of vulnerability, and consequently threaten the functions and ecosystem services they provide to society. In addition, there is currently no effective transfer of information on the scope of the regulations to help citizens and authorities understand the importance of their application in order to maintain the functions and services provided by these ecosystems. Therefore, in addition to the new regulations created, it is necessary to generate coordinated networks including academic institutions, social, educational, and governmental organizations, and to propose strategies to be included in decision-making. One alternative for the protection of peatlands in Chilean Patagonia is their inclusion as part of the so-called Natural Climate Solutions (Natural Climate Solutions or Nature-Based Solutions [47]). The Intergovernmental Panel on Climate Change [29] determined that peatlands represent an immediate impact alternative for climate change mitigation. These strategies seek, through conservation, restoration and proper land management, to increase carbon storage capacity in landscapes, as well as to decrease greenhouse gas emissions (CO_2, CH_4) in different ecosystems such as forests, peatlands and grasslands, thus mitigating climate change [19]. Therefore, it is imperative to promote the protection, conservation and restoration of peatlands, with the aim of maintaining or increasing their capacity to act as carbon sinks, an urgent function of national importance to achieve carbon neutrality by 2050.

4   Conclusions and Recommendations

(i) Peatlands harbor vast biodiversity, with a high proportion of species endemic to southern South America; they regulate multiple levels of ecosystem functioning and provide innumerable benefits and services that improve the quality of life of human societies.

(ii) Natural peatlands throughout western and eastern Patagonia are unique millennial archives for understanding ecological, environmental and climatic changes in the Southern Hemisphere. These ecosystems should be considered millennial heritage ecosystems.

(iii) Peatlands are one of the most important carbon sinks in Chile, so both restoration and conservation of pristine peatlands represent effective alternative mechanisms to achieve the country's projected carbon neutrality by 2050. However, peatlands are being directly impacted by human extractive activities and roads, as well as by global climate change, and therefore could behave as carbon emitters rather than carbon sinks.

(iv) The current regulations that regulate and supervise the management of both moss harvesting and peat extraction divide ecosystems into different components, causing different institutions and regulations to act in parallel according to their own interests, and fail to recognize the interaction of the multiple levels at which the ecosystem functions and the landscape in which they are inserted. In view of this, laws should be modified or redesigned.

(v) The current management of peat extraction from natural ecosystems is not sustainable, because it leads to the drastic modification of hydrology, topography, and natural biogeochemical cycles and loss of biota, making the recovery of complex food webs and their role in the various ecosystem functions impossible. Adequate management and conservation of biodiversity and ecosystem services generates multiple benefits to society and can bring greater benefits to human beings than land use change, which causes the loss of natural habitats.

(vi) Most current natural peatland restoration practices are costly and lack a clear objective (e.g. social, economic, ecological). There are no successful examples of restoration of peatlands and their diverse ecosystem functions in Chile.

We also provide the following recommendations:

- Undertake a national inventory that includes the description, conservation status, identification of critical areas and implementation of their monitoring, as far as possible under standard international methodologies. This would be an essential step to obtain baseline information and thus better address decision-making regarding the management and conservation of pristine peatland ecosystems in climate change mitigation.

- Incorporation of these ecosystems into international networks and treaties, to increase the interaction of management and conservation experiences and to have a regulatory policy consistent with the role of peatlands as global climate regulators. The millenary climate records provided by peatlands are a powerful tool to be included in global climate change projection models.
- Integrate and strengthen the institutions that regulate and govern national peatland activities and property rights. The legal dismemberment of peat, *sphagnum* moss, and water in different institutions and with different legal frameworks, currently prevents a single governance to manage holistically the goods, services and ecosystem functions of peatlands. The principle of progressive improvement should be advanced order to maintain the well-being of key ecosystems such as peatlands as a human right to a healthy environment.
- Maintain the maximum number of natural peatlands in a pristine state and avoid land use change as much as possible, together with greater dissemination of the multiple natural benefits that these ecosystems provide to society in the medium- and long-term, which are greater than those derived from the exploitation of peat, which generates only short-term economic benefits and that do not necessarily even benefit local communities.
- Promote and expand environmental education and special interest tourism (based on cultural and environmental identity) for the incorporation of the communities of the region. Peatlands are part of the environmental culture of Chilean Patagonia and an asset for tourism throughout their distribution. Through environmental education, we must seek to transform the habits that make our socioeconomic development model incompatible with the conservation of these ecosystems.

Acknowledgements CAM is grateful to PAI project 77180002 and FONDECYT 11220705. RM is grateful to PAI project 79170119. Both projects awarded by the National Agency for Research and Development of Chile, Government of Chile.

References

1. Amesbury, M. J., Gallego-Sala, A., & Loisel, J. (2019). Peatlands as prolific carbon sinks. *Nature Geoscience, 12*(11), 880–881.
2. Arroyo, M. T., Pliscoff, P., Mihoc, M., & Arroyo-Kalin, M., (2005). The Magellanic moorland. In L. H. Fraser & P. A. Keddy (Eds.), *The world's largest wetlands, ecology and conservation* (pp. 424–445). Cambridge University Press.
3. Biester, H., Kilian, R., Franzen, C., Woda, C., Mangini, A., & Schöler, H. F. (2002). Elevated mercury accumulation in a peat bog of the Magellanic moorlands, Chile (53°S) an anthropogenic signal from the southern hemisphere. *Earth and Planetary Science Letters, 201*(3), 609–620.

4. Bonn, A., Allott, T., Evans, M., Joosten, H., & Stoneman, R. (2016). Peatland restoration and ecosystem services. In A. Bonn, T. Allott, M. Evans, H. Joosten, & R. Stoneman (Eds.), *Ecological reviews* (pp. 1–16). Cambridge University Press.

5. Borken, W., Horn, M. A., Geimer, S., Aguilar, N. A. B., & Knorr, K. H. (2016). Associative nitrogen fixation in nodules of the conifer *Lepidothamnus fonkii* (Podocarpaceae) inhabiting ombrotrophic bogs in southern Patagonia. *Scientific Reports, 6*(1), 39072.

6. Bragina, A., Oberauner-Wappis, L., Zachow, C., Halwachs, B., Thallinger, G. G., Müller, H., & Berg, G. (2014). The Sphagnum microbiome supports bog ecosystems functioning under extreme conditions. *Molecular Ecology, 23*(18), 4498–4510.

7. Cardone, I., & Worman, G. (2008). Los turbales y su relación con la comunidad. Vox populi Centro de Estudios Patagónicos.

8. Contador, T., Kennedy, J., & Rozzi, R. (2012). The conservation status of southern South American aquatic insects in the literature. *Biodiversity and Conservation, 21*, 2095–2107.

9. Corporación Nacional Forestal (CONAF). (2006). *Catastro de uso de suelo y vegetación. Región de Magallanes y Antártica Chilena.* Retrieved from: http://sit.conaf.cl/

10. Corporación Nacional Forestal (CONAF). (2017). *Superficie de uso de suelo regional.* Retrieved from: https://www.conaf.cl/nuestros-bosques/bosques-en-chile/catastro-vegetacional/

11. Díaz, M. F., Larraín, J., Zegers, G., & Tapia, C. (2008). Caracterización florística e hidrológica de turberas de la isla grande de Chiloé Chile. *Revista Chilena de Historia Natural, 81*(4), 455–468.

12. Domínguez, E., Bahamonde, N., & Muñoz-Escobar, C. (2012). Efectos de la extracción de turba sobre la composición y estructura de una turbera de Sphagnum explotada y abandonada hace 20 años, Chile. *Anales del Instituto Patagonia, 40*(2), 37–45.

13. Domínguez, E. (2014). *Manual de buenas prácticas para el uso sostenido del musgo Sphagnum magellanicum en Magallanes, Chile.* Boletín INIA No. 276. Instituto de Investigaciones Agropecuarias. Centro Regional de Investigación Kampenaike.

14. Domínguez, E., Vega-Valdés, D., Dollenz, O., Villa-Martínez, R., Aravena, J. C., Henríquez, J. M., & Muñoz-Escobar, C. (2015a). Flora y vegetación de turberas de la Región de Magallanes. In E. Domínguez & D. Vega-Valdés (Eds.), *Funciones y servicios ecosistémicos de las turberas en Magallanes* (pp. 149–195). Instituto de Investigaciones Agropecuarias. Centro Regional de Investigación Kampenaike. Colección de libros INIA No. 33.

15. Domínguez, E., Doorn, M., Silva, R., & Flaneigs, L. (2015b). *Bases comerciales para el desarrollo sostenible del musgo Sphagnum en Magallanes, Chile.* Boletín INIA No. 309. Instituto de Investigaciones Agropecuarias. Centro Regional de Investigación Kampenaike.

16. Fernández, L. D., Lara, E., & Mitchell, E. A. D. (2015). Checklist, diversity and distribution of testate amoebae in Chile. *European Journal of Protistology, 51*(5), 409–424.

17. Frolking, S., Talbot, J., Jones, M., Treat, C., Kauffman, J., Tuittila, E. S., & Roulet, N. (2011). Peatlands in the Earth's 21st century climate system. *Environmental Reviews, 19*, 371–396.

18. Graham, L., Graham, J., Knack, J., Trest, M., Piotrowski, M., & Arancibia-Avila, P. (2017). A sub-Antarctic peat moss metagenome indicates microbiome resilience to stress and biogeochemical functions of early Paleozoic terrestrial ecosystems. *International Journal of Plant Sciences, 178*, 8.

19. Griscom, B. W., Adams, J., Ellis, P. W., Houghton, R. A., Lomax, G., Miteva, D. A., Schlesinger, W. H., Shoch, D., Siikamäki, J. V., Smith, P., Woodbury, P., Zganjar, C., Blackman, A., Campari, J., Conant, R. T., Delgado, C., Elias, P., Gopalakrishna, T., Hamsik, M. R., … Fargione, J. (2017). Natural climate solutions. *Proceedings of the National Academy of Sciences, USA, 114*(44), 11645–11650.

20. Groeneveld, E., Massé, A., & Rochefort, L. (2007). *Polytrichum strictum* as a nurse-plant in peatland restoration. *Restoration Ecology, 15*, 709–719.

21. Grootjans, A., Iturraspe, R., Fritz, C., Moen, A., & Joosten, H. (2014). Mires and mire types of península Mitre, Tierra del Fuego, Argentina. *Mires and Peat, 14*(1).

22. Guzmán, J. (2015). Roedores de las turberas en Magallanes. In E. Domínguez & D. Vega-Valdés (Eds.), *Funciones y servicios ecosistémicos de las turberas en Magallanes* (pp. 279–393). Instituto de Investigaciones Agropecuarias. Centro Regional de Investigación Kampenaike. Colección de libros INIA No. 33.
23. Henn, J, J., Anderson, C.B., & Martínez Pastur, G. (2016). Landscape-level impact and habitat factors associated with invasive beaver distribution in Tierra del Fuego. *Biol. Invasions, 18,* 1679–1688. https://doi.org/10.1007/s10530-016-1110-9
24. Heusser, C. (2003). Ice age of Southern Andes. A chronicle of palaeoecological events. In J. Rose (Ed.), *Developments in quaternary science*, Vol. 3, No. 12 (pp. 154–170). Elsevier.
25. Hoyos-Santillán, J., Miranda, A., Lara, A., Rojas, M., & Sepúlveda-Jauregui, A. (2019). Protecting Patagonian peatlands in Chile. *Science, 366*(6470), 1207–1208.
26. Hoyos-Santillán, J., & Mansilla, C. A. (2021). Dinámica del carbono en turberas de la Patagonia chilena. In E. Domínguez & M.P. Martínez (Eds.), *Funciones y servicios ecosistémicos de las turberas de Sphagnum en la Región de Aysén* (pp. 65–89). Instituto de investigaciones Agropecuarias, Centro Regional de Investigación Tamel Aike. Colección de libros INIA No. 4.
27. Ibarra, J., Anderson, C., Altamirano, T., Rozzi, R., & Bonacic, C. (2010). Diversity and singularity of the avifauna in the austral peat bogs of the Cape Horn Biosphere Reserve, Chile. *Ciencia e Investigación Agraria, 37*.
28. Instituto Forestal de Chile. (2020). *Productos forestales no madereros (marzo)*. Boletín de Productos Forestales No Madereros (PFNM) No. 35. Instituto Forestal de Chile.
29. Intergovernmental Panel on Climate Change. (2019). Summary for Policymakers. In P. R. Shukla, J. Skea, E. Calvo Buendia, V. Masson-Delmonte, H. Portner, D. C. Roberts, P. Zhai, R. Slade, S. Connors, R. van Diemen, M. Ferrat, E. Haughey, S. Luz, S. Neogi, M. Pathak, J. Petzold, J. Portugal Pereira, P. Vyas, E. Huntley, & J. Malley (Eds.), *Climate change and Land: an IPCC special report on climate change, desertification, land degradation, sustainable land management, food security, and greenhouse gas fluxes in terrestrial ecosystem* (pp. 1–36). Intergovernmental Panel on Climate Change. Retrieved from: https://www.ipcc.ch/srccl/chapter/summary-for-policymakers/Iturraspe, R. (2016). Patagonian peatlands (Argentina and Chile). In: C. Finlayson, G. Milton, R. Prentice, & N. Davidson (Eds.), *The Wetland book* (pp. 1–10). Springer.
30. Iturraspe, R. (2016). Patagonian peatlands (Argentina and Chile). In C. Finlayson, G. Milton, R. Prentice & N. Davidson (Eds.), *The wetland book* (pp. 1–10). Springer.
31. Jeréz, V., & Muñoz-Escobar, C. (2015). Coleópteros y otros insectos asociados a turberas del pá-ramo magallánico en la Región de Magallanes, Chile. In E. Domínguez & D. Vega-Valdés (Eds.), *Funciones y servicios ecosistémicos de las turberas en Magallanes* (pp. 199–228). Instituto de Investigaciones Agropecuarias. Centro Regional de Investigación Kampenaike. Colección de libros INIA No. 33.
32. Jofré, J., Goffinet, B., Marino, P., Raguso, R., Nihei, S., Massardo, F., & Rozzi, R. (2011). First evidence of insect attraction by a southern hemisphere Splachnaceae: The case of *Tayloria dubyi* Broth. in the Biosphere Reserve Cape Horn, Chile. *Nova Hedwigia, 92*, 317–326.
33. Joosten, H., & Clarke, D. (2002). *Wise use of mires and peatlands: Background and principles including a framework for decision-making. International Mire Conservation Group and International Peat Society*. Retrieved from: http://www.gret-perg.ulaval.ca/fileadmin/fichiers/fichie rsGRET/pdf/Doc_generale/WUMP_Wise_Use_of_Mires_and_Peatlands_book.pdf
34. Kleinebecker, T., Hölzel, N., & Vogel, A. (2007). Gradients of continentality and moisture in south Patagonian ombrotrophic peatland vegetation. *Folia Geobotanica, 42*(4), 363–382.
35. Leifeld, J., & Menichetti, L. (2018). The underappreciated potential of peatlands in global climate change mitigation strategies. *Nature Communications, 9*(1), 1071.
36. Liebner, S., Zeyer, J., Wagner, D., Schubert, C., Pfeiffer, E. M., & Knoblauch, C. (2011). Methane oxidation associated with submerged brown mosses reduces methane emissions from Siberian polygonal tundra. *Journal of Ecology, 99*(4), 914–922.
37. Loisel, J., & Yu, Z. (2013). Holocene peatland carbon dynamics in Patagonia. *Quaternary Science Reviews, 69*, 125–141.

38. Luebert, F., & Pliscoff, P. (2018). *Sinopsis bioclimática y vegetacional de Chile*. Editorial Universitaria.
39. Mackenzie, R., Lewis, L. R., & Rozzi, R. (2016). Nuevo registro de *Sphagnum falcatulum* Besch (Sphagnaceae) en isla Navarino, Reserva de la Biósfera Cabo de Hornos. *Anales del Instituto de la Patagonia, 44*(1), 79–84.
40. Malvarez, A. I., Kandus, P., & Carbajo, A. (2004). Distribución regional de los turbales de la Patagonia. In: D. E. Blanco & V. M. de la Balze (Eds.), *Los turbales de la Patagonia. Bases para su inventario y la conservación de su biodiversidad* (pp. 22–29). Wetland International.
41. Markgraf, V., Whitlock, C., & Haberle, S. (2007). Vegetation and fire history during the last 18,000 cal yr B.P. in southern Patagonia: Mallín Pollux, Coyhaique, province Aysén (45°41 30 S, 71°50 30 W, 640 m elevation). *Palaeogeography, Palaeoclimatology, Palaeoecology, 254*(3), 492–507.
42. Marquet, P. A., Buschmann, A. H., Corcoran, D., Díaz, P. A., Fuentes-Castillo, T., Garreaud, R., Pliscoff, P., & Salazar, A. (2023). *Global change and acceleration of anthropic pressures on Patagonian ecosystems*. Springer.
43. Mathijssen, P. J. H., Gałka, M., Borken, W., & Knorr, K. H. (2019). Plant communities' control long term carbon accumulation and biogeochemical gradients in a Patagonian bog. *Science of the Total Environment, 684*, 670–681.
44. McCulloch, R. D., Blaikie, J., Jacob, B., Mansilla, C. A., Morello, F., De Pol-Holz, R., San Román, M., Tisdall, E., & Torres, J. (2020). Late glacial and Holocene climate variability, southernmost Patagonia. *Quaternary Science Reviews, 229*, 106131.
45. Ministerio del Medio Ambiente y Centro de Ecología Aplicada. (2011). *Diseño del inventario nacional de humedales y el seguimiento ambiental*. Ministerio del Medio Ambiente Centro de Ecología Aplicada. Retrieved from: http://bibliotecadigital.ciren.cl/bitstream/handle/123456 789/6276/HUM-CEA-001.pdf?sequence=1&isAllowed=y
46. NDC - Contribución Determinada a Nivel Nacional. (2020). *Contribución Determinada a Nivel Nacional de Chile. Actualización 2020*. Retrieved from: https://mma.gob.cl/wp-content/upl oads/2020/07/Espanol-21-julio.pdf
47. Nesshöver, C., Assmuth, T., Irvine, K. N., Rusch, G. M., Waylen, K. A., Delbaere, B., Haase, D., Jones-Walters, L., Keune, H., Kovacs, E., Krauze, K., Külvik, M., Rey, F., van Dijk, J., Vistad, O. I., Wilkinson, M. E., & Wittmer, H. (2017). The science, policy and practice of nature-based solutions: An interdisciplinary perspective. *Science of the Total Environment, 579*, 1215–1227.
48. Oficina de Estudios y Políticas Agrarias, ODEPA. (2015). *Exportaciones de musgos secos, distintos de los usados para ramos y adornos y de los medicinales*. Retrieved from: https://www.odepa.gob.cl/series-anuales-por-producto-de-exportaciones-importaciones
49. Ortiz, J. C. (2015). Anfibios de las turberas del extremo austral de Chile. In: E. Domínguez & D. Vega-Valdés,(Eds.), *Funciones y servicios ecosistémicos de las turberas en Magallanes* (pp. 229–240). Instituto de Investigaciones Agropecuarias. Centro Regional de Investigación Kampenaike. Colección de libros INIA No. 33.
50. Pacheco-Cancino, P., & Carrillo-López, R. (2017). *Manual de producción artificial del musgo Sphagnum magellanicum* Brid. Universidad de La Frontera. Retrieved from: https://isbn.cloud/9789562363228/manual-de-produccion-artificial-del-musgo-Sphagnum-magellanicum-brid/
51. Page, S., Rieley, J., & Banks, C. (2011). Global and regional importance of the tropical peatland carbon pool. *Global Change Biology, 17*(2), 798–818.
52. Pan, Y., Birdsey, R., Fang, J., Houghton, R., Kauppi, P., Kurz, W., Phillips, O. L., Shvidenko, A., Lewis, S. L., Canadell, J. G., Ciais, P., Jackson, R. B., Pacala, S. W., McGuire, A. D., Piao, S., Rautiainen, A., Sitch, S., & Hayes, D. (2011). A large and persistent carbon sink in the world's forests. *Science, 333*, 988–993.
53. Pisano, E. (1977). Fitogeografía de Fuego-Patagonia chilena. *Anales del Instituto de la Patagonia, 8*, 121–250.
54. Poulter, B., Bousquet, P., Canadell, J. G., Ciais, P., Peregon, A., Saunois, M., Arora, V. K., Beerling, D. J., Brovkin, V., Jones, C. D., Joos, F., Gedney, N., Ito, A., Kleinen, T., Koven, C. D., McDonald, K., Melton, J., R., Peng, C., Prigent, C., Schroeder, R., Riley, W. J., Riley, W. J.,

Saito, M., Spahni, R., Tian, H., Taylor, L., Viovy, N., Wilton, D., Wiltshire, A., Xu, X., Zhang, B., Zhang, Z., & Zhu, Q. (2017). Global wetland contribution to 2000–2012 atmospheric methane growth rate dynamics. *Environmental Research Letters, 12*, 094013.
55. Primack, R., Rozzi, R., Feinsinger, P., Dirzo, R., & Massardo, F. (2001). *Fundamentos de conservación biológica. Perspectivas latinoamericanas* (2nd ed.). Fondo de cultura Económica.
56. Riffo, J. (2019). *Curso biodiversidad y conservación de humedales en la Región de Magallanes.* Versión XXII. INIA. Retrieved from: https://www.inia.cl/wp-content/uploads/2019/04/Jose-Riffo_Legislación.pdf
57. Riveros, G. A., Kush, A., Cárcamo, J., & Domínguez, E. (2015). Avifauna en turberas Fuego-Patagónicas. In: E. Domínguez & D. Vega-Valdés (Eds.), *Funciones y servicios ecosistémicos de las turberas en Magallanes* (pp. 245–275). INIA No. 33.
58. Saatchi, S. S., Harris, N. L., Brown, S., Lefsky, M., Mitchard, E. T. A., Salas, W., Zutta, B. R., Buermann, W., Lewis, S. L., Hagen, S., Petrova, S., White., L., Silman, M., & Morel, A. (2011). Benchmark map of forest carbon stocks in tropical regions across three continents. *Proceedings of the National Academy of Sciences, USA, 108*(24), 9899–9904.
59. Saavedra, B., & Villarroel, G. (2019). *Chile, país de humedales: 40 mil reservas de vida.* Retrieved from: https://chile.wcs.org/Portals/134/Libro%20Humedales%20WCS.pdf?ver=2019-02-08-203952-653
60. Scharlemann, J. P. W., Tanner, E. V. J., Hiederer, R., & Kapos, V. (2014). Global soil carbon: Understanding and managing the largest terrestrial carbon pool. *Carbon Management, 5*(1), 81–91.
61. Sjögersten, S., Black, C., Evers, S., Hoyos-Santillan, J., Wright, E., & Turner, B. (2014). Tropical wet-lands: ¿a missing link in the global carbon cycle? *Global Biogeochemical Cycles, 28*, 1371–1386.
62. Vaccarezza, F. (2012). *Gestión ambiental de turberas en Magallanes (Chile).* TDX (Tesis Doctoral). Universitat de Barcelona. Retrieved from: http://www.tdx.cat/handle/10803/96650
63. Veblen, T., Burns, B., Kitzberger, T., Lara, A., & Villalba, R. (1995). The ecology of the conifers of southern South America. In N. Enright & R. Hill (Eds.), *Ecology of the southern conifers* (pp. 120–155). Melbourne University Press.
64. De Vleeschouwer, F., Vanneste, H., Mauquoy, D., Piotrowska, N., Torrejón, F., Roland, T., Stein, A., & Le Roux, G. (2014). Emissions from pre-hispanic metallurgy in the South American atmosphere. *PLoS One, 9*(10), e111315.
65. Yu, Z., Loisel, J., Brosseau, D. P., Beilman, D. W., & Hunt, S. J. (2010). Global peatland dynamics since the last glacial maximum. *Geophysical Research Letters, 37*(13), 1–5.

Claudia A. Mansilla Ph.D. Environmental Sciences, University of Stirling, Scotland, UK. Assistant Professor, GAIA-Antarctic Research Center, Universidad de Magallanes, Chile.

Erwin Domínguez Master in Botany, Universidad de Concepción. Researcher in natural grasslands and natural resources, Instituto de Investigaciones Agropecuarias—INIA, Kampenaike, Chile.

Roy Mackenzie Biologist, Pontificia Universidad Católica de Valparaíso. Ph.D. in Microbiology, Universitat Autònoma de Barcelona. Assistant Professor, Universidad de Magallanes, Chile.

Jorge Hoyos-Santillan Ph.D. Environmental Sciences, University of Nottingham, U.K. Associate Professor, University of Nottingham, U.K. Researcher, Laboratory of Environmental Biogeochemistry in Extreme Ecosystems, Universidad de Magallanes, Chile.

Juan Marcos Henríquez D. in Biological Sciences, Universidad de Concepción, Chile. Associate Professor, Universidad de Magallanes, Chile.

Juan Carlos Aravena B.Sc. and M.Sc. in Biological Sciences, Universidad de Chile. Ph.D. in Environmental Sciences, University of Western Ontario, Canada. Associate Professor and Director of the GAIA-Antarctic Research Center, Universidad de Magallanes, Chile.

Rodrigo Villa-Martínez M.Sc. in Biological Sciences and Ph.D. in Evolutionary Biology, Universidad de Chile. Associate Professor, GAIA-Antarctic Research Center, Universidad de Magallanes, Chile.

Steppe Ecosystems in Chilean Patagonia: Distribution, Climate, Biodiversity, and Threats to Their Sustainable Management

7

Sergio Radic-Schilling, Paulo Corti, René Muñoz-Arriagada, Nicolás Butorovic, and Laura Sánchez-Jardón

Abstract

Steppe ecosystems in Chilean Patagonia are located in the Andean rain shadow zones in the Aysén and Magallanes administrative regions, on the eastern slopes of the Andes along the border with Argentina, and in the north-central sector of Tierra del Fuego island. This chapter describes the distribution, climatic conditions, biodiversity, and changes in vegetation coverage of the steppe in Chile. Steppe biodiversity faces threats due to the negative impacts of multiple anthropogenic activities, particularly the development of the hydrocarbon sector, grazing pressure, wildfires, and the introduction of exotic species, as well as global climate change. Conservation of the Patagonian steppe in Chile depends both on protection within state parks or reserves and on cultural changes in the management of the unprotected area; economic uses and conservation must be integrated through sustainable management strategies.

Keywords

Patagonia · Chile · Grasslands · Meadows · Tussock grasses · Management

S. Radic-Schilling (✉) · R. Muñoz-Arriagada · N. Butorovic · L. Sánchez-Jardón
Universidad de Magallanes, Punta Arenas, Chile
e-mail: sergio.radic@umag.cl

P. Corti
Universidad Austral de Chile, Los Ríos, Chile
e-mail: pcorti@uach.cl

© Pontificia Universidad Católica de Chile 2023
J. C. Castilla et al. (eds.), *Conservation in Chilean Patagonia*, Integrated Science 19,
https://doi.org/10.1007/978-3-031-39408-9_7

1   Introduction

The steppe is a dry, cold terrestrial ecosystem, where herbaceous, graminoid, and shrub species dominate. In the south of South America they extend through the Argentine Patagonia, from the coasts of the Atlantic Ocean to Chile, where they are restricted to the rain shadow zone to the east of the Andes Range in Aysén and Magallanes, except in Tierra del Fuego where they occupy the driest central-northern sector of the island [27]. These steppes develop in a semi-arid continental climatic regime, in which despite summer rainfall, evaporation usually exceeds precipitation [1]. The cold steppe ecosystems, dominated by grasses of the genera *Stipa*, *Festuca*, and *Poa*, would have reached their greatest development during the Late Pleistocene, due to the cold, dry climate that prevailed during that era [36]. Paleoecological records suggest that the steppe was the predominant ecosystem in vast areas of the planet during the Last Glacial Maximum (*ca.* 18,000 years BP) [36]. Steppe ecosystems have continental-type climates, as most of these environments are associated with large territories and mountainous areas with little oceanic influence [85]. Steppes have high seasonal climatic variability,summers are generally dry and warm, but winters are long and cold [85]. Being arid ecosystems, they have in common water as a limiting factor for the development of vegetation, with rainfall that generally does not exceed 400 mm yr^{-1} [85].

Stressors to an organism can affect its distribution; humidity and temperature are the main limiting stressors for vegetation in the steppes [2]. Water availability and temperature interact to determine boundaries for different plant communities globally,steppes occupy a specific place in a scheme of global biomes according to the interaction of these two physical factors (Fig. 1).

The incidence of livestock in managed steppe is diverse and ranges from extensive pastoral production systems to intensified systems [21]. Both the Northern Hemisphere steppes and the Patagonian steppe sustain extensive livestock use where large volumes of livestock are concentrated, especially sheep and goats [21, 50, 58]. Livestock management in the steppes of southern Chile is sedentary, although with winter and summer rotations [58]. There are differences and similarities between the steppes of Chilean Patagonia and those of other parts of the world, related to climatic variables, soil type, and dominant vegetation (Table 1).

2   Scope and Objectives

The objective of this chapter is to analyze the geographic distribution of the steppe in southern South America, along with the physical (soil and climate), and biological (vegetation and fauna) characteristics of these ecosystems represented in Chilean Patagonia. The chapter describes how the environmental functions and benefits of this ecosystem have been exploited to date for human beings, especially for livestock development, and discusses the main threats to its integrity and future, primarily due to the consequences of climate change, frequency of fires, intense livestock use, expansion of tourism, and invasive species. Comparisons of

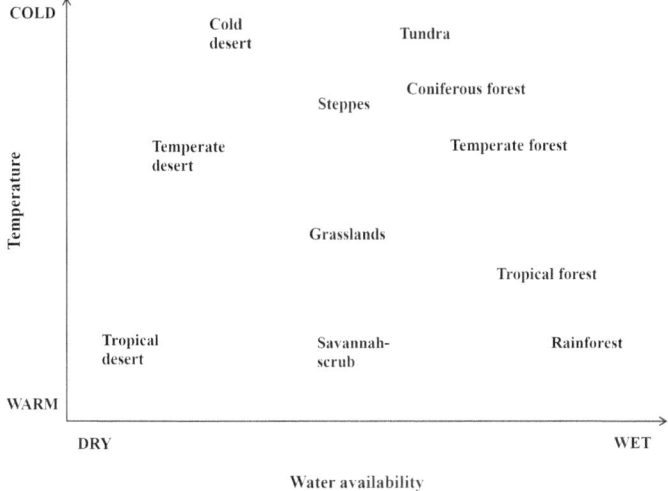

Fig. 1 General relationship between the types of plant communities or biomes and the conditions of temperature and water availability, indicating the position of the steppes. *Source* modified from Bidwell [2]

climatic and vegetation conditions are presented in the main precipitation and temperature gradients that distinguish the northern and southern, western and eastern zones of the geographic range of the Chilean Patagonian steppe.

3 Physical and Biological Characteristics of Steppe Ecosystems, Considering Threats and Ecosystem Services

3.1 Spatial Distribution of the Patagonian Steppe in Chile

The Patagonian steppe is a biome—though also classified as an ecoregion [54]—that is widely distributed in south-central Argentina, and found only in some restricted geographic areas of extreme southern Chile, where it is always in the rain shadow east of the Andes Range [58, 27, 67].

The Chilean steppe ecosystem is located in the regions of Aysén del General Carlos Ibáñez del Campo (Aysén Region) and Magallanes y de la Antártica Chilena (Magallanes Region), on the eastern slope of the Andes Mountains. In the Magallanes Region it is also found in the northern sector of the large island of Tierra del Fuego [58, 67]. The Chilean Patagonian steppe extends from 44° 43' S–68° 42' W to 54° 06' S–73° 80' W, in the eastern fringe of the binational steppe macrozone, which is the wettest area of its distribution.

Different authors have estimated different surface areas for the Chilean Patagonian steppe (e.g. [51, 57], because different variables such as climate, soils, vegetation, or combinations of these can be taken into account [56]. The main

Table 1 Similarities and differences in climate and flora characteristics between the Chilean Patagonian steppe and the Eurasian, Spanish, and North American steppes

Features	STEPPE			
	Chilean Patagonia	Eurasia	Spain	North America
Temperature (min–max °C)	−10–25 (1;20)	−30–25 (2;27)	8,9–17,9 [3]	−3,7–18,6 [4;23]
Annual rainfall (mm)	200–588 (5;18;20)	300–580 [2;25]	190–396 [3;24]	125–620 [4;22;23]
Predominant soil type	Mollisols [6;20]	Mollisols [7;21]	Mollisols [8]	Mollisols [8]
Soil pH	5.7–7.7 (19;20)	6.6–8.2 [2;26]	7.7–8.0 [10]	7.0–8.0 [11]
Shrub species	*Empetrum rubrum, Lepidophylletum cupresiforme, Chiliotrichium diffusum, Berberis buxifolia, Junellia tridens, Adesmia boronioides* [9;12]	*Artemisia pauciflora, Spiraea hypericifolia, Atriplex cana, Halocnemum strobilaceo, Nitraria schoberi* [13]	*Periploca laevigata, Maytenus senegalensis, Ziziphus lotus, Stipa tenacissima, Launaea arborescens, Genista pumila, Artemisia* spp. (14)	*Chrysothamnus, Artemisia arbuscula, A.cana, A. tripartita, Atriplex spinosa,Salsola kali* (15)
Poaceae species	*Festuca gracillima, F. magellanica and F. pallescens, Luzula alopecurus, Stipa humilis, Hordeum publiflorum, Stipa neaei, Poa dusenii* (9;16)	*Poaceae barnhart, Stipa lessingiana, S. sareptana, S. pennata, S. dasiphilla, and Festuca sulcata* [13]	*Stipa tenacissima, Poa ligulata, Festuca hystrix, Bromus hordeaceus* [14]	*Bromus tectorum, Sitanion hystrix, Sandberg's bluegrass, Bouteloua gracilis* (15;17)

Sources cited: (**1**) Santana et al. [72], (**2**) Maher et al. [48], (**3**) Blasco [3], (**4**) Williams et al. [86], (**5**) Covacevich and Ruz [18], (**6**) Borrelli & Oliva [5], (**7**) IUSS Working Group [41], (**8**) USDA [77], (**9**) Pisano [56], (**10**) Rey et al. [64], (**11**) Hironaka et al. [33], (**12**) Cruz & Lara [19], (**13**) Kudrevatykh et al. [45], (**14**) Ollero & van Staalduinen [53], (**15**) Brandt & Rickard [6], (**16**) Valle et al. [79], (**17**) Hoffman et al. [35], (**18**) Pisano [58], (**19**) Sáez [68], (**20**). Hepp & Stolpe [31], (**21**) Driessen & Deckers [23], (**22**) NCDC*, (**23**) Bailey [1], (**24**) Cirera & García [14], (**25**). Chendev et al. [12], (**26**) Torn et al. [75], (**27**) Boonman & Mikhalev [4]. *National Climatic Data Center (NCDC). Climate of New Mexico. https://www.ncdc.noaa.gov/climatenormals/clim60/states/Clim_NM_01.pdf, accessed May 2020

characteristic of the steppe is its dominant graminoid, tough herbaceous plant community without the presence of trees, although it may include shrubs [58]. The dominant biomass of the herbaceous vegetation is provided by the perennial grasses that form dense clumps called tussock grasses (in Spanish *coirones*). The characteristic species in Aysén is *Festuca pallescens*, while in Magallanes,

F. gracillima dominates [58], both species are present transversely in the different soil and climatic conditions mentioned above. Tussock grasses are usually present in all steppe environments, sometimes with low dominance values, including areas where hygrophilous species dominate and coirones may be absent, due to topographic characteristics that favor more humid conditions (Fig. 2).

The area occupied by steppe is marginal in the Aysén Region (108,490 km²; [31], with an approximate area of 2,562 km² (2.36%, Table 2). The steppe in Aysén is concentrated on the border with Argentina, with some valleys that intrude to the west where they form an ecotone between the open steppe and the forest and mountain areas of the Andes [58]. The steppe valleys that enter from Argentina into Chile are found in Alto Río Cisnes, Coyhaique Alto, Balmaceda, the areas adjacent to the General Carrera and Cochrane lakes and the Chacabuco River valley [67]. The commune of Coyhaique has the highest representation of steppe in this region (Table 2).

The Magallanes Region (1,382,033 km²) is subdivided into two sectors, one continental, including archipelagos and channels (132,033 km²) and the other Antarctic (1,250,000 km²) [18]. The region presents a much wider steppe distribution than Aysén, covering an estimated area of 24,434 km² (18.5% of the region's continental area, Table 2), which is physiognomically similar to the Argentine steppe, which it is a continuum of [67].

The steppe is also distributed on the eastern slope of the Andes, in Puerto Natales, and on both sides of the easternmost portion of the Strait of Magellan, close to the Atlantic Ocean, reaching a large area in the northern part of the island of Tierra del Fuego [58, 67]. 91.55% of the steppe area of Chilean Patagonia is located in the provinces of Tierra del Fuego and Magallanes (47.70% and 43.85%, respectively), while only 8.45% is in the province of Última Esperanza (Table 2). This province is a steppe zone with spatial and topographic patterns more similar to the steppe of the Aysén Region. The elevation gradient is similar to that of the Aysén Region.

The altitudinal gradient decreases from north to south; the Aysén steppe has average altitudes of over 400 m above sea level (m), while the southernmost portions, located in the Magallanes Region, have average altitudes of less than 200 m. (Fig. 3). Changes in vegetation occur as a result of this elevation gradient, which generates an altitudinal variant of the Chilean Patagonian steppe that includes high, semi-arid, and non-forested territories with more or less dense herbaceous vegetation [46, 57].

3.2   Climate of the Patagonian Steppe

The Patagonian steppe of Chile, unlike other cold steppes, maintains more stable climatic conditions between summer and winter seasons [85]. The colder steppe climate in Chilean Patagonia also has a more pronounced thermal amplitude and relatively low rainfall, with higher annual amounts at its western boundary,

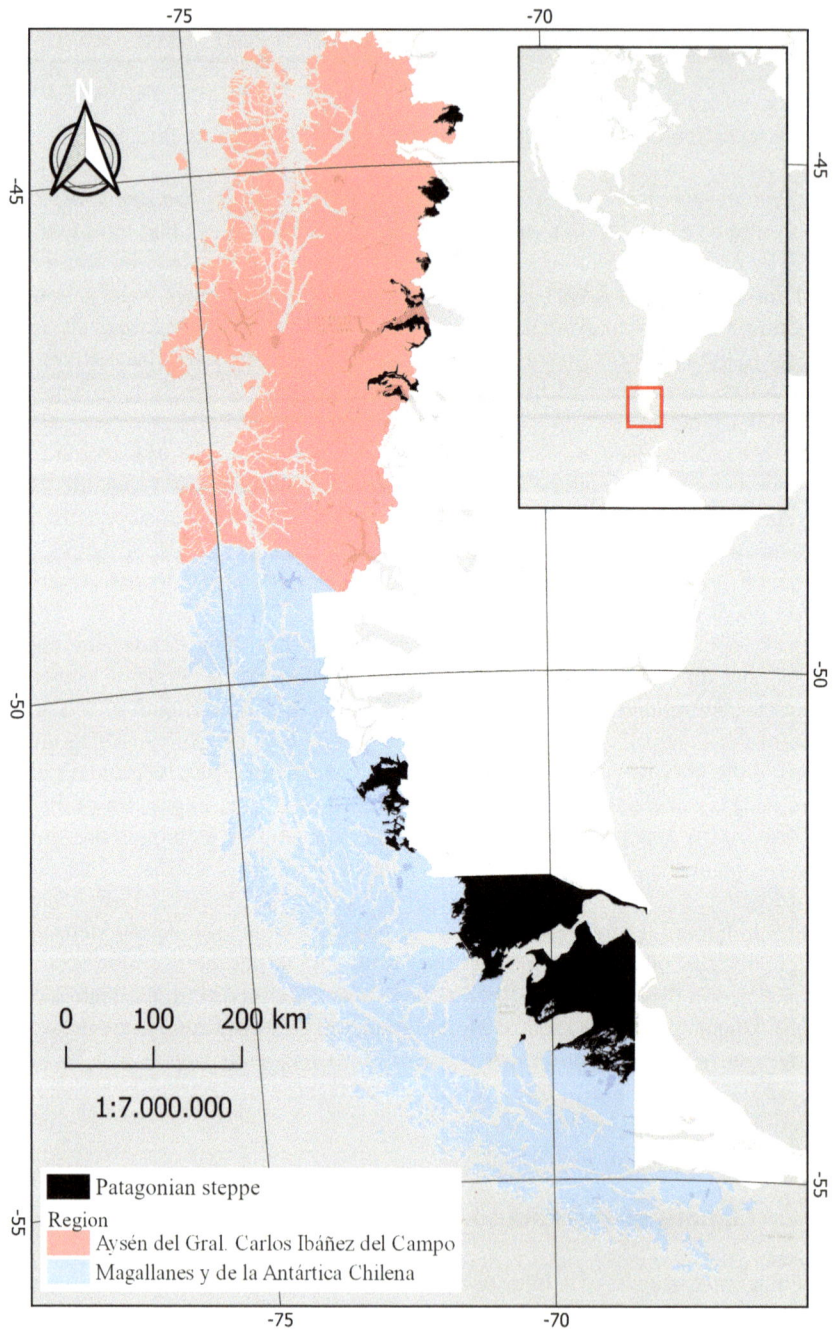

Fig. 2 Area and location of the steppes in Chilean Patagonia. *Source* Own elaboration, based on Pisano [56], Moore [51], Pliscoff & Fuentes-Castillo [59]. The approximation was generated from photo interpretation using high-resolution imagery sources available from Google satellite, Bing and Yandex satellite, and auxiliary data from the ASTER GDEM Digital Elevation Model

Table 2 Patagonian steppe distribution according to political-administrative divisions

Region	Province	Commune	Area (hectares)
Aysén	Capitan Prat	Cochrane	52,471.4
	Coyhaique	Coyhaique	82,737.0
		Lago Verde	43,828.8
	General Carrera	Chile Chico	52,496.3
		Ibañez River	24,661.8
Total			**256,195**
Magallanes	Magallanes	Laguna Blanca	296,309.0
		Punta Arenas	72,429.8
		Rio Verde	34,807.4
		San Gregorio	667,991.0
	Tierra del Fuego	Porvenir	677.422.0
		Primavera	377,306.0
		Timaukel	110,826.0
	Ultima Esperanza	Natales	38,299.0
		Torres del Paine	168,030.0
Total			**2,443,420**

Source Prepared by the authors, based on: [51, 56, 59]

decreasing to the east and north, reaching 200 mm yr^{-1} [11], the potential evapotranspiration fluctuates between 470 and 680 mm yr^{-1}, depending on the area of the region and the year [61]. Winters are shorter and summers are rather cold compared to the steppes of Central Asia or North America [85].

The Andes Range forms a barrier that stops the air masses coming from the Pacific Ocean in its Patagonian and Fuegian sections, and is responsible for local climatic modifications [24]. In the Aysén Region the Andes Range is in the east,the steppe is found in areas of lower altitude in the east that appear as intrusions from Argentina. There are four main areas, separated by elevated platforms: Alto Río Cisnes, Ñirehuao, Coyhaique Alto, and Balmaceda. All of these are depositional planes bordered by gentle hills (Regional Secretariat of Planning and Coordination, [73]. The Andes Range is on the western edge of the Magallanes Region, almost facing the Pacific Ocean, approximately north–south in its northern part, twisting its course to a northwest-southeast direction in its southern part [11] the steppe is located at the eastern end. There is a strong predominance of winds from the western quadrant (westerlies) [50], which upon reaching the American continent at its southern end are modified locally by the geomorphology, altitude and distance from the sea, giving rise to an enormous variety of climates in the Magallanes Region [24].

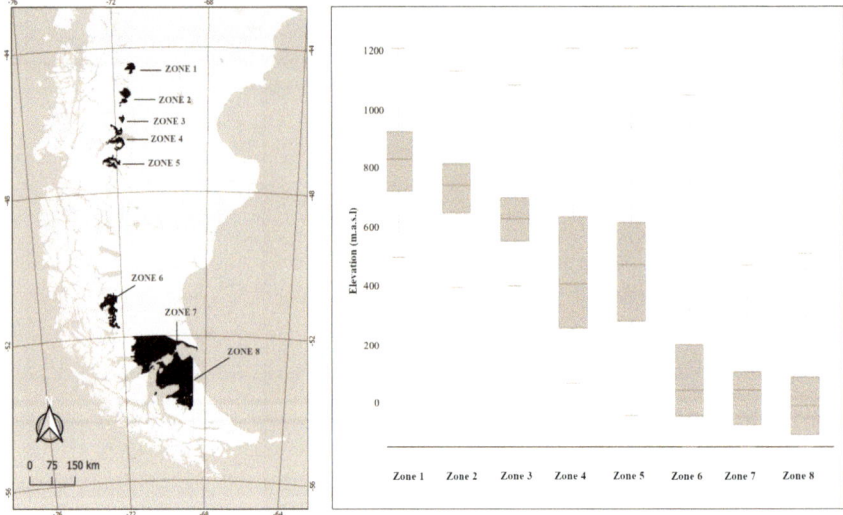

Fig. 3 Distribution of the Patagonian steppe according to zones. Zones 1 to 5 correspond to the Aysén Region and zones 6 to 8 correspond to the Magallanes Region. The box plot illustrates the elevational variation of steppe zones in Chilean Patagonia. The lines inside the boxes indicate the medians, while the boxes represent the 1st and 3rd quartiles. The lines above and below the boxes are the extreme values. *Source* Own elaboration; elevations were calculated from 7.5 M of data from the ASTER GDEM digital elevation model

The climatic conditions have a longitudinal gradient in the steppe of the Aysén Region; the climate is more arid than in the western strip, receiving strong influence from the dominant bioclimates in the east. It also has some continental tendency, with greater annual thermal amplitude, along with a progressive decrease in precipitation and relative humidity [22, 46]. In the Magallanes Region the climatology is determined by various atmospheric factors of general circulation, such as the position of the Pacific anticyclone and the Humboldt Current [11]. These atmospheric factors are associated with a strong zonal component of winds from the west and the proximity of the Antarctic continent, with frequent displacement of the polar front towards mid-latitudes [11]. Therefore, several factors interact to define a main climate type for the Magellanic steppe.

There is a very marked Foëhn effect, which occurs when moisture-laden air masses precipitate on the western slope of the Andes (windward), while precipitation drops sharply on the eastern edge (leeward) [11]. This happens because of the strong adiabatic gradient that causes the temperature of the air mass to be lower on the western edge and higher on the eastern edge [11]. This is the main effect responsible for the uneven distribution of precipitation in Patagonia, both in Aysén [31] and Magallanes [11].

Precipitation occurs throughout the year in the steppe zone of Aysén, with an approximate annual average between 444 mm yr^{-1} [58] and 588 mm yr^{-1} [31] and a range of 11–70 mm between the driest and wettest months, respectively,

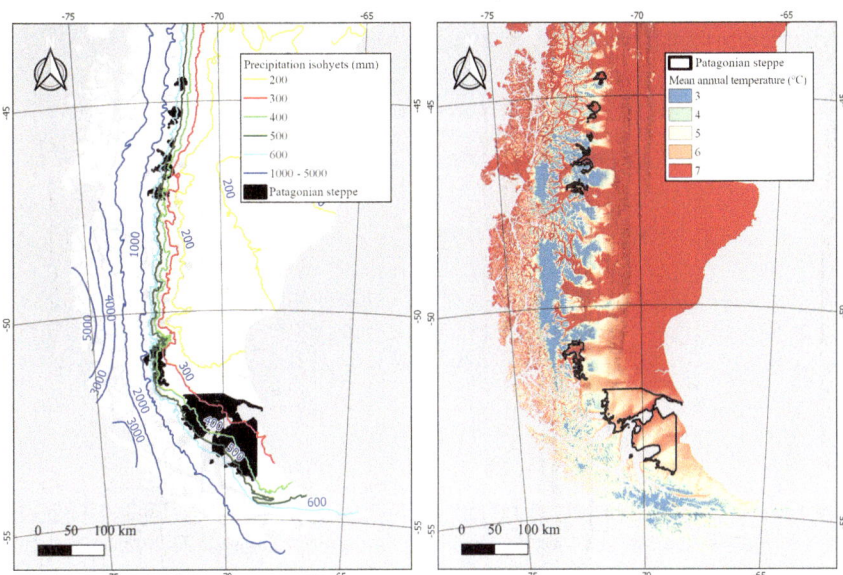

Fig. 4 Precipitation isohyets (mm-yr^{-1}) and isotherms (annual mean °C) for the Aysén and Magallanes regions. *Source* modified from Fick and Hijmans [25]

so there would not be marked seasonality [58], Fig. 4). However, precipitation is somewhat lower in the Magallanes Region, approximately 328 mm yr^{-1}, with a range of 16–39 mm [58], Fig. 4). Although scarce, precipitation is more evenly distributed over the seasons [56]. In both regions, much of the precipitation occurs as snow in the winter months (June–August, [58]), although in the Magallanes Region the annual snow cover is variable [83].

Average annual temperatures in the steppe zone in Aysén are less than 6.5 °C [31, 58], with ranges between 2.2 °C (average annual minimum) and 11.6 °C (average annual maximum) [31], while in the steppe zone of Magallanes the average is 4.7 °C with a range of 0–9.2 °C [58]. Average temperatures in inland areas of Magallanes in the winter months do not exceed 0 °C and extremes can reach − 25° to −30 °C [11]. In summer the extreme night temperatures reach −11 °C, and can reach 30 °C during the day [11].

The following climatic groups have been defined for the steppe zone in the Aysén and Magallanes Regions according to the Köppen-Geiger classification, which considers ranges of precipitation and temperature [44]: (i) boreal Andean climate (Cfc) (Aysén [31, 44]), (ii) trans-Andean climate with steppe degeneration (Dfk) (Aysén—[57], Magallanes—[57]), (iii) cold steppe climate (BSk), the dominant climate in the Aysén steppe and in the extensive area of eastern Magallanes and Tierra del Fuego [31, 44, 57]. The climate stations most representative of the cold steppe climate in the Aysén and Magallanes Regions are indicated in

Table 3 Mean precipitation (mm yr^{-1}) of different localities of the Aysén and Magallanes regions (Mag) in the Patagonian steppe zone, indicating the number of years with records

	Location	Precipitation	Years-data
1	Coyhaique Alto (Aysén)	327.9	6
2	Balmaceda (Aysén)	588.0	48
3	San Felipe Bay (Mag)	292.1	3
4	Side River, in Cerro Sombrero (Mag)	287.2	8
5	San Sebastián (Mag)	372.5	3
6	Onaissin, in Maria Cristina (Mag)	241.2	8
7	Pampa Guanaco (Mag)	336.2	19
8	San Gregorio (Mag)	268.1	7
9	Villa Tehuelche (Mag)	317.6	7
10	Aymond Mount (Mag)	216.4	7

Source own elaboration, based on data from Hepp and Stolpe [31] and from the automatic stations of the Dirección General de Aguas (DGA) (2019). Real-time hydrological data. Ministry of Public Works (in Spanish Ministry of Public Works). https://dga.mop.gob.cl/Paginas/hidrolineasatel.aspx. Accessed December 2019

Table 3. In both regions, precipitation is within the range of 200 to 400 mm yr^{-1} (Magallanes) and between 327 and 588 mm yr^{-1} (Aysén).

Guanaco and sheep grazing in steppe dominated by coironales. Mount Aymond sector, Magallanes and Chilean Antarctica Region.
Photograph by Paulo Corti.

Guanaco and sheep grazing in steppe dominated by coironales. Mount Aymond sector, Magallanes and Chilean Antarctica Region. Photograph by Paulo Corti

3.3   Soils in the Patagonian Steppe

Most of the soils of the Fuegian-Patagonian area have developed on moraine systems of the second and third glaciation or on mixed moraine sediments of the same epochs [56]. Both types of soils contain abundant ash from Pleistocene-Quaternary volcanism mixed with fluvial and/or colluvial sand and silt [56]. A flat or slightly undulating landscape predominates, dominated by sedimentary plateaus from the Tertiary period without a well-defined drainage towards the sea,rainfall is channelled into temporary lagoons or large internal lowlands [5].

Soils in the steppe of the Aysén Region have formed in topographic positions that vary from alluvial terraces to undulating hills with volcanic ash-generating materials on glacial and fluvioglacial deposits [31]. The aridity of the eastern plains of the Magallanes Region conditions the type of vegetation and soil development [56], from semi-arid, more weathered, less leached and less acidic, to those located further west, with a shallow surface horizon [56]. These soils have an upper layer of fine sand with organic matter [5]. Textures change to clay loam in depth,they are stony throughout the profile, with a pH ranging from slightly acidic to moderately alkaline [5]. The coironales in this area have an average soil bulk density of 0.71 g cm^{-3}, ranging from 0.58 to 0.87 g cm^{-3} [62], and low nitrogen content [68].

There are mollisol and inceptisol soils in the small valleys inserted in semi-arid and xerophytic scrubland areas that occupy the steppe in the eastern part of the Aysén Region [31], according to the soil classification of the United States Department of Agriculture (USDA). Mollisols and aridisols predominate in the Magallanes steppe 5. Mollisols, also called chernozems and kastanozems (or brown soils) according to the soil classification of the Food and Agriculture Organization of the United Nations (FAO), are formed under grassland vegetation in climates with moderate to marked water deficit [47] and have a horizon with concentrations of secondary carbonates,calcium is the dominant cation in the surface horizons [47]. These soils appear with increasing precipitation (*ca.* 300 mm) in well-drained areas [56]. Aridisols do not have water available when the temperature is suitable for the growth of plant species [47], while inceptisols are incipient with slight morphological development in the subsoil, including structure formation or brownish color [31].

There are mosaics of other types of soils with azonal characteristics in the steppes [29], generated by humid depressions, and hydromorphic soils of the mallín type, which present considerable accumulations of incompletely humidified organic matter that forms saturated humus in soils with little differentiation between horizons [58], where acidity tends to decrease with depth [79]. According to FAO soil taxonomy, these humid depressions or vegas are classified as histosols and fluvisols,there is considerable variability in their organic matter content and

Table 4 Macronutrients and general characteristics of soil groupings in the steppe zone of the Aysén Region, compared to chestnut soils, grasslands, meadows, meadows and murtillares in the steppe zone of the Magallanes Region

Macronutrient averages of the different soil groupings					
Soil grouping or vegetal communities	Phosphorus	Sulfur	Potassium	Calcium	Magnesium
	Olsen (mg kg^{-1})	Extractable (mg kg^{-1})	(mg kg^{-1})	(meq 100 g^{-1})	
Steppe Aysén region	13.9	1.6	342	9.6	2.7
Chestnut	5.0	4.0	406	8.2	3.1
Grasslands					
pH > 6	16.0	9.0	512	13.9	4.2
pH 5.9–5.7	13.0	10.0	339	9.4	3.7
Meadows					
Saline-sodium	9.0	600	378	32.8	9.1
Salinas	1.0	117	606	32.3	10.7
Non-saline sodium	1.0	29	638	6.1	4.2
Non-saline	12.0	21	596	19.0	5.7
Organic	13.0	153	309	26.1	6.7
Murtillar of Magallanes	6.0	6.8	6.8	1.7	1.3
Averages of general characteristics of different soil groupings					
	pH water	Aluminum saturation (%)	Organic matter (%)	Extractable aluminium (mg-kg^{-1})	
Steppe Aysén region	6.0	0.3	8.8	–	
Chestnut	6.2	0.3	6.1	53	
Grasslands					
pH > 6	6.2	0.2	9.1	77	
pH 5.9–5.7	5.8	0.6	9.2	171	
Meadows					
Saline-sodium	7.7	0	14.4	42	
Saline	7.3	0	11.1	5	
Non-Saline Sodium	6.1	2.1	6.8	144	
Non-Saline	6.4	0.5	10.8	106	
Organic	5.7	6.2	40.1	116	
Murtillar of Magallanes	4.8	55.0	30.7	748	

Source modified from Sáez [68], Hepp and Stolpe [31] and Valle et al. [79]

pH, tending to salinize in arid conditions [79]. They have a soil bulk density of 0.28 g cm^{-3}, with a range of 0.20–0.39 g cm^{-3} [62].

The vegas are the soils with the highest fertility within the Patagonian steppe [68]. These humid depressions have been classified into five types for the Magallanes Region, where it is indicated that the soils of non-saline vegas have low phosphorus (P) availability and low retention capacity,but they have high content of calcium bases (Ca), magnesium (Mg), potassium (K), sulfur (S), and micronutrients (Table 4).

Steppe (chestnut) soils have P values as low as 5 mg kg^{-1}, 6.0 mg kg^{-1} for the d deedle-Dee (in Spanish murtilla) or heath community and reaching 16 mg kg^{-1} in the more alkaline grasslands (Table 4); they possess an average phosphorus buffering capacity of 13.6 kg P ppm^{-1} Olsen [31]. They have high K and Mg content, reaching values of 512 mg kg^{-1} and 4.2 meq 100 g^{-1} dry soil respectively, but with the lowest values for both nutrients in the murtilla or heathland community (Table 4). S availability is 1.6 and 4.0 mg kg^{-1} for the steppe soils of the Aysén Region and the drier sectors of Magallanes, respectively (Table 4). The soils under the vegetation community of vegas have S values higher than 12 mg kg^{-1}, a threshold figure [66], since an agronomic response to grassland fertilization above this value is not expected. S availability is 6.8 mg kg^{-1} for the murtilla or heathland community, which is considered low.

Although steppe soils respond to fertilization with N, P, and S, increasing the productivity of the grassland, annual production would remain low due to the arid conditions in which the naturalized grassland is found. Therefore the application of fertilizers to increase pasture productivity is not profitable [68]. The exception would be the vega communities, which stand out for their higher levels of nutrients and moisture compared to soils under other plant communities such as coironales and murtillares [79].

The pH tends to be slightly acidic to alkaline in most of the steppe soils, with values ranging from 5.7 to 7.7 (Table 4). The pH values of the soils under the main steppe plant communities range from 4.8 - 6.2 for the murtilla and coironal communities, respectively (Table 4). The murtilla (*Empetrum rubrum*) community has this pH because this species has the ability to lower soil pH [15], Borelli and Oliva, 5. The average values of aluminum (Al) saturation are lower than 6.2% (Table 4), which according to Rodríguez [66] is considered low, the exception is the murtilla community, which has 55% saturation.

The highest extractable Al value found was 171 mg kg^{-1} (Table 4), which has no effect on phosphorus retention, since a value lower than 400 mg kg^{-1} is considered very low [66]. Once again, the exception is the murtilla community, which has 748 mg kg^{-1}.

3.4   Predominant Vegetation of the Patagonian Steppe

The representative vegetation of the Patagonian steppe is a community of graminoids and hard herbs, without the presence of trees [58]. Gajardo [27] and

Fig. 5 Coiron grasslands (*Festuca* spp.) in the steppe of Tierra del Fuego

Luebert and Pliscoff [46] suggested that the steppes in Aysén and Magallanes belong to two distinct sub-regions, differentiated mainly by the greater presence of the shrub component in Aysén and by the dominance of different species of coirones. The steppe in the Aysén Region is primarily associated with coiron grasslands, similar to those found in the provinces of Magallanes and Tierra del Fuego, but covering a significantly smaller area [31], and with the difference that they are mainly dominated by *Festuca pallescens* [57] and *Stipa* (*Stipa* spp.) [31]. Coironales in the steppe of the Magallanes Region occupy approximately 24% of the total area destined for livestock use [74], including several plant communities with variable physiognomy, from the typical hard grass steppe of *F. gracillima* (Fig. 5) to shrub steppe and shrublands, and also includes vegas or hygromorphic cespitose communities [56]. The climatic differences between the steppe of the Aysén Region and that of Magallanes would be the main cause of the floristic differences [58].

The steppe of the Aysén Region is characterized by the white or sweet coirón (*Festuca pallescens*), which forms robust plants with a strong and deep root system; it is accompanied by other perennial grasses of the genera *Festuca, Agrostis, Stipa, Poa, Bromus,* and *Deschampsia* that grow in aggregate form as tussock [32]. Other xerophytic shrubs common in the Aysén steppe are the green bush (*Nardophyllum obtusifolium*), different species of yareta (*Azorella* spp.), and senecio (*Senecio* sp.), fachine (*Chiliotrichum diffusum*), neneo (*Mulinum spinosum*), white hawthorn (*Discaria chacaye*), and christmas bush (*Baccharis magellanica*) [76]. The proportion of endemic species in the dominant families is very high, with up to 60% endemism in Leguminosae and 33% in Compositae [20]. Other outstanding elements are the wide diversity of lichen and herb species with showy flowers, such as the Chilean oxalis (*Oxalis adenophylla*), the Darwin's slipper (*Calceolaria uniflora*), and a variety of orchids such as the porcelain orchid (*Chloraea magellanica*) and the peatland orchid (*C. chica*) [76].

Fig. 6 Community of plant species distributed according to exposure and water availability in the steppe of Tierra del Fuego: **a** area with greater exposure to the wind where murtilla or diddle-dee (*Empetrum rubrum*) and balsam-bog (*Bolax gummifera*) predominate; **b** area that is protected from the wind and with greater accumulation of water in the soil, where the fachine bush (*Chiliotrichium diffusum*) predominates; **c** area at the base of the hill, where there is a greater accumulation of water compared to the other communities, is dominated by reeds (*Marsippospermum grandiflorum*)

The drier Magellanic steppe has an extensive coiron grassland, which contributes 30–70% of the total cover [5]. The spaces between coirones and shrubs are occupied by a large group of small native grasses, mainly perennials with rhizomes and stolons, dicotyledons, and naturalized species that are a key contribution to the diversity and pastoral value of this community, which is preferred by sheep to the detriment of coiron [13, 74]. There are also native herbaceous plants, mainly annuals [27]. This combination produces a plant formation that develops with variable vigor depending on the characteristics of the site [74], especially water availability [2, 58, 74] (Fig. 6).

Three main plant communities can be distinguished in the steppe of the Magallanes Region, according to the edaphic conditions [51, 58], topography [51], drainage, wind exposure [58] and microclimate of the sites [51]. Frequently there is a combination of these communities (Fig. 7, [51, 57]: **(a) Natural grasslands:** represented mainly by three types: (i) coirón grasslands (*Festuca gracillima* and *F. magellanica*) accompanied by other grasses and herbaceous species [51] which have local hygrophytic, mesophytic and xerophytic expressions [58], (ii) mesic grasslands, which are the most humid within this plant community [51] and are composed of meadows and hygrophytic grasslands [58], (iii) salt meadows, with azonal vegetation that develops in inland depressions or marine coasts [29] where evaporation exceeds water flow, accumulating salts [51]. The coironales are found mainly in the great plains of the northern sector of the provinces of Magallanes [58] and Tierra del Fuego [51], while the meadows are generated by specific conditions in depressed sectors of the site [51, 58], which are concentrators of

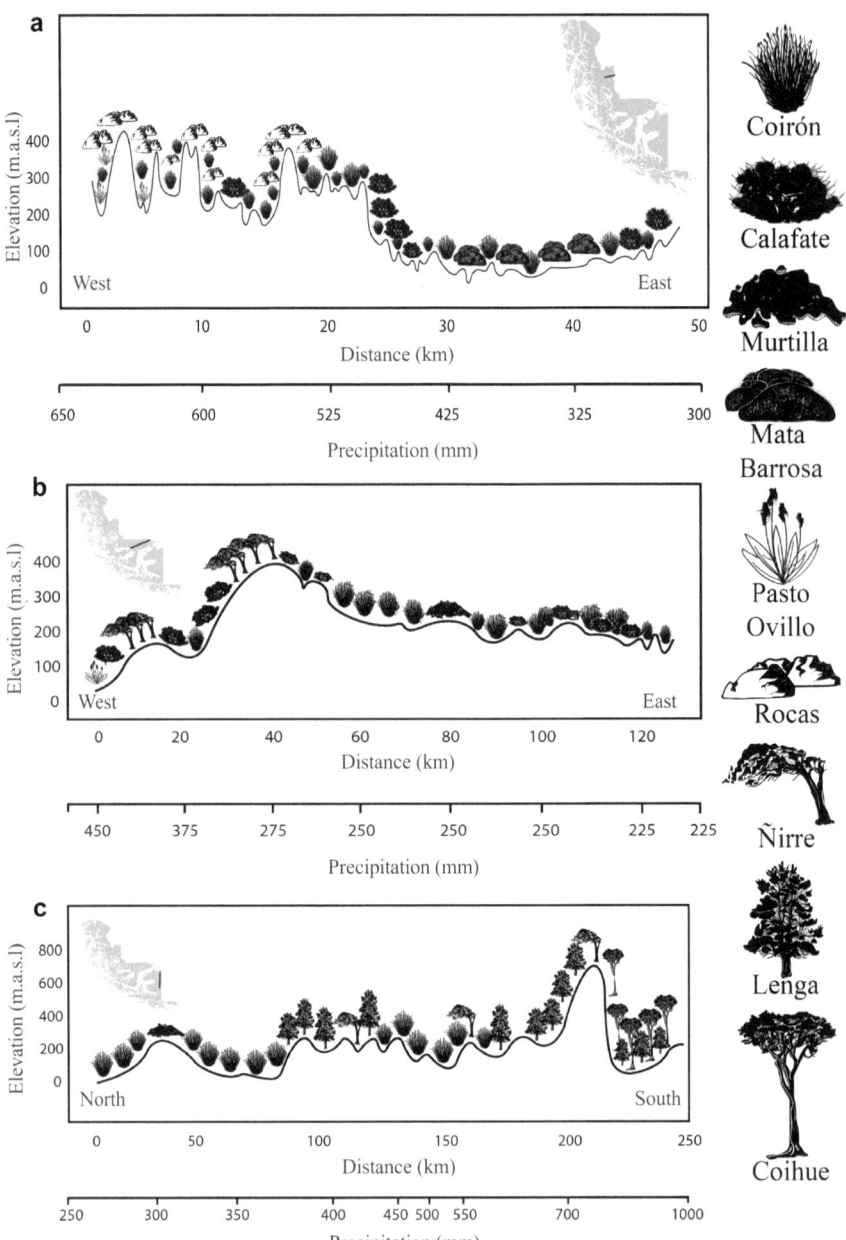

Fig. 7 Vegetation profile of three transects in the provinces of: **a** Última Esperanza, **b** Magallanes, **c** Tierra del Fuego. The elevation, precipitation (mm yr^{-1}), and main components of vegetation within the transect are indicated. The transects are shown with a black line in region maps. *Source*: own elaboration, based on the ASTER GDEM model and [25]. [Coirón: *Festuca gracillima;* Calafate: *Berberis microphylla*; Murtilla: *Empetrum rubrum*; Mata barrosa: *Mulinum spinosus*; Pasto ovillo: *Dactylis glomerata*; Rocas: rocks; Ñirre: *Nothofagus antarctica*; Lenga: *Nothofagus pumilio*; Coihue: *Nothofagus betuloides*]

surface runoff [58]. **(b) Shrubs:** represented mainly by fachine [51, 58] and other species of more restricted distribution such as the green bush, black bush (*Mulguraea tridens*), paramela (*Adesmia boronioides*) [58], mata barrosa (*Mulinum spinosum*) [78, 80] and barberry (*Berberis microphylla*) [51, 56]. Fachine appears in the most humid sectors of the Patagonian steppe [58], with a dispersion in the region above 350 mm yr^{-1}, where the coirón gradually gives way to this shrub community [51] **(c) Subshrubs and/or heaths:** mainly composed of murtilla or diddle-dee (*Empetrum rubrum*) [51, 58], accompanied by other creeping shrub species such as christmas bush *Baccharis magellanica* [51, 78], dwarf barberry (*Berberis empetrifolia*) [56, 78] and species that form cushions, tufted azorella (*Azorella caespitosa*), nardófilo (*Nardophyllum bryoides*) [51, 78], balsam-bog (*Bolax gummifera*) [78, 80] and yaretilla (*Azorella trifurcata*) [56, 78]. Murtilla is generally found on flat, very exposed terraces, with stony, nutrient-poor [5], thin [51], coarse-textured, acidic soils [15].

3.5   Fauna in the Patagonian Steppe

The fauna that currently inhabit the Patagonian steppe come mainly from species that persisted during the cold periods of the Pleistocene (*ca.* 2.6 million years BP) and that managed to survive the harsh climate conditions [35]. Paleontological evidence indicates that animal species that survived in periglacial areas expanded rapidly once favorable warmer environments were established in more recent times [35]. Thus the fauna of the Patagonian steppe in Chile derives from colonization in an east–west direction as the ice retreated.

In contrast to the steppes of the Northern Hemisphere, where there is a great diversity of animal species, the Patagonian steppe has a rather reduced faunal diversity, but with important endemisms [67]. Mammals inhabiting the Patagonian steppe are represented by seven orders, 17 families and at least 14 genera [42]. Few large mammals are found, with the exception of the guanaco (*Lama guanicoe*), a camelid widely distributed in the steppe in both Aysén and Magallanes.

Most are herbivorous mammals, rodents of the subfamily Sigmodontinae such as the Edwards's long-clawed mouse (*Notiomys edwardsii*), and the family Ctenomyidae, such as the Magellanic tuco tuco (*Ctenomys magellanicus*) [42], and in Coyhaique *Ctenomys coyhaiquensis* [76], or of medium size, such as Chinchillidae (e.g. Wolffsohn's viscacha, *Lagidium wolffsohni*, [42]. These last two families have species with fossorial habits,others form colonies [42], all are groups endemic to South America. Two of the small- and medium-sized species belong to the family Dasypodidae or armadillos: the dwarf armadillo (*Zaedyus pichiy*) and the larger hairy armadillo (*Chaetophractus villosus*) [42].

There are two canids among the carnivores of the Chilean Patagonian steppe: the culpeo (*Lycalopex culpaeus*) and grey (*L. griseus*) foxes; the former is larger [42]. There are also three species of felids: the Geoffroy's cat (*Leopardus geoffroyi*), the pampas cat (*L. colocolo*), and the puma (*Puma concolor*),the last is the largest carnivore in the steppe [42]. There are also smaller carnivores of the family Mustelidae, the Patagonian skunk (*Conepatus humboldtii*), the lesser grison (*Galictis cuja*) and the Patagonian ferret (*Lyncodon patagonicus*) [42].

Introduced mammal species are also found, which have a negative impact on the steppe ecosystem due to predation and grazing on native species [40, 50]. One carnivorous mustelid, the American mink (*Neovison vison*), is found in both Aysén and Magallanes [40]. Two rodents, the muskrat (*Ondatra zibethicus*) and the American beaver (*Castor canadensis*), are present only in Magallanes [40]. Two lagomorphs, the rabbit (*Oryctolagus cuniculus*) and the European hare (*Lepus europaeus*) are present in both regions, and two ungulates, the red deer (*Cervus elaphus*) and the wild boar (*Sus scrofa*), still only present in Aysén, although there is a red deer farm on the island of Tierra del Fuego [40]. There are also feral domestic species such as cattle and horses, as well as cats and dogs, which roam near urban areas, degrading the ecosystem and affecting native bird and mammal species [40, 42].

Birds are represented in the Chilean Patagonian steppe by several orders. Among the most visible are those that group together flightless birds, the Rheiformes, represented by the Darwin's rhea (*Rhea pennata*) and others with limited flight, and the Tinamiformes with the elegant crested tinamou (*Eudromia elegans*), adapted to open and arid environments [84]. The large flying carrion-eating Andean condor (*Vultur gryphus*) is a member of the order Cathartiforme, birds of prey include the orders Accipitriforme, with the black-chested eagle (*Geranoaetus melanoleucus*) and the variable hawk (*G. polyosoma*), and the Falconiformes, with the crested caracara (*Caracara plancus*) and the aplomado falcon (*Falco femoralis*) [84]. Water birds of the order Anseriformes include the black-necked swan (*Cygnus melan-coryphus*), the coscoroba swan (*Coscoroba coscoroba*), the upland goose (*Chloephaga picta*), the red shoveler (*Spatula platalea*) and the spectacled ducks (*Speculanas specularis*), and the wigeon (*Lophonetta specularioi des*),the order Phoenicopteriforme is represented by the Chilean flamingo (*Phoenicopterus chilensis*). These use permanent bodies of water or small temporary lagoons in the steppe, which are highly productive in plant matter [84]. Several smaller birds belonging to the order Passeriformes and seabirds of the order Charadriiformes use the steppe as a nesting and feeding ground [84].

The herpetofauna is mainly represented by lizards of the genus *Liolaemus*, with the Magellan's tree iguana (*L. magellanicus*; [8]), a and few amphibians [17]. The latter are the Patagonian toad (*Nannophryne variegata*), the portezuelofrog (*Atelognathus salai*) and the large four-eyed toad (*Pleurodema bufonina*), which tolerate the limited humidity conditions of the steppe [17].

3.6   Ecosystem Services of the Chilean Patagonian Steppe

Ecosystem services are biophysical processes that generate resources, functions, or goods that are useful for human well-being [30, 39]. Ecosystem services acquire different values depending on the type of stakeholder and the rate of ecosystem provision [69]. Some human groups focus their activity on livestock production, others on water quality and access for people and animals [69], nutrient cycling, cultural services and climate regulation, or pollination functions [30]. There has also been increasing interest in protected areas that ensure the viability of biodiversity and its ecological processes and benefits. However, the demand for ecosystem services depends on the level of education, income, values, culture and geographic location of the human groups that use them [69].

Steppe ecosystem services support over one billion people globally [30]. If we consider the benefits provided by steppe ecosystems for diverse human groups [69], including native peoples, the value of the Patagonian steppe in Chile is clearly underestimated and focused in relation to one of the main economic activities, extensive sheep and cattle ranching [49]. This steppe cattle ranching began in Magallanes over 130 years ago [49] and about 100 years ago in Aysén [70], and continues to this day. An important ecosystem service of the steppe is the capacity of soil organic matter to sequester significant volumes of carbon [43]. The movement of carbon from the atmosphere to the soil and vegetation, where it is stored, is called carbon sequestration, and is an important ecosystem contribution to global climate change mitigation through fixation via photosynthesis [10, 50]. The type of grazing management of steppe grasslands directly influences carbon accumulation. Intensive grazing (Fig. 8) and grazing exclusion are the management types that reduce the amount of carbon sequestered and lead to lower species richness in plant biomass, compared to moderate grazing [69]. These results could vary in the steppe, depending on soil type, vegetation and climate [10].

The type of management to which the steppe is subjected, especially when it involves production, can affect the supply of other ecosystem services such as the

Fig. 8 Example of forage and biomass production depending on the type of grazing. **a** area with intensive grazing; **b** area with moderate grazing

Fig. 9 Progression of resource management in natural ecosystems since the first human settlements. *Source* modified from Briske [9]

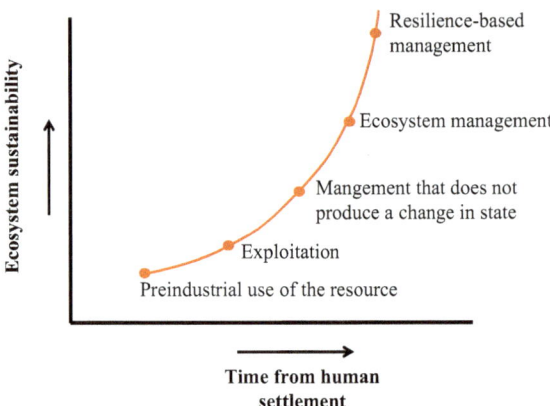

quality and quantity of available water, the conservation of biodiversity of plant and animal species, and carbon sequestration [69]. Therefore, it is important to promote policies that provide incentives and benefits that support conservation and recognize the value of the multiple ecosystem services provided by steppes, and thus ensure management that facilitates their resilience and sustainability [30]. Resilience-based management generates the greatest ecosystem sustainability (Fig. 9), as it recognizes the inevitability of change and seeks to channel it to maximize multiple ecosystem services [9]. This type of management implies that the natural resource used is developed in such a way as to allow its recovery to the initial state prior to intervention. An example of this type of management is grazing with stocking rate adjusted to the production of the site, which allows the steppe vegetation to recover within its annual cycle.

3.7 Conservation and Threats

There are several threats to the conservation of the Patagonian steppe; the most relevant ones related to anthropogenic impacts are mining [38] and the oil industry [29], which generate significant negative effects [29, 38]. The area of hydrocarbon exploitation in the Magallanes Region coincides with the steppe ecosystem [37]. Overgrazing of grasslands by livestock leads to aridization [26] and allows the invasion of exotic species of plants and animals [7, 40]. Tourism [37, 38], despite being an activity with sustainable development purposes, has generated changes that increase pressures on the territory, with a tendency towards intensive exploitation that could be detrimental to steppe systems [37]. Added to these threats are the fires that are frequent in this ecosystem [55, 81, 82] and the change in climate conditions due to global change, which will produce warmer winters, with precipitation mostly in the form of rain instead of snow [50, 85].

Because it is considered an ecosystem of agricultural importance [58], the representation of the steppe in the National System of Protected Areas (in Spanish

SNASPE) is low in the regions of Aysén and Magallanes [59, 60]. It is worth mentioning that these two regions have 69% of the total area of SNASPE in Chile [16]. There are few state protected areas that have total or partial coverage of steppe surface within their boundaries. The Patagonia National Park is located in the Aysén Region (421 km^2), Torres del Paine National Park (401.9 km^2) and Pali Aike National Park (44.2 km^2) are in the Magallanes Region. Only Pali Aike National Park has exclusive protection of steppe environments, since in the other areas only small portions of isolated steppes are protected within a matrix mainly composed of deciduous forests in mountainous areas. The total area of Patagonian steppe in southern Chile is approximately 26,996.9 km^2,the percentage of steppe under some form of state protection is only 3.2% of the total [60]. A similar situation occurs in the Argentine steppe, where only 0.6% of its area is protected [29].

Fires are one of the most aggressive and sudden disturbances that can affect native ecosystems, strongly depleting and modifying their cover, structure and composition [55], as well as reducing carbon and nutrient storage. The more severe the calcination of biomass and organic soil, the more severe the impact on the ecosystem's resilience, i.e. its ability to return to the state prior to the disturbance. Fires in the Patagonian steppe have had different degrees of severity, causing a decrease in the cover of all functional groups of vegetation and an increase in exotic species in sectors where the fire reached greater severity [28]. However, fire regimes have changed in response to climate variability, vegetation type and changes in land use,therefore, they are expected to continue to change during the twenty-first century, increasing fire frequency as a consequence of climate change [50] and increasing anthropogenic impact [52].

In the ecotone, i.e. the transition zone of the Patagonian steppe with the deciduous forests of lenga (*Nothofagus pumilio*) and ñirre (*N. antarctica*) present in Aysen Region in just a few decades the development of livestock pastures has produced a mosaic-type anthropized landscape, in which large tracts of semi-natural grasslands are found with scattered remnant fragments of native forests that still persist [32, 70]. It is here that intentional fires have been an important disturbance, which together with extensive livestock activity, explain the biodiversity and ecological functioning of these ecosystems [71].

An example of the effect of fires can be seen in Torres del Paine National Park, where they have affected nearly 47,000 hectares over the last 30 years. These events have damaged pre-Andean scrubland, steppe, and forest ecosystems [81], This park has been affected by 57 fires of varying sizes since 1980, which have impacted its ecosystems [55]. The Olguín fire (in 2011) was the most extensive and devastating, burning about 7% of the park's surface, of which 59.7% was Patagonian steppe, 28.6% scrub or shrub steppe, 9.7% native forest, and 1.9% other vegetation [55].

Another alteration of the original landscape of the Patagonian steppe is agricultural and livestock production, which has altered the floristic composition and original plant productivity of its communities [58]. In the Magellanic steppes sheep grazing and high stocking rates without monitoring have transformed many grass communities into degraded grasslands [15]. Pasture management by intensive grazing or the exclusion of grazing generate the greatest decrease in species richness [69]. Maintaining soil cover is a vitally important issue in steppe grasslands in order to avoid soil erosion, grassland degradation, loss of primary and secondary productivity, and biodiversity [13].

The invasion of exotic plant species is another frequent threat in Chilean Patagonia [7]. One of the plant species of exotic origin that has had the greatest impact on the steppe ecosystem of the Magallanes Region is mouse-ear hawkweed (*Hieracium pilosella,* [63]). The mouse-ear hawkweed can colonize bare soil in degraded grasslands,once its established, the species spreads more by stolons than seed production, which is an effective strategy to occupy space before resident plant species [63]. Between 2004–2018 this species invaded 52% of the area of a property [65] at a rate of 63 ± 15 ha yr^{-1}.

4   Recommendations

The following are recommendations for biodiversity conservation, ecosystem processes, and sustainable management of Chilean Patagonian grasslands:

- Given the heterogeneity of the steppe system and the various intensities of management and impacts, development of a monitoring plan using satellite images and remote sensing techniques is advisable in order to determine quantitatively the spatio-temporal trends in the different types of steppe that occur in the environments of Chilean Patagonia (Fig. 2). For example, coironales dominated by perennial grasses (*Festuca* spp.), areas with shrub cover and more humid areas such as meadows. In addition, a monitoring plan of water availability, temperature changes, and soil fertility, the main site conditions that regulate the dominance of one plant community over another, is recommended in order to prevent or mitigate degradation processes in steppe areas subject to intensive livestock grazing or affected by fires.
- Degraded pastures can be improved through the application of amendments that add deficient nutrients, such as nitrogen, phosphorus, and sulphur. In addition, in the specific case of Azonal soils such as the meadows (in spanish vegas) if the conditions of greater water availability and cover are maintained, these would be areas of greater resilience to climate change. Thus it is recommended that their management be differentiated from the rest of the steppe plant communities, so that their characteristics are maintained over time.
- It is necessary to implement a comprehensive management plan that considers all possible threats to Patagonian steppe biodiversity and that takes into account the possible adverse effects of localized mineral and hydrocarbon extraction that degrade and erode these environments. Other threats to the steppe ecosystem that must be prevented or mitigated include the introduction of exotic animal and plant species that cause damage, which have been poorly assessed (mainly mouse-ear hawkweed, beaver, mink, and lagomorph species), the serious impact

of fires, especially in areas of climatic transition with native forest, unregulated tourism, and agricultural activity that does not consider sustainable planning in terms of animal load and impacts. The conservation of the Chilean Patagonian steppe depends both on the protection of the ecosystem in state parks or reserves, which cover only a small portion of the steppe [60], and on the implementation of management plans and management of disturbances such as fires and exotic species in the extensive unprotected area.

- The possibility of increasing the value of conserved steppe through the market using carbon credits captured in the Patagonian soil and vegetation should be studied. In Chilean Patagonia this is an undervalued ecosystem service, which is applied in other similar grasslands around the world, and is a consensus tool that could contribute to the conservation of the steppes and their biodiversity. To this end, it is necessary to prioritize forms of management that maximize carbon sequestration in the soil and vegetation (see Fig. 8).

- There is a need to improve scientific knowledge on the management and impact of the abundant exotic species of animals and plants introduced in the Chilean Patagonian steppe. This objective requires integrated management with Argentine authorities in the border areas, since most of these invasive species are found in both countries and have migrated spontaneously from one country to the other. It is necessary to develop concrete plans to reduce the environmental and productive damage caused by these species. It is urgent to prevent the introduction of new alien species for any purpose, since due to limited knowledge, the Patagonian steppe for both countries has been used as an experimental ground because of its scarce population and the undervaluation of its ecosystem services. Accordingly, greater valuation and knowledge of the steppe ecosystem and its resources is essential to establish control or mitigation measures prior to the introduction of a species for economic purposes, especially in areas subject to use by the local community. Management based on ecosystem resilience should consider the ecological-social component in an integral manner, with the participation of the main regional stakeholders in the design and evaluation of management actions, which could reduce the degradation of steppes and enable their sustainable use.

Acknowledgements The authors thank Lorena Bahamonde, Mirna Navarro, and Cinthya Glucevic for their support during the development of the chapter. Paulo Corti and Sergio Radic thank the FONDECYT 1171039 project.

References

1. Bailey, R. (1980). *Description of the ecoregions of the United States.* Miscellaneous publication, N° 1391. U. S. Department of Agriculture.
2. Bidwell, R. (1993). Physiology and distribution of plant communities. In R. Bidwell (Ed.), *Plant physiology* (pp. 687–722). A.G.T. Editor, S.A.
3. Blasco, M. (2012). Climate and water deficit in the steppe ecosystem of Lécera, Spain. *Journal of Agricultural Sciences, 29,* 5–15.

4. Boonman, J., & Mikhalev, S. (2005).The Russian steppe. In J. M., Suttie, S. G. Reynolds, & Batello, C. (Eds.). *Grasslands of the world* (pp. 381–416). Food and Agriculture Organization of the United Nations.

5. Borrelli, P., & Oliva, G. (2001). Effect of animals on rangelands. In Borrelli, P., Oliva, G. (Eds.), *Chapter 4, Sustainable sheep farming in southern Patagonia. Santa Cruz Agricultural Experimental Station* (pp. 99–128). National Institute of Agricultural Technology (INTA).

6. Brandt, C., & Rickard, W. (1994). *Alien taxa in the North American shrub-steppe four decades after cessation of livestock grazing and cultivation agriculture.* Environmental Sciences Department, Pacific Northwest Laboratory.

7. Bravo-Monasterio, P., Pauchard, A., & Fajardo, A. (2016). *Pinus contorta* invasion into treeless steppe reduces species richness and alters species traits of the local community. *Biological Invasions, 18*, 1883–1894.

8. Breitman, M., Avila, L., Sites, J. N., & Morando, M. (2011). Lizards from the end of the world: Phylogenetic relationships of the *Liolaemus lineomaculatus* section (Squamata: Iguania: Liolaemini). *Molecular Phylogenetics and Evolution, 59*, 364–376.

9. Briske, G. (2017). Rangeland systems: Foundation for a conceptual framework. In Briske, D. (Ed.), *Rangeland systems: Processes, management and challenges* (pp. 1–21). Springer Series on Environmental Management.

10. Brown, J. (2008). Carbon sequestration and sink resources in grazed lands. In O'Rourke, J. (Ed.), *People and policy in rangeland management* (pp. 240–247). International Rangeland Congress.

11. Butorovic, N. (2019). *Comportamiento de las variables precipitación y temperatura del aire en la ciudad de Punta Arenas durante el período Enero-Julio 2019.* Report requested by Empresa Pecket-Energy.

12. Chendev, Y., Sauer, T., Hernández, G., & Lee Burras, C. (2015). History of east European Chernozem soil degradation; protection and restoration by tree windbreaks in the Russian steppe. *Sustainability, 7*, 705–724.

13. Cipriotti, P., Collantes, M., Escartín, C., Cabeza, S., Rauber, R., & Braun, K. (2014). Experiencias de largo plazo para el manejo de una hierba invasora de pastizales: El caso de *Hieracium pilosella* L. en la estepa fueguina. *Ecología Austral, Special Section, 24*, 135–144.

14. Cirera, J., & García, B. (2006). *Estepas ibéricas el paisaje olvidado.* SEO/BirdLifeand Fundació Territori i Paisatge.

15. Collantes, M., Anchorena, J., & Cingolani, A. (1999). The steppes of Tierra del Fuego: Floristic and growth form patterns controlled by soil fertility and moisture. *Plant Ecology, 140*, 61–75.

16. Corporació Nacional Forestal (CONAF). (2019). Parques nacionales. http://www.conaf.cl/parques-nacionales/parques-de-chile/. Accessed 28 Apr 2020

17. Correa, C., Cisternas, J., & Correa-Solís, M. (2011). Lista comentada de las especies de anfibios de Chile (Amphibia: Anura). *Boletín De Biodiversidad De Chile, 6*, 1–21.

18. Covacevich, N., & Ruz, E. (1996). Praderas en la zona austral: XII Región (Magallanes). In I. Ruiz (Ed.), *Praderas para Chile* (pp. 639–655). Instituto de Investigación Agropecuaria; Ministerio de Agricultura.

19. Cruz, G., & Lara, A. (1987). *Regiones naturales del área de uso agropecuario de la XII Región, Magallanes y de la Antártica Chilena.* INIA Kampenaike-Intendencia de la XII Región.

20. Davis, S., Heywood, V., & Hamilton, A. (1997). *Centres of plant diversity: a guide and strategy for their* conservation (Vol. 3). IUCN Publications Unit.

21. Derner, J., Hunt, L., Filho, K., Ritten, J., Capper, J., & Han, G. (2017). Livestock production systems. In Briske, D. (Ed.). *Rangeland systems: Processes, management and challenges* (pp. 347–372). Springer Series on Environmental Management.

22. Di Castri, F., & Hajek, E. (1976). *Bioclimatología de Chile.* Vicerrectoría Académica, Universidad Católica de Chile.

23. Driessen, P., & Deckers, J. (2001). *Lecture notes on the major soils of the world.* Food and Agriculture Organization of the United Nations (FAO).

24. Endlicher, W., & Santana, A. (1988). El clima del sur de la Patagonia y sus aspectos ecológicos. Un siglo de mediciones climatológicas en Punta Arenas. *Anales Instituto Patagonia. Serie Ciencias Naturales (chile), 18*, 57–86.
25. Fick, S., & Hijmans, R. (2017). WorldClim 2: New 1 km spatial resolution climate surfaces for global land areas. *International Journal of Climatology, 37*(12), 4302–4315.
26. Gaitán, J., Bran, D., Oliva, G., Aguiar, M., Buono, G., Ferrante, D., Nakamatsu, V., Ciari, G., Salomone, J., Massara, V., Martínez, G., & Maestre, F. (2018). Aridity and overgrazing have convergent effects on ecosystem structure and functioning in Patagonian rangelands. *Land Degradation and Development, 29*, 210–218.
27. Gajardo, R. (1994). *La vegetación natural de Chile: Clasificación y distribución geográfica.* Editorial Universitaria.
28. Ghermandi, L., González, S., Lescano, M., & Oddi, F. (2013). Effects of fire severity on early recovery of Patagonian steppes. *International Journal of Wildland Fire, 22*(8), 1055–1062.
29. Green, L., & Ferreyra, M. (2011). *Flores de la estepa Patagónica, guía para el reconocimiento de las principales especies de plantas vasculares de la estepa.* Vásquez Mazzini Editores.
30. Havstad, K. (2008). Ecosystem functions of grazing lands. People and policy in rangeland management. In *XXI international rangeland congress* (pp. 216–221). IGC.
31. Hepp, C., & Stolpe, N. (2014). *Characterization and properties of soils in western Patagonia (Aysén).* Instituto de Investigaciones. Agriculture and Livestock, INIA Research Center.
32. Hepp, C. (1988). Praderas en la zona austral XI Región (Aysén). In Ruiz Núñez I. (Ed.), *Praderas para Chile* (pp. 624–638). Instituto de Investigaciones Agropecuarias.
33. Hironaka, M., Fosberg, M., & Winward, A. (1983). *Sagebrush-grass habitats of southern Idaho.* Forest, Wildlife and Range Experiment Station, University of Idaho, Moscow, United States. Bulletin No. 35.
34. Hewitt, G. (2000). The genetic legacy of the Quaternary ice ages. *Nature, 405*, 907–913.
35. Hoffman, A., Perrettaa, H., Lemoinea, N., & Smith, M. (2019). Blue grama grass genotype affects palatability and preference by semi-arid steppe grasshoppers. *Acta Oecologica, 96*, 43–48.
36. Hopkins, D. M. (1982). Aspects of the paleogeography of Beringia during the late Pleistocene. In D. M. Hopkins, J. V. Mathews, C. E. Schwaeger, & S. B. Young (Eds.), *Paleoecology of Beringia* (pp. 3–28). Academic Press.
37. Inostroza, L. (2012). Patagonia, antropización de un territorio natural. *Cuadernos De Investigación Urbanística, 83*, 86.
38. Inostroza, L., Zasada, I., & König, H. (2016). Last of the wild revisited: Assessing spatial patterns of human impact on landscapes in southern Patagonia, Chile. *Regional Environmental Change, 16*, 2071–2085.
39. Intergovernmental Science-Policy Platform on Biodiversity and Ecosystem Services (IPBES) (2019). *Summary for policymakers of the global assessment report on biodiversity and ecosystem services of the Intergovernmental Science-Policy Platform on Biodiversity and Ecosystem Services.* In S. Díaz, J. Settele, E. S. Brondízio, H. T. Ngo, M. Guèze, J. Agard, A. Arneth, P. Balvanera, Brauman, K. A., Butchart, S. H. M. M., Chan, K. M. A., Garibaldi, L. A., Ichii, K., Liu, J., Subramanian, S. M., Midgley, G. F., Miloslavich, P., Molnár, Z., Obura, D., Pfaff, A., Polasky, S., Purvis, A., Razzaque, J., Reyers, B., Roy Chowdhury, R., Shin, Y. J., Visseren-Hamakers, I. J., Willis, K. J., & Zayas, C. N. (Eds.). IPBES Secretariat.
40. Iriarte, J., Lobos, G., & Jaksic, F. (2005). Invasive vertebrate species in Chile and their control and monitoring by governmental agencies. *Revista Chilena De Historia Natural, 78*, 143–154.
41. IUSS Working Group (2006). *World reference base for soil resources. A framework for international classification correlation and communication.* World Soil Resources Reports No. 103. FAO. Retrieved from http://www.fao.org/3/a-a0510e.pdf
42. Johnson, W., Franklin, W., & Iriarte, J. (1990). The mammalian fauna of the northern Chilean Patagonia: A biogeographical dilemma. *Mammalia, 54*, 457–470.
43. Kenne, G., & Kloot, R. (2019). The carbon sequestration potential of regenerative farming practices in South Carolina, USA. *American Journal of Climate Change, 8*, 157–172.

44. Kottek, M., Grieser, J., Beck, C., Rudolf, B., & Rubel, F. (2006). World map of the Köppen-Geiger climate classification updated. *Meteorologische Zeitschrift, 15*, 259–263.
45. Kudrevatykh, I., Kalinin, I., & Alekseev, O. (2019). Biogenic accumulation of chemical elements by plants of genus *Poaceae* Barnhart and Genus *Artemisia* L. in the dry steppe and semidesert zones of the south of the Russian plain. *Contemporary Problems of Ecology, 12*, 377–385.
46. Luebert, F., & Pliscoff, P. (2018). *Sinopsis bioclimática y vegetacional de Chile* (2nd ed.). Editorial Universitaria.
47. Luzio, W., Casanova, M., & Vera, W. (2006). Génesis de suelos. In W. Luzio, & M. Casanova (Eds.), *Avances en el conocimiento de los suelos de Chile* (pp. 21–41). Universidad de Chile and Servicio Agrícola y Ganadero (SAG).
48. Maher, B., Alekseev, A., & Alekseeva, T. (2003). Magnetic mineralogy of soils across the Russian steppe: Climatic dependence of pedogenic magnetite formation. *Palaeogeography, Palaeoclimatology, Palaeoecology, 201*, 321–341.
49. Martinic, M. (1980). Ocupación del ecúmene en Magallanes, 1843–1930. La colonización en áreas marginales. *Anales Del Instituto De La Patagonia, 7*, 7–46.
50. Marquet, P. A., Buschmann, A. H., Corcoran, D., Díaz, P. A., Fuentes-Castillo, T., Garreaud, R., Pliscoff, P., & Salazar, A. (2023). *Global change and acceleration of anthropic pressures on Patagonian ecosystems.* Springer.
51. Moore, D. (1983). *Flora of Tierra del Fuego.* Livesey Limited.
52. Moreno, P., Vilanova, I., Villa-Martínez, R., & Francois, J. (2018). Modulation of fire regimes by vegetation and site type in southwestern Patagonia since 13 ka. *Frontiers in Ecology and Evolution, 6*(34), 1–10.
53. Ollero, H., & van Staalduinen, M. (2012). Iberian steppes. Department of Ecology, Universidad Autónoma de Madrid, Cantoblanco. In M. Werger, M. van Staalduinen (Eds.), *Eurasian steppes. Ecological problems and livelihoods in a changing world* (pp. 273–288). Springer.
54. Olson, D. M., & Dinerstein, E. (2002). The global 200: Priority ecoregions for global conservation. *Annals of the Missouri Botanical Garden, 89*, 199–224.
55. Peña, A., & Ulloa, J. (2017). Mapping burned vegetation recovery using pre- and post-fire spectral index classification. *Journal of Remote Sensing, 50*, 37–48.
56. Pisano, E. (1977). Fitogeografía de Fuego-Patagonia chilena I. Comunidades vegetales entre las latitudes 52 y 56° S. *Anales Del Instituto De La Patagonia, 8*, 121–248.
57. Pisano, E. (1981). Bosquejo fitogeográfico de Fuego-Patagonia. *Anales Del Instituto De La Patagonia, 12*, 159–171.
58. Pisano, E. (1985). La estepa patagónica como recurso pastoril en Aysén y Magallanes. Sección Botánica. Anales Instituto Patagonia. *Ambiente y Desarrollo, 1*(2), 45–59.
59. Pliscoff, P., & Fuentes-Castillo, T. (2008). *Análisis de representatividad ecosistémica de las áreas protegidas públicas y privadas en Chile: creación de un sistema nacional integral de áreas protegidas para Chile.* Final report. MMA/GEF-UNDP.
60. Pliscoff, P., Martínez-Harms, M. J., & Fuentes-Castillo, T. (2023). *Representativeness assessment and identification of priorities for the protection of terrestrial ecosystems in Chilean Patagonia.* Springer.
61. Radic, S., Opazo, S., Mihovilovic, E., & Olave, C. (2011). Determinación de evapotranspiración potencial con los métodos de Penman-Monteith, Ivanov y Turc. In *Proceedings annual congress of the Chilean society of animal production (SOCHIPA). XXXVI annual meeting* (pp 141–142). INIA-UMAG.
62. Radic, S., Fernández, A., Opazo, S., McAdam, J. H., & Ivelic, J. (2013). Soil bulk density from grasslands in the Magallanes region, Chile. In *10th international conference of agrophysics* (p. 105). ICA.
63. Radic, S., Ivelic, J., Gross, P., Ruiz, R., & Muñoz, R. (2020). *Guía para el control de Hieracium pilosella para la Región de Magallanes y Antártica Chilena.* Mejoramiento competitivo de la cadena de valor de la lana y la carne ovina en la Región de Magallanes y Antártica Chilena. Programa Territorial Integrado (PTI).

64. Rey, A., Pegoraro, E., Oyonarte, C., Were, A., Escribano, P., & Raimundo, J. (2011). Impact of land degradation on soil respiration in a steppe (*Stipa tenacissima L.*) semi-arid ecosystem in the SE of Spain. *Soil Biology & Biochemistry, 43*, 393–403.
65. Reyes, D., Muñoz, R., & Radic, S. (2017). Mapping of the affected surface by the presence of *Hieracium pilosella* in a livestock farm of the Region de Magallanes y Antárctica Chilena, using Sentinel-2 data. In 2017 first IEEE international symposium of geoscience and remote sensing (GRSS-CHILE) (pp. 1–4). IEEE.
66. Rodríguez, J. (1991). *Manual de fertilización*. Ediciones Universidad Católica, Facultad de Agronomía.
67. Roig-Juñent, S., Griotti, M., Domínguez, M., Agrain, F., Campos-Soldini, P., Carrara, R., Cheli, G., Fernández-Campón, F., Flores, G., Katinas, L., Muzón, J., Neita-Moreno, J., Pessacq, P., San Blas, G., Scheibler, E., & Crisci, J. (2018). The Patagonian steppe biogeographic province: Andean region or South American transition zone? *Zoologica Scripta*. 1–7
68. Sáez, C. (1994). *Caracterización de la fertilidad de los suelos de la Región de Magallanes. Características fisicoquímicas y nutricionales de los suelos*. Informe Final: Proyecto Fundación Fondo Investigaciones Agropecuarias, Universidad de Magallanes. Escuela de Ciencias y Tecnologías en Recursos Agrícolas y Acuícolas.
69. Sala, O., Yahdjian, L., Havstad, K., Aguiar, M. (2017). Rangeland ecosystem services: nature's supply and humans' demand. In Briske, D. (Ed.), *Rangeland systems: processes, management and challenges* (pp. 467–490). Springer Series on Environmental Management.
70. Sánchez-Jardón, L., Acosta, B., del Pozo, A., Casado, M., Ovalle, C., Elizalde, H., Hepp, C., & de Miguel, J. (2010). Grassland productivity and diversity on a tree cover gradient in *Nothofagus pumilio* in NW Patagonia. *Agriculture, Ecosystems & Environment, 137*, 213–218.
71. Sánchez-Jardón, L., Acosta, B., del Pozo, A., Casado, M. A., Ovalle, C., & de Miguel, J. (2014). Variability of herbaceous productivity along *Nothofagus pumilio* forest-open grassland boundaries in northern Chilean Patagonia. *Agroforestry Systems, 88*, 397–411.
72. Santana, A., Olave, C., & Butorovic, N. (2010). Estudio climatológico con registros de alta resolución temporal en campamento Posesión (ENAP). Magallanes, Chile. *Anales Del Instituto De La Patagonia, 38*, 5–34.
73. Regional Secretariat of Planning and Coordination (SERPLAC) (2005). *Atlas Región de Aysén*. Ministerio de Desarrollo Social, Gobierno de Chile. LOM Ediciones Ltda. Retrieved from http://www.desarrollosocialyfamilia.gob.cl/btca/txtcompleto/mideplan/planreg.ordenam.territorial-aysen.pdf
74. Servicio Agrícola y Ganadero (SAG) (2003). *El pastizal de Tierra del Fuego*. Punta Arenas, Región de Magallanes y Antártica Chilena, Chile: Government of Chile.
75. Torn, M., Lapenis, A., Timofeev, A., Fischer, M., Babikov, B., & Harden, J. (2002). Organic carbon and carbon isotopes in modern and 100-year-old-soil archives of Russian steppe. *Global Change Biology, 8*, 941–953.
76. Torres-Mura, J. C., & Rojas, V. G. (2004). *Reserva Nacional Lago Jeinimeni: historia natural*. Proyecto Biodiversidad de Aysén—Conaf XI Región.
77. United State Department of Agriculture (1999). *Soil taxonomy a basic system of soil classification for making and interpreting soils surveys*. United States Department of Agriculture. Natural Resources Conservation Service. Retrieved from https://www.nrcs.usda.gov/Internet/FSE_DOCUMENTS/nrcs142p2_051232.pdf
78. Uribe, I. (2004). *Manual de terreno para la identificación de especies en pastizales de la XII Región*. Servicio Agrícola y Ganadero.
79. Valle, S., Radic, S., & Casanova, M. (2015). Soils associated with three important grazing plant communities in southern Patagonia. *Revista Agro Sur, 43*(2), 89–99.
80. Vidal, O. (2006). *Flora Torres del Paine: Guía de campo*. Edición Fantástico Sur.
81. Vidal, O. (2012). Torres del Paine, ecoturismo e incendios forestales: Perspectivas de investigación y manejo para una biodiversidad erosionada. *Revista Bosque Nativo Revista Bosque Nativo, 50*, 33–39.
82. Vidal, O., Aguayo, M., Niculcar, R., Bahamonde, N., Radic, S., San Martín, C., Kusch, A., Latorre, J., & Félez, J. (2015). Plantas invasoras en el Parque Nacional Torres del Paine

(Magallanes, Chile): Estado del arte, distribución post-fuego e implicancias en restauración ecológica. *Anales Del Instituto De La Patagonia, 43*(1), 75–96.

83. Villa, M., Opazo, S., Moraga, C., Muñoz-Arriagada, R., & Radic, S. (2020). Patterns of vegetation and climatic conditions derived from satellite images relevant for sub-Antarctic rangeland management. *Rangeland Ecology & Management, 73*, 552–559.

84. Vuilleumier, F. (1998). François avian biodiversity in forest and steppe communities of Chilean Fuego-Patagonia. *Anales Del Instituto De La Patagonia, 26*, 41–57.

85. Wesche, K., Ambarli, D., Kamp, J., Török, P., Treiber, J., & Dengler, J. (2016). The Palaearctic steppe biome: A new synthesis. *Biodiversity and Conservation, 25*, 2197–2231.

86. Williams, M., Paige, G., Thurow, T., Hild, A., & Gerow, K. (2011). Songbird relationships to shrub-steppe ecological site characteristics. *Rangeland Ecology and Management, 64*, 109–118.

Part IV

Marine Ecosystems

Coastal-Marine Protection in Chilean Patagonia: Historical Progress, Current Situation, and Challenges

8

David Tecklin, Aldo Farías, María Paz Peña, Xiomara Gélvez, Juan Carlos Castilla, Maximiano Sepúlveda, Francisco A. Viddi, and Rodrigo Hucke-Gaete

Abstract

Chilean Patagonia offers a unique opportunity at both the national and international levels to establish an integrated system of coastal-marine protection of enormous value for biodiversity and society. This chapter describes the creation, current status, and principal geographic characteristics of the different forms of coastal-marine protection in the region in order to provide an overview of progress and challenges. Current coverage of marine protected areas, which have been the focus of most work to date, is limited to 6% (11,218 km^2) of Patagonia's coastal-marine zone. However, the interior waters within national parks and national reserves that make up the National Protected Area System cover an additional 35% of the coastal zone (63,933 km^2) and represent 85% of the legally protected marine area. In addition, requests by Indigenous communities to establish Indigenous People's Coastal Marine Spaces (in Spanish Espacios Costeros Marinos de Pueblos Originarios, ECMPO) now total 62.931 km^2 across 65 different areas and present an important potential complementary

D. Tecklin (✉) · A. Farías · M. P. Peña · X. Gélvez
Austral Patagonia Program, Faculty of Economics and Administrative Sciences, Universidad Austral de Chile, Valdivia, Chile
e-mail: david.tecklin@gmail.com

J. C. Castilla
Department of Ecology, Faculty of Biological Sciences and Interdisciplinary Center for Global Change. Pontificia Universidad Católica de Chile, Santiago, Chile

M. Sepúlveda
The Pew Charitable Trusts, Av. Andrés Bello, 2233 Santiago, Chile

F. A. Viddi
Blue Whale Center, Universidad Austral de Chile, Valdivia, Chile

R. Hucke-Gaete
Institute of Marine and Limnological Sciences, Universidad Austral de Chile, Valdivia, Chile

© Pontificia Universidad Católica de Chile 2023
J. C. Castilla et al. (eds.), *Conservation in Chilean Patagonia*, Integrated Science 19,
https://doi.org/10.1007/978-3-031-39408-9_8

conservation tool. This study thus suggests the need to expand our understanding of marine biodiversity conservation in Patagonia with a recognition of all forms of marine protection as well as complementary areas such as ECMPOs. Finally, we provide recommendations for priority strategies to consolidate a large-scale integrated coastal-marine conservation system for Chilean Patagonia. These include strengthening the effective management of the marine portion of national parks and reserves, developing a protocol for the recognition of ECMPOs as marine protected areas when requested by their proponents, the creation of public and public–private funding mechanisms, technical assistance for all forms of protection, and the importance of integrated sea-land planning and management.

Keywords

Patagonia • Chile • Marine protected areas • Interior waters of national parks • Indigenous coastal-marine spaces

1   Introduction

The geography of the Chilean Patagonian coastline presents an unusual situation of marine conservation and sea–land interface, with a much greater conservation potential than has been recognized and achieved to date. The Patagonian continental coast of Chile extends for approximately 1,600 linear km, with a succession of fjords, channels and archipelagos between Reloncaví Sound and the Diego Ramírez Islands (41° 42′ S 73° 02′ W 56° 29′ S 68° 44′ W). The Patagonian inland sea (maritory), delimited by these archipelago systems, is exposed to constant and diverse flows of water, nutrients, and energy between terrestrial and marine systems, which produce a great heterogeneity of coastal-marine environments and an important associated marine biodiversity [19, 23, 37].[1] At the same time, this geographical configuration of the coastal zone gives rise to a unique institutional, sociocultural, and economic situation in Chile. South of the Reloncaví Sound, most of the fjords along the continental coastline are adjacent to public lands (*tierras fiscales*) or to national parks and reserves that are part of the National Protected Areas System (in Spanish SNASPE). The large Patagonian archipelagos also largely fall within the SNASPE. Despite their remoteness and the sparse coastal population south of Chiloé Island (*ca.* 42° S), these archipelagos represent ancestral maritories of Indigenous peoples that have been inhabited and used by humans for thousands of years [4, 35]. Currently, these coastal-marine environments, particularly in northern Patagonia, are subject to multiple uses and pressures from fishing, industrial aquaculture (salmon and mussels), and tourism [5, 17, 27]. In

---

[1] Without detracting from the interesting debate and information regarding the concept of maritorio or maritory in Chile (see School of Architecture [1, 9, 14]), we use the term broadly to describe coastal-marine spaces in the interior of Chilean Patagonia in both their physical and institutional dimensions.

this context, public policies and explicit marine conservation efforts are relatively recent.

A broad and holistic vision of the scope of marine protection in the region has been lacking to date. Studies, reports, and agency communications related to marine protection tend to be limited to marine protected areas (MPAs) such as the marine reserves and marine parks that are established under the General Fisheries and Aquaculture Law No. 18,892 of 1991[2] (in Spanish LGPA), and the Multiple-Use Marine and Coastal Protected Areas (MU-MCPAs) and Nature Sanctuaries recognized under the General Environmental Framework Law No. 19,300 of 1994 (in Spanish LBGMA). A few publications have recognized the existence of marine areas within national parks and reserves administered by the National Forestry Corporation (in Spanish Corporación Nacional Forestal, CONAF) in Patagonia [3, 18, 30], but generally the literature tends to take for granted that national parks and reserves are only terrestrial in nature.

Without detracting from the fundamental contribution of MPAs to Patagonian marine conservation, in this chapter we propose broadening the perspective of marine protection to integrate all existing legal categories that contribute, or could contribute, to the protection of Patagonian coastal-marine ecosystems. Taking this broader view than has been used to date, this analysis serves to evaluate the scope of legal protection and provides perspectives for consolidating an effective coastal-marine conservation system in Patagonia.

2   Scope and Objectives

The main contribution of this chapter is the compilation, updating, and analysis of information regarding the creation and current distribution of coastal-marine protection in Chilean Patagonia. An important objective is to highlight the major opportunity to configure an integrated coastal-marine conservation system that is widely distributed across the region. To this end, we review progress in the creation and establishment of MPAs and other legal categories that contribute to marine protection. Finally, we discuss the principal challenges, needs, and opportunities that arise from a more integrated paradigm of marine protection.

3   Methods

Our review of the coastal-marine protected areas of Chilean Patagonia is based on an exploration of official documents and secondary information, as well as geographic analysis using official cartographic information analyzed in a geographic information system (GIS) using ArcGIS 10.5 [13]. Within the geographic area

---

[2] Promulgated by Decree No. 430 of 1991 and published in the Official Gazette in 1992. https://www.leychile.cl/Navegar?idNorma=13315&idVersion=Diferido.

of Chilean Patagonia, analysis is circumscribed to the inland coastal zone as it
is defined in the country's legal framework, which establishes that the coastal
zone coincides with Chile's territorial waters, thus extending from the highest tide
line to 12 nautical miles. With respect to the limits of Patagonia's biogeographic
ecoregions and the geographic spaces known as northern Patagonia and southern
Patagonia, we rely on Hucke-Gaete et al. [23].

To assess the potential level of marine protection, we used as the unit of analysis
all protected areas (PAs) with marine representation and/or scope. This includes
Marine Parks (MP), Marine Reserves (MR), Multiple-Use Marine and Coastal
Protected Areas (MU-MCPA), Nature Sanctuaries (NS) and Ramsar sites, as well
as the coastal-marine portion of the National Parks and National Reserves of
the SNASPE. In addition, the analysis of complementary forms of conservation
included the Benthic Resources Management and Exploitation Areas (in Spanish Areas de Manejo y Explotación de Recursos Bentónicos, AMERB). This is
a category of coastal-marine administration that includes fishing and conservation management objectives. It also included Indigenous Peoples Coastal Marine
Spaces (in Spanish Espacios Costeros Marinos de Pueblos Originarios, ECMPO)[3]
that have been decreed or are under review. In the latter case, the ECMPO
requested and accepted for formal review as admissible by the Undersecretariat
for Fisheries and Aquaculture (in Spanish, Subsecretaria de Pesca y Acuicultura,
SUBPESCA) as of January 2020 were included. ECMPO requests not yet declared
admissible at that date (Chaitén-Desertores and Yagán), but for which official cartography was available, were also included. For some ECMPO requests, for the
purposes of this analysis we grouped together different sectors or portions of an
ECMPO that have been requested by the same organization but are individualized
in the official databases for various administrative reasons.

Official information from the SUBPESCA data viewer and the National Regis-
ter of Protected Areas of the Ministry of the Environment (in Spanish Ministerio
del Medio Ambiente, MMA) was used to review the spatial coverage of the MPAs
(Table 1). There are no official cartographic data for the marine portion of the
SNASPE, since its reports and statistics only count its land area. Official maps
from the Ministry of National Assets (in Spanish Ministerio de Bienes Nacionales,
MBN) were used to address this limitation. We identified 3 NRs, Katalalixar,
Las Guaitecas, and Kawésqar, and 4 NPs, Isla Magdalena, Laguna San Rafael,
Bernardo O'Higgins, and Alberto de Agostini as including coastal-marine areas.
Cape Horn NP was excluded from the calculations for this study because, although
its creation decree indicates that it includes coastal-marine areas, its perimeter
could not be specified. We also analyzed the potential complementary contribu-
tion to marine conservation from ECMPOs and AMERBs. For this purpose, we

---

[3] Indigenous People's Coastal Marine Spaces: delimited marine space, whose administration is
given to indigenous communities or associations of them, whose members have exercised the
customary use of such space" (Article 2°, Law 20,249).

Table 1 Summary of variables analyzed, and sources of information consulted

Variables	Data	Sources of information
Surface area of study area	Land and marine areas (up to 12 nautical miles)	INE map service: http://www.censo2017.cl/map-service/
PA surfaces and complementary areas	SNASPE	CONAF's territorial information system. https://sit.conaf.cl/
		Cadastre of the Ministry of National Assets http://www.catastro.cl
	MP, MR, MU-MCPA	SUBPESCA data viewer https://mapas.subpesca.cl/ideviewer/
	ECMPO, AMERB	
	NS	National registry of protected areas MMA. http://areasprotegidas.mma.gob.cl/
	Ramsar sites	
Shoreline	Perimeter of coastal zone	INE map service: http://www.censo2017.cl/map-service/
MPA establishment	Decrees to establish areas	National registry of protected areas MMA. http://areasprotegidas.mma.gob.cl/
	Conservation objectives	

PA Protected Areas of the National Protected Area System, *MP* Marine Park, *MR* Marine Reserve, *MU-MCPA* Multiple-Use Marine Coastal Protected Area, *ECMPO* Indigenous Peoples' Coastal Marine Spaces, *NS* Nature Sanctuaries, *AMERB* Benthic Resources Management and Exploitation Areas, *INE* Instituto Nacional de Estadística

counted only the net contribution, excluding areas of overlap with existing PAs (SNASPE and MPAs).

To complement the area analysis of each category, we also calculated the extension of the shoreline included in each area. We used the National Institute of Statistics map (in Spanish Instituto Nacional de Estadística, INE) of regional limits and measured the linear extent of islands, islets, and the continental coast. The sources of primary and secondary information are summarized in Table 1. This mainly includes national maps and decrees, management reports, articles, and other gray literature sources. Similarly, the historical evolution of the establishment of MPAs based on a review of legislation, area creation decrees, and available bibliography.

4   Results

4.1 Creation and Evolution of Marine Protected Areas in Chilean Patagonia

The designation of MPAs in Patagonia has been much more recent and institutionally heterogeneous compared to terrestrial PAs. The first Patagonian MPA, Estero Quitralco Nature Sanctuary, was decreed in the municipality of Aysén in 1996. The creation of areas grew from then on, with the declaration of 11 MPAs under different categories (Figs. 1 and 2; Table 2). The first legislative framework to explicitly include the protection and conservation of marine areas was the General Fisheries and Aquaculture Law of 1991. This legislation originated in response to fishing management problems and is focused in part on the sustainability of artisanal fishing. However, it also establishes a framework for marine protection through the categories of MP[4] and MR,[5] that are intended to safeguard hydrobiological resources and protect reproduction areas,[6] and whose declaration corresponds to SUBPESCA and administration to the National Fisheries Service (in Spanish SERNAPESCA) [38]. SUBPESCA is thus a key entity in determining marine conservation policies within the national public institutional framework. Despite multiple studies and proposals to establish a network of MPAs in Chile (e.g., [36]), this has not materialized to date.

The first MR and MP were declared in Chilean Patagonia in 2004 (Fig. 1). Francisco Coloane MP was declared that year in the Magallanes region with the goal of preserving the feeding sites of humpback whales and other aquatic communities present in the area [26]. Two coastal MRs were declared in the province of Chiloé in 2004: (i) Pullinque, with the purpose of safeguarding a natural shoal of Chilean oyster (*Ostrea chilensis*), and (ii) Putemún, to conserve a natural shoal of giant mussel (*Choromytilus chorus*) [29]. In Chile, the importance and attention given to MPAs grew with the establishment of the country's national environmental institutions (Fig. 1), in particular through the General Environmental Framework Law, which created the National Environmental Commission (in Spanish Comisión Nacional del Medio Ambiente, CONAMA) and its subsequent modification in 2010 through Law No. 20,417, which transformed this commission into the Ministry of the Environment (MMA). The MMA has responsibility for generating policies and standards for PAs, including MP and MR, although the

---

[4] "The Marine Parks will be under the guardianship of the Service and no type of activity may be carried out in them, except those authorized for observation, research or study purposes" (Title II, art. 3, letter d. LGPA).

[5] "Marine Reserve: area of protection of hydrobiological resources to protect reproduction zones, fishing grounds and areas of repopulation by management. These areas will be under the control of the Service and extractive activities may only be carried out in them for transitory periods, subject to a well-founded resolution of the Undersecretariat" (Title I, art. 2, number 36. LGPA).

[6] There are currently five marine reserves in Chile (totaling 80.3 km2) and 10 marine parks (totaling 859,964.9 km2) mostly created after 2016 (http://mapas.subpesca.cl/ideviewer).

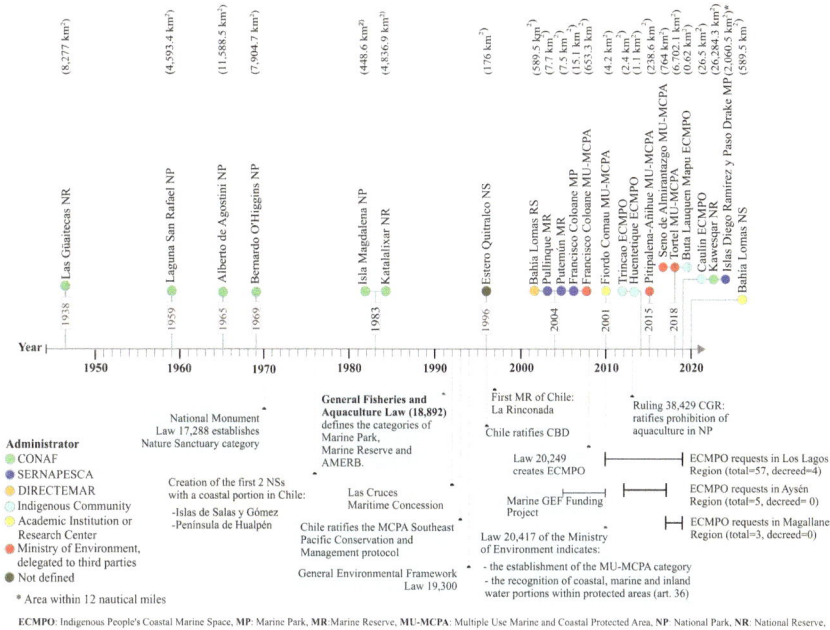

Fig. 1 Evolution of coastal-marine protection in Chilean Patagonia. The marine-coastal protection area is indicated. For ECMPO, total = number of applications; decreed = number of decreed areas

management responsibility for these areas varies according to the legal category (National Biodiversity Strategy years 2003 and 2017).[7] The new environmental institutional framework particularly reinforced the figure of MU-MCPA, a category whose management corresponds to the MMA[8] [34]. However, as the MMA has not had the capacity to administer and manage MPAs to date, this has been delegated to third parties in the form of agreements or concessions over PAs.

The first stage of the declaration of MPAs in Patagonia was driven primarily by government actions within the framework of the project "Conservation of Globally Important Biodiversity along the Chilean Coast" (GEF-Marine Project). In 2004, this project led to the declaration of the Francisco Coloane MP as the first MP in the country, and around which the Francisco Coloane MU-MCPA (in Spanish Area Marina Costera Protegida de Multiples Usos, AMCP-MU) was also

---

[7] The 2003 strategy established the goal of protecting at least 10% of terrestrial and marine ecosystems by 2015. The 2030 Strategy is expected to implement a network with 80% of MPAs with management plans in place.

[8] THE MU-MCPA category originated under the Permanent Commission for the South Pacific (CPPS) and are defined as areas "that include portions of water and seabed, rocks, beaches and fiscal beach lands, flora and fauna, historical and cultural resources that are set aside by law or other efficient means to protect all or part of the environment so delimited" [34].

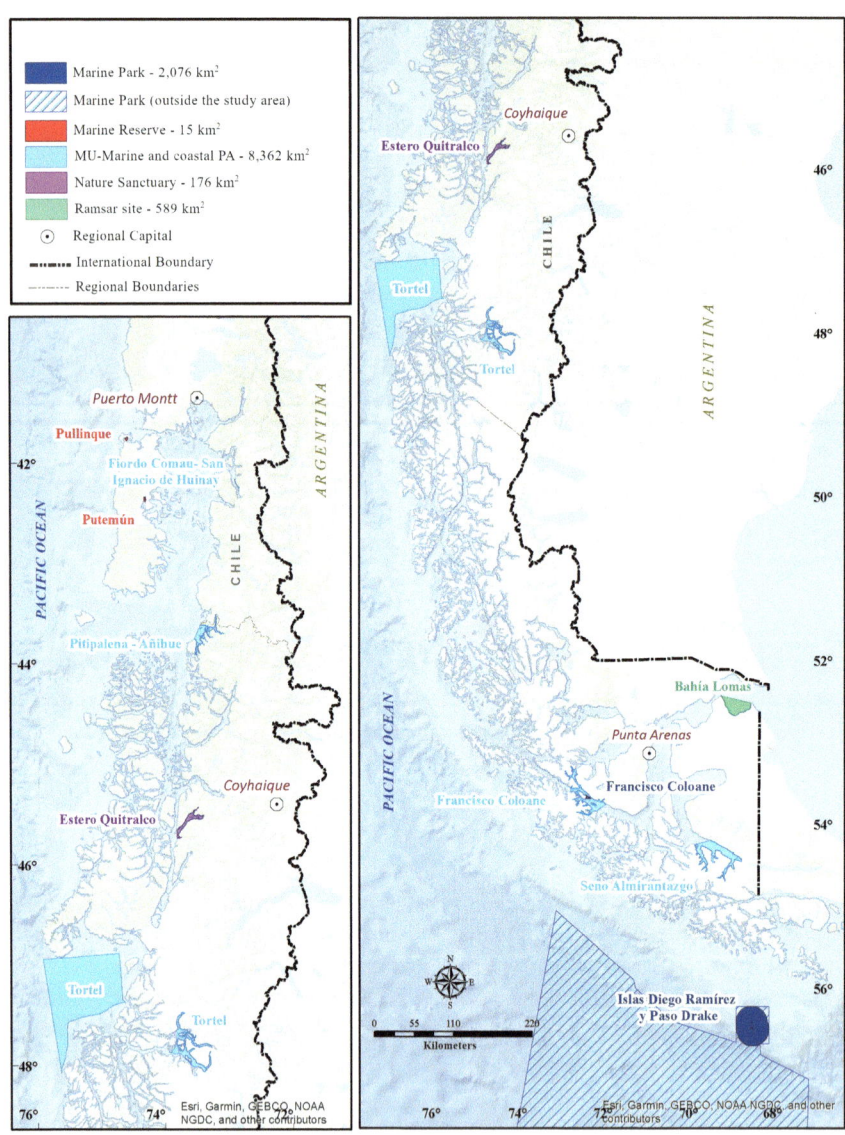

Fig. 2 Marine protected areas declared in Chilean Patagonia as of December 2019

established (Fig. 1), both of these MPAs are located in southern Patagonia. The subsequent stages in the creation of MPAs were largely promoted by conservation non-governmental organizations (NGOs). The first relevant initiative, which established a pattern in terms of geographic and institutional approaches, was promoted for the Chiloense marine ecoregion in northern Patagonia [23]. Here, in 2003, several NGOs led by the Blue Whale Center (CBA), the Austral University of

Table 2 List of areas with coastal-marine protection in Chilean Patagonia, indicating the total area of maritory within 12 nautical miles (own elaboration)

Protection figure	Name of the area	Marine surface in Patagonia (km^2)	Year of creation	Region	Official management instrument
National park	Isla Magdalena	448.61	1983	Aysén	No
National park	Laguna San Rafael	4,593.40		Aysén	Yes
National park	Alberto de Agostini	11,588.48	1965	Magallanes	No
National park	Bernardo O'Higgins	7,904.76	1969	Aysén and Magallanes	No
National reserve	Katalalixar	4,836.98	1983	Aysén	No
National reserve	Las Guaitecas	8,277.03	1938	Aysén	No
National reserve	Kawésqar	26,284.29	2019	Magallanes	No
SubTotal SNASPE		**63,933.55 km^2**			
Marine park	Francisco Coloane	15.06	2004	Magallanes	No
Marine park	Islas Diego Ramírez—Paso Drake	2,060.51	2018	Magallanes	No
Marine reserve	Pullinque	7.73	2004	Los Lagos	Yes
Marine reserve	Putemún	7.53	2004	Los Lagos	Yes
Marine coastal protected area	Fiordo de Comau—San Ignacio de Huinay	4.15	2003	Los Lagos	No
Marine coastal protected area	Francisco Coloane	653.27		Magallanes	No
Marine coastal protected area	Pitipalena—Añihué	238.62	2015	Aysén	Yes

(continued)

Table 2 (continued)

Protection figure	Name of the area	Marine surface in Patagonia (km²)	Year of creation	Region	Official management instrument
Marine coastal protected area	Seno Almirantazgo	764.00	2018	Magallanes	No
Marine coastal protected area	Tortel	6,702.10	2018	Aysén	No
Nature sanctuary	Estuario Quitralco	176.00		Aysén	No
RAMSAR site/Nature sanctuary[a]	Bahía Lomas	589.46	2004/ 2020	Magallanes	No information
SubTotal AMP		**11,218.43**			
Total marine protection		**75,151.98**			

[a] Bahía Lomas was decreed as a Nature Sanctuary, maintaining the same surface area as the RAMSAR site

Chile (UACh) and the World Wildlife Fund (WWF-Chile), with support from the Regional Government of Los Lagos, promoted the creation of an MU-MCPA for the Gulf of Corcovado, the exposed coast of Chiloé and the north of Las Guaitecas, under the concept of a "large maritory with general zoning and regulation of uses" [22]. Although the area was not finally decreed, the initiative generated baseline research and a proposal for core protection areas[9] in Patagonia. In 2015, the Pitipalena-Añihue area was finally decreed as an MU-MCPA (Fig. 1), with support from the Raúl Marín Balmaceda community, the Melimoyu Foundation and other NGOs. In the same year, a MP was declared for the area from Tic-Toc Bay to the west of the Gulf of Corcovado.

Between 2016 and 2018, there was a last phase of MPA creation with the designation of three new conservation areas: (i) Tortel MU-MCPA in the Aysén Region, promoted by the Municipality of Tortel, with support from the NGO Oceana [28], (ii) Seno de Almirantazgo MU-MCPA in Tierra del Fuego, with support from the Wildlife Conservation Society [41], (iii) the Islas Diego Ramírez y Paso Drake MP south of Cape Horn, presented by SUBPESCA with technical support from the Subantarctic Biocultural Conservation Program of the Puerto Williams University Center [32, 37]. An important antecedent to the designation of the Islas Diego Ramírez y Paso Drake MP, was the earlier establishment of the Cabo de Hornos

---

[9] Potential areas to be designated MPAs due to their important natural and cultural heritage values.

Biosphere Reserve with a maritime area of 29,727.9 km^2, which is of particular relevance as the first Biosphere Reserve to integrate marine and terrestrial areas [33].

4.2 Marine Protection in National Parks and National Reserves in Chilean Patagonia

Beginning in 1938, with the Las Guaitecas NR, but primarily in the 1960s, most of the large Patagonian archipelagos were designated as NP or NR (73% in the Aysén Region and 86% in the Magallanes Region[10]). However, due to the precariousness and ambiguities in the legal framework of the SNASPE and because it is under the administration of CONAF, an institution historically focused on terrestrial management, the recognition of the marine portion of these protected archipelagos has not been consistent or widely accepted by public institutions. However, in legal terms, over the last decade, the coastal-marine waters contained in the NPs and NRs have been increasingly recognized through a series of administrative and legislative acts [18, 30]. This recognition was first reinforced by the 2002 modification (Law No. 19,800) of the General Environmental Framework Law, which established in Article 158 the prohibition of "all extractive fishing and aquaculture activities in the NPs", while also noting that these activities are exceptionally allowed in the NRs with the appropriate authorization. However, the marine scope of NPs and NRs was more clearly established with the 2010 amendment to the General Environmental Framework Law (no. 19,300), which establishes in Article 36 that PAs include the "portions of sea, beach lands, sea beaches, lakes, lagoons, glaciers, reservoirs, watercourses, marshes and other wetlands, located within their perimeter." This recognition is of great relevance to Chilean Patagonia, where the large archipelagos were designated as parks or reserves with perimeters that encompass the islands along with their channels and fjords. Such areas include, from north to south, Las Guaitecas NR, Isla Magdalena NP, Laguna San Rafael NP, Katalalixar NR, Bernardo O'Higgins NP, Kawésqar NR, Alberto de Agostini NP, and Cabo de Hornos NP (Fig. 3).

Despite this legislation, the protected nature of the SNASPE marine area has not been fully supported by the other State agencies, an issue that is beginning to be clarified through a series of rulings by the Comptroller General of the Republic (CGR) for disputes over the expansion of salmon farming in Chilean Patagonia. The first ruling (No. 28,757 of 2007) followed protest by environmental organizations in Aysén regarding the granting of salmon farming concessions in Las Guaitecas NR [18]. The CGR recognized the protected nature of the waters of the NR but allowed the granting of concessions in the reserve. Additional jurisprudence in support of the protected nature of these maritories began

---

[10] According to our own calculations, GIS laboratory Austral Patagonia Program, Universidad Austral de Chile.

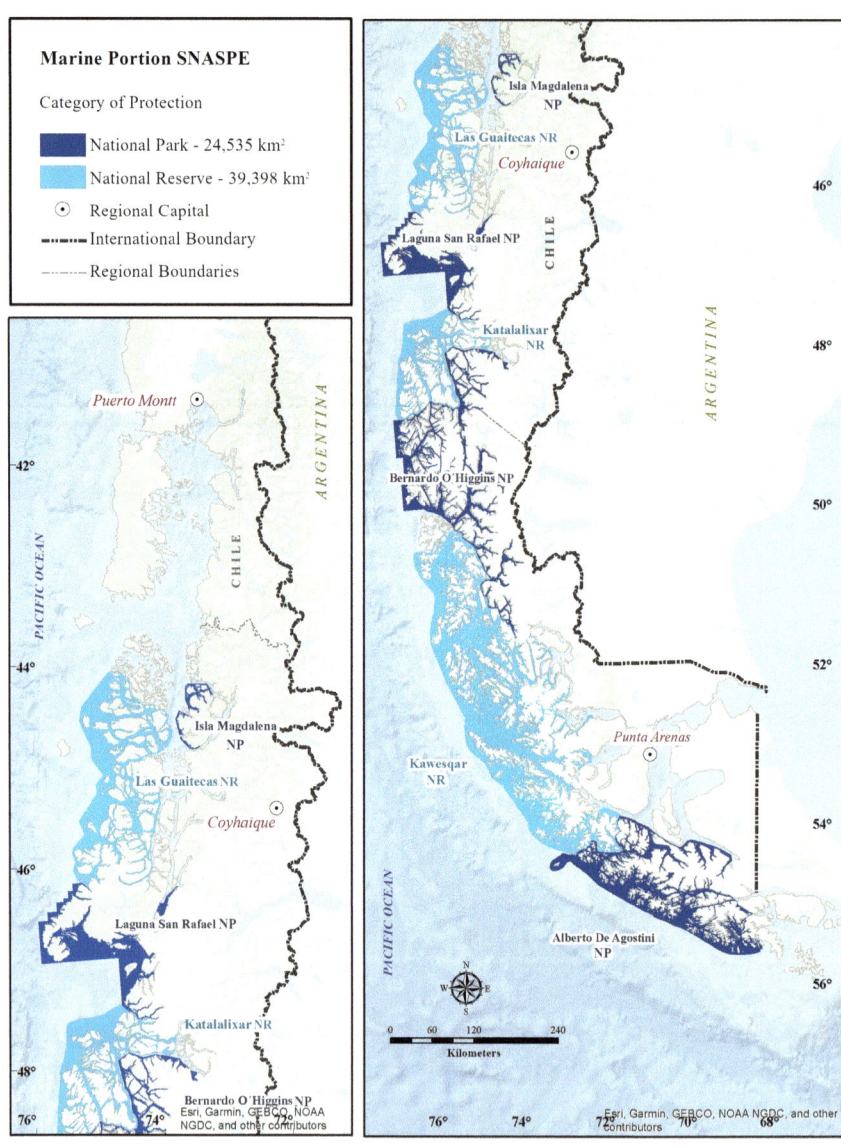

Fig. 3 National parks and reserves with marine portions in Chilean Patagonia as of December 2019

to emerge in 2012. This was the result of a controversy between CONAF and SUBPESCA regarding the feasibility of granting salmon farming concessions in Bernardo O'Higgins NP and Alberto de Agostini NP. The Regional Comptroller's Office of Magallanes and Chilean Antarctica, through Ruling No. 1,326 of 2012, affirmed the relevance of the coastal-marine portions of the NPs and the prohibition of aquaculture in them. This controversy originated in the opposition of the Kawésqar Community of Puerto Edén to the installation of salmon farming in Bernardo O'Higgins NP, which triggered the administrative confrontation.

In 2013, the Ministry of Economy, Development, and Tourism requested the reconsideration of Ruling No. 1,326 of 2012, but the CGR reaffirmed the regional decision.[11] In 2019, the legal feasibility of protecting the maritory within the SNASPE was further reinforced with the establishment of the Kawésqar NR ($26,284.29$ km^2) over the entire inland waters of the former Alacalufes Forest Reserve, thus creating the first 100% marine NR under CONAF's administration. This declaration had its origin in the Indigenous consultation with Kawésqar communities for the reclassification of the Alacalufes Forest Reserve to NP, which documented the communities' interest in effectively protecting the inland waters of a future National Park and led to a government commitment to evaluate this option [25].

4.3 Complementary Conservation Areas in Chilean Patagonia: Indigenous Peoples' Marine Coastal Spaces

Law No. 20,249 was enacted in 2008, creating the category of Indigenous Peoples Marine Coastal Spaces (EMCPO). Also known as the "Lafquenche Law", this legislation's objective is to "safeguard the customary use of these spaces in order to maintain the traditions and use of natural resources by the communities linked to the coastline" (Article 3).[12] Although ECMPOs are not recognized as MPAs, Article 5 of the law states that the administration of the ECMPO "must ensure the conservation of the natural resources included in it." Since its enactment, Indigenous organizations have used the law for various purposes [2], and it lies with the organization requesting the area to propose the degree of protection to be assigned to each ECMPO.

While this issue remains the subject of debate, we consider ECMPOs as comparable to multiple-use MPAs when they are requested with explicit conservation objectives that are then included in a management plan.[13] The ECMPO Law

---

[11] Ruling No. 38,429 of 2013, of the Comptroller General of the Republic, concluding that "From the harmonic interpretation of Articles 158 of Law 18,892 and 36 of Law 19,300, it is evident that it is not possible to develop aquaculture activities in maritime waters that are part of an NP, which is also consistent with the Washington Convention, under which our country is obliged not to exploit the existing resources in that category of protection for commercial purposes (applies criteria of Ruling No. 56,465 of 2008)".

[12] "Delimited marine space, whose administration is given to Indigenous communities or associations of them, whose members have exercised the customary use of such space" (Art. 2, letter e, Law 20,249).

[13] It should be noted that some ECMPOs, such as the case of the Chaitén-Islas Desertores application, in addition to being applied for conservation objectives, have followed planning processes based on open standards for conservation practice, a methodology adopted by the MMA for MPAs.

establishes a formal administrative review process for evaluating requests from Indigenous communities, but in practice, review have taken much longer than stipulated by the law. Since 2010, 62 ECMPOs have been requested in northern Patagonia (Los Lagos and northern Aysén regions) and three in southern Patagonia, but to date, only four of the areas requested have been decreed, all of which are in Los Lagos Region (Figs. 4 and 5). The potential role of ECMPOs in conservation has been increasingly highlighted in the literature, although to date, this has not been echoed in public policy [2, 4, 20, 21, 27, 40] .

4.4 Total Coverage and Distribution of Coastal-Marine Protection in Chilean Patagonia

The total area of the coastal zone of Chilean Patagonia is 183,073 km^2, with 41% of this area under some form of legal protection (75,151 km^2, Table 2). The SNASPE represents 35% of the total coastal-marine area of Patagonia, while MPAs represent only 6% (Fig. 6). With regard to the proportion of protected area in Patagonia, the SNASPE represents 85%, and MPAs 15%. The most important of the latter are the MU-MCPAs and MPs, with 11 and 3% of the total protected marine area, while the MR, NS and Ramsar Sites account for 1% of the total protected area (Figs. 7 and 8).[14]

4.5 Protection Applied to the Coastline of Chilean Patagonia

To complement the analysis of coastal-marine water surfaces, we calculated the complete length of the coastline of Chilean Patagonia (100,627 km) and the coastline included in each category of coastal protection and management. The SNASPE represents 79% (79,365 km) of the Patagonian coastline, and 97% of the coastline under official protection. The MPAs cover 2.2% (2,187 km) of the Patagonian coastline, and only 2.7% of the protected coastline (Fig. 9).

4.6 Potential Complementary Protection by Other Categories of Coastal-Marine Administration

The current decreed area of ECMPOs in Chilean Patagonia covers 30.7 km^2 (Fig. 6), and the largest of these areas is Caulín with *ca.* 26 km^2 (Fig. 4). There is also 62,931 km^2 of requested ECMPOs, most of which is concentrated in the Magallanes Region (Figs. 4 and 5). A portion of these ECMPOs already have regional and national approval but with decrees pending. Existing requests overlap with

---

[14] An important issue to analyze in future work is the 560 km^2 overlap between Alberto de Agostini NP and the recently created AMCP-MU Seno Almirantazgo.

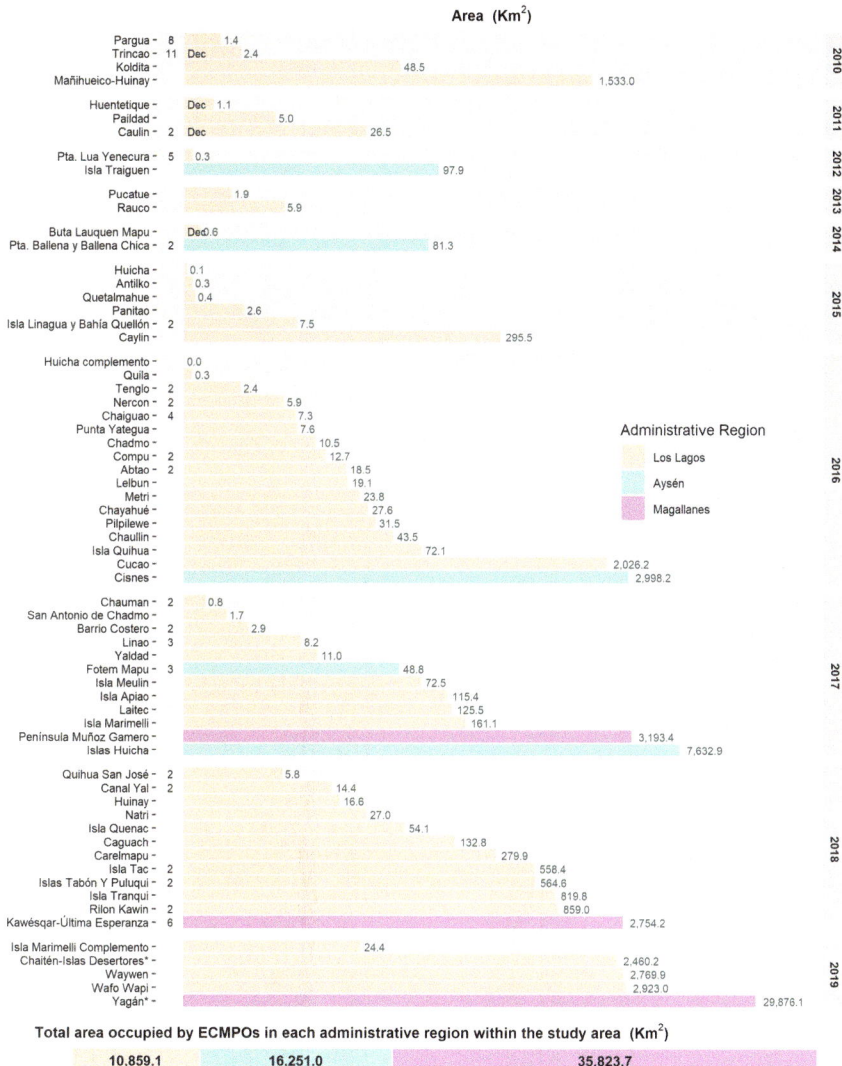

Fig. 4 List and total area of ECMPOs requested in Chilean Patagonia between 2010 and 2019 by administrative region. ECMPOs with more than one zone are indicated in the 2nd column, those lacking acceptance for formal review as of January 2020 with an (*) and those decreed with a (Dec)

six established PAs: Las Guaitecas NR (with an overlap of 4,874 km^2), Kawésqar NR (3,871 km^2), Alberto D'Agostini NP (7,643 km^2), Bernardo O'Higgins NP (143 km^2), Islas Diego Ramírez y Paso Drakes MP (2,060 km^2) and the MU-MCPA Fiordo Comau-San Ignacio de Huinay (2 km^2). The total ECMPO area that does not overlap with existing PAs is 44,368 km^2, which is a remarkable additive

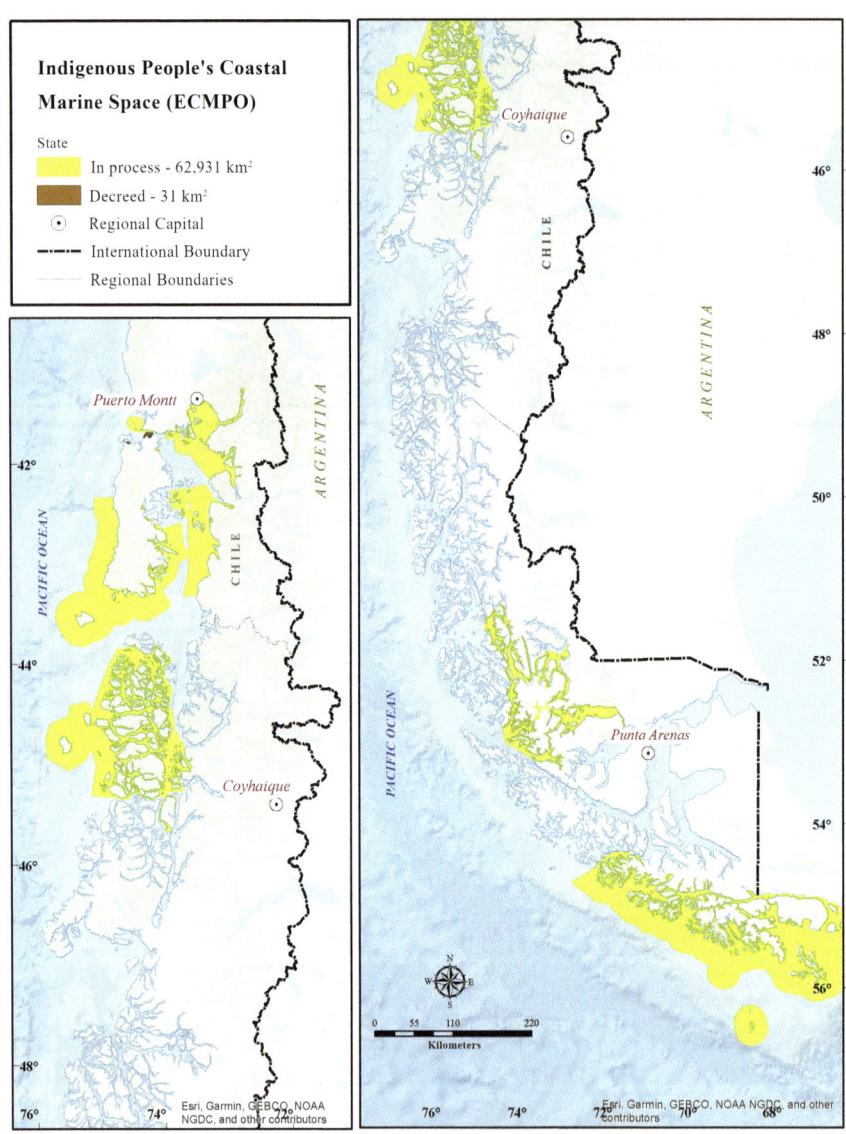

Fig. 5 Distribution of ECMPOs requested and decreed in Chilean Patagonia to 2020

potential in terms of marine protection. Therefore, if the ECMPOs are considered protected areas, they would contribute 24% of the marine protected area.

There are 344 decreed and pending Benthic Resource Management and Exploitation Areas (AMERB) in Chilean Patagonia, covering 735 km of the coastline and 521 km² of coastal waters (<1% of the total in both cases). We have not analyzed them in this chapter because of their reduced presence in the region,

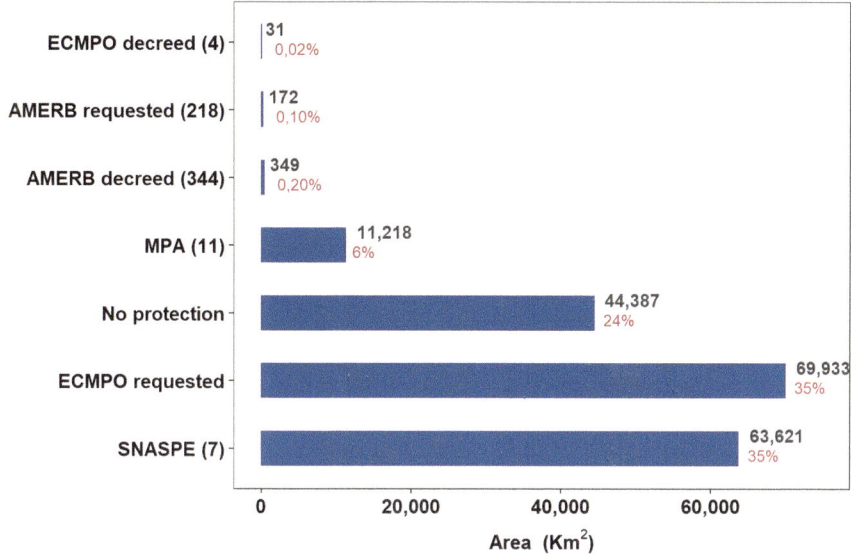

Fig. 6 Distribution of the coastal zone of Chilean Patagonia by protection or administration. The surface areas calculated for AMERB and ECMPO exclude areas that overlap with PAs (SNASPE and MPA)

however, there is evidence of the important role played by these areas at the national level both for biodiversity conservation and for the development of fishing communities [6, 8, 10, 16].

4.7 Level of Protection and Restrictions Applicable to Marine Areas Under Protection in Chilean Patagonia

Legal protection (on paper) is a first step toward conservation, but its real effect on conservation in decreed areas is highly variable. Current legislation and its application by the authorities generally allows multiple uses, including extractive fishing in the vast majority of the protected area (Table 3). The presence of strictly protected areas is reduced in the Patagonian coastal zone; it only includes the Francisco Coloane MP (15.06 km^2) and the Islas Diego Ramírez y Paso Drake MP, of which only 2,060 km^2 of its total 144,390 km^2 correspond to the coastal zone. MU-MCPAs have no a priori restriction on any use (due to the lack of a regulation in force). However, to date, their creation decrees have prohibited intensive aquaculture, except for those farms that were installed prior to the declaration of the Pitipalena-Añihué and Fiordo Comau MU-MCPAs. Environmental legislation establishes that in PAs (but not in ECMPOs or AMERBs), the development of economic activities, such as the establishment of salmon or mussel farms, requires an Environmental Impact Assessment (EIA); however, this requirement has not been

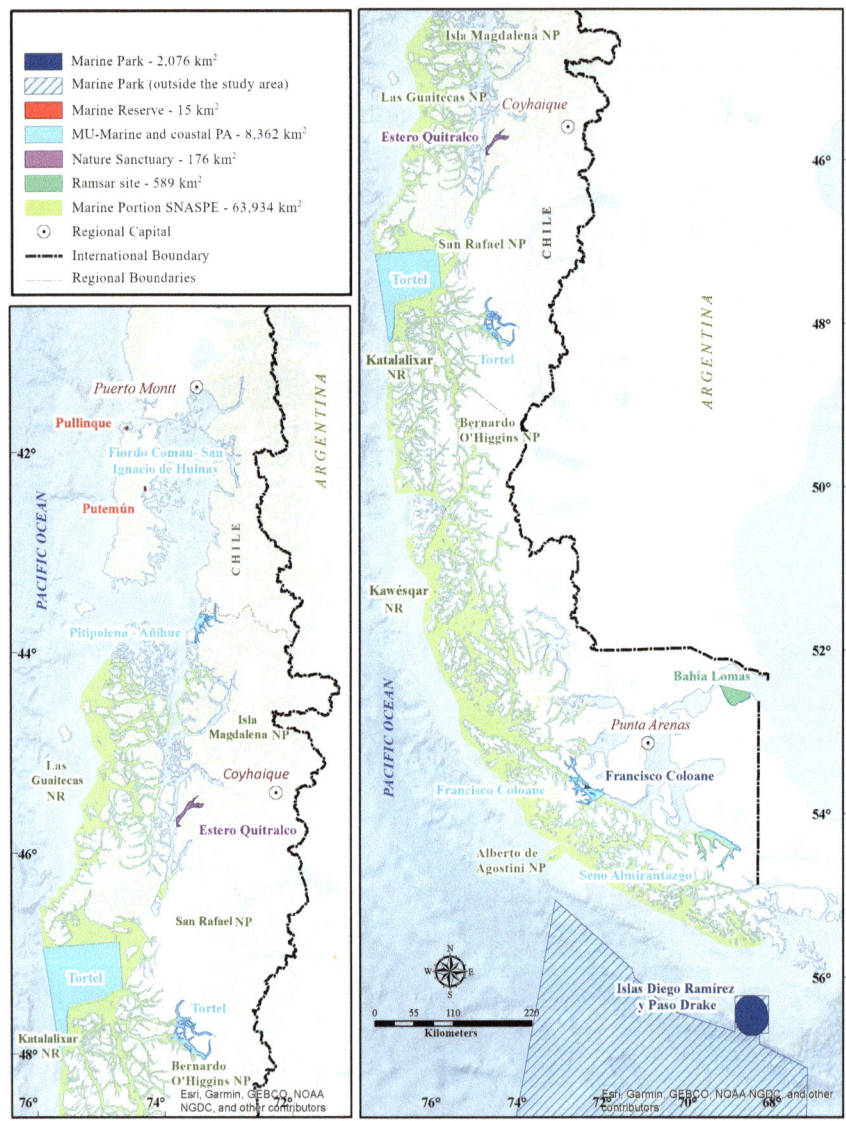

Fig. 7 Marine protection was decreed in Chilean Patagonia by 2020

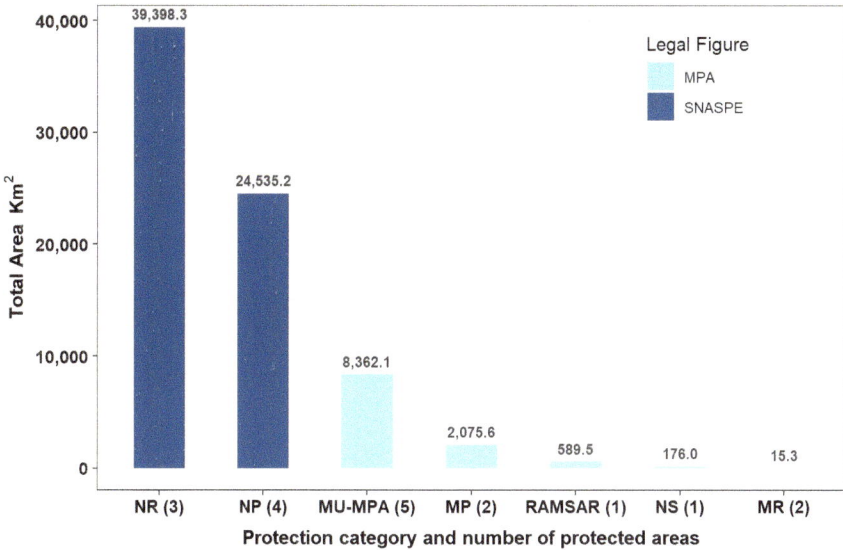

Fig. 8 Coastal-marine protection area according to legal figure for Chilean Patagonia

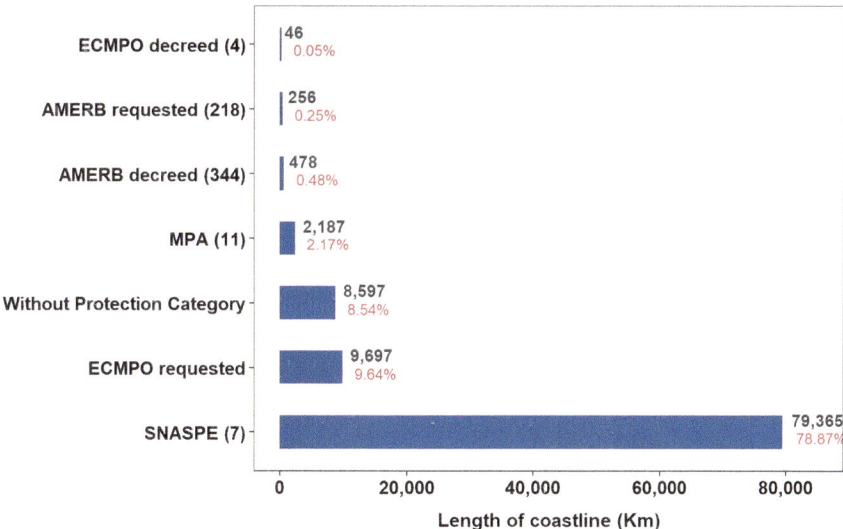

Fig. 9 Distribution of the coastline of Chilean Patagonia according to protection and administration figures

Table 3 Level of restriction applicable according to legal categories for the coastal zone of Chilean Patagonia (-) activities permitted with management plan, (-) activities prohibited

Legal Figure	Legal basis	Administrative Entity	Equivalence IUCN Category	Artisanal fishing	Industrial fishing	Aquaculture	Mining	Hydrocarbons	Tourism	Research
MP	Law 18.892/1989	SERNAPESCA	Ia	-	-	-	-	-	-	-
MR	Law 20.417/2010 S.D. 238/2004		IV, VI	-	-	-	-	-	-	-
MU-MCPA	Decree 827/1995 Law 20.417/2010	MMA, delegated to third parties	VI	-	-	-	-	-	-	-
NS	Law 17.288/1970 Law 19.300/1994 Law 20.417/2010 Law 20.417/2010	MMA, delegated to third parties	III, IV	-	-	-	-	-	-	-
NP	Law 18.892/1989	CONAF	II	-	-	-	-	-	-	-
NR	Law 20.417/2010 Opinion 38.429/2013		IV, VI	-	-	-	-	-	-	-
Ramsar	D.S. 771/1981 Agreement 287/2005	MMA, delegated to third parties		-	-	-	-	-	-	-
ECMPO	Law No. 20.249 D.S. 143/2008	Applicant association or community		-	-	-	-	-	-	-
AMERB	Law 18.892/1989 S.D. 355/1995 S.D. 96/2016	Artisanal fishermen's organizations		-	-	-	-	-	-	-

MP Marine Park, *MR* Marine Reserve, *MU-MCPA* Multiple-Use Marine Coastal Protected Area, *NS* Nature Sanctuary, *NP* National Park, *NR* National Reserve, *Ramsar* Ramsar site, *ECMPO* Indigenous Peoples' Coastal Marine Space, *AMERB* Benthic Resources Management and Exploitation Areas

respected in NR.[15] In addition, in Quitralco NS the installation of salmon farms was permitted without requiring an EIA.

---

[15] Article 11, letter (d) of the General Environmental Framework Law, which establishes the obligation of projects within PAs to submit to the EIA system, and there is an accumulation of jurisprudence regarding the obligation to do so through a full EIA rather than a Declaration of Environmental Impact (see discussion in [24]).

The SNASPE situation is even more complex and differs between NPs and NRs. The 2002 modification (No. 19,800) of the General Environmental Framework Law establishes in Article 158 the prohibition of "all extractive fishing and aquaculture activities in the NPs", while this is permitted in the NRs with the appropriate authorization. The Washington Convention, an international treaty ratified by Chile, prohibits aquaculture in NP and at least restricts it in the NR, although the environmental agencies have not complied with such restrictions.[16] Since 2013, the authorities have effectively prohibited aquaculture in NPs; however, such measures do not affect concessions granted prior to that date, allowing, for example, 19 salmon farming concessions previously granted in Alberto de Agostini NP to remain in effect. It is important to note that artisanal fishing is compatible with NR status. In the NPs, despite the general prohibition established by the General Environmental Framework Law, the CGR's jurisprudence (Ruling No. 41,121 of 2014) indicated the need to evaluate the situation on a case-by-case basis, as artisanal fishing is considered a "consolidated" use in Bernardo O'Higgins NP, i.e., that it could not be prohibited because the use was installed prior to legal recognition of the park's inland waters.[17] While the law allows aquaculture activity in the NR, this must demonstrate compatibility with the objectives of the NR and its management plan, and these latter conditions should also be applied to extractive fishing [24]. However, due to the ambiguities described above, 319 salmon farming concessions have been granted in Las Guaitecas NR. In addition, in Kawésqar NR there are 68 salmon farming concessions granted and another 62 applications under review, in addition to 53 concessions in process for mussels and algae [38]. Table 3 summarizes the legal restrictions applicable to uses in the different categories of marine protection. It should be noted, however, that several of these activities could also be prohibited or regulated through management plans for individual areas.

---

[16] Article 2 of the Washington Convention states, "The following shall be understood as NR:" Regions.

[17] Ruling No. 41,121 of 2014, National Comptroller of the Republic.

Scientific expedition in Kawesqar National Park, Magallanes Region. Photo by Nicolás Muñoz (UACh-CBA)

5   Discussion

The evolution of marine protection in Chilean Patagonia has shown important advances both through the declaration of MPAs and the legal recognition of the marine portion of the SNASPE's large parks and reserves. There is also growing harmonization between MPA and SNASPE planning due to the adoption by both the MMA and CONAF of planning methodologies based on the Open Standards for Conservation [12, 39]. However, the accumulated progress in the declaration of areas is not the result of a coordinated public policy to optimize efforts but rather of an accumulation of independent efforts, many of which required 5–10 years, and today, the vast majority of these areas exist as "paper protected areas".

The contribution of the SNASPE units to the total area protected, in addition to their wide latitudinal distribution and concomitant diversity of environments, is particularly noteworthy. However, this has not been recognized in public strategies and policies. Similarly, attention should be given to the advances in the recognition of Indigenous rights in Chilean Patagonia and the rapid growth of ECMPO applications, whose contribution to marine conservation could be significant, both in the coverage of threatened marine ecosystems that are underrepresented in the current system and in generating new local management models. At the same time, a general regulation for the integrated management of the system of marine and terrestrial PAs is still pending, as is the regulation of the figures established by the LGBMA. Our emphasis on integrated management refers, first, to the need for greater recognition of the various forms of coastal-marine protection in use; second, to their integration with terrestrial PAs; and finally, to the search for ways to articulate marine-terrestrial protection with other conservation and local development strategies promoted by civil society in each Patagonian region. The Biodiversity and Protected Areas Service Bill was introduced in the National Congress in 2011 and could contribute to this end. If approved, it will consolidate the management of marine and terrestrial PAs under the MMA, but this legislative process remains incomplete. Therefore, economies of scale are not taken advantage of in the designation and management of areas, and each area requires major efforts to define the governance arrangement, management plans, and financing.

A detailed analysis of the level of effective management of the different forms of protection and specific units is beyond the scope of this chapter, but, with few exceptions where public–private efforts have advanced in setting up local management structures, Patagonian marine protection remains only in legal terms (on paper). In general, management plans have not been established, no budgetary resources have been allocated, and therefore, field activities, such as monitoring and enforcement, are practically nonexistent.[18] In budgetary terms, MPAs suffer from large financial gaps that jeopardize the viability of the conservation objectives for which they were created [7].

A specific line item of $200 million Chilean pesos for MPA management was established for the first time within the MMA's national budget in 2018.[19] Moving forward, MPA implementation could be supported by the goals for management of MPAs in the recently adopted national climate commitments (National Determined Contributions, NDC) within the Paris Agreement [11]. To date, there is no coordinated effort in the marine management of the SNASPE. While this lack is perhaps somewhat less critical in NPs, due to their higher level of restriction, the two largest NRs (Kawésqar and Las Guaitecas) currently suffer from intense pressures due to aquaculture uses (e.g., salmon farming). This situation represents a critical gap that requires regularization through the planning and management tools available for the SNASPE and the cooperation of other sectoral regulatory authorities, in particular SUBPESCA and the Navy's General Directorate of Maritime Territory and Merchant Marine. This is not trivial, considering that Chilean Patagonia still has relatively pristine conditions, maintains an important capacity as a carbon sink and can be considered a biodiversity refuge in the face of climate change [15, 22]. Marine protection in Patagonia must be understood in an integrated manner when establishing MPAs or "refuges" that can buffer the multiple threats to species and their need to migrate/adapt, but especially, that allow the maintenance of coastal ecosystem functions. Well-implemented and managed MPAs help marine ecosystems adapt to various types of impacts and, therefore, increase their resilience [31]. This is the challenge we must solve as a society before thresholds or points of no return are exceeded.

---

[18] The absence of MPA management was recorded in the Audit Report of the Office of the Comptroller General of the Republic No. 825 of 2018.

[19] Indication to the 2019 public sector budget bill n° 258–366.

6 Conclusions and Recommendations

Chilean Patagonia is a region with a high potential for coastal-marine conservation that is unique in the country and the world. A combination of factors explains this situation: the particular geography of the Patagonian coastal zone, with its extensive archipelagos, inland seas, and marine-terrestrial interconnections; the advances in the establishment of MPAs during the last two decades; the recognition of the marine portions of the NPs and NR that are part of the SNASPE; the development of proposals for ECMPOs by Indigenous organizations; and the growing involvement of diverse local and national stakeholders in proposing protection measures for the coastal zone. However, there is a gap between the current scenario and many public and private conservation strategies and policies, which still focus exclusively on conventional MPAs. Based on this scenario, we recommend four priority strategies:

- The effective management of the marine portion of the SNASPE should be strengthened. In the short term, ensure the completion, with the appropriate phases of citizen participation and Indigenous consultation, of the management plans for the coastal-marine portions of the NPs and NRs that contain inland waters in Patagonia. Over the long term, ensure an appropriate management formula for the large, protected areas located in archipelagos, supported by newly available technologies and co-management between CONAF and other public and/or local entities with authority over coastal-marine space or resources.
- Develop a protocol and legal procedures to recognize the MPA status of ECMPOs when their proponents request it and generate a system of governmental support for their management, as well as for the preparation, implementation, and monitoring of their management plans.
- Generate public and public–private financing and technical support mechanisms for the installation of management systems for all categories of marine protection, including conventional MPAs, SNASPE, and ECMPOs. Considering the scope of the challenge, it is crucial to seek economies of scale through equipment and management capabilities that can serve multiple units.
- Encourage integrated planning and management between terrestrial and marine environments to optimize conservation efforts and promote the transfer of capacities from terrestrial to marine environments.

Acknowledgements The authors thank The Pew Charitable Trusts for providing financial support for this work. In addition, the suggestions and contributions of the anonymous reviewers who helped to improve the chapter are sincerely appreciated. Juan Carlos Castilla is grateful for the support of the Facultad de Ciencias Biológicas, Universidad Católica de Chile. We are grateful to CONAF Aysén and Magallanes for their collaboration with the revision of cartographic data.

References

1. Álvarez, R., Ther-Ríos, F., Skewes, J., Hidalgo, C., Carabias, D., & García, C. (2019). Reflexiones sobre el concepto de maritorio y su relevancia para los estudios de Chiloé contemporáneo. *Revista Austral de Ciencias Sociales, 36*, 115–126.
2. Araos, F., Godoy, C., Andrade, R., Ther, F., Gelcich, S., & Salas, C. (2017). Conservación marina y costera en Chile: trayectorias institucionales, innovaciones locales y recomendaciones para el futuro locales. In L. Ferreira, L. Schmidt, M. Pardo, J. Calvimontes, & E. Viglio (Eds.), *Clima de tensão. Ação humana, biodiversidade e mudanças climáticas* (pp. 529–544). UNICAMP.
3. Aravena, J. C., Vela-Cruz, G., Torres, J., Huencoy, C., & Tonko, J. C. (2018). Parque Nacional Bernardo O'Higgins/territorio Kawésqar Waes: Conservación y gestión en un territorio ancestral. *Magallania, 46*(1), 49–63.
4. Aylwin, J., Arce, L., Guerra, F., Núñez, D., Álvarez, R., Mansilla P., Alday, D., Caro, L., Chiguay, C., & Huenucoy, C. (2023). *Conservation and indigenous people in Chilean Patagonia.* Springer.
5. Buschmann, A. H., Niklitschek, E. J., & Pereda, S. (2023). *Aquaculture and its impacts on the conservation of Chilean.* Springer.
6. Castilla, J. C. (1996). La futura red chilena de parques y reservas marinas y los conceptos de conservación, preservación y manejo en la legislación nacional. *Revista Chilena Historia. Natural, 69*, 253–270.
7. Castilla, J. C. (2008). El océano y la conservación en Chile: Los eternos olvidados. *Centro de Estudios Públicos, 112*, 206–217.
8. Castilla, J. C. (2010). Fisheries in Chile: Small-pelagics, management, rights, and sea zoning. *Bulletin of Marine Science, 86*(2), 221–234.
9. Castilla, J. C. (2014). Chile es Mar. In Colección Santander Museo Chileno de Arte Precolombino (Eds.), *El Mar de Chile* (pp. 13–40). Edición Carlos Aldunate del Solar. Ograma Impresores. http://www.precolombino.cl/archivos_biblioteca/publicaciones-en-pdf/libros-de-arte/mar-de-chile/mar-de-chile.pdf
10. Castilla, J. C., Manríquez, P., Alvarado, J., Rossón, A., Pino, C., Espoz, C., Soto, R., Oliva, D., & Defeo, O. (1998). Artisanal Caletas: As units of production and co-managers of benthic invertebrates in Chile. *Canadian Journal of Fisheries and Aquatic Sciences (Special Publication), 125*, 407–413.
11. Contribución Determinada Nacional de Chile. (2020). *Contribución Determinada Nacional de Chile (NDC).* Actualización 2020. https://mma.gob.cl/wp-content/uploads/2020/04/NDC_Chile_2020_espan%CC%83ol-1.pdf
12. Corporación Nacional Forestal. (2017). *Manual para la planificación del manejo de las Áreas Protegidas del SNASPE.* Santiago de Chile. https://www.conaf.cl/wp-content/files_mf/1515526054CONAF_2017_MANUALPARALAPLANIFICACI%C3%93NDELASAREASPROTEGIDASDELSNASPE_BajaResoluci%C3%B3n.pdf
13. Environmental Systems Research Institute. (2016). *ArcGIS desktop: Release 10.5.*
14. Escuela de Arquitectura Universidad Católica de Valparaíso. (1971). Maritorios de los archipiélagos de la Patagonia occidental. In Escuela de Arquitectura Universidad Católica de Valparaíso (UCV) (Ed.), *Fundamentos de la Escuela de Arquitectura* (pp. 1–18). Escuela de Arquitectura UCV. https://wiki.ead.pucv.cl/images/a/a9/OFI_1971_Maritorios.pdf
15. Farías, L., Ubilla, K., Aguirre, C., Bedriñana, L., Cienfuegos, R., Delgado, V., Fernández, C., Fernández, M., Gaxiola, A., González, H., Hucke-Gaete, R., Marquet, P., Montecino, V., Morales, C., Narváez, D., Osses, M., Peceño, B., Quiroga, E., Ramajo, L., Sepúlveda, H., Soto, D., Vargas, E., Viddi, F., & Valencia, J., (2019) *Nueve soluciones basadas en el océano para contribuir a los NDC de Chile.* Comité científico COP25, Mesa Océanos, Ministerio de Ciencia, Tecnología, Conocimiento e Innovación. http://www.cr2.cl/wp-content/uploads/2019/12/Nueve-soluciones-para-NDC.pdf

16. Gelcich, S., Godoy, N., Prado, L., & Castilla, J. C. (2008). Add-on conservation benefits of marine territorial user rights fishery policies in central Chile. *Ecological Applications, 18*, 273–281.
17. Guala, C., Veloso, K., Farías, A., & Sariego, F. (2023). *Analysis of tourism development linked to protected areas*. Springer.
18. Guerra, F. (2015). ¿Se puede intervenir en las áreas silvestres protegidas del Estado? Una aproximación al contexto chileno a partir del dictamen N° 38.429 de la Contraloría General de la República. *Revista Justicia Ambiental, 7*, 189–202. https://cl.boell.org/sites/default/files/libro_fima_interior_y_tapas.pdf
19. Häussermann, V., Försterra, G., & Laudien, J. (2023). *Hard bottom macrobenthos of Chilean Patagonia: Emphasis on conservation of sublittoral invertebrate and algal forests*. Springer.
20. Hiriart-Bertrand, L., Troncoso, J. M., Correo, A., & Vargas, C. (2019). Del reconocimiento del derecho consuetudinario a la implementación de acciones de resguardo: el caso de los espacios costeros marinos de pueblos originarios en Chile. In M. Ruiz, R. Oyanadel, & B. Monteferri (Eds.), *Mar, costas y pesquerías: Una mirada comparativa desde Chile, México y Perú* (pp.137–150). SPDA.
21. Hiriart-Bertrand, L., Silva, J. A., & Gelcich, S. (2020). Challenges and opportunities of implementing the marine and coastal areas for indigenous peoples policy in Chile. *Ocean and Coastal Management, 193,* 105233.
22. Hucke-Gaete, R., Álvarez, R., Gálvez, M., Farías, A., Lo Moro, P., Montecinos, Y., Navarra, M., & Ruíz, J. (2010). *Conservando el mar de Chiloé, Palena y Las Guaitecas. Síntesis del estudio "Investigación para el desarrollo de Área Marina Costera Protegida Chiloé, Palena y Guaitecas"*. Universidad Austral de Chile, Imprenta América.
23. Hucke-Gaete, R., Viddi, F. A., & Simeone, A. (2023). *Marine mammals and seabirds of Chilean Patagonia: Focal species for the conservation of marine ecosystems*. Springer.
24. Martínez, I., & Paredes, C. (2020). Régimen jurídico del desarrollo de la salmonicultura en áreas protegidas. *Fundación Terram. 35*. https://www.terram.cl/descargar/recursos_naturales/salmonicultura/app_-_análisis_de_políticas_públicas/Conservando-o-Cultivando-Regimen-jurídico-del-desarrollo-de-la-salmonicultura-en-areas-protegidas.pdf
25. Ministerio de Bienes Nacionales. (2017). *Sistematización del proceso de consulta indígena al pueblo Kawésqar por la ampliación y re-clasificación de la Reserva Nacional Alacalufes*. http://www.bienesnacionales.cl/wp-content/uploads/2017/12/InformeFinal_pci_revltv_nba_lpa_final_05102017_f2.pdf
26. Ministerio del Medio Ambiente. (2018). *Área Marina Costera Protegida de Múltiples Usos y Parque Marino Francisco Coloane*. Plan Estratégico para el manejo de conservación y desarrollo de actividades sostenibles. 2018–2027. http://catalogador.mma.gob.cl:8080/geonetwork/srv/spa/resources.get?uuid=c9ea050a-f6aa-491e-897d-388f35c5671b&fname=Plan%20Estrategico%20para%20el%20manejo%20de%20conservaci%C3%B3n%20Francisco%20Coloane_WCS_Versi%C3%B3n%20WEB.pdf&access=public
27. Molinet, C., & Niklitschek, E. J. (2023). *Fisheries and marine conservation in Chilean Patagonia*. Springer.
28. Oceana. (2009). *Área Marina Costera Protegida de Múltiples Usos - Tortel 1–80*. https://chile.oceana.org/sites/default/files/reports/oceana_2010final.pdf
29. Praus, S., Palma, M., & Domínguez, R. (2011). *La situación jurídica de las actuales áreas protegidas de Chile*. Fondo del Medio Ambiente Mundial.
30. Recordón, H. J. (2013). Áreas protegidas en el espacio marino y costero nacional. In H. D. J. Bermúdez (Ed.), *Justicia ambiental, derecho e instrumentos de gestión del espacio marino costero* (pp. 277–306). LOM.
31. Roberts, C. M., O'Leary, B. C., McCauley, D. J., Cury, P. M., Duarte, C. M., Lubchenco, J., Pauly, D., Sáenz-Arroyo, A., Sumaila, U. R., Wilson, R. W., Worm, B., & Castilla, J. C. (2017). Marine reserves can mitigate and promote adaptation to climate change. *Procedings of the National Academy of Sciences, 114*, 6167–6175.
32. Rozzi, R., Massardo, F., Mansilla, A., Squeo, F. A., Barros, E., Poulin, E., Frangopulos, M., Rosenfeld, S., Contador, T., González-Weaver, C., MacKenzie, R., Crego, R. D., Viddi, F.,

Naretto, J., Gallardo, M. R., Jiménez, J. E., Pérez, C., Rodríguez, J. P., Méndez, F., Barroso, O., Rendoll, J., Schüttler, E., Morello, F., Carvajal, D., Kennedy, J., Convey, P., Russell, S., Berchez, F., Sumida, P. Y. G., Müller, E., Izurieta, A., Cruz, F., Rozzi, A., Armesto, J., Kalin-Arroyo, M., & Martinic, M. (2017). *Informe técnico para la propuesta de creación Parque Marino Cabo de Hornos – Diego Ramírez.* https://issuu.com/umag9/docs/ebook_low_parque_marino_umag_fin_di

33. Rozzi, R., Rosenfeld, S., Armesto J. J., Mansilla, A., Núñez-Ávila, M., & Massardo, F. (2023). *Ecological connections across the marine-terrestrial interface in Chilean Patagonia.* Springer.

34. Sierralta, L., Serrano, R., Rovira, J., & Cortés, C. (2011). *Las áreas protegidas de Chile. Antecedentes, institucionalidad, estadísticas y desafíos.* División de Recursos Naturales Renovables y Biodiversidad, Ministerio del Medio Ambiente. http://bibliotecadi-gital.ciren.cl/bit stream/handle/123456789/6990/HUM2-0008.pdf?sequence=1&isAllowed=y

35. Skewes, J. C., Álvarez, R., & Navarro, M. (2012). Usos consuetudinarios, conflictos actuales y conservación en el borde costero de Chiloé insular. *Magallania (Punta. Arenas), 40,* 109–125.

36. Subsecretaría de Pesca y Acuicultura. (2006). *Identificación de zonas representativas de los ecosistemas marinos nacionales susceptibles de ser declaradas como Áreas Marinas Protegidas asociadas al ámbito del sector pesquero.* SUBPESCA.

37. Subsecretaría de Pesca y Acuicultura. (2017). *Antecedentes técnicos Parque Marino Isla Diego Ramírez - Paso Drake, Región de Magallanes y Antártica chilena. Informe Técnico (R. Pesq.) n°220/2017.* SUBPESCA. http://www.umag.cl/vcm/wp-content/uploads/2017/11/SUB PESCA-R.Pesq-N%C2%B0220-2017-PM-Islas-Diego-Ramírez_Paso-Drake.pdf

38. Subsecretaría de Pesca y Acuicultura. (2020). *Aplicación de visualización de mapas de la Subsecretaría de Pesca y Acuicultura.* http://mapas.subpesca.cl/ideviewer/

39. Tacón, A., Tecklin, D., Farías, A., Peña, M. P., & García, M. (2023). *Terrestrial protected areas in Chilean Patagonia: Characterization, historical evolution, and management.* Springer.

40. Tecklin, D. (2015). La apropiación de la costa chilena: ecología política de los derechos privados en torno al mayor recurso público del país. In M. Prieto, B. Bustos, & J. Barton (Eds.), *Ecología política en Chile: Naturaleza, propiedad, conocimiento y poder* (pp. 121–142). Editorial Universitaria.

41. Vila, A., Püschel, N., Rodríguez, M., Guijón, R., & Kusch, A. (2017). Propuesta Área Marina Costera Protegida de Múltiples Usos "Seno Almirantazgo" Tierra del Fuego, Región de Magallanes y de la Antártica chilena. Informe técnico. Wildlife Conservation Society. https://programs.wcs.org/DesktopModules/Bring2mind/DMX/Download.aspx?EntryId=33731&Por talId=134&DownloadMethod=attachment

David Tecklin BA, Swarthmore College, M.A. University of California, Berkeley and Ph.D. in Geography, University of Arizona, USA. Research Associate, Austral Patagonia Program, Universidad Austral de Chile. Principal Officer, The Pew Charitable Trusts.

Aldo Farías Forestry Engineer, Universidad Austral de Chile. Executive Coordinator, Austral Patagonia Program, Universidad Austral de Chile, Valdivia.

María Paz Peña Biologist in Natural Resources and Environment, Pontificia Universidad Católica de Chile. Ph.D. in Sciences, mention Systematics and Ecology, Universidad Austral de Chile. Researcher, Austral Patagonia Program, Universidad Austral de Chile.

Xiomara Gélvez Environmental engineer, University of Antioquia, Colombia. Specialist in cartography. Austral Patagonia Program, Universidad Austral de Chile, Valdivia.

Juan Carlos Castilla Professor of Natural Sciences and Chemistry, Pontificia Universidad Católica de Chile. Ph.D. Marine Biology and D.Sc., Bangor University, Wales, U.K. Full Professor and Emeritus, Pontificia Universidad Católica de Chile. In 2010 received the National Award for Applied and Technological Sciences.

Maximiliano Sepúlveda Veterinarian, Universidad de Chile. M.Sc. Universidad Austral de Chile. Ph.D. University of Minnesota, USA. Works in research and management in biological conservation with emphasis on protected areas and endangered species.

Francisco Viddi Marine Biologist, Universidad Austral de Chile. Ph.D. in Environmental Sciences, Macquarie University, Australia. Specialist in marine mammal ecology and marine conservation. Member of the IUCN Cetacean Specialist Group.

Rodrigo Hucke-Gaete Marine Biologist, Universidad Austral de Chile. Doctor in Sciences and professor at Universidad Austral de Chile. Marine ecologist specialized in marine mammals and marine conservation. Member of the IUCN Cetacean Specialist Group.

Marine Mammals and Seabirds of Chilean Patagonia: Focal Species for the Conservation of Marine Ecosystems

9

Rodrigo Hucke-Gaete, Francisco A. Viddi, and Alejandro Simeone

Abstract

Chilean Patagonia boasts 100,627 km of coastline, including approximately 40,050 islands and numerous fjords and channels, which generate a high degree of geomorphological and hydrographic complexity and make it one of the largest mega-estuarine environments on Earth. These characteristics, among others, generate structurally and functionally unique marine ecosystems, as well as biodiversity hotspots. In this chapter we perform a comprehensive literature review and highlight the use of marine mammals and seabirds as focal species to guide conservation initiatives aimed at minimizing anthropogenic impacts as well as safeguarding the ecosystem integrity of Patagonia as a climate refuge. Given their characteristics as ecologically relevant umbrella, indicator, and sentinel species, we suggest that focal species are highly useful in guiding the prioritization of management and conservation initiatives to achieve world-class conservation standards in Chilean Patagonia.

Keywords

Patagonia · Chile · Climate change refugia · Marine protected areas · Marine spatial planning · Threats · Ecosystem integrity · Resilience

R. Hucke-Gaete (✉) · F. A. Viddi
Institute of Marine and Limnological Sciences, Universidad Austral de Chile, Campus Isla Teja, Valdivia, Chile
e-mail: rhucke@uach.cl

NGO Centro Ballena Azul, Valdivia, Chile

A. Simeone
Department of Ecology and Biodiversity, Faculty of Life Sciences, Universidad Andrés Bello, Santiago, Chile

J. C. Castilla et al. (eds.), *Conservation in Chilean Patagonia*, Integrated Science 19, https://doi.org/10.1007/978-3-031-39408-9_9

1   Introduction

There is worldwide consensus that the oceans are in crisis, that fisheries are in decline, that the numbers of endangered species are increasing, and that different ecosystems have been damaged or destroyed due to anthropogenic pressures [69]. However, according to [21], over the coming decades there are still opportunities to rebuild marine life and ecosystem functions if the main pressures affecting the oceans can be mitigated by 2050. Very few places on Earth can be considered totally pristine, not even remote Chilean Patagonia. This area has sustained local livelihoods for thousands of years [7], has had episodes of excessive exploitation in the near past, and today maintains an important artisanal and industrial fishing sector [62], intense maritime transport and is the neuralgic center of the Chilean salmon and mussel aquaculture industry [59].

Chilean Patagonia's marine ecosystems have been described as biodiversity hotspots, particularly for high trophic level predator species such as marine birds and mammals. These groups have been recommended as focal species for the development of management and conservation proposals [41] due to their emblematic character, ecological roles, potential as key species in the habitats where they are found or because of their characteristics as umbrella species. These groups of animals can act as indicators of ecosystem health, since changes in their distribution, abundance, behavioral patterns and/or trophic ecology may reflect important changes in the environment, whether these are of natural or anthropogenic origin [11].

Marine protected areas have been proposed as a means of conserving Chilean Patagonia's marine ecosystems. However, it is not yet clear if these initiatives are sufficient to represent biodiversity adequately or if they are the optimum tool for conservation. This is in a region where a series of actors, often with conflicting interests, coexist and exert intense pressure on ecosystems and marine resources [34, 62]. The approach proposed in this chapter is to use seabirds and marine mammals as focal species to achieve broad conservation objectives. Focal species have been defined as those that warrant conservation interest because they possess characteristics that identify them as functionally important, key, umbrella, indicator, flagship, vulnerable or sensitive species, and therefore are useful for consideration both in the selection and delimitation of conservation initiatives and in planning and management, including research and monitoring.

2   Scope and Objectives

The objective of this chapter is to evaluate the state of knowledge and conservation of birds and marine mammals that inhabit Chilean Patagonia in the context of the history of exploitation and threats In this region. The aim is to identify gaps, challenges and opportunities for improvement to promote appropriate and effective management actions using the concept of focal species.

3 Methods

The mainstream and gray literature available for Chilean Patagonia between Reloncaví Sound and the Diego Ramírez islands (41° 42'S 73° 02'W; 56° 29'S 68° 44'W) was reviewed to identify the most relevant aspects of the biology of seabird and marine mammal species, the threats that affect them and the alternatives available to conserve these groups in their ecosystems. The authors' experience of more than two decades on these issues, both in Patagonia and in other parts of Chile, is added to this search. We distinguished two major Patagonian marine areas for this study: (i) the Chiloense Marine Ecoregion or northern Patagonia, *ca.* 41°–47°S [42], and (ii), the Channels and Fjords Marine Ecoregion of southern Chile or southern Patagonia, *ca.* 47°–56°S [101].

4 Results

4.1 Patagonia's Marine Ecosystems: a Hotspot for Focal Species

Chilean Patagonia extends linearly for more than 1,600 km of continental coastline, including 100,627 km of coastline and 40,050 islands [93], with a high degree of geomorphological and hydrographic complexity [74, 89]. These factors, added to the high variability of meteorological conditions, create ecosystems that are considered structurally and functionally unique. The oceanographic characteristics of the marine ecosystems of Chile's Patagonian fjords and channels generally have a permanent influx of Subantarctic oceanic water through channels and gulfs, which has higher temperature, nutrient concentration and salinity than the water in the interior zone. This oceanic water mixes in the coastal zone with freshwater generated by high precipitation, glacial melt and coastal runoff, thus producing a mega-estuarine system of positive circulation [73], Pickard and [74, 88]. The freshwater body is generally devoid of nutrients (except silicic acid), but contains high concentrations of particulate and dissolved organic matter [30]. The large Patagonian ice fields (North, South, Muñoz-Gamero, Santa Inés and Darwin) between 46°S and 48°S are considered valuable freshwater reservoirs of global importance and have a profound influence on the functioning of the marine ecosystems [68, 81]. Estuarine areas serve as habitat for many marine species during some phase of their lives. This includes several commercially important fish species that spawn on the open coasts of the Chonos Archipelago and Guafo Island, and whose eggs and larvae have been detected in the inland waters of the fjords and channels, which are thought to serve as nurseries in this initial life phase [8].

Several studies have reported the high productivity of inland Patagonian waters, particularly during spring, as reflected by the high growth rates of phytoplankton [49], which in turn favor the abundance of planktonic herbivores and carnivores [67]. Planktonic crustaceans, particularly copepods and euphausiids (krill), predominate in abundance in Patagonian fjords and channels. *Euphausia valentinii* is the most abundant euphausiid in Chilean Patagonia [67] and is considered a key species, as it establishes an ecological link between microplankton and higher trophic levels (i.e. fish, penguins and whales; [31]. The squat lobster (*Munida gregaria/subrugosa*) may constitute more than 50% of the macrobenthic invertebrate biomass [6], it is the most abundant decapod in the coastal waters of Tierra del Fuego, with abundance as high as 27 individuals/m^2 [33]. This species is preyed upon by a variety of higher order predators, including whales, dolphins, sea lions, birds, fish, spider crabs and octopuses. It is hypothesized that they are a direct trophic link between the detritus and the larger predators [83].

Due to this primary and secondary productivity, Chilean Patagonia is home to important populations of higher order predators such as seabirds and marine mammals. Some of these species are migratory, such as blue and humpback whales, as well as numerous seabirds (albatrosses, shearwaters, terns), while others are resident and maintain an annual presence in the area, such as sea lions, otters, dolphins, porpoises, black-browed albatrosses, imperial cormorants and Magellan penguins, among others [42, 98, 100]. Approximately 56 species of marine mammals have been recorded in Chile, representing 42% of the species richness of this functional group globally. A total of 32 species of cetaceans have been recorded in Chilean Patagonia, out of approximately 44 species present throughout the country [3], and 6 species of marine carnivores (sea lions, fur seals, seals and otters, [95] (Table 1). Until very recently, most of the information available in the literature on marine mammals in Chilean Patagonia was data collected during the whaling season through opportunistic sightings, strandings, range updates and osteological material, most of which is scattered in technical reports, conference reports and unpublished scientific papers [98].

Among the most outstanding features that have been reported recently for marine mammals in northern Patagonia is the presence of an important feeding and nursing area for blue whales (*Balaenoptera musculus*) [39]. Historical information from 1907 [94] indicates that masses of blue whales were common in the Gulf of Corcovado. However, the extraordinary presence of this species in the area was soon forgotten, and it was almost 100 years before the return of this species to this historic site was observed. An important feeding area for humpback whales (*Megaptera novaeangliae*) has been described in the Strait of Magellan (southern Patagonia, around Carlos III Island), and is the first such area recognized for the entire southeastern Pacific [26]. An additional area was later described in northern Patagonia [43]. Other species of large cetaceans frequently observed in feeding and/or transit behavior in Chilean Patagonia include sei or Rudolphi's whales (*Balaenoptera borealis*), fin whales (*Balaenoptera physalus*), southern right whales (*Eubalena australis*), common and Antarctic minke (*Balaenoptera bonaerensis, B. acutorostrata*) and sperm whales (*Physeter macrocephalus*) [3, 40, 42, 98].

There are at least 19 species of small cetaceans in the region (dolphins, ziphids and porpoises), including the Chilean dolphin (*Cephalorhynchus eutropia*), which is the only cetacean species endemic to Chile [100]. Four species of pinnipeds (sea lions and seals) have also been recorded,the most abundant are the common sea lion or fur seal (*Otaria byronia*) and the southern fur seal (*Arctocephalus australis*), which reproduce in the area. Although important knowledge gaps remain

Table 1 List of bird and marine mammal species recorded for Chilean Patagonia, including their conservation status according to the Red List of the International Union for Conservation of Nature* and the Chilean Ministry of the Environment** (in Spanish Ministerio del Medio Ambiente, MMA)

	Species	Common name	Conservation Category	
			IUCN	MMA
Class MAMMALIA				
Order CETACEA				
Family Balaenopteridae	Balaenoptera musculus	Blue whale	EN	EN
	Balaenoptera physalus	Fin whale	VU	CR
	Balaenoptera borealis	Sei whale	EN	CR
	Balaenoptera bonaerensis	Antarctic minke whale	NT	LC
	Balaenoptera acutorostrata	Common minke whale	LC	LC
	Megaptera novaeangliae	Humpback whale	LC	VU
Family Balaenidae	Eubalaena australis	Southern right whale	CR	EN
Family Neobalaenidae	Caperea marginata	Pygmy right whale	LC	DD
Family Physeteridae	Physeter macrocephalus	Sperm whale	VU	VU
Family Ziphiidae	Berardius arnuxii	Arnoux's beaked whale	DD	DD
	Hyperoodon planifrons	Southern bottle-nosed whale	LC	LC
	Ziphius cavirostris	Cuvier's beaked whale	LC	LC
	Tasmacetus shepherdi	Shepherd's beaked whale	DD	DD
	Mesoplodon densirostris	Blainville's beaked whale	DD	DD
	Mesoplodon layardii	Strap-toothed beaked whale	DD	DD
	Mesoplodon grayi	Gray's beaked whale	DD	DD
	Mesoplodon hectori	Hector's beaked whale	DD	DD
	Mesoplodon traversii	Spade-toothed whale	DD	DD

(continued)

Table 1 (continued)

	Species	Common name	Conservation Category	
			IUCN	MMA
	Mesoplodon bowdoini	Andrews' beaked whale	DD	DD
Family Delphinidae	*Orcinus orca*	Orca	DD	DD
	Pseudorca crassidens	False killer whale	NT	DD
	Globicephala melas	Long-finned pilot whale	LC	DD
	Lagenorhynchus obscurus	Dusky dolphin	LC	LC
	Lagenorhynchus australis	Peale's dolphin	LC	LC
	Lagenorhynchus cruciger	Hourglass dolphin	LC	LC
	Tursiops truncatus	Bottlenose dolphin	LC	LC
	Grampus griseus	Risso's dolphin	LC	LC
	Lissodelphis peronii	Southern right-whale dolphin	LC	DD
	Cephalorhynchus eutropia	Chilean dolphin	NT	NT
	Cephalorhynchus commersonii	Commerson's dolphin	LC	EN
Family Phocoenidae	*Phocoena spinipinnis*	Burmeister's porpoise	NT	DD
	Phocoena dioptrica	Spectacled porpoise	LC	DD
Order CARNIVORA				
Family Otariidae	*Otaria flavescens*	South American sea lion	LC	LC
	Arctocephalus australis	South American fur seal	LC	NT
Family Phocidae	*Mirounga leonina*	Southern elephant seal	LC	VU
	Hydrurga leptonyx	Leopard seal	LC	LC
Family Mustelidae	*Lontra felina*	Marine otter	EN	VU
	Lontra provocax	Southern river otter	EN	EN

(continued)

Table 1 (continued)

	Species	Common name	Conservation Category	
			IUCN	MMA
Class BIRDS				
Order ANSERIFORMES				
Family Anatidae	*Tachyeres patachonicus*	Flying steamer-duck	LC	LC
	Tachyeres pteneres	Fuegian steamer-duck	LC	NT
	Chloephaga hybrida	Kelp goose	LC	VU
Order SPHENISCIFORMES				
Family Spheniscidae	*Aptenodytes patagonicus*	King penguin	LC	–
	Spheniscus humboldti	Humboldt penguin	VU	VU
	Spheniscus magellanicus	Magellanic Penguin	NT	–
	Eudyptes chrysolophus	Macaroni penguin	VU	–
	Eudyptes chrysocome	Southern rockhopper penguin	VU	–
Order PROCELLARIIFORMES				
Family Diomedeidae	*Diomedea epomophora*	Southern Royal albatross	VU	–
	Diomedea sanfordi	Northern Royal albatross	EN	–
	Diomdea exulans	Wandering albatross	VU	–
	Diomedea antipodensis	Antipodes albatross	EN	–
	Thalassarche chrysostoma	Gray-headed albatross	EN	NT
	Thalassarche bulleri	Buller's albatross	NT	–
	Thalassarche melanophris	Black-browed albatross	LC	LC
	Thalassarche cauta	White-capped albatross	NT	–

(continued)

Table 1 (continued)

	Species	Common name	Conservation Category	
			IUCN	MMA
	Thalassarche salvini	Salvin's albatross	VU	–
	Thalassarche eremita	Chatham albatross	VU	–
	Phoebetria palpebrata	Light mantled albatross	NT	–
Family Procellariidae	*Macronectes giganteus*	Southern giant petrel	LC	–
	Macronectes halli	Northern giant petrel	LC	–
	Fulmarus glacialoides	Southern fulmar	LC	–
	Daption capense	Cape petrel	LC	–
	Aphrodroma brevirostris	Kerguelen petrel	LC	–
	Pachyptila desolata	Antarctic prion	LC	–
	Pachyptila belcheri	Thin-billed prion	LC	–
	Halobaena caerulea	Blue petrel	LC	–
	Procellaria westlandica	Westland petrel	EN	–
	Procellaria aequinoctialis	White-chinned petrel	VU	–
	Ardenna grisea	Sooty shearwater	NT	–
	Ardenna gravis	Great shearwater	LC	–
	Ardenna creatopus	Pink-footed shearwater	VU	EN
	Puffinus puffinus	Manx shearwater	LC	–
	Pelecanoides urinatrix	Common diving-petrel	LC	–
	Pelecanoides magellani	Magallanic diving-petrel	LC	–
Family Oceanitidae	*Fregetta tropica*	Black-bellied storm-petrel	LC	–

(continued)

Table 1 (continued)

	Species	Common name	Conservation Category	
			IUCN	MMA
	Oceanites oceanicus	Wilson's storm-petrel	LC	–
	Oceanites pincoyae	Pincoya storm-petrel	DD	–
Order SULIFORMES				
Family Phalacrocoracidae	*Phalacrocorax gaimardi*	Red-legged cormorant	NT	NT
	Phalacrocorax brasilianus	Neotropic cormorant	LC	–
	Phalacrocorax magellanicus	Rock cormorant	LC	–
	Phalacrocorax atriceps	Imperial cormorant	LC	–
Order PELECANIFORMES				
Family Pelecanidae	*Pelecanus thagus*	Peruvian pelican	NT	–
Order CHARADRIIFORMES				
Family Stercorariidae	*Stercorarius chilensis*	Chilean skua	LC	–
	Stercorarius parasiticus	Parasitic jaeger	LC	–
Family Laridae	*Chroicocephalus maculipennis*	Brown-hooded gull	LC	–
	Leucophaeus pipixcan	Franklin's gull	LC	–
	Leucophaeus scoresbii	Dolphin gull	LC	–
	Larus dominicanus	Kelp gull	LC	–
	Sterna hirundinacea	South American tern	LC	–
	Sterna paradisaea	Arctic tern	LC	–

* https://www.iucnredlist.org/
** https://clasificacionespecies.mma.gob.cl/

regarding the ecology of marine mammals and the marine systems on which they depend in this region, this gap is slowly being filled by systematic studies that have reported on the distribution, abundance, habitat modeling, behavioral and movement patterns, as well as the ecological determinants of these different processes [9, 10, 35, 39, 40, 43, 98–100]. These studies demonstrated that seasonal and spatial primary productivity is an important indicator of the meso-scale distribution and movement patterns of whales. At a finer scale, studies (particularly in dolphins) have shown how certain oceanographic processes (e.g. tidal fronts and currents), the influence of rivers and freshwater, as well as habitats formed by macroalgal/kelp forests, are of great importance for habitat selection and essential biological behaviors such as reproduction and feeding.

One hundred and nine seabird species have been recorded in Patagonia [85], which represents 30% of the national species richness. Chilean Patagonia is home to nearly 50% of the seabirds recorded in Chile (Table 1). These figures make the Patagonian region an area of great importance in terms of seabird species richness for Chile and the world. An important number of seabirds that inhabit or visit Chilean Patagonia are high trophic level predators. The most common albatross species in Patagonia is the black-browed albatross (*Thalassarche melanophris*), which reaches its highest abundance in summer at breeding sites located in southern Patagonia. At least six colonies of this species have been documented between 51–56°S, totaling > 134,000 pairs [5, 55, 58, 82]. The sooty shearwater (*Ardenna grisea*) is the most abundant species in northern Patagonia during the summer months, when it arrives in large numbers to breed [79]. It is frequently observed in flocks of thousands of individuals, especially during their migrations. Breeding colonies of this species have been identified on the Metalqui, Guamblin, and Guafo islands. The last of these has an estimated population of over 4 million breeding pairs [79], and colonies have up to 300,000 pairs on the Wollaston and Hermite islands [86]. Another species that maintains an important population in Chilean Patagonia is the Magellanic penguin (*Spheniscus magellanicus*). Boersma et al. [12] estimated that there are at least 23 colonies of this species with > 144,000 pairs between 41–55°S. [76] estimated at least 12 nesting sites of southern rockhopper penguin (*Eudyptes chrysocome*) in southern Patagonia, with > 396,000 pairs. [19] mentioned 12 colonies of macaroni penguin (*Eudyptes chrysolophus*) in southern Patagonia, but the population size in this area is undetermined and apparently declining.

Other seabird species that visit Chilean Patagonia include the wandering albatross (*Diomedea exulans*), northern royal albatross (*Diomedea sanfordi*), southern royal albatross (*Diomedea epomophora*), Salvin's albatross (*Thalassarche salvini*) and the Westland petrel (*Procellaria westlandica*). The Antarctic giant petrel (*Macronectes giganteus*), the southern fulmar (*Fulmarus glacialoides*), the Magellanic diving-petrel (*Pelecanoides magellani*) and the Wilson's storm-petrel (*Oceanites oceanicus*) are other relatively common Procellariiformes at certain times of the year, many of which nest in the region [15, 46, 86].

In summary, Chilean Patagonia contains a high diversity of focal species of birds and marine mammals. This diversity could be explained by the significant heterogeneity of Patagonia's environment, its primary and secondary productivity, and the processes that sustain them. Compared to other areas of Chile, and certainly the world, this vast region is home to emblematic animal groups, many of which are classified as vulnerable or endangered (IUCN, 2018), and which are potentially key to the functioning of the ecosystems located here (Table 1). The presence of these species groups is a great opportunity to boost conservation efforts under a focal or umbrella species approach.

4.2 Areas Identified as Relevant for Marine Biodiversity in Chilean Patagonia

One of the most interesting prioritization exercises, due to its large geographic scope, was carried out for the Chiloense Marine Ecoregion; it identified 13 ecologically important areas suitable for recommendation as MPAs (for details see Fig. 1 in [42]. This exercise, the first of its kind in Chile, was performed using MARXAN software [104], it incorporated the best available information on ecological aspects (*e.g.* species, bio-oceanographic processes, ecosystems) and human aspects (e.g. costs). The identification of sites of conservation importance was guided by three main criteria, to: (i) represent the critical biodiversity of the Chiloense ecoregion; (ii) reflect the threats in the area and (iii) incorporate the working scale of ecoregions. A second exercise was conducted for the southern Patagonia region [101], it identified 33 ecologically important areas (Fig. 1).

4.3 Current and Potential Threats: Challenges and Obstacles for the Conservation of Marine Ecosystems in Chilean Patagonia

Chilean Patagonia has been occupied by humans for over 10,000 years. Until the early nineteenth century, this occupation included only subsistence uses by the five native peoples that inhabited this region [7]. Subsequently, human occupation in Patagonia was encouraged through processes of intensive natural resource extraction by people who saw this area as a place to obtain profit and then leave. Beginning at the end of the nineteenth century, sea lions (common and fur seals) and otters (chungungo or marine otter and huillín or southern river otter) were an important focus of exploitation [61]. Heavy hunting pressure on these species brought them to the brink of extinction and was followed closely and in parallel by the hunting of large cetaceans, primarily right, blue, humpback and sperm whales in the Gulf of Corcovado and exposed coast of Chiloé [78].

Although there is no current economic activity based on hunting these species, there is strong pressure on the proper functioning and sustainability of marine ecosystems, as well as on the species that inhabit them. Aquaculture, industrial and artisanal fishing, as well as coastal development projects, tourism, and transportation stand out as major threats [32, 59]. All these activities produce a number of ecological impacts or effects on marine mammal and bird species (Figures 1 and 2). The population status of more than 95% of the major fishery resource species in Chile is uncertain or clearly overexploited [70]. Six species of high commercial importance in Patagonia are considered overexploited or collapsed fisheries: southern hake, hoki and southern blue whitting (*Merluccius australis*,

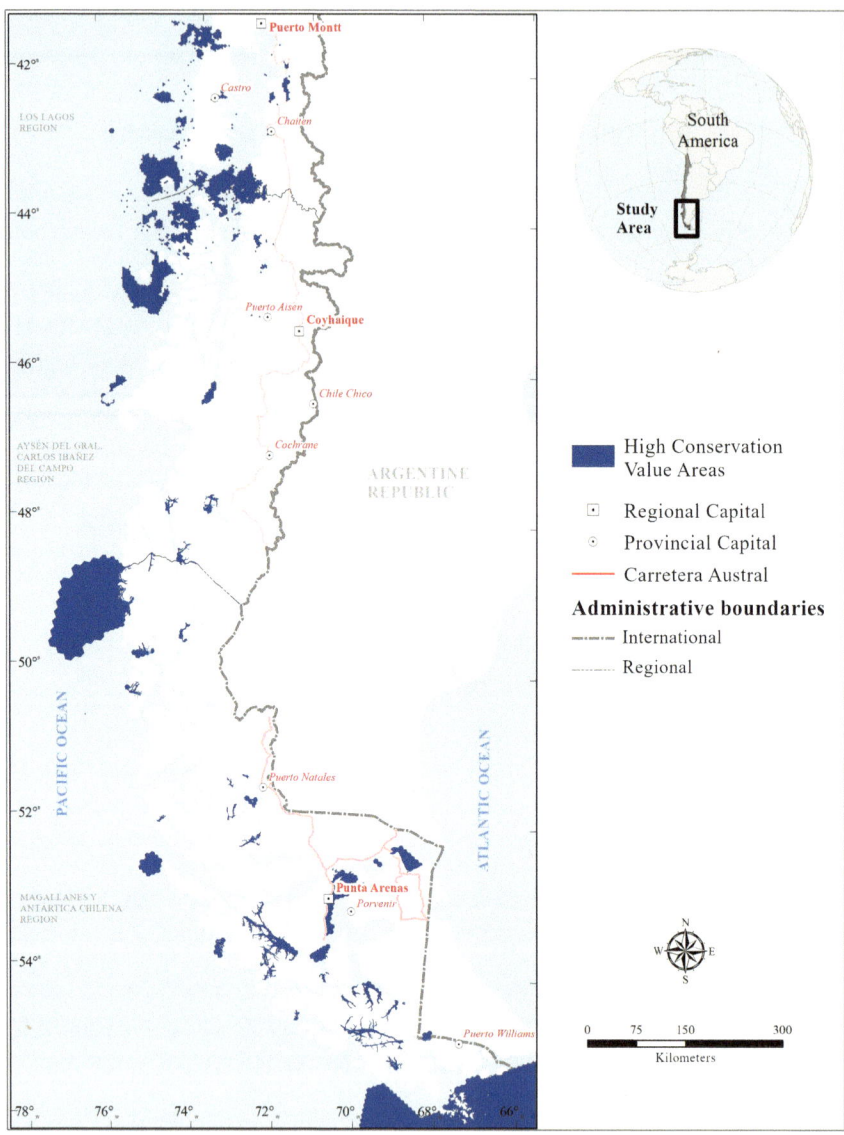

Fig. 1 Map of high conservation value areas in chilean patagonia (modified from [42] and [101])

Macruronus magellanicus, and *Micromesistius australis*, respectively), sea urchin (*Loxechinus albus*), loco (*Concholepas concholepas*) and deep-sea cod (*Dissostichus eleginoides*) [62]. The negative interaction between fisheries and non-target marine fauna is little studied in Chile. These interactions have only been evaluated in a few cases, primarily between mammals and seabirds and fisheries (*e.g.* [20, 40, 45, 63, 87].

Mortality of the white-chinned petrel (*Procellaria aequinoctialis*) by artisanal longline fleets of southern hake and deep-sea cod has been reported in northern Patagonia. [91] indicated that mortality of Magellan penguin and sooty shearwater in gill nets is common in this area. Of great concern is a recent study by the [47], which estimated that between 2015 and 2018, more than 10,000 (95% CI = 6,898- 16,670) black-browed albatrosses died as result of interactions with trawlers in the southern austral demersal fishery. Since its large-scale implementation in Chilean waters in the early 1980s, the aquaculture industry has increased its initial production more than 140 times, especially in the Los Lagos Region (northern Patagonia) where > 90% of national production is located. Chile is currently the second largest producer of salmonids in the world (485,000 t/year). The production of the blue mussel (*Mytilus edulis chilensis*) (58,000 t/year), although less important than that of salmon, is considered one of the most significant such industries in the Southern Hemisphere.[1] Mussel aquaculture occurs massively in coastal waters and does not require nets, cages, or supplementary feed. However, cultivation of these mollusks occupies large areas and can cause significant organic enrichment, mainly on the seafloor, due to high bio-deposition rates (fecal and pseudo-fecal), as well as the frequent detachment of mussels from suspended systems. These events significantly alter the chemical composition of the sediment and reduce the amount of available oxygen [18]. Little is known about how these crops impact birds and marine mammals, with the exception of the habitat displacement and spatial disturbance that the crop structures impose on Chilean dolphins [80], and the contrasting potential benefit as a food and resting source for the flightless steamer-duck (*Tachyeres pteneres*) [60].

Intensive salmon farming in Chile has considerable impacts on the marine environment [13], since the activity is based on supplementary feeding (food rich in phosphorus and nitrogen), the use of significant quantities of antibiotics and other chemicals (e.g. pesticides, disinfectants, antifoulants), as well as the presence of cages, anchorages and nets, and the constant re-supply by sea. This industry has different impacts on the marine ecosystems of Patagonia [13]. Interactions between aquaculture and marine mammals are often negative, as mammals are affected by habitat loss, gunfire used to deter approaches (mainly sea lions) and accidental entanglement in sea lion protection nets or anchoring lines [43, 77, 80]. However, the indirect negative effect that the industry generates on ecosystems is probably much more relevant, with the massive escapes of these exotic and eurytrophic species, the spread of parasites and diseases to native species, eutrophication and anoxia of entire fjords, among many other impacts [59, 62].

As human activities intensify in Chilean Patagonia (particularly salmon farming), so does maritime traffic, which has been widely recognized as an important factor affecting seabird and marine mammal populations. The risk of collision represents a danger to these species [54], and also the underwater noise generated by

---

[1] http://www.sernapesca.cl/informacion-utilidad/anuarios-estadisticos-de-pesca-Y-acuicultura.

cavitation can generate changes in behavior, distribution, abundance and population dynamics [36]. Main shipping routes are located between Puerto Montt and Aysén Fjord as result of the increased transport of cargo, fuel, tourist activities, aquaculture and fishing. A recent study by [10] identified three potentially conflictive zones due to the overlap between important areas for blue whales, salmon farming concessions and the density of maritime traffic. These are the Gulf of Ancud in the Chiloé inland sea, the Corcovado Gulf and the Moraleda Channel (Fig. 2). Collisions with blue whales and sei whales have been recorded recently, both species whose conservation status is of concern, and which are probably unable to sustain much mortality in addition to natural mortality (Fig. 2) [41, 10].

Other environmental impacts resulting from oil spills have not been investigated and are only scarcely monitored. Examples include the May 2001 spill from the Panamanian-flagged oil tanker José Fuchs, which released 440 t of crude oil along 120 km of coastline in the southern area of the Moraleda Channel, and the July, 2019 spill of 40 thousand liters of diesel oil off Guarello Island by a mining company north of the Kawésqar National Reserve [16].

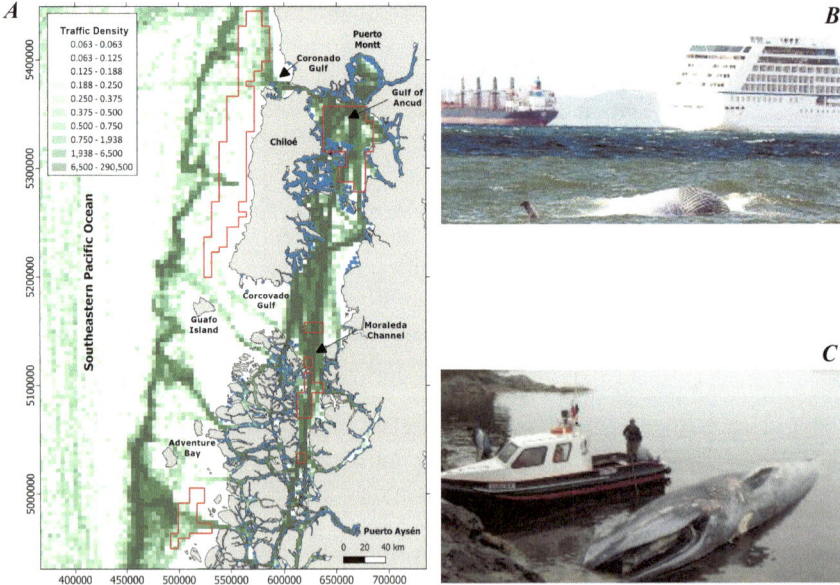

Fig. 2 **a** Maritime traffic in northern Patagonia and its overlap with areas that include 20% of the highest predicted densities for blue whales [10] (red polygons). The raster gradient (from light gray to green) indicates the density of ship positions per km^2 according to Automatic Identification System (AIS) data averaged from 2012 to 2016. (*Source* www.globalfishingwatch.org). Dots in dark blue indicate salmon farming concessions in 2013. (*Source* www.subpesca.cl). **b** Dead blue whale floating near Puerto Montt with fractured jaw and pectoral fin (*Source* El Llanquihue, front page, 13 February, 2014). **c** Blue whale stranded in Melimoyu bay (Commune of Puerto Cisnes) with the caudal fin severed at its base, most likely by a propeller (*Source* El Diario de Aysén, 24 February 2017)

In addition to increased boat traffic, one of the most pervasive and long-lasting human impacts is the generation of pollution, including plastic accumulation and fragmentation. Almost 80% of global floating marine debris comes from coastal human settlements, while the remaining 20% comes from vessels and ocean platforms [16]. Floating garbage is a threat to hundreds of species of birds, mammals, sea turtles and fish, which tend to become entangled, drown or suffer damage to their digestive systems [53]. It is common in Patagonia to observe large amounts of garbage, including plastic bags, ropes and net debris (Fig. 3).

Hinojosa [37] reported that between 1 and 50 items/km^2 of floating marine debris were recorded in northern Patagonia during seven Maritime Training and Instruction Center cruises between 2002 and 2005. This is substantially higher than the numbers reported for open coastal waters (0.01–25 items/km^2) and very close to values reported for semi-enclosed bays in highly populated regions around

Fig. 3 Examples of beaches with an accumulation of anthropogenic garbage near Puerto Aguirre, northern Patagonia. Plastic debris found throughout the area's beaches come from salmon farming activities, as well as household and fishing waste (© R. Hucke-Gaete)

the World (40 items/km^2). This figure increases considerably in Chiloe's inner sea, where the maximum abundance of garbage was found: 250 items/km^2; 80% of this was extruded polystyrene foam (styrofoam) and the rest included plastic fragments, plastic bags, ropes and salmon feed sacks. The problem persists, and it has been estimated using multispectral satellite imagery that more than 50 t of marine debris could be found along a 100 km stretch of Patagonian coastline [1], much of which can cause entanglement or obstruction of the respiratory and/or gastric tracts in species such as birds and marine mammals. This threat becomes increasingly complex to manage as plastic degrades into microparticles and fibers, which have already been recorded in sea lion feces [71], and crab stomachs in areas as isolated as Cape Horn [4].

Climate change will affect the physical, biological and biogeochemical properties of the oceans and coasts at different spatial and temporal scales, modifying their structure and ecological functions [48, 59]. These changes, in turn, will cause feedback in the climate system. The environmental stress in which the oceans find themselves, due to a combination of various factors, will affect the resilience of some marine ecosystems to climate change. Variations in the exchange of freshwater and matter between oceanic and terrestrial and coastal systems in Chilean Patagonia, triggered by climate change or direct human activities, are projected to affect the cycling of nutrients and carbon and therefore the health of coastal fjord ecosystems [50]. Harmful algal blooms (HABs) worldwide have increased in frequency, magnitude, intensity and geographic extent in recent decades [65]. This is especially critical in Chilean Patagonia, with historically recurrent but increasingly intense HAB events [59]. The largest mass mortality of sei whales ever recorded in the world (more than 343 individuals) was observed south of the Taitao Peninsula during the summer of 2015; the event was attributed to an intense HAB during an El Niño event [33]. Another event was observed in the summer of 2016 that caused massive mortality of invertebrates and fish, generating losses greater than US$ 800 million and sanitary problems due to more than 40,000 t of decomposing biomass [56]. In addition to HABs, an increase in populations of predatory gelatinous organisms (cnidarians and ctenophores) has been detected in recent decades in various marine ecosystems, attributed to climate change and/or fisheries that have eliminated natural predators of these organisms [75]. Gelatinous organisms are voracious predators that can affect the structure and dynamics of pelagic communities by consuming a wide variety of herbivorous zooplankters and fish in early stages [67]. For example, the massive proliferation of gelatinous filter feeder (subantarctic salp, *Ilhea magalhanica*) in the Chiloé Inland Sea caused fish mortality and a drastic decrease in phytoplankton cells, as well as a historical low in surface chlorophyll concentration [27].

5   Discussion

5.1   Integrated Conservation of Marine Ecosystems in Chilean Patagonia—Gaps, Challenges, and Opportunities

5.1.1   The Current Scenario of Marine Protection in Chilean Patagonia

Given the accumulation of evidence that marine protected areas (MPAs) contribute to the conservation of habitats and populations (Halpern, 2003) and that generates a positive spillover effect, which can maintain and even increase the overall yield of adjacent fisheries [25], the United Nations Environmental Program-World Conservation Monitoring Centre (UNEP-WCMS, 2018) has generated scenarios for their accelerated development. The global MPA surface was approximately 2 million km^2 (0.7% of the oceans) in 2000. This area has increased to *ca.* 27 million km^2 (*ca.* 7.5% of the oceans) since then, with more than 15,000 MPAs established around the world [97]. In Chile, however, unlike terrestrial environments, the history of MPA creation is recent and sporadic. The 176 km^2 Estero Quitralco National Sanctuary was created in Chilean Patagonia in 1996; it can be considered the first MPA in Chile.

MPAs covered less than 0.5% of the total sea area of Chile's Exclusive Economic Zone (EEZ) until 2009. In 2010, with the creation of the Motu Motiro Hiva Marine Park (150,000 km^2), Chile began to play a leading role in the creation of large oceanic MPAs [23]. In the following years the country designated more than 1.4 million km^2 of marine areas as MPAs, reaching more than 43% of the EEZ. This area is larger than the entire surface of continental Chile (750,000 km^2) and seven times larger than the surface of the terrestrial protected areas created in the country in the last 100 years. However, more than 90% of the area protected lies in the waters of the EEZ around oceanic islands (in territories beyond 12 nautical miles from the continental coast or territorial sea): Rapa Nui and Salas y Gómez Islands, Juan Fernández Archipelago, Desventuradas Islands and Diego Ramírez Islands.

This leaves an important gap in biological representation and coverage [93] and becomes even more relevant if we consider that there are areas of high biodiversity that have no MPA coverage [42, 101].

Humpback whale diving in a Patagonian feeding ground, Carlos III MPA

5.1.2 Focal Species and Their Use in Conservation in Chilean Patagonia

Due to their large biomass and historical abundance, several species of marine mammals are important consumers of productivity at different trophic levels, and are considered key focal species that play an essential ecological role in maintaining the integrity of the community structure and dynamics and the flow of nutrients and energy [38]. Robertson et al. [82] suggested that large cetaceans could play a role analogous to marine upwellings, by lifting nutrients from the depths and releasing them to the surface as fecal material that tends to disperse rather than sink [38]. Proposed that the role of large cetaceans could be an important and little-considered essential piece to understand the high productivity of certain areas of Chilean Patagonia holistically.

Patagonian marine ecosystems have highly seasonal primary production, which results in an efficient carbon sink through sedimentation during the spring [29]. This also results in the transport and exchange of significant amounts of organic matter between terrestrial and marine systems, being the main contributor to the carbon flux of coastal marine ecosystems [84, 102]. Chilean fjord lands have recently been identified as carbon sinks [50], and it is suggested that Chilean Patagonia likely captures more CO_2 than is released on the coast of northern Chile [96]. Because considerable aggregations of whales feed throughout Chilean Patagonia during austral summer and autumn, the potential influence on the dynamics of primary productivity and ecological processes facilitated by this megafauna in the biogeochemical carbon cycle should be explored [22, 57]. Proposed that Chilean Patagonia be considered a climatic refuge where the recovery and maintenance of the integrity of marine ecosystems is promoted, pressures on them are reduced and thus ecosystem functions and services are strengthened. By promoting this, Chilean Patagonia would reinforce global efforts to mitigate the impact of climate change as a nature-based solution.

Important advances have been made in Chilean Patagonia during the last decade in the identification of significant habitats, or nuclei, for some important behaviors of marine mammals and birds [42, 86]. The results to date have helped to determine habitat selection, movement and distribution patterns associated with environmental and oceanographic conditions and factors that trigger these processes. The approach is to establish conservation and management efforts in those areas in such a way that the relatively well-understood focal species function as umbrella species, so that the conservation of their habitat also extends protection to less visible species. These species should also be considered indicators and incorporated into MPA management plans. According to the Commission for the Conservation of Antarctic Marine Living Resources,[2] which manages Southern Ocean fishery resource species from an ecosystem perspective, an indicator species must show a measurable response to changes in the availability of exploited species. This could, for example, include variations in population size, reproductive success, body mass or foraging behavior. This same concept can be used to measure the effectiveness of measures implemented in MPAs through the monitoring of carefully selected indicators (e.g. duration of feeding trips, growth rate of young, reproductive success, changes in diet, condition and survival of adults). This approach can improve the cost-effectiveness and standardization of monitoring to achieve the stated management objectives, by documenting key ecosystem parameters rather than attempting to obtain full understanding of complex processes before taking appropriate adaptive measures.

5.1.3   Is the Establishment of MPAs in Chilean Patagonia the Best Marine Conservation Tool? Recommendations for a Conservation Model Under a Multi-Sectoral Approach and Marine Spatial Planning

To date, there are 34 officially decreed MPAs in Chile under different categories, representing approximately 43% of the surface area of the EEZ. With this coverage, Chile has taken an important step toward meeting the Aichi goals (protection of 10% of the sea) and has undoubtedly become a major player worldwide in the creation of MPAs, particularly in large oceanic areas. However, most of these areas do not have a management plan. The Aichi targets not only address surface area but also require that these areas be effectively and equitably managed, ecologically representative and well- connected [14]. On that basis, we consider the State of

---

[2] CCAMLR: https://www.ccamlr.org/es/science/programa-de-seguimiento-del-ecosistema-de-la-ccrvma-cemp.

Chile to be far from achieving this international goal. Facing the challenges for adequate and effective management of MPAs is perhaps the greatest problem that Chile and many other countries have today [28].

Notwithstanding the existence of tools for the implementation of MPAs in Chile, the formal establishment of MPAs is complex, as the range of MPA categories is under the administrative wing of different government agencies, which has a direct impact on governance systems. The agencies do not necessarily coordinate or may even be in conflict over the jurisdiction of protected areas. This also leads to ineffective use of resources and replication of actions. The long-awaited Biodiversity and Protected Areas Service could be a major solution as a coordinating entity and for the effective management of MPAs [93].

The mere fact of establishing MPAs does not guarantee success in biodiversity conservation. When MPAs are simply decreed, but resources are insufficient for effective design and management, these areas become paper parks [103]. Availability of resources is an important determinant for the success or failure of an MPA, and also the lack of social involvement and lack of coordination between government agencies trigger a flawed and ineffective governance and management system [72]. Thus, the establishment of MPAs can generate the dangerous illusion of protection when in fact this is not occurring [2]. It appears there are no longer options for additional large-scale MPAs in Chile, so the core of marine conservation guidelines in Chilean Patagonia should focus on the appropriate design of a network of small and medium-sized MPAs (100–1,000 km^2). Comprehensive protection based on ecosystem and landscape management that holistically includes terrestrial and marine systems, which are so intertwined in Chilean Patagonia, is urgently needed [84, 92].

Chilean Patagonia is a geographically complex region, both because of its oceanographic, geological, cryosphere and ecological processes and because of the multiple spectra of interests and uses of marine ecosystems [42, 61, 62, 93]. It is therefore a region where decisions on the management, conservation and uses of natural resources are complex. Single solutions such as MPA designation are insufficient,attention should be channeled to multisectoral approaches. An example of such a process-oriented approach is marine spatial planning (MSP) [2], with appropriate attention to ecosystem services and human welfare in Patagonia [66]. It is important to highlight and foment the processes of macro- and micro-level coastal zoning of the administrative regions that are part of Patagonia, especially because this was established as a goal in the National Biodiversity Strategy 2017–2030. There are other tools that could contribute enormously if well implemented and carried out as part of zoning and MSP. [24] mentioned complementary or auxiliary alternatives for scaling up marine biodiversity conservation, which include business model innovations for biodiversity benefits through territorial fishing use rights (i.e. Management and Exploitation Areas for Benthic Resources) and the creation of municipal conservation areas. Both tools have demonstrated cross-cutting results in biodiversity conservation, improvement of livelihoods and the recovery of fishing stocks.

Another tool is the designation of Indigenous People's Coastal Marine Spaces (in Spanish ECMPOs), defined spaces whose administration is given to indigenous communities or associations that have exercised customary use of the space, ascertained by the National Corporation for Indigenous Development. All of the original peoples of Patagonia had and still maintain a maritime connection, including the incorporation of marine mammals into their daily lives for shelter, navigation, food, hunting and social bonding as well as in their mythology and cosmovision. Because of its recent implementation, we know little about successes or failures of ECMPOs as a conservation tool. The inclusion of these ECMPOs is certainly paramount in an MSP approach, given the high number of requests submitted to the Undersecretariat of Fisheries and Aquaculture [93]. One of their roles would be to act by moderating or safeguarding areas for the use of Patagonian marine resource species, both for the benefit of the Indigenous communities and for the many other activities that take place in the region. It is important to ensure that the State Protected Areas on land in Patagonia protect the marine space of their inland waters. An interesting recent case is Kawésqar National Park (former Alacalufes Forest Reserve, *ca.* 28,000 km^2) and the concomitant Kawésqar National Reserve (*ca.* 26,000 km^2), which aim to conserve the marine portion of the waters adjacent to the national park. The particularity of this case is that the National Forestry Corporation (in Spanish CONAF) is in charge of both areas, and therefore the elaboration and implementation of their respective management plans [93].

The prior planning processes carried out in Chilean Patagonia [42, 101], although important for identifying possible conservation areas, should now be complemented by new processes that fill knowledge gaps through new research efforts that cover broader spatiotemporal scales, consistent with the life history of focal species. Finally, we believe that the strategic use of mammals and seabirds as focal and sentinel species in Chilean Patagonia has great conservation potential. This makes sense from an ecological perspective given the characteristics already indicated above, and because seabirds and marine mammals are emblematic groups that generate empathy in the public and are part of ecosystems that provide services for human welfare [66] through special interest tourism, for example.

6   Conclusions and Recommendations

Chilean Patagonia presents areas of great importance for marine biodiversity and is potentially a refuge from climate change [22, 59]. The generation of comprehensive conservation processes must first involve eradicating or minimizing the threats that affect this region [21]. The conservation tools to be applied in Chilean Patagonia should consider addressing the shortcomings pointed out in this study, especially those related to the current ineffectiveness of MPAs due to low or nonexistent management, governance, organization, social involvement, resources, and available funds. These conservation tools will undoubtedly be relevant if they cease to be only paper-based and are implemented along with the application of tools emanating from MSP. Marine conservation efforts in Chilean Patagonia must find a balance between social needs, current uses and the need to protect biodiversity, including political, social, private sector, academic and NGO involvement. The following are considered priority recommendations:

- Focal species such as seabirds and marine mammals can be very useful to guide prioritization in management and conservation processes, ideally under an MSP approach, given their characteristics as umbrella, ecologically important, indicator and sentinel species. We consider it essential to minimize the anthropogenic stressors that negatively affect their populations. It is important to develop or update abundance estimates to monitor this, as well as to develop indicators of changes in the ecosystems. This should use standardized methodologies that allow us to establish trends and thus measure the effectiveness or failure of the measures implemented under an ecosystem approach.
- We recommend narrowing the various management and knowledge gaps to permit the adequate management and conservation of marine ecosystems in Chilean Patagonia. Special emphasis should be placed on the gaps in representativeness and challenges in the adequate implementation of MPAs. It is also essential to advance the priority research topics identified by [77] related to aquaculture and its impacts, as well as to understand and promote the maintenance and restoration of the effects of marine vertebrates as essential components of ecological processes that promote carbon capture as a nature-based solution to the effects of climate change [38], see measure 4 in [22].
- MPAs are good alternatives to promote marine conservation processes, but they are not the only tool. It is also important to promote MSP initiatives in zoning processes both nationally and regionally, and to include additional tools such as ECMPOs along with achieving effective coordination with government agencies. Of relevance are the new processes for generating management plans for terrestrial protected areas (through CONAF) whose administrative boundaries include portions of the sea (inland waters, such as canals and fjords) [93].
- Within MSP processes it is essential to build opportunities for coordination and cooperation to define the function and role of the private sector. This is a determining aspect in the fulfillment of the objectives of different conservation tools, particularly in relation to adequate financing, which will make it possible to sustain adaptive and world-class management plans.
- We recommend that current and future MPAs aim to include IUCN-derived standards and promote their certification through the "Green List", an initiative that aims to recognize and increase the number of protected areas that function equitably, are well managed globally and deliver long-term conservation success.

Acknowledgements The authors would like to thank the Programa Austral Patagonia (UACH) and The Pew Charitable Trusts for making this work possible. Our gratitude also goes to the coordinators of this process, especially to Dr. Juan Carlos Castilla. Additionally, special thanks to Mr. Aldo Farías for the preparation of Fig. 1 and to Dr. Luis Bedriñana-Romano for the preparation of Fig. 3. We thank an anonymous referee for constructive comments on an earlier version of this chapter.

References

1. Acuña-Ruz, T., Uribe, D., Taylor, R., Amézquita, L., Guzmán, M. C., Merrill, J., Martínez, P., Voisin, L., Mattar, B., & C. (2018). Anthropogenic marine debris over beaches: Spectral characterization for remote sensing applications. *Remote Sensing of Environment, 217*, 309–322.

2. Agardy, T., di Sciara, G. N., & Christie, P. (2011). Mind the gap: Addressing the shortcomings of marine protected areas through large scale marine spatial planning. *Marine Policy, 35*(2), 226–232.

3. Aguayo-Lobo, A., Torres, D. N., & Acevedo, J. R. (1998). Los mamíferos marinos de Chile: I. *Cetacea. Serie Científica INACH, 48*, 19–159.

4. Andrade, C., & Ovando, F. (2017). First record of microplastics in stomach content of the southern king crab *Lithodes santolla* (Anomura: Lithodidadae), Nassau Bay, cape Horn, Chile. *Anales del Instituto de la Patagonia, 45*(3), 59–65.

5. Arata, J., Robertson, G., Valencia, J., & Lawton, K. (2003). The Evangelistas islets, Chile: A new breeding site for black-browed albatrosses. *Polar Biology, 26*, 687–690.

6. Arntz, W. E., & Gorny, M. (1996). Cruise report of the joint Chilean-German-Italian Magellan 'Victor Hensen' Campaign in 1994. *Berichte zür Polarforschung, 190*, 1–113.

7. Aylwin, J., Arce, L., Guerra, F., Núñez, D., Álvarez, R., Mansilla P., Alday, D., Caro, L., Chiguay, C., and Huenucoy, C. (2023). *Conservation and Indigenous People in Chilean Patagonia.* Springer.

8. Balbontín, F., & Bernal, R. (1997). Distribución y abundancia de ictioplancton en la zona austral de Chile. *Ciencia y Tecnología Marina, 20*, 155–163.

9. Bedriñana-Romano, L., Viddi, F. A., Torres-Florez, J. P., Ruiz, J., Nery, M. F., Haro, D., Montecino, Y., & Hucke-Gaete, R. (2014). At-sea abundance and spatial distribution of South American sea lion (*Otaria byronia*) in Chilean northern Patagonia: How many are there? *Mammalian Biology, 79*(6), 384–392.

10. Bedriñana-Romano, L., Hucke-Gaete, R., Viddi, F. A., Morales, J., Williams, R., Ashe, E., Garcés-Vargas, J., Torres-Florez, J. P., & Ruiz, J. (2018). Integrating multiple data sources for assessing blue whale abundance and distribution in Chilean northern Patagonia. *Diversity and Distributions, 24*(7), 991–1004.

11. Boersma, D. (2008). Penguins as marine sentinels. *BioScience, 58*(7), 597–607.

12. Boersma, P. D., Garcia Borboroglu, P., Frere, E., Godoy, C., Kane, O., Pozzi, L. M., Pütz, K., Raya Rey, A., Rebstock, G. A., Simeone, A., Smith, J., van Buren, A., & Yorio, P. (2015). Pingüino de Magallanes. In: P., García, y P. D., Boersma (Eds). *Pingüinos, historia natural y conservación*, pp. 253–285. Buenos Aires, Argentina: Vázquez Mazzini editores.

13. Buschmann, A. H., Niklitschek, E. J., & Pereda, S. (2021). Acuicultura y sus impactos en la conservación de la Patagonia chilena. In: J. C., Castilla, J. J., Armesto y M. J., Martínez-Harms (Eds.), *Conservación en la Patagonia chilena: evaluación del conocimiento, oportunidades y desafíos*, pp. 367–387. Santiago, Chile: Ediciones Universidad Católica de Chile.

14. Convention on Biological Diversity (2018). *Decision adopted by the Conference of the Parties to the Convention on Biological Diversity at its 14th meeting.* Decision 14/8. Protected areas and other effective area-based conservation measures. Retrieved from: www.cbd.int/doc/decisions/cop-14/cop-14-dec-08-en.pdf.

15. Clark, G. S., Von Meyer, A. P., Nelson, J. W., & Watt, J. N. (1984). Notes on the sooty shearwater and other avifauna of the Chilean offshore island of Guafo. *Notornis, 31*, 225–231.

16. Coe, J. M., & Rogers, D. (2012). *Marine debris: Sources, impacts, and solutions.* Springer.

17. Cooke, J. G. (2018). *Balaenoptera musculus* (errata 2019). IUCN, 2018. *Red list of threatened species 2018*: e.T2477A156923585. Retrieved from: https://doi.org/10.2305/IUCN.UK.2018-2.RLTS.T2477A156923585.en.

18. Cranford, P. J., Strain, P. M., Dowd, M., Hargrave, B. T., Grant, J., & Archambault, M. C. (2007). Influence of mussel aquaculture on nitrogen dynamics in a nutrient enriched coastal embayment. *Marine Ecology Progress Series, 347*, 61–78.
19. Crossin, G. T., Trathan, P. N., & Crawford, R. J. M. (2015). Pingüino macaroni *(Eudyptes chryso- lophus)*. In: P., García, y P.D., Boersma (Eds). *Pingüinos, historia natural y conservación*, pp. 197–224. Vázquez Mazzini editores. Buenos Aires, Argentina.
20. De la Torriente, A., Quiñones, A., Miranda, D., & Echevarría, F. (2010). South American sea lion and spiny dogfish predation on artisanal catches of southern hake in fjords of Chilean Patagonia. *Journal of Marine Science, 67*, 294–303.
21. Duarte, C. M., Agusti, S., Barbier, E., Britten, G. L., Castilla, J. C., Gattuso, J. P., Fulweiler, R. W., Hughes, T. P., Knowlton, N., Lovelock, C. E., Ltze, H. K., Predragovic, M., Poloczanska, E., Roberts, C., & Worm, B. (2020). Rebuilding Marine Life. *Nature, 580*(7801), 39–51.
22. Farías, L., Ubilla, K., Aguirre, C., Bedriñana, L., Cienfuegos, R., Delgado, V., Fernández, C., Fernández, M., Gaxiola, A., González, H., Hucke-Gaete, R., Marquet, P., Montecino, V., Morales, C., Narváez, D., Osses, M., Peceño, B., Quiroga, E., Ramajo, L., Sepúlveda, H., Soto, D., Vargas, E., Viddi, F., & Valencia, J. (2019). *Nueve soluciones basadas en océano para contribuir a los NDC de Chile*. Comité Científico COP25, Mesa Océanos, Ministerio de Ciencia, Tecnología, Conocimiento e Innovación. Retrieved from: http://www.cr2.cl/wp-con tent/uploads/2019/12/Nueve-soluciones-para-NDC.pdf.
23. Fernández, M., Rodríguez, M., Gelcich, S., Hiriart-Bertrand, L., & Castilla, J. C. (2021). Advances and challenges in marine conservation in Chile: A regional and global comparison. *Marine and Freshwater Ecosystems, 1–12,*. https://doi.org/10.1002/aqc.3570.
24. Gelcich, S., Peralta, L., Donlan, C. J., Godoy, N., Ortiz, V., Tapia-Lewin, S., Vargas, C., Kein, A., Castilla, J. C., Fernández, M., & Godoy, F. (2015). Alternative strategies for scaling up marine coastal biodiversity conservation in Chile. *Maritime Studies, 14*(1), 5.
25. Gell, F. R., & Roberts, C. M. (2003). Benefits beyond boundaries: The fishery effects of marine reserves. *Trends in Ecology and Evolution, 18*(9), 448–455.
26. Gibbons, J., Capella, J., & Valladares, C. (2003). Rediscovery of a humpback whale, *Megaptera novaeangliae*, summering ground in the strait of Magellan, Chile. *Journal of Cetacean Research and Management, 5*, 203–208.
27. Giesecke, R., Clement, A., Garcés-Vargas, J., Mardones, J., González, H. E., Caputo, L., & Castro, L. (2014). Massive salp outbreaks in the inner sea of Chiloé island (southern Chile): Possible causes and ecological consequences. *Latin American Journal of Aquatic Research, 42*(3), 604–621.
28. Gill, D. A., Mascia, M. B., Ahmadia, G. N., Glew, L., Lester, S. E., Barnes, M., Craigie, I., Darling, E. S., Free, C. M., Geldmann, J., Holst, S., Jensen, O. P., White, A. T., Basurto, X., Coad, L., Gates, R. D., Guannel, G., Mumby, P. J., Thomas, H., … Fox, H. E. (2017). Capacity shortfalls hinder the performance of marine protected areas globally. *Nature, 543*(7647), 665–669.
29. González, H. E., Calderón, M. J., Castro, L., Clement, A., Cuevas, L., Daneri, G., Iriarte, J. L., Lizárraga, L., Martínez, R., Menschel, E., Silva, N., Carrasco, C., Valenzuela, C., Vargas, C. A., & Molinet, C. (2010). Primary production and its fate in the pelagic food web of the Reloncaví fjord and plankton dynamics of the interior sea of Chiloé, northern Patagonia, Chile. *Marine Ecology Progress Series, 402*, 13–30.
30. González, H. E., Castro, L. R., Daneri, G., Iriarte, J. L., Silva, N., Tapia, F., Teca, E., & Vargas, C. A. (2013). Land-ocean gradient in haline stratification and its effects on plankton dynamics and trophic carbon fluxes in Chilean Patagonian fjords (47–50°S). *Progress in Oceanography, 119*, 32–47.
31. González, H. E., Graeve, M., Kattner, G., Silva, N., Castro, L., Iriarte, J. L., Osman, L., Daneri, G., & Vargas, C. A. (2016). Carbon flow through the pelagic food web in southern Chilean Patagonia: Relevance of *Euphausia vallentini* as a key species. *Marine Ecology Progress Series, 557*, 91–110.
32. Guala, C., Veloso, K., Farías, A., & Sariego, F. (2021). Caracterización del desarrollo turístico asociado a las áreas silvestres protegidas de la Patagonia chilena. In: J. C., Castilla, J.

J., Armesto y M. J., Martínez-Harms (Eds.), *Conservación en la Patagonia chilena: evaluación del conocimiento, oportunidades y desafíos*, pp. 575–598. Santiago, Chile: Ediciones Universidad Católica de Chile.

33. Gutt, J., Helsen, E., Arntz, W., Buschmann, A. (1999). Biodiversity and community structure of the mega-epibenthos in the Magellan area (South America). Scientia Marina, 63(S1), 155–170. Häussermann, V., Gutstein, C. S., Bedington, M., Cassis, D., Olavarría, C., Dale, A. C., Valenzuela-Toro, A. M., Pérez-Álvarez, M. J., Sepúlveda, H. H., McConnell, K. M., Horwitz, F. E. and Försterra, G. (2017). Largest baleen whale mass mortality during strong El Niño event is likely related to harmful toxic algal bloom. *PeerJ, 5*, e3123.

34. Häussermann, V., Försterra, G., & Laudien, J. (2023). *Hard Bottom Macrobenthos of Chilean Patagonia: Emphasis on Conservation of Subltitoral Invertebrate and Algal Forests.* Springer.

35. Heinrich, S., Genov, T., Fuentes-Riquelme, M., & Hammond, P. S. (2019). Fine-scale habitat partitioning of Chilean and peale's dolphins and their overlap with aquaculture. *Aquatic Conservation Marine and Freshwater Ecosystems., 29*(S1), 212–226.

36. Hildebrand, J. A. (2009). Anthropogenic and natural sources of ambient noise in the ocean. *Marine Ecology Progress Series, 395*, 5–20.

37. Hinojosa, I. A., & Thiel, M. (2009). Floating marine debris in fjords, gulfs and channels of southern Chile. *Marine Pollution Bulletin, 58*(3), 341–350.

38. Hucke-Gaete, R.Whales might also be an important component in Patagonian fjord Ecosystems: Comment to Iriarte, et al. (2011). *Ambio, 40*(1), 104–105.

39. Hucke-Gaete, R., Osman, L. P., Moreno, C. A., Findlay, K. P., & Ljungblad, D. K. (2003). Discovery of a blue whale feeding and nursing ground in southern Chile. In: Proceedings of the Royal Society of London. Series B (Suppl.) *Biology Letters, 271*, S170-S173.

40. Hucke-Gaete, R., Moreno, C. A., & Arata, J. A. (2004). Operational interactions of sperm whales and killer whales with the Patagonian toothfish industrial fishery off southern Chile. *CCAMLR Science, 11*, 127–140.

41. Hucke-Gaete, R., Viddi, F. A., & Bello, M. E. (2006). Marine conservation in southern Chile: the importance of the Chiloe-Corcovado area for blue whales, biological diversity and sustainable development. Valdivia, Chile: Imprenta América.

42. Hucke-Gaete, R., Lo Moro, P., & Ruiz, J. (2010). Conservando el mar de Chiloé, Palena y Guaitecas. Síntesis del estudio Investigación para el desarrollo de Área Marina Costera Protegida Chiloé, Palena y Guaitecas. Valdivia, Chile: Imprenta América.

43. Hucke-Gaete, R., Haro, D., Torres-Flórez, J. P., Montecinos, Y., Viddi, F. A., Bedriñana, L., & Ruiz, J. (2013). A historical feeding ground for humpback whales in the eastern south Pacific revisited: The case of northern Patagonia, Chile. *Aquatic Conservation: Marine and Freshwater Ecosystems, 23*, 858–867.

44. Hucke-Gaete, R., Bedriñana-Romano, L., Viddi, F. A., Ruiz, J. E., Torres-Florez, J. P., & Zerbini, A. N. (2018). From Chilean Patagonia to Galapagos, Ecuador: Novel insights on blue whale migratory pathways along the Eastern South Pacific. *PeerJ, 6*, e4695.

45. Hückstädt, L. A., & Krautz, M. C. (2004). Interaction between southern sea lions *Otaria flavescens* and jack mackerel *Trachurus symmetricus* commercial fishery off central Chile: A geostatistical approach. *Marine Ecology Progress Series, 282*, 285–294.

46. Imberti, S. (2005). Distribución otoñal de aves marinas y terrestres en los canales chilenos. *Anales Instituto de la Patagonia (Chile), 33*, 21–30.

47. Instituto de Fomento Pesquero (2019). Programa de investigación del descarte y captura de pesca inci- dental y programa de monitoreo y evaluación de los planes de reducción del descarte y la captura de incidental en las pesquerías demersales 2018–2019. Informe final, sección II. Retrieved from: https://www.ifop.cl/wp-content/contenidos/uploads/Repositoriol fop/InformeFinal/2019/P-581141b.pdf.

48. Intergovernmental Panel on Climate Change (2014). *Climate Change 2014: Synthesis report.* Contribution of Working Groups I, II and III to the Fifth Assessment Report of the Intergovernmental Panel on Climate Change, core writing team, Pachauri, R. K. & Meyer, L. A. (Eds.). Geneva, Switzerland: IPCC.

49. Iriarte, J. L., González, H. E., Liu, K. K., Rivas, C., & Valenzuela, C. (2007). Spatial and temporal variability of chlorophyll and primary productivity in surface waters of southern Chile (41.5- 43°S). *Estuarine, Coastal and Shelf Science, 74*, 471–480.

50. Iriarte, J. L., González, H. E., & Nahuelhual, L. (2010). Patagonian fjord ecosystems in southern Chile as a highly vulnerable region: Problems and needs. *Ambio, 39*(7), 463–466.

51. International Union for Conservation of Nature (2018). Red list of threatened species 2018: e.T2477A156923585.

52. King, M. C. and Beazley, K. F. (2005). Using focal species for marine protected area network planning in the Scotia-Fundy region of Atlantic Canada. *Aquatic Conservation: Marine and Freshwaters Ecosystems 15*, 367–385. Retrieved from: https://doi.org/10.2305/IUCN. UK.2018-2.RLTS.T2477A156923585.en.

53. Laist, D. W. (1997). Impacts of marine debris: entanglement of marine life in marine debris including a comprehensive list of species with entanglement and ingestion records. In: J. M. Coe y D. B. Rogers (Eds.), *Marine debris: sources, impacts, and solutions*, pp. 99–139. New York, USA: Springer.

54. Laist, D. W., Knowlton, A. R., Mead, J. G., Collet, A. S., & Podesta, M. (2001). Collisions between ships and whales. *Marine Mammal Science, 17*(1), 35–75.

55. Lawton, K., Robertson, G., Valencia, J., Wienecke, B., & Kirkwood, R. (2003). The status of black-browed albatrosses *Thalassarche melanophrys* at Diego de Almagro island, Chile. *Ibis, 145*, 502–505.

56. León-Muñoz, J., Urbina, M. A., Garreaud, R., & Iriarte, J. L. (2018). Hydroclimatic conditions trigger record harmful algal bloom in western Patagonia (summer 2016). *Scientific Reports, 8*(1), 1–10.

57. Lutz, S. J., & Martin, A. H. (2014). *Fish carbon: Exploring marine vertebrate carbon services*. GRID-Arendal, Arendal.

58. Marín, M., & Oehler, D. (2007). Una nueva colonia de anidamiento para el albatros de ceja negra (*Thalassarche melanophrys*) para Chile. *Anales Instituto Patagonia (Chile), 35*, 29–33.

59. Marquet, P. A., Buschmann, A. H., Corcoran, D., Díaz, P. A., Fuentes-Castillo, T., Garreaud, R., Pliscoff, P., & Salazar, A. (2023). *Global Change and Acceleration of Anthropic Pressures on Patagonian Ecosystems*. Springer.

60. Medina-Vogel, G., Pons, D. J., & Schlatter, R. P. (2019). Relationships between off-bottom bivalve aquaculture and the Magellanic steamer duck *Tachyeres pteneres* in southern Chile. *Aquaculture Environment Interactions, 11*, 321–330.

61. Molinet, C., Solari, M. E., Díaz, M., Marticorena, F., Díaz, P. A., Navarro, M., & Niklitschek, E. (2018). Fragmentos de la historia ambiental del sistema de fiordos y canales nor-patagónicos, sur de Chile: Dos siglos de explotación. *Magallania, 46*(2), 107–128.

62. Molinet, C., and Niklitschek, E. J. (2023). *Fisheries And Marine Conservation In Chilean Patagonia*. Springer.

63. Moreno, C. A., Arata, J. A., Rubilar, P., Hucke-Gaete, R., & Robertson, G. (2006). Artisanal longline fisheries in southern Chile: Lessons to be learned to avoid incidental seabird mortality. *Biological Conservation, 127*, 27–36.

64. Myers, R. A., & Worm, B. (2003). Rapid worldwide depletion of predatory fish communities. *Nature, 423*(6937), 280–283.

65. O'Neil, J. M., Davis, T. W., Burford, M. A., & Gobler, C. J. (2012). The rise of harmful cyanobacteria blooms: The potential roles of eutrophication and climate change. *Harmful Algae, 14*, 313–334.

66. Outeiro, L., Häussermann, V., Viddi, F., Hucke-Gaete, R., Försterra, G., Oyarzo, H., Kosiel, K., & Villasante, S. (2015). Using ecosystem services mapping for marine spatial planning in southern Chile under scenario assessment. *Ecosystem Services, 16*, 341–353.

67. Palma, S., & Silva, N. (2004). Distribution of siphonophores, chaetognaths, euphausiids and oceanographic conditions in the fjords and channels of southern Chile. *Deep Sea Research Part II: Topical Studies in Oceanography, 51*, 513–535.

68. Pantoja, S., Iriarte, J. L., & Daneri, G. (2011). Oceanography of the Chilean Patagonia. *Continental Shelf Research, 31*(3), 149–153.

69. Pauly, D., Christensen, V., Dalsgaard, J., Froese, R., & Torres, F. (1998). Fishing down marine food webs. *Science, 279*(5352), 860–863.
70. Pérez-Matus, A., and Buschmann, A. H. (2003). *Sustentabilidad e incertidumbre de las principales pesquerías chilenas.* Santiago, Chile: Publicaciones Oceana. Retrieved from: http://www.navarro.cl/historico/ambiente/pesca/MATERIAL%20PESCA/principales_pesquerias_chilenas.pdf
71. Pérez-Venegas, D. J., Toro-Valdivieso, C., Ayala, F., Brito, B., Iturra, L., Arriagada, M., Seguel, M., Barrios, C., Sepúlveda, M., Oliva, D., Cárdenas-Alayza, S., Urbina, M. A., Jorquera, A., Castro-Nallar, E., & Galbán-Malagón, C. (2020). Monitoring the occurrence of microplastic ingestion in otariids along the Peruvian and Chilean coasts. *Marine Pollution Bulletin, 153*, 110966.
72. Petit, I. J., Campoy, A. N., Hevia, M. J., Gaymer, C. F., & Squeo, F. A. (2018). Protected areas in Chile: Are we managing them? *Revista Chilena de Historia Natural, 91*(1), 1.
73. Pickard, G. L. (1971). Some physical oceanographic features of inlets of Chile. *Fisheries Research Board of Canada, 28*, 1077–1106.
74. Pickard, G. L., & Stanton, B. R. (1980). Pacific fjords: a review of their water characteristics. In: H.J. Freeland, D.M. Farmer, and Levings C.D. (Eds.). *Fjord Oceanography*, pp. 1–51. Boston, Massachusetts, USA: Springer.
75. Purcell, J. E., Uye, S., & Lo, W. T. (2007). Anthropogenic causes of jellyfish blooms and their direct consequences for humans: A review. *Marine Ecology Progress Series, 350*, 153–174.
76. Pütz, K., Raya Rey, A., & Otley, H. (2015). Pingüino penacho amarillo del sur. In: García, P., & Boersma, P. D. (Eds.). *Pingüinos, historia natural y conservación*, pp. 121–139. Buenos Aires, Argentina: Vázquez Mazzini editores.
77. Quiñones, R., Fuentes, M., Montes, R. M., Soto, D., & León-Muñoz, J. (2019). Environmental issues in Chilean salmon farming: A review. *Reviews in Aquaculture, 11*, 375–402.
78. Quiroz, D. (2014). Etnografía histórica de la planta ballenera de isla Guafo [1921-1937]. *Magallania (Punta Arenas), 42*, 81–107.
79. Reyes-Arriagada, R., Campos-Ellwanger, P., Schlatter, R. P., & Baduini, C. (2007). Sooty shearwater (*Puffinus griseus*) on Guafo Island: The largest seabird colony in the world? *Biodiversity Conservation, 16*, 913–930.
80. Ribeiro, S., Viddi, F. A., Cordeiro, J. L., & Freitas, T. R. O. (2007). Fine-scale habitat selection of Chilean dolphins (*Cephalorhynchus eutropia*): Interactions with aquaculture activities in southern Chiloé island, Chile. *Journal of the Marine Biological Association of the United Kingdom, 87*, 119–128.
81. Rivera, A., Aravena, J. C., Urra, A., & Reid, B. (2023). *Chilean Patagonian Glaciers And Environmental Change.* Springer.
82. Robertson, G., Moreno, C. A., Lawton, K., Arata, J., Valencia, J., and Kirkwood, R. (2007). An estimate of the population sizes of black-browed (Thalassarche melanophrys) and grey-headed (T. chrysostoma) albatrosses breeding in the Diego Ramírez archipelago, Chile. Emu, 107, 239–244. Roman, J. and McCarthy, J. J. (2010). The whale pump: Marine mammals enhance primary productivity in a coastal basin. *PLoS ONE, 5*(10), e13255.
83. Romero, M. C., Lovrich, G. A., Tapella, F., & Thatje, S. (2004). Feeding ecology of the crab *Munida subrugosa* (Decapoda: Anomura: Galatheidae) in the Beagle channel, Argentina. *Journal of the Marine Biological Association of the United Kingdom, 84*, 359–365.
84. Rozzi, R., Rosenfeld, S., Armesto J. J., Mansilla, A., Núñez-Ávila, M., & Massardo, F. (2023). *Ecological Connections Across the Marine-Terrestrial Interface in Chilean Patagonia.* Springer
85. Schlatter, R., & Simeone, A. (1999). Estado del conocimiento y conservación de las aves en mares chilenos. *Estudios Oceanológicos (Chile), 18*, 25–33.
86. Scofield, R. P., & Reyes-Arriagada, R. (2013). A population estimate of the sooty shearwater *Puffinus griseus* in the Wollaston and Hermite island groups, cape Horn archipelago, Chile, and concerns over conservation in the area. *Revista de Biología Marina y Oceanografía, 48*, 623–628.

87. Sepúlveda, M., Pérez, M. J., Sielfeld, W., Oliva, D., Durán, L. R., Rodríguez, L., Araos, V., & Buscaglia, M. (2007). Operational interaction between south american sea lions *Otaria flavescens* and artisanal (small-scale) fishing in Chile: Results from interview surveys and on-board observations. *Fisheries Research, 83*(2), 332–340.
88. Silva, N., Calvete, C., & Sievers, H. (1998). Masas de agua y circulación general para algunos canales australes entre Puerto Montt y laguna San Rafael, Chile (Crucero CIMAR-Fiordo 1). *Ciencia y Tecnología Marina, 21*, 17–48.
89. Silva, N., and Palma, S. (2008). The CIMAR Program in the austral chilean channels and fjords. In: Progress in the Oceanographic knowledge of Chilean interior waters, from Puerto Montt to cape Horn (pp. 11–15). Valparaíso: Comité Oceanográfico Nacional. Pontificia Universidad Católica de Valparaíso. Retrieved from: http://www.cona.mil.cl/revista/english/1.1%20Nelson%20Silva%20-%20Sergio%20Palma.pdf.
90. Spalding, M. D., Fox, H. E., Allen, G. R., Davidson, N., Ferdaña, Z. A., Finlayson, M., Halpern, B. S., Jorge, M. A., Lombana, A., Lourie, S. A., Martin, K. D., McManus, E., Molnar, J., Recchia, C. A., & Robertson, J. (2007). Marine ecoregions of the world: A bioregionalization of coastal and shelf areas. *BioScience, 57*(7), 573–583.
91. Suazo, C. G., Schlatter, R. P., Arriagada, A. M., Cabezas, L. A., & Ojeda, J. (2013). Fishermen's perceptions of interactions between seabirds and artisanal fisheries in the Chonos archipielago, Chilean Patagonia. *Oryx, 47*, 184–189.
92. Tacón, A., Tecklin, D., Farías, A., Peña, M. P., & García, M. (2023). *Terrestrial Protected Areas in Chilean Patagonia: Characterization, Historical Evolution, and Management.* Springer.
93. Tecklin, D., Farías, A., Peña, M. P., Gélvez, X., Castilla, J. C., Sepúlveda, M., Viddi, F. A., & Hucke-Gaete, R. (2023). *Coastal-Marine Protection in Chilean Patagonia: Historical Progress, Current Situation, and Challenges.* Springer.
94. Tønnessen, J. N., & Johnsen, A. O. (1982). *The history of modern whaling.* University of California Press.
95. Torres, D. N., Aguayo, A., & Acevedo, J. (2000). Mamíferos marinos de Chile II. *Carnivora. Serie Científica INACH, 50*, 25–103.
96. Torres, R., Pantoja, S., Harada, N., González, H. E., Daneri, G., Frangopulos, M., Rutllant, J., Duarte, C. M., Ruiz-Halpern, S., Mayol, E., & Fukasawa, M. (2011). Air-sea CO_2 fluxes along the coast of Chile: from CO_2 outgassing in central northern upwelling waters to CO_2 uptake in southern Patagonian fjords. *Journal of Geophysical Research Oceans, 116*(C09006).
97. United Nation Environmental Program-World Conservation Monitoring Centre—IUCN—NGS (2018). *Protected planet report 2018.* UNEP-WCMC, IUCN and NGS: Cambridge UK; Gland, Switzerland; and Washington D. C., USA. Retrieved from: https://livereport.protectedplanet.net/pdf/Protected_Planet_Report_2018.pdf.
98. Viddi, F. A., Hucke-Gaete, R., Torres-Flórez, J. P., & Ribeiro, S. (2010). Spatial and seasonal variability in cetacean distribution in the fjords of northern Patagonian, Chile. *ICES Journal of Marine Science, 67*, 959–970.
99. Viddi, F. A., Harcourt, R. G., Hucke-Gaete, R., & Field, I. C. (2011). Fine-scale movement patterns of the sympatric Chilean and peale's dolphins in the northern Patagonian fjords, Chile. *Marine Ecology Progress Series, 436*, 245–256.
100. Viddi, F. A., Harcourt, R. G., & Hucke-Gaete, R. (2015). Identifying key habitats for the conservation of Chilean dolphins in the fjords of southern Chile. *Aquatic Conservation: Marine and Freshwater Ecosystems, 26*(3), 506–516.
101. Vila, A., Falabella, V., Gálvez, M., Farías, A., Droguett, D., & Saavedra, B. (2016). Identifying high-value areas to strengthen marine conservation in the channels and fjords of the southern Chile ecoregion. *Oryx, 50*(2), 308–316.
102. Walsh, J. J. (1991). Importance of continental margins in the marine biogeochemical cycling of carbon and nitrogen. *Nature, 350*, 53–55.
103. Watson, J. E. M., Dudley, N., Segan, D. B., & Hockings, M. (2014). The performance and potential of protected areas. *Nature, 515*(7525), 67–73.

104. Watts, M. E., Ball, I. R., Stewart, R. S., Klein, C. J., Wilson, K., Steinback, C., Lourival, R., Kircher, L., & Possingham, H. P. (2009). Marxan with zones: Software for optimal conservation-based land and sea-use zoning. *Environmental Modelling and Software, 24*(12), 1513–1521.

Rodrigo Hucke-Gaete Marine Biologist, Universidad Austral de Chile. Doctor in Sciences and professor at Universidad Austral de Chile. Marine ecologist specialized in marine mammals and marine conservation. Member of the IUCN Cetacean Specialist Group.

Francisco A. Viddi Marine Biologist, Universidad Austral de Chile. Ph.D in Environmental Sciences, Macquarie University, Australia. Specialist in marine mammal ecology and marine conservation. Member of the IUCN Cetacean Specialist Group.

Alejandro Simeone Biologist, Universidad Austral de Chile. Ph.D in Natural Sciences, University of Kiel, Germany. Associate Professor, Universidad Andrés Bello, Chile. Seabird specialist.

Hard Bottom Macrobenthos of Chilean Patagonia: Emphasis on Conservation of Sublitoral Invertebrate and Algal Forests

10

Vreni Häussermann, Günter Försterra, and Jürgen Laudien

Abstract

The region of the fjords, channels, and islands of Chilean Patagonia (41° 42'S 73° 02'W; 56° 29'S 68° 44'W) has one of the most rugged coasts and is one of the least studied marine systems worldwide. Over the last two decades, we have collected samples (diving down to 35 m) at more than 500 stations, taken underwater photographs, recorded videos from remotely operated vehicles and carried out a comprehensive literature review on benthic (hard bottom) macroinvertebrates and macroalgae. Based on this research, we propose a subdivision of Chilean Patagonia into three biogeographical provinces and 13 ecoregions. The inventory developed indicates the occurrence of rich and extensive sublittoral associations formed by 13 bioengineering species, conforming 11 types of submarine invertebrates, and two types of macroalgal forests. According to the national inventory of wildlife species by conservation status, six of the invertebrate species thriving in Patagonia belong to one of the categories of threatened species. The main local threats to these habitat-forming species are aquaculture, infrastructure and industrialization projects, fishing, and invertebrate harvesting, as well as threats from climate change and volcanic activities. Finally, we identify knowledge gaps and provide recommendations for the protection

V. Häussermann (✉)
Escuela de Ingeniería en Gestión de Expediciones y Ecoturismo, Facultad de Ciencias de la Naturaleza, Universidad San Sebastián, Lago Panguipulli, 1390 Puerto Montt, Chile
e-mail: Verena.haussermann@uss.cl

G. Försterra
Facultad de Recursos Naturales, Escuela de Ciencias del Mar, Universidad Católica de Valparaíso, Avda. Brasil, 2950 Valparaíso, Chile

J. Laudien
Alfred-Wegener-Institut Helmholtz-Zentrum für Polar- und Meeresforschung, Am Alten Hafen 26, 27568 Bremerhaven, Germany

© Pontificia Universidad Católica de Chile 2023
J. C. Castilla et al. (eds.), *Conservation in Chilean Patagonia*, Integrated Science 19,
https://doi.org/10.1007/978-3-031-39408-9_10

and conservation of the biodiversity and ecosystem services provided by these species' associations.

Keywords

Patagonia • Chile • Biogeography • Sublittoral forests • Macroinvertebrates • Macroalgae • Threatened species • Conservation • Marine protected areas

1   Introduction

Chilean Patagonia covers a marine area of 121,948 km^2 located between Reloncaví Sound and the Diego Ramírez Islands (41° 42'S 73° 02'W; 56° 29'S 68° 44'W); the linear extension of the continental coastline, including fjords, channels, islands, islets and rocky areas, is 100,627 km [69]. This generates marine coastal ecosystems that are among the most biologically structured in the world [13, 31, 38]. The region is characterized by pronounced physical and chemical gradients, which together with the factors mentioned above result in highly diverse sublittoral habitats, considered "hotspots" of marine diversity [21, 31], which are among the least studied in the world [4].

There were no taxonomic surveys and biodiversity inventories of hard or consolidated sublittoral bottoms in Chilean Patagonia until about 20 years ago. Over the last 20 years, our team has used SCUBA diving (self-contained underwater breathing apparatus) to conduct numerous on-site inventories of macrobenthicepifauna, mainly on hard bottoms and rocky walls, down to 35 m depth. The influence of the low salinity layer (LSL) is very pronounced in the first 10–15 m in these environments but is no longer evident below those depths [39]. According to our studies, the benthic communities of Chilean Patagonia can be divided into three biogeographic provinces with 13 ecoregions (adapted from [31]). Approximately 50% of Chilean Patagonia's land area is protected, but the 11 marine protected areas (MPAs) represent only 6% of marine Patagonia (excluding areas that are part of the National System of State Protected Wild Areas, see [69]). These MPAs have little or no protection, and there is no integrated conservation system for marine Patagonia [38, 69].

2   Scope and Objectives

Our objectives for this chapter, based on our publications, literature review and scientific observations are: (i) to describe the biocenoses or forests of hard-bottom benthic macroinvertebrates and macroalgae, which are important bioengineering species; (ii) to provide information on the biodiversity of these species for each of the biogeographic provinces; (iii) to identify the main anthropogenic and natural impacts on these benthic systems; (iv) to identify the main knowledge gaps and research opportunities; and (v) to provide recommendations for the protection and conservation of these biocenoses.

3   Methods

Between 1989 and 2019, in the area of Chilean Patagonia between Valdivia (*ca.*, 40°S) and the Beagle Channel (*ca.*, 55°S), we: (i) collected samples of sublittoral benthic macroinvertebrates (>5 mm length) from hard bottoms at 405 sites to a maximum depth of 35 m using SCUBA diving (Fig. 1a); (ii) recorded videos of many of these macroinvertebrates at 93 sites between 0–500 m depth with a remotely operated vehicle; (iii) conducted photo-transects at 53 sites down to 28 m (49 photos at seven different depths at each site); (iv) have created species occurrence lists of 26 taxa from ten phyla at 167 sites since 2011, using a predefined list of 70 species that can be reliably identified from high-quality underwater photos; and (v) photographed, collected, documented and preserved 12,181 specimens of these macroinvertebrates. We also analyzed the literature for reliable taxonomic records and compiled a list of 1,811 species of benthic macroinvertebrates for Chilean Patagonia [31]. The information is contained in the PAtagonia MArine DAtabase (PAMADA), which is a collection of biological (mainly benthic), physico-chemical and oceanographic data, currently containing 20,159 species occurrence points. This chapter contains a summary of this information. We have also reviewed the Chilean National Inventory of Species Conservation Category according to the Wildlife Species Classification Regulations of the Ministry of the Environment, (in Spanish Ministerio del Medio Ambiente, MMA) (https://clasificacionespecies.mma.gob.cl/).

4   Results

4.1 Biogeographical Subdivisions in Chilean Patagonia and Latitudinal Trends in the Number of Hard-Bottom Sublittoral Macroinvertebrate Species

Our studies of sublittoral hard-bottom marine macroinvertebrates at more than 400 dive sites in Chilean Patagonia allow us to propose a latitudinal division of Chilean Patagonia into three marine biogeographic provinces: (i) Northern Patagonia (NP): 42°–47°S; (ii) Central Patagonia (CP): 47°–54°S; and (iii) Southern Patagonia (SP): 54°–56°S (Fig. 1a). Each province includes three main ecoregions: Fjords, channels and exposed coast. Due to geomorphological aspects and other oceanographic conditions, at least four other ecoregions are proposed: (i) the east coast of Chiloé Island, with extensive muddy bottoms; (ii) the Corcovado Gulf; (iii) the Gulf of Penas; (iv) the large semi-enclosed inland seas in SP, such as the Otway and Skyring Sounds [31].

In the Taitao Peninsula (47°S), which separates the NP and CP provinces, a part of the West Wind Drift Current meets the continent, and the division between the Cape Horn Current and the Humboldt Current occurs. The boundary between the CP and SP provinces is in the deep Straits of Magellan (54°S), where a strong component of the circumpolar current predominates, flowing through the

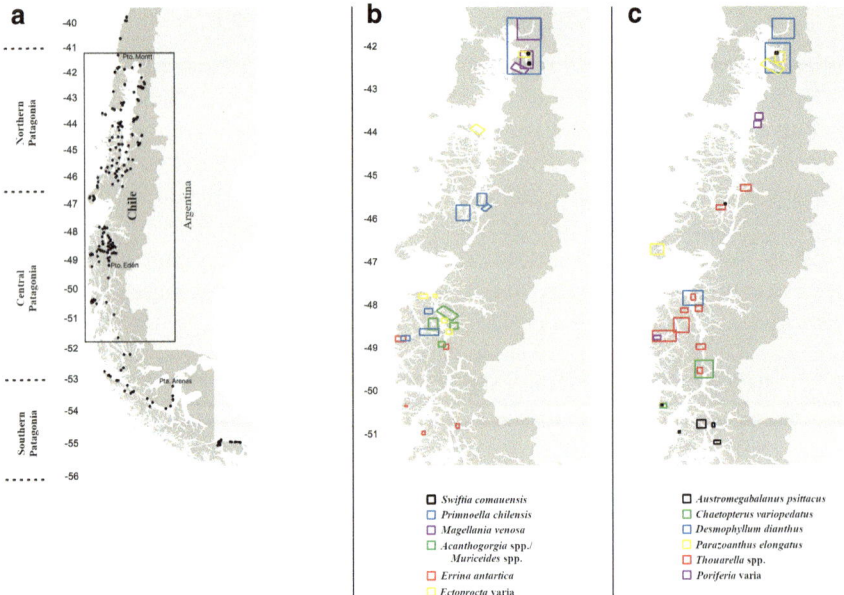

Fig. 1 a Map of Chilean Patagonia showing the boundaries of the provinces of Northern Patagonia (NP), Central Patagonia (CP) and Southern Patagonia (SP) and the sites surveyed (black dots) between 1998 and 2019. **b** and **c** Maps of the most prominent sublittoral marine invertebrate forests. No sublittoral invertebrate forests were observed in SP

strait from west to east, thus hindering latitudinal dispersal of organisms with planktonic larvae [58]. In comparison to the glacier-free fjords of NP, the intra-continental fjords in CP and SP are strongly influenced by glacial erosion. In the NP province, the sea anemone *Anthothoe chilensis* is abundant on mussel banks, other species restricted to this region are the encrusting anemone *Parazoanthus elongatus* and gorgonians of the genus *Swiftia*; there are also several bioengineering benthic invertebrates (sensu [42], such as stony corals (Fig. 2a), ectoprocts, brachiopods (Fig. 2b), mussels, barnacles and gorgonians (*Primnoella chilensis* and *Thouarella* spp.) [31].

The fjords in the CP province have subtidal walls with steep slopes and low invertebrate colonization. The soft coral *Alcyonium glaciophilum* and the gorgonians *Acanthogorgia* sp. 1 and sp. 2 are species restricted to this province. The presence of hydrocorals, gorgonians (*Acanthogorgia* spp., *Muriceides* spp., *Thouarella* spp. (Fig. 2c), *Primnoella chilensis*), ectoprocts and sponge gardens is outstanding in the CP channels. In this province, limestone archipelagos such as Madre de Dios harbor a low number of anthozoans, and there are hydrocoral reefs in channels with strong currents.

Brown macroalgal forests are common in shallow coastal sites in the SP Province (Fig. 2d). Anthozoan and decapod species are generally very scarce.

Fig. 2 Photographs of sublittoral benthic forests in Chilean Patagonia. **a** Forest of *Desmopyhllum dianthus* stony corals. **b** *Magellania venosa* brachiopod forest. **c** *Thouarella brucei* gorgonian forest. **d** *Macrocystis pyrifera* macroalgal forest

However, the anemones *Stomphia coccinea* and *Bunodactis octoradiata,* the decapod *Pagurus forceps* and the bivalves *Aequiyoldia eightsii, Cuspidiaria patagonica, Policordia radiata* and *Cyclochlamys multistriata* are typical for this province [31]. Our current database for species numbers of five hard-bottom invertebrate taxa from Chilean Patagonia shows that (i) anthozoans initially increase in species numbers south of Puerto Montt (42°S) and then decrease toward the extreme south: 15 species at 40°S; 41 at 45°S; 37 at 50°S and 15 at 55°S; (ii) gastropods decrease toward higher latitudes: 34 species at 40°S; 28 at 45°S; 25 at 50°S and 24 at 55°S; (iii) bivalves and pycnogonids show relative stability in the number of species throughout Patagonia: 35 species of bivalves and six species of pycnogonids at 40°S; 39 and 11 species at 45°S; 38 and 9 species at 50°S; 37 and 10 species at 55°S, respectively; (iv) decapods decrease toward higher latitudes across Patagonia: 59 species at 40°S; 48 at 45°S; 42 at 50°S and 14 at 55°S.

4.2 Sublittoral Benthic Hard-Bottom Species of Outstanding Importance

4.2.1 Sublittoral Forest-Forming Species of Macroinvertebrates and Brown Macroalgae

Our inventory of benthic biodiversity allowed us to identify 13 communities of different macroinvertebrates and brown macroalgae that form sublittoral forests

(Figs. 1 and 2). These habitat-forming species or ecosystem bioengineers [42] modulate the abiotic and biotic environment and therefore maintain a self-organized habitat, which is used by diverse associated fauna [62]. Forest-forming invertebrates and algae are subdivided into five subsets: (i) species with massive endoskeletons: cold-water stony corals, hydrocorals and Ectoprocta (ii) species with massive exoskeletons: bivalves, brachiopods and barnacles; (iii) species with scattered calcified structures or with spicules: gorgonians and sponges; (iv) non-calcifying invertebrate species; (v) macroalgal species (see details in Table 1).

(i) **Sublittoral forests of cold-water stony corals, hydrocorals and ectoprocts.**
The cosmopolitan stony coral *Desmophyllum dianthus* (Scleractinia) has been described in Chilean Patagonia from 7 to 2,460 m depth, with a distribution between 42°–56°S [23]. *D. dianthus* grows on steep rock faces with a slope > 80° and on lower surfaces of boulders and below the influence of the LSL. In NP, it can form banks with a density of up to 1,500 individuals/m² and has a maximum length of up to 40 cm [31].

Individuals can grow on top of others with up to five individuals, forming colony-like structures. These coral banks are found between 20–400 m in Comau [22], Reloncaví and Reñihue fjords [23]. A patch of 100 m² was also found in Pitipalena Fjord (Fig. 1b). The species is scarce in the rest of Patagonia, with accumulations of small individuals at the mouth of the Messier Channel. The species is a sessil predator [37] and grows relatively fast [41], with marked seasonal reproduction and high fecundity [20]. Hundreds of species are associated with these forests, including sponges, echinoderms, snails, anemones, ectoprocts, polychaetes, scleractinian corals, brachiopods *(Magellania venosa),* bivalves *(Aulacomya atra* and *Acesta patagonica)* and the fish *Sebastes oculatus.*

The hydrocoral *Errina antarctica* (Stylasterina) is distributed in semi-exposed and exposed environments below the LSL from south of Chiloé (43°S) to the Subantarctic islands between 10–771 m. In CP, it is locally abundant in some channels with strong currents; its colonies can grow fan-shaped on steep walls or as bushes on the bottom of channels, reaching up to 40 cm in diameter and creating reef-like formations [33]. Hydrocoral reefs provide habitat and shelter for numerous species (58 associated taxa have been identified, [78] and serve as substrate for sedentary filter-feeding species such as the crinoid *Gorgonocephalus chilensis* and the feather star *Florometra magellanica.* The dead areas of these colonies are used by a large number of sessile, sedentary, mobile and burrowing organisms [30]. Aggregations and sublittoral forests of several species of calcifying ectoprocts such as *Aspidostoma giganteum*, *Adeonella* spp. and *Microporella hyadesi* occur in exposed or semi-exposed sites with strong currents. Individuals reach a diameter of up to 30 cm and create highly structured habitats for semi-sessile and mobile organisms such as the hermit crab *Pagurus comptus*, the gastropod *Calliostoma consimilis* and ophiuroids *(e.g. Ophiacantha rosea).*

Table 1 Sublittoral forests of macroinvertebrates and brown algae in the coastal zone of Chilean Patagonia. CP: Central Patagonia

Name	Ecosystem bioengineering species	Ecosystem bioengineer category	Bathymetric distribution of species	Geographic distribution of species	Main habitat
Stony coral forests	*Desmophyllum dianthus*	Bioengineers with massive endoskeleton	7–2.460 m	42°–54°S Cosmopolitan	Aggregations on vertical walls. Very numerous in Reloncaví, Comau and Reñihue fjords. Small patch in Pitipalena
	Caryophyllia huinayensis		11–800 m	36°–53°30′S	On almost vertical rocky substrate
	Tethocyathus endesa		11–240 m	36°–48°30′S	Substrates from 71° to 145°; withstands some sedimentation
Hydrocoral forests	*Errina antarctica*	Bioengineer with massive endoskeleton	10–500 m	43°–56°S Also in Subantarctic islands	Channels with strong currents, mainly in CP
Aggregations of bryozoans (Ectoprocta)	*Adeonella* spp.	Bioengineers with massive endoskeleton	10–33 m	42°–56°S	Channels with strong currents; exposed areas
	Microporella hyadesi		14–27 m	43°–56°S	
	Aspidostoma giganteum		4–45 m	42°–56°S	
Mussel banks	*Mytilus chilensis*	Bioengineers with massive exoskeleton	Intertidal–37 m	20°–56°S	Fjords and channels at exposed coast
	Perumytilus purpuratus		Intertidal	18°–56°S	
	Aulacomya atra		Subtidal–<30 m	18°–56°S	

(continued)

Table 1 (continued)

Name	Ecosystem bioengineering species	Ecosystem bioengineer category	Bathymetric distribution of species	Geographic distribution of species	Main habitat
Brachiopod banks	*Magellania venosa*	Bioengineer with massive exoskeleton	2–1.362 m	30°–55°S	Fjords: Reloncaví, Comau, Reñihue, Pitipalena and Magdalena Sound
Barnacle aggregations	*Austromegabalanus psittacus*	Bioengineer with massive exoskeleton	Intertidal–35 m	18°–56°S	Channels with strong currents; exposed areas
Gorgonian gardens	*Primnoella chilensis*	Bioengineers with calcified structures	down to 320 m	41° - 55°S	Fjords and channels. *Acanthogorgia* and *Muriceides* spp. in CP support some sedimentation
	Thouarella spp.		15–1.500 m	41°–56°S	
	Acanthogorgia spp.		20–28 m	48°–51°S	
	Muriceides spp.		18–32 m	48°–49°S	
	Swiftia comauensis		15–60 m	42°10'–42° 30'S: Comau and Reñihué Fjords	
Sponge fields	Demospongiae and Calcarea	Bioengineers with calcified structures	Intertidal and subtidal		Exposed coasts
Polychaete fields	*Chaetopterus variopedatus*	Bioengineer without calcified structures	1–485 m	27°–56°S	Channels
Ascidian aggregations	*Sycozoa sigillinoides*	Bioengineer without calcified structures	10–548 m	44°–54°S Aggregation: 54° 59'S; 68° 20'W	Exposed channels and shorelines

(continued)

Table 1 (continued)

Name	Ecosystem bioengineering species	Ecosystem bioengineer category	Bathymetric distribution of species	Geographic distribution of species	Main habitat
Aggregations of encrusting anemones	Parazoanthus elongatus	Bioengineer without calcified structures	10–35 m	41–47°S	Channels and fjords: Comau, Reñihué and Slight Estuary
Brown macroalgae forests	Durvillaea incurvata	Bioengineers without calcified structures	Intertidal–15 m	22–43°S	Exposed coasts
	Durvillaea antarctica		Intertidal–15 m	44°–55°S and Subantarctic	Exposed coasts
	Lessonia flavicans (syn. L. vadosa)		Intertidal–40 m	Magallan area (little-known)	Exposed and protected coasts
	Lessonia spicata*		Intertidal and shallow subtidal	30–48°S	Semi-exposed and exposed coasts
	Lessonia berteroana*		Intertidal and shallow subtidal	17°–30°S	Semi-exposed and exposed coasts
	Lessonia trabeculata		Down to 40 m	14–42°S	Protected and exposed coasts
	Macrocystis pyrifera		Down to 40 m	14–56°S and Subantarctic	Protected and exposed coasts
Beds of rhodoliths	Melobesiodeae: unidentified species	Bioengineers with massive exo-skeletons	Intertidal–270 m	Guarello Island: 50° 20' S 75° 23' W Melinka: 43° 53' S 73° 44' W. Amita Island: 44° 4' S 73° 52' W	Channels with clear water

[* Formerly Lessonia nigrescens, which remains a valid species because no individuals have been found in its type locality of Cape Horn].

(ii) **Sublittoral forests of bivalves, brachiopods and barnacles.** Numerous sites in Patagonia's rocky intertidal zone are dominated by bivalves such as *Mytilus chilensis* and *Brachidontes purpuratus*; the former being more abundant in fjords and channels and the latter on the exposed coast. *Aulacomya atra* is abundant in the subtidal zone down to 20 m. The mussel beds can be up to 30 cm thick and host numerous invertebrate species, especially when the beds have various age structures, such as the anemones *Anthothoe chilensis* (NP), *Metridium senile* (a species introduced to Patagonia, being on the rise throughout North and South Patagonia, [35], and creating huge problems for sea urchin fisherfolks, gastropods of the genus *Crepidula,* echinoderms, sponges, polychaetes and crustaceans, which typically live inside the matrix of the beds.

In some fjords of NP (Reloncaví, Comau, Reñihué, Pitipalena) and in Magdalena Sound, there are forests of the brachiopod *Magellania venosa* [6, 7]. This species can numerically dominate the benthos in Comau Fjord on steep rocky walls between 15–35 m, where densities of up to 416 ind./m^2 have been observed [6]. The species is also observed in soft bottoms at the mouths of some fjords at depths from 150 to 200 m (*e.g.* Comau Fjord). Shell growth is rapid, which may explain its high population density and coexistence with mussels [6]. Aggregations of the giant barnacle *Austromegabalanus psittacus* have been observed on the exposed coast and semi-exposed channels below the LSL with individuals growing on top of each other. At sites with higher and more stable salinity and moderate wave intensity, individuals grow at rates between 0.06 and 0.13 mm/day [45] and can reach heights of up to 30 cm, forming large banks.

(iii) **Sublittoral forests of gorgonians and sponges.** The sea whip *Primnoella chilensis* dominates on moderately steep slopes of fjords and channels in NP; in this province the gorgonia *Swiftia comauensis* is restricted to Comau and Reñihué Fjord. Although present in some channels of NP, gorgonians of the genus *Thouarella* dominate in the channels of CP. The branching gorgonians of the genera *Acanthogorgia* and *Muriceides* are restricted to fjords and channels of CP that are impacted by fine sediment. Gorgonian forests provide habitat for numerous species such as the anemone *Dactylanthus antarcticus* and the nudibranch *Tritonia odhneri,* which prey or graze on gorgonians of the genera *Primnoella* and *Thouarella* [31]. Throughout Chilean Patagonia, in some semi-exposed and exposed sites below 5–10 m, sponges of the classes Demospongiae and Calcarea with calcified spicules form sponge gardens (*e.g. Tedania mucosa* and *Mycale magellanica*). These gardens are especially common in the channels of CP, where *Amphimedon maresi* have three-dimensional structures which provide refuges and habitats for numerous other species. Encrusting sponges often dominate the benthic fauna, especially in the first few meters of wave-exposed sites; a significant percentage of these species have not yet been identified. Approximately 70% of the sponge species collected in our surveys had not yet been described [31]. Some invertebrates such as the starfish *Poraniopsis echinaster* and the gastropods *Fissurellidea* sp. and *Buchanania onchidioides* may prey on sponges.

(iv) **Sublittoral forests of polychaetes, ascidians, and encrusting anemones.** Forests of the polychaete *Chaetopterus variopedatus* are present in fjords and channels of CP, with dense fields of up to 100 m^2 at depths between 10–20 m. The tubes of these polychaetes can reach up to 50 cm length and provide habitat for a great variety of species, such as hydrozoans and sponges (*Halisarca magellanica*), decapods (*Campylonothus vagans*) and echinoderms (*Gorgonocephalus chilensis*). Forests of other non-calcifying invertebrate species are very scarce in the biogeographic provinces of Patagonia; some patches of encrusting anemones *Parazoanthus elongatus* (Comau and Reñihué Fjord and Slight Estuary) are found in NP, and an accumulation of the ascidian *Sycozoa sigillinoides* in the Beagle Channel (SP).

(v) **Sublittoral brown macroalgal forests.** The combination of high nutrient load and a salinity between 33 and 34 along the exposed rocky coasts of Chilean Patagonia creates the conditions for the formation of sublittoral forests of brown macroalgae that provide habitat, shelter and food for a large diversity of species [2, 8, 15, 28]. The main forests are formed by: (i) *Macrocystis pyrifera*, which can reach up to 15 m in length and is found in shallow intertidal and subtidal habitats down to 15–18 m [51, 54, 76, 77], (ii) *Durvillea incurvata* (30°–43°S) and *D. antarctica* (44°–55°S) [26], which are commonly found in shallow, exposed or very exposed inter- or subtidal habitats; and (iii) the genus *Lessonia*, which in Patagonia includes two species: *L. flavicans* (47°–55°S; [65]), which forms inter- and subtidal forests in the Magallanes Region, and *L. spicata* (30°–48°S), which forms intertidal forests at semi-exposed and exposed coasts [61]. Ojeda and Santelices [54] described the autecology and population dynamics of *M. pyrifera*, while [64] summarized the ecological relationships of invertebrate and algal communities in these Patagonian forests. Trophic webs of up to 122 invertebrate and algal species have been described for these forests [1, 3, 15, 52, 59] and in mixed forests of *M. pyrifera* and *Lessonia* spp. [27]. Generalist top predators such as the starfish *Cosmasterias lurida* [74], the crab *Peltarion spinosolum*, carnivorous anemones and predatory nemerteans are prominent in the trophic webs. Vásquez et al. [75] studied the trophic webs of the sea urchins (*Loxechinus albus, Pseudechinus magellanicus, Arbacia dufresnei* and *Austrocidaris canaliculata*) living in these forests and found that there was no trophic competition among adults, and that predation on adult urchins is not an important factor in regulating their densities.

(vi) **Sublittoral rhodolith beds.** Red coralline algae form rhodolith beds (unidentified species of Melobesiodeae; [60]. Rhodoliths extend from shallow zones to the maximum depth of the photic zone (down to 270 m), creating habitat, refuge, or settlement sites for a large number of marine species. For Chilean Patagonia, there is only one report of rhodolith beds in the Guaitecas and Madre de Dios archipelagos [46].

4.3 Sublittoral Benthic Macroinvertebrate Hard-Bottom Species in Chilean Patagonia Included in the Ministry of Environment's List "Conservation Category According to the Wildlife Classification Regulation"

The list of species whose continued existence is considered problematic, published by the Chilean Ministry of Environment,[1] is based on expert opinions. This list includes six sublittoral macroinvertebrate species: (i) the gorgonian *Swiftia comauensis,* only observed in Comau and Reñihué Fjord, 42°10' to 42° 30'S 72°E (Fig. 3c) is considered to be "critically endangered"; (ii) the sea whip *Primnoella chilensis* (41°–55°S), also present in Brazil and Argentina (Fig. 3b) is categorized as "endangered"; (iii) the shallow water ecotype of the hydrocoral *Errina antarctica* (43°–54°S), also present in the southwest Atlantic and Subantarctic islands (Fig. 3a), is considered "vulnerable"; (iv) the cosmopolitan stony coral *Desmophyllum dianthus* (42°–53°S), also present in the Juan Fernández archipelago (Fig. 3c) is considered to be "near threatened"; (v) the rock shrimp, *Campylonothus vagans* (41°–56°S), which is also present in the southwest Atlantic (Fig. 3B) is categorized under "minor concern"; (vi) for the non-retractile anemone, *Bolocera kerguelensis,* 41°–54°S; also present in the Antarctic and southwest Atlantic, and the deep-water ecotype of the hydrocoral *Errina antarctica* the category "data deficient" applies.

4.4 Threats to Marine Macroinvertebrate Biodiversity in Chilean Patagonia from Local Stressors

Patagonian marine ecosystems are subject to a vast anthropogenic transformation that can, and in many cases already has, affected biological conservation. An important number of species that form invertebrate forests are unique, fragile and sensitive to increased sedimentation, eutrophication, use of various chemicals, overfishing and deforestation [11, 12, 40, 50]. The main threats to the benthic communities of Patagonian fjords and channels are aquaculture, infrastructure and industrialization projects, fishing and invertebrate harvesting [25, 55, 57]. The steady increase in salmonid production over the last two decades has had an important impact on the ecology of Chilean Patagonia, for example, eutrophication and increased sedimentation poses a threat to benthic species [36]. The combination of high solar radiation, increased surface temperature and decreased precipitation associated with climate change, especially in El Niño years, has also resulted in a more pronounced vertical stratification of the water column. Because of the above factors, harmful algal blooms (HABs) are occurring more frequent and widespread in Patagonia [43, 47]. When the microalgae in these blooms die, they sink, and there is increased oxygen consumption by bacterial flora. This leads to hypoxia events, which can cause stress, mortality, and changes in food webs [48, 56].

---

[1] See: https://clasificacionespecies.mma.gob.cl/.

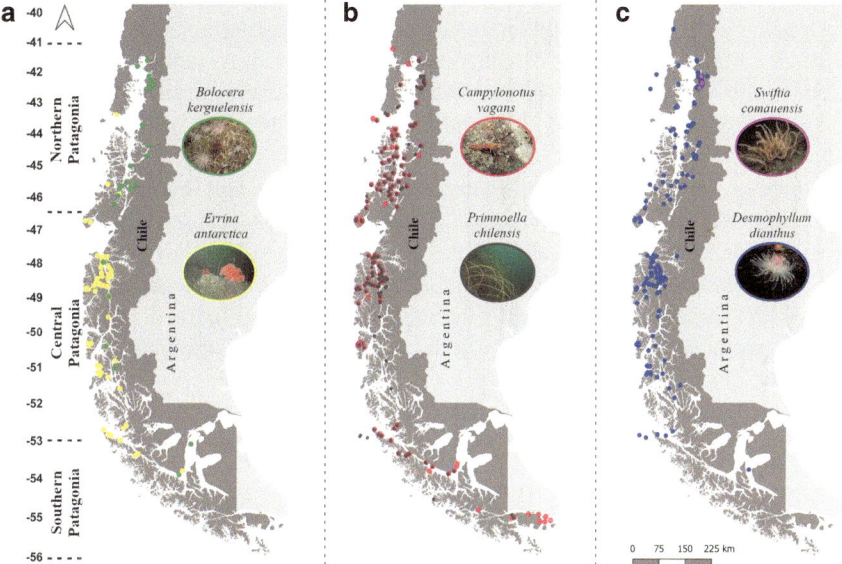

Fig. 3 Distribution map of benthic invertebrate species of sublittoral hard bottoms of Chilean Patagonia included in the list of conservation categories of endangered wildlife; according to the Wildlife Species Classification Regulations (MMA). **a** *Bolocera kerguelensis* ("data deficient"), *Errina antarctica* (shallow water ecotype: "vulnerable", deep water ecotype: "data deficient") **b** *Campylonotus vagans* ("least concern"), *Primnoella chilensis* ("endangered") **c** *Swiftia comauensis* ("critically endangered"), *Desmophyllum dianthus* ("near threatened")

For example, in 2012, 99% of the specimens of the coral *D. dianthus* died along approximately 15 km coastline of Comau Fjord, possibly due to a hypoxia event in combination with elevated hydrogen sulfide levels [18, 24]. The biological changes observed in Comau Fjord between 2003 and 2013, such as the decrease in the abundance of gorgonians, ectoprocts and long-lived anemones [32], could also be a consequence of the constant increase in the production and growth of salmonids in this fjord during these years [9, 10]. The continued use of chemical products by the industry, including antibiotics and crustacean control products [10], is harmful to Patagonian ecosystems. The abundance of some decapods has declined significantly in some Patagonian locations, as drugs against salmon lice have lethal and sublethal effects on crustacean larvae [29, 32].

The amount of waste remaining in the water and boat traffic have increased considerably during recent decades [32]. Several infrastructure and industrialization projects accompany this development, such as the construction or expansion of coastal roads and harbors. This kind of construction increases the exposure of benthic fauna to sediment and can destabilize the terrestrial vegetation layer, leading to landslides, which in turn can cause local tsunamis [66]. Artisanal fishing has increased significantly in Chilean Patagonia over the last 25 years [50], and the heavy extraction of species such as the gastropod *Concholepas concholepas* (loco), mussels, giant squid and algae may cause modifications in the benthic ecosystem, including macroinvertebrate forests and marine algae [32]. There is evidence of the direct impact of fisheries on Patagonian ecosystems. For example, the mussel banks of Reloncaví Fjord and Comau Fjord were considerably reduced between 2003 and 2013 [32, 49], and at some locations, *e.g.* in Comau Fjord, anemones monopolized the freed space. [17] demonstrated that the extraction of the urchin *L. albus* affects both the exploited populations and the associated benthic communities. Volcanic eruptions in Patagonia also have caused adverse effects on marine ecosystems. For example, the eruption of the Chaitén Volcano in 2008 released large amounts of sediments that affected the filtration processes of benthic organisms (Rogers, 1990). Volcanic activity can also affect pH and alkalinity and increase methane and sulfide levels in the water, negatively affecting the survival of benthic organisms [73].

5 Discussion

Our analyses of the distribution and diversity of the hard-bottom sublittoral macroinvertebrate fauna of Chilean Patagonia suggest the existence of three biogeographic provinces with different species assemblages: NP, CP, and SP. These coincide in latitude with what [63] classified as three ecoregions: Chiloé-Taitao, Kawésqar and Magallanes. Our results also indicate the existence of three ecoregions in each province: fjords, channels and exposed coasts (nine ecoregions in total) and at least four additional ecoregions (see results) characterized by variables such as salinity, water temperature, currents, slope and substrate [71].

This chapter describes 13 Patagonian macroinvertebrate and macroalgal communities that form sublittoral forests. These bioengineering species maintain self-organized habitats, host hundreds of species and are fundamental to Patagonian ecosystems. Only six of these species of benthic macroinvertebrates are included in the Ministry of Environment's list "Conservation Categories according to the Wildlife Species Classification Regulations", mainly due to the low number of existing scientific studies. More research and long-term monitoring are required to understand the threats to benthic macroinvertebrates and to develop conservation strategies.

Our records show that the number of species of anthozoans, gastropods and decapods decreases from low to high Patagonian latitudes; for decapods this agrees with the observations of Thatje and Arntz [68]. For bivalves and pycnogonids, a stable number of species was observed across latitudes. However, Valdovinos et al. [72], who analyzed all described Patagonian mollusks, reported a decrease in species numbers from low to high latitudes. Our approach of using only valid species with records confirmed by a specialist, which are observed atleast five sites, may account for the disparity of these results.

In contrast to the brown macroalga *M. pyrifera* forests in the Northern Hemisphere (California, Alaska), where sea urchins control kelp abundance and distribution, sea urchins do not play such a structuring ecological role in the Magallanes Region (Beagle Channel) [14]. While the sea otter *Enhydra lutris nereis* is a key predator that controls the density of urchins *Strongylocentrotus* spp. in Californian kelp forests, there is no such key predator of *L. albus* in the Magellanic forests [14, 15], see also [16].

The marine ecosystems of Chilean Patagonia are threatened by local anthropogenic (salmon farming, transport, HABs, and others), natural (volcanic eruptions) and global (climate change) stressors, natural stressors being of inferior importance. The NP biogeographic province is the most threatened of Patagonian marine ecosystems [47, 50]. However, salmon farming, artisanal and industrial fishing, and transport activities are expanding in CP and SP [50, 53]. A synthesis of some of the major local (eutrophication of waters, HAB) and global (climate change) threats in the marine areas of Chilean Patagonia can be found in [12, 34, 47, 67, 70].

The alteration or disappearance of species in marine ecosystems is imperceptible to the public and authorities without constant and long-term monitoring, as they occur below the surface. Patagonian marine ecosystems are at high risk of losing species and diverse ecosystem services [12, 40], those species forming sublittoral forests of macroinvertebrates and macroalgae should be protected as a priority. Marine Protected Areas (MPAs) are one of the most widely used tools for this purpose worldwide. Current knowledge suggests that between 25 and 75% of the global coastal zone would need to be protected to avoid serious ecological crises [5]. Tecklin et al. [69] described the current situation of MPAs in Chilean Patagonia in detail, pointing out that excluding the coastal-marine extensions of the large Patagonian National Parks, only 6% of the marine area is under (paper) protection. Edgar et al. [19] compared MPAs around the world and concluded that four of five characteristics must be met for an MPA to be effective: large size, isolation (from impacted areas), age older than 10 years, good surveillance, and high level of protection. No MPA in Chilean Patagonia meets these conditions. In addition to creating an integrated system of MPAs in Patagonia, including some with high protection or without intervention (no-take), efforts must be made to regulate fisheries [50] and revise current salmon farming practices [11, 47].

6   Conclusions and Recommendations

We provide the following recommendations addressing knowledge gaps, urgent conservation needs and threats to the Patagonian hard-bottom benthic communities and ecosystems that make up sublittoral forests.

- The greatest gaps in knowledge are found in the channels and exposed coasts between 50° and 56°S, which are generally difficult to access due to climate instability and the high cost of expeditions (Fig. 1a). However, there is an urgent need to promote and fund basic research to understand ecosystem structures and

Fig. 4 A cold-water coral bank dominated by *Desmophyllum dianthus* in Comau Fjord, 30 m depth

dynamics in all sublittoral ecoregions, especially of western Patagonia and to develop long-term time series.

- The status of sublittoral macroinvertebrate species classified by the MMA as having conservation problems is still poorly understood. Further research efforts are needed to obtain more thorough assessments and to register further species with conservation concerns in the Red List of the International Union for Conservation of Nature.

- Conservation efforts for the existing MPAs in Patagonia, most of which do not have management plans, monitoring, park rangers or funds to manage them, should be strengthened. The creation of an integrated land–sea system of MPAs should also be accelerated, including a set of reference or non-intervention marine-coastal areas for each of the biogeographic ecoregions. In addition, oceanographic data collected by the aquaculture industry must be made accessible.

- The negative impacts of various anthropogenic, global and natural stressors on Patagonian benthic ecosystems should be studied and evaluated as soon as possible, to determine the carrying capacity of the ecosystems. Regarding artisanal and industrial fisheries and aquaculture (especially in NP), it is necessary to apply precautionary fishing and ecosystem principles and introduce sustainable guidelines, such as reducing the nutrient loads entering the fjords and channels to avoid hypoxia. In addition, the carbon footprint of these activities needs to be significantly reduced, to lower the global impact.

Acknowledgements We are very grateful to the reviewers and to Juan Carlos Castilla for suggestions that significantly helped to improve the chapter. We thank all former research assistants and interns at the Huinay Scientific Field Station for their support over the years. We would also like to thank all the cooperating scientists, without whom we would not have been able to identify the species mentioned in this chapter, especially Diego Zelaya and Roland Melzer for reviewing the information on mollusks and arthropods. We are very grateful to Stacy Ballyram and Paulina Kurpet for creating the maps. Thanks also to Thomas Heran, Juan Pablo Espinoza and Serena Marchant for their support. Funding was provided by Fondecyt project N° 1131039 and Conicyt project N° PII20150106 for VH and N° 1201717 and 1150843 for GF.

References

1. Adami, M. L., & Gordillo, S. (1999). Structure and dynamics of the biota associated with *Macrocystis pyrifera* (Phaeophyta) from the Beagle Channel. *Tierra del Fuego. Scientia Marina, 63*(S1), 183–191
2. Almanza, V., Buschmann, A. H., Hernández-González, M. C., & Henríquez, L. A. (2012). Can giant kelp (*Macrocystis pyrifera*) forests enhance invertebrate recruitment in southern Chile? *Marine Biology Research, 8*(9), 855–864
3. Andrade, C., Ríos, C., Gerdes, D., & Brey, T. (2016). Trophic structure of shallow-water benthic communities in the sub-Antarctic Strait of Magellan. *Polar Biology, 39*(12), 2281–2297
4. Arntz, W. E. (1999). Magellan-Antarctic: ecosystems that drifted apart. Summary review. In: Arntz, W. E., & Ríos, C. (Eds.), *Magellan-Antarctic: ecosystems that drifted apart,* pp. 503–511. Madrid: Instituto de Ciencias del Mar C.S.I.C
5. Baillie, J., & Ya-Ping, Z. (2018). Space for nature. *Science, 361*(6407), 1051
6. Baumgarten, S., Laudien, J., Jantzen, C., Häussermann, V., & Försterra, G. (2014). Population structure, growth and production of a recent brachiopod from the Chilean fjord region. *Marine Ecology, 35*(4), 401–413
7. Betti, F., Bavestrello, G., Bo, M., Enrichetti, F., Loi, A., Wanderlingh, A., Pérez-Santos, I., & Daneri, G. (2017). Benthic biodiversity and ecological gradients in the Seno Magdalena (Puyuhuapi Fjord, Chile). *Estuarine, Coastal and Shelf Science, 198,* 269–278
8. Bruno, D. O., Victorio, M. F., Acha, E. M., & Fernández, D. A. (2018). Fish early life stages associated with giant kelp forests in sub-Antarctic coastal waters (Beagle Channel, Argentina). *Polar Biology, 41*(2), 365–375
9. Buschmann, A. H., Riquelme, V. A., Hernández-González, M. C., Varela, D., Jiménez, J. E., Henríquez, L. A., Vergara, P. A., Guíñez, R., & Filún, L. (2006). A review of the impacts of salmonid farming on marine coastal ecosystems in the southeast Pacific. *ICES Journal of Marine Science: Journal du Conseil, 63*(7), 1338–1345
10. Buschmann, A. H., Cabello, F., Young, K., Carvajal, J., Varela, D. A., & Henríquez, L. (2009). Salmon aquaculture and coastal ecosystem health in Chile: Analysis of regulations, environmental impacts and bioremediation systems. *Ocean & Coastal Management, 52*(5), 243–249
11. Buschmann, A. H., Niklitschek, E. J., and Pereda, S. (2023). *Aquaculture and Its Impacts on the Conservation of Chilean Patagonia.* Springer
12. Cabello, F. C., & Godfrey, H. P. (2016). Harmful algal blooms (HABs), marine ecosystems and human health in the Chilean Patagonia. *Revista Chilena de Infectología, 33*(5), 561–562
13. Camus, P. A. (2001). Biogeografía marina de Chile continental. *Revista Chilena de Historia Natural, 74*(3), 587–617
14. Castilla, J. C., and Moreno, C. A. (1982). Sea urchins and *Macrocystis pyrifera*: experimental test of their ecological relations in southern Chile. In: J. M., Lawrence. (Ed.), *Proceedings of the International Echinoderm Conference*, pp. 257–263. Tampa Bay, Florida, USA: International Echinoderm Conference
15. Castilla, J. C. (1985). Food webs and functional aspects of the kelp, *Macrocystis pyrifera*, community in the Beagle Channel, Chile. In: Siegfried, W. R., Condy, P. R., & Laws, R. M. (Eds.), *Antarctic Nutrient Cycles and Food Webs*, pp. 407–414. Berlin, Heidelberg: Springer

16. Contreras, S., & Castilla, J. C. (1987). Feeding behavior and morphological adaptations in two sympatric sea urchin species in central Chile. *Marine Ecology Progress Series, 38,* 217–224

17. Contreras, C., Niklitschek, E., Molinet, C., Díaz, P., & Díaz, M. (2019). Fishery-induced reductions in density and size truncation of sea urchin Loxechinus albus affects diversity and species composition in benthic communities. *Estuarine, Coastal and Shelf Science, 219,* 409–419

18. Castrillón, A. L. (2017). Respuesta metabólica de tres corales escleractinios de aguas frías (*Desmophyllum dianthus, Caryophyllia huinayensis* y *Tethocyathus endesa*) cuando se someten a condiciones de hipoxia y sulfuro de hidrógeno. [Thesis]. Universidad Católica del Norte, Facultad de Ciencias del Mar, Chile

19. Edgar, G. J., Stuart-Smith, R. D., Willis, T. J., Kininmonth, S., Baker, S. C., Banks, S., Barrett, N. S., Becerro, M. A., Bernard, A. T. F., Berkhout, J., Buxton, C. D., Campbell, S. J., Cooper, A. T., Davey, M., Edgar, S. C., Försterra, G., Galván, D. E., Irigoyen, A. J., Kushner, D. J., … Thomson, R. J. (2014). Global conservation outcomes depend on marine protected areas with five key features. *Nature, 506*(7487), 216–220

20. Feehan, K. A., Waller, R. G., & Häussermann, V. (2019). Highly seasonal reproduction in deep-water emergent *Desmophyllum dianthus* (Scleractinia: Caryophylliidae) from the northern Patagonian fjords. *Marine Biology, 166*(4), 52

21. Fernández, M., Jaramillo, E., Marquet, P. A., Moreno, C. A., Navarrete, S. A., Ojeda, P. F., Valdovinos, C. R., & Vásquez, J. A. (2000). Diversity, dynamics and biogeography of Chilean benthic nearshore ecosystems: An overview and guidelines for conservation. *Revista Chilena de Historia Natural, 73*(4), 797–830

22. Fillinger, L., & Richter, C. (2013). Vertical and horizontal distribution of *Desmophyllum dianthus* in Comau fjord, Chile: A cold-water coral thriving at low pH. *PeerJ, 1,* e194

23. Försterra, G., Beuck, L., Häussermann, V., & Freiwald, A. (2005). Shallow-water *Desmophyllum dianthus* (Scleractinia) from Chile: characteristics of the biocoenoses, the bioeroding community, heterotrophic interactions and (paleo)-bathymetric implications. In: Freiwald, A., & Roberts, J. M. (Eds), *Cold-water corals and ecosystems,* pp. 937–977. Berlin, Heidelberg: Springer

24. Försterra, G., Häussermann, V., Laudien, J., Jantzen, C., Sellanes, J., & Muñoz, P. (2014). Mass die-off of the cold-water coral *Desmophyllum dianthus* in the Chilean Patagonian fjord region. *Bulletin of Marine Science, 90*(3), 895–899

25. Försterra, G., Häussermann, V., & Laudien, J. (2017). Animal forests in the Chilean fjords: discoveries, perspectives and threats in shallow and deep waters. In: Rossi, S., Bramanti, L., Gori, A., & Orejas Saco del Valle, C. (Eds.), *Marine animal forests: the ecology of benthic biodiversity hotspots,* pp. 1–35. Switzerland: Springer

26. Fraser, C. I., Morrison, A., & Olmedo Rojas, P. (2020). Biogeographic processes influencing Antarctic and sub-Antarctic seaweeds. *Antarctic Seaweeds: Diversity, Adaptation and Ecosystem Services,* 43–57

27. Friedlander, A. M., Ballesteros, E., Bell, T. W., Giddens, J., Henning, B., Hüne, M., Muñoz, A., Salinas-de-León, P., & Sala, E. (2018). Marine biodiversity at the end of the world: Cape Horn and Diego Ramírez islands. *PLoS ONE, 13*(1), e0189930

28. Friedlander, A. M., Ballesteros, E., Bell, T. W., Caselle, J. E., Campagna, C., Goodell, W., Hüne, M., Muñoz, A., Salinas-de-León, P., Sala, E., & Dayton, P. K. (2020). Kelp forests at the end of the earth: 45 years later. *PLoS ONE, 15*(3), e0229259

29. Gebauer, P., Paschke, K., Vera, C., Toro, J. E., Pardo, M., & Urbina, M. (2017). Lethal and sublethal effects of commonly used anti-sea lice formulations on non-target crab *Metacarcinus edwardsii* larvae. *Chemosphere, 185,* 1019–1029

30. Häussermann, V., & Försterra, G. (2007). Extraordinary abundance of hydrocorals (Cnidaria, Hydrozoa, Stylasteridae) in shallow water of the Patagonian fjord region. *Polar Biology, 30*(4), 487–492

31. Häussermann, V., & Försterra, G. (2009). *Marine Benthic Fauna of Chilean Patagonia.* Puerto Montt, Chile: Nature in Focus

32. Häussermann, V., Försterra, G., Melzer, R., & Meyer, R. (2013). Gradual changes of benthic biodiversity in Comau fjord, Chilean Patagonia–lateral observations over a decade of taxonomic research. *Spixiana, 36*(2), 161–171
33. Häussermann, V., & Försterra, G. (2014). Vast reef-like accumulation of the hydrocoral *Errina antarctica* (Cnidaria, Hydrozoa) wiped out in Central Patagonia. *Coral Reefs, 33*(1), 29
34. Häussermann, V., Gutstein, C. S., Beddington, M., Cassis, D., Olavarria, C., Dale, A. C., Valenzuela-Toro, A. M., Pérez-Álvarez, M. J., Sepúlveda, H. H., McConnell, K. M., Horwitz, F. E., & Försterra, G. (2017). Largest baleen whale mass mortality during strong El Niño event is likely related to harmful toxic algal bloom. *PeerJ, 5*, e3123
35. Häussermann, V., Molinet, C., Díaz Gómez, M., Försterra, G., Henríquez, J., Espinoza Cea, K., Matamala Ascencio, T., Hüne, M., Cárdenas, C., Glon, B. T., & N., & Subiabre Mena, D. (2022). Recent massive invasions of the circumboreal sea anemone *Metridium senile* in North and South Patagonia. *Biological Invasions, 24*, 3665–3674
36. Hargrave, B. T. (2010). Empirical relationships describing benthic impacts of salmon aquaculture. *Aquaculture Environment Interactions, 1*, 33–46
37. Höfer, J., González, H. E., Laudien, J., Schmidt, G. M., Häussermann, V., & Richter, C. (2018). All you can eat: The functional response of the cold-water coral *Desmophyllum dianthus* feeding on krill and copepods. *PeerJ, 6*, e5872
38. Hucke-Gaete, R., Viddi, F. A., & Simeone, A. (2021). *Marine Mammals and Seabirds of Chilean Patagonia: Focal Species for the Conservation of Marine Ecosystems*. Springer
39. Iriarte, J. L., Pantoja, S., Iriarte, L., & Daneri, G. (2014). Oceanographic processes in Chilean fjords of Patagonia: From small to large-scale studies. *Progress in Oceanography, 129*, 1–7
40. Iriarte, J. (2018). Natural and human influences on marine processes in Patagonian subantarctic coastal waters. *Frontiers in Marine Sciences, 5*, 360
41. Jantzen, C., Laudien, L., Sokol, S., Försterra, G., Häussermann, V., Kupprat, F., & Richter, C. (2013). In situ short-term growth rates of a cold-water coral. *Marine and Freshwater Research, 64*(7), 631–641
42. Jones, C. G., Lawton, J. H., & Shachak, M. (1994). Organisms as ecosystem engineers. *Oikos, 69*, 373–386
43. León-Muñoz, J., Urbina, M. A., Garreaud, R., & Iriarte, J. L. (2018). Hydroclimatic conditions trigger record harmful algal bloom in western Patagonia (summer 2016). *Scientific Reports, 8*, 1330
44. López, D. A., López, B. A., Arriagada, S. E., González, M. L., Mora, O. A., Bedecarratz, P. C., Pineda, M. O., Andrade, L. I., Uribe, J. M., & Riquelme, V. A. (2012). Diversification of Chilean aquaculture: The case of the giant barnacle *Austromegabalanus psittacus* (Molina,1782). *Latin American Journal of Aquatic Research, 40*(3), 596–607
45. López, D. A., Espinoza, E. A., López, B. A. & Santibañez, A. F. (2008). Molting behavior and growth in the giant barnacle *Austromegabalanus psittacus* (Molina, 1782). *Revista de biología marina y oceanografía 43*(3), 607–613
46. Macaya, E., Riosmena-Rodríguez, R., Melzer, R., Meyer, R., Försterra, G., & Häussermann, V. (2014). Rhodolith beds in the south-east Pacific. *Marine Biodiversity, 45*, 153–154
47. Marquet, P. A., Buschmann, A. H., Corcoran, D., Díaz, P. A., Fuentes-Castillo, T., Garreaud, R., Pliscoff, P., and Salazar, A. (2023). *Global Change and Acceleration of Anthropic Pressures on Patagonian Ecosystems*. Springer
48. Mayr, C., Rebolledo, L., Schulte, K., Schuster, A., Zolitschka, B., Försterra, G., & Häussermann, V. (2014). Responses of nitrogen and carbon deposition rates in Comau fjord (42°S, southern Chile) to natural and anthropogenic impacts during the last century. *Continental & Shelf Research, 78*, 29–38
49. Molinet, C. A., Díaz, M. A., Arriagada, C. B., Cares, L. E., Marín, S. L., Astorga, M. P., & Niklitschek, E. J. E. (2015). Spatial distribution pattern of *Mytilus chilensis* beds in the Reloncaví fjord: Hypothesis on associated processes. *Revista Chilena de Historia Natural, 88*, 1–12
50. Molinet, C., and Niklitschek, E. J. (2023). *Fisheries and Marine Conservation in Chilean Patagonia*. Springer

51. Mora-Soto, A., Palacios, M., Macaya, E. C., Gómez, I., Huovinen, P., Pérez-Matus, A., Young, M., Golding, N., Toro, M., Yaqub, M., & Macias-Fauria, M. (2020). A high-resolution global map of giant kelp (*Macrocystis pyrifera*) forests and intertidal green algae (Ulvophyceae) with Sentinel-2 Imagery. *Remote Sensing, 12*, 694

52. Moreno, C. A., & Jara, F. (1984). Ecological studies on fish fauna associated with *Macrocystis pyrifera* belts in the south of Fueguian islands. *Chile. Marine Ecology Progress Series, 15*(1), 99–107

53. Niklitschek, E. J., Soto, D., Lafon, A., Molinet, C., & Toledo, P. (2013). Southward expansion of the Chilean salmon industry in the Patagonian fjords: Main environmental challenges. *Reviews in Aquaculture, 5*(3), 72–195

54. Ojeda, F. P., & Santelices, B. (1984). Invertebrate communities in holdfasts of the kelp *Macrocystis pyrifera* from southern Chile. *Marine Ecology Progress Series, 16*(1), 65–73

55. Pantoja, S., Iriarte, J. L., & Daneri, G. (2011). Oceanography of the Chilean Patagonia. *Continental Shelf Research, 31*(3), 149–153

56. Pérez-Santos, I., Castro, L., Ross, L., Niklitschek, E., Mayorga, N., Cubillos, L., Gutiérrez, M., Escalona, E., Castillo, M., Alegría, N., & Daneri, G. (2018). Turbulence and hypoxia contribute to dense biological scattering layers in a Patagonian fjord system. *Ocean Science, 14*(5), 1185–1206

57. Quiñones, R. A., Fuentes, M., Montes, R. M., Soto, D., & León-Muñoz, J. (2019). Environmental issues in Chilean salmon farming: A review. *Reviews in Aquaculture, 11*(2), 375–402

58. Riemann-Zürneck, K. (1986). Zur Biogeographie des Südwestatlantik mit besonderer Berücksichtigung der Seeanemonen (Coelenterata: Actinaria). *Helgoländer Meeresuntersuchungen, 40*, 91–149

59. Ríos, C., Arntz, W. E., Gerdes, D., Mutschke, E., & Montiel, A. (2007). Spatial and temporal variability of the benthic assemblages associated to the holdfasts of the kelp *Macrocystis pyrifera* in the Strait of Magellan, Chile. *Polar Biology, 31*(1), 89–100

60. Riosmena-Rodríguez, R., Nelson, W., & Aguirre, J. (Eds.). (2016). *Rhodolith/maërl beds: A global perspective*. Springer International Publishing

61. Rosenfeld, S., Méndez, F., Calderón, M. S., Bahamonde, F., Rodríguez, J. P., Ojeda, J., Marambio, J., Gorny, M., & Mansilla, A. (2019). A new record of kelp *Lessonia spicata* (Suhr) Santelices in the subantarctic channels: Implications for the conservation of the "huiro negro" in the Chilean coast. *PeerJ, 7*, e7610

62. Rossi, S. (2013). The destruction of the 'animal forests' in the oceans: Towards an oversimplification of the benthic ecosystems. *Ocean and Coastal Management, 84*, 77–85

63. Rovira, J., & Herreros, J. (2016). Clasificación de ecosistemas marinos chilenos de la zona económica exclusiva. *Departamento de Planificación y Políticas en Biodiversidad. División de Recursos Naturales y Biodiversidad. Ministerio del Medio Ambiente*

64. Santelices, B. (1992). Marine phytogeography of the Juan Fernandez Archipelago: A new assessment

65. Searles R. B. (1978). The genus *Lessonia* Bory (Phaeophyta, Laminariales) in southern Chile and Argentina. *British Phycological Journal, 13*, 361–381

66. Sepúlveda, S. A., Náquira, M. V., & Arenas, M. (2011). Susceptibility of coastal landslides and related hazards in the Chilean Patagonia: The case of Hornopirén area (42°S). *Investigaciones Geográficas, 43*, 35–46

67. Smale, D. A., Wernberg, T., Oliver, E. C., Thomsen, M., Harvey, B. P., Straub, S. C., Burrows, M. T., Alexander, L. V., Benthuysen, J. A., Donat, M. G., Feng, M., Hobday, A. J., Holbroo, N. J., Perkins-Kirkpatrick, S. E., Scannell, H. A., Gupta, A. S., Payne, B. L., & Moore, P. J. (2019). Marine heatwaves threaten global biodiversity and the provision of ecosystem services. *Nature Climate Change, 9*(4), 306–312

68. Thatje, S., & Arntz, W. E. (2004). Antarctic reptant decapods: More than a myth? *Polar Biology, 27*(4), 195–201

69. Tecklin, D., Farías, A., Peña, M.P., Gelvez, X., Castilla, J.C., Sepúlveda, M., Viddi, F. A., & Hucke-Gaete, R. (2023). *Coastal-Marine Protection in Chilean Patagonia: Historical Progress, Current Situation, And Challenges*. Springer

70. Torres, R., Pantoja, S., Harada, N., González, H. E., Daneri, G., Frangopulos, M., Rutllant, J. A., Duarte, C. M., Rúiz-Halpern, S., Mayol, E., & Fukasawa, M. (2011). Air-sea CO_2 fluxes along the coast of Chile: From CO_2 outgassing in central northern upwelling waters to CO_2 uptake in southern Patagonian fjords. *Journal of Geophysical Research, 116*, C09006
71. Tyberghein, L., Verbruggen, H., Pauly, K., Troupin, C., Mineur, F., & De Clerck, O. (2012). Bio-ORACLE: A global environmental dataset for marine species distribution modelling. *Global Ecology and Biogeography, 21*, 272–281
72. Valdovinos, C., Navarrete, S. A., & Marquet, P. A. (2003). Mollusk species diversity in the Pacific: Why are there more species towards the pole? *Ecography, 26*(2), 139–144
73. Vaquer-Sunyer, R., & Duarte, C. M. (2010). Sulfide exposure accelerates hypoxia-driven mortality. *Limnology and Oceanography, 55*, 1075–1082
74. Vásquez, J., & Castilla, J. C. (1984). Some aspects of the biology and trophic range of *Cosmasterias lurida* (Asteroidea, Asteriinae) in belts of *Macrocystis pyrifera* at Puerto Toro. *Chile. Medio Ambiente, 7*(1), 47–51
75. Vásquez, J. A., Castilla, J. C., & Santelices, B. (1984). Distributional patterns and diets of four species of sea urchins in giant kelp forest (*Macrocystis pyrifera*) of Puerto Toro, Navarino island. *Chile. Marine Ecology Progress Series, 19*(1), 55–63
76. Villegas, M. J., Laudien, J., Sielfeld, W., & Arntz, W. E. (2008). *Macrocystis integrifolia* and *Lessonia trabeculata* (Laminariales; Phaeophyceae) kelp habitat structures and associated macrobenthic community off northern Chile. *Helgoland Marine Research, 62*, 33–43
77. Villegas, M., Laudien, J., Sielfeld, W., and Arntz, W. (2019). Effect of foresting barren ground with *Macrocystis pyrifera* (Linnaeus) C. Agardh on the occurrence of coastal fishes off northern Chile. *Journal of Applied Phycology*, 31(3), 2145–2157
78. Winkler, M. (2013). Macroepibenthic communities associated with the hydrocoral *Errina antarctica* from the Chilean fjord region: Does bathymetry influence community structure? Alfred-Wegener-Institut Helmholtz-Zentrum für Polar- und Meeresforschung and Universität Koblenz-Landau, Landau: Bachelor thesis

Vreni Häussermann Diploma in Biology and Ph.D. in Zoology, Ludwig-Maximilians University, Munich, Germany. Assistant Professor, Universidad San Sebastián, Chile. Pew Marine Fellow, Humboldt Scholar, and Rolex Laureate. Studies on the marine biodiversity of Chilean Patagonia since 1998.

Günter Försterra Diploma in Biology, Ludwig-Maximilians-University, Munich, Germany. Researcher at Universidad Católica de Valparaíso. He has been working for more than 20 years on marine biodiversity in Chilean Patagonia.

Jürgen Laudien Ph.D. in Marine Biology, University of Bremen, Germany. Senior Researcher, Alfred-Wegener-Institut Helmholtz-Zentrum für Polar-und Meeresforschung, Bremerhaven, Germany. Researcher in Chilean Patagonia since 2009.

Fisheries and Marine Conservation in Chilean Patagonia

11

Carlos Molinet and Edwin J. Niklitschek

Abstract

The Patagonian fjords and channels system has been subjected to intense fishing pressure over the last 50 years. There are currently more than 40,000 registered fishermen and official landings in 2019 were 192,891 tons (t) of benthic species, 9600 t of demersal species and 14,400 t of pelagic species (www.ser napesca.cl). These quantities do not include unknown levels of illegal fishing, under-reporting and bycatch. In this chapter we present an overview of the recent developments in benthic, demersal and industrial fisheries, their current and future environmental impacts and efforts to achieve ecosystem-based fishery management. The management of these resource species has been focused in a mono-specific and hierarchical approach of limited effectiveness, because: (i) the lack of regulatory adherence by users; (ii) the vast size and isolation of the territory; (iii) the institutional weaknesses of the management, monitoring, and enforcement systems. The few fisheries in the area that have enough scientific capture and effort information to be evaluated have been declared in a state of over-exploitation. The effects of the large removals of macroalgae, invertebrates, and fish on biological diversity and ecosystem structure and function have yet to be evaluated. This is a difficult task given the absence of distribution, abundance, and diversity baselines, and the overall limited information available for local ecosystems. There is, however, preliminary evidence showing that recent exploitation levels have affected both benthic biodiversity and the

C. Molinet (✉)
Instituto de Acuicultura, Universidad Austral de Chile, Campus Pelluco, Puerto Montt, Chile
e-mail: cmolinet@uach.cl

C. Molinet · E. J. Niklitschek
Programa de Investigación Pesquera, Universidad Austral de Chile-Universidad de Los Lagos, Puerto Montt, Chile

E. J. Niklitschek
Centro i~Mar, Universidad de los Lagos, Puerto Montt, Chile

© Pontificia Universidad Católica de Chile 2023
J. C. Castilla et al. (eds.), *Conservation in Chilean Patagonia*, Integrated Science 19, https://doi.org/10.1007/978-3-031-39408-9_11

availability of prey for threatened species of birds and mammals. After reviewing all previous issues, we propose four basic lines of action: (i) move quickly towards new models of governance, locally based, for fisheries management; (ii) move towards multi-specific and/or ecosystem-based management models of local fisheries; (iii) carry out a multi-year systematic effort to study and monitor local biodiversity, aimed at defining and evaluating areas and conservation measures; and iv) design, establish, and monitor an extended network of Patagonian marine protected areas that can serve, among other aims, as reference non-altered marine environments.

Keywords

Patagonia • Chile • Artisanal and industrial fisheries • Overfishing • Illegal fishing • Co-management • Ecosystem approach • Monitoring • Biodiversity

1   Introduction

Chile's Patagonian fjord and channel system (PFCS) is one of the most extensive estuarine ecosystems in the world [42, 43]. It is located on the western edge of Patagonia and covers the Los Lagos, Aysén and Magallanes administrative regions. For the purposes of this chapter, its geographic scope is defined as extending from the Reloncaví Sound to Cape Horn (41° 42′ S 73° 02′W; 55° 58′ S 67° 17′W). Its continental coastline extends approximately 1,600 km from north to south, however, if the >40,000 islands in this zone are included, it extends 100,627 km [22, 58]. This system can be divided into three eco-geographic regions: (i) Chiloé-Taitao, between the Chacao Channel and the Taitao Peninsula (42°–47°S); (ii) Kawésqar (47°–54°S); (iii) Magallanes (54°–56°S) [47]. PFCS habitats are highly heterogeneous in time and space, which is driven by the geomorphological configuration and the local and external oceanographic and meteorological forcings that regulate them [43, 54, 55]. These characteristics promote dynamic processes of retention and transport of nutrients, particles, eggs and larvae, which sustain high levels of diversity and biological production and extend the latitudinal and bathymetric ranges of distribution of some species [34].

The human relationship with the Patagonian coastline has evolved from a subsistence economy until the mid-nineteenth century to a commercial and export-oriented economy, in which fisheries experienced a rapid expansion from the mid-1970s to the mid-1980s [28]. The PFCS presently concentrates important artisanal fisheries, whose landings in 2019 reached 192,891 t of benthic species, 9600 t of demersal species and 14,400 t of pelagic species, the last composed almost exclusively of austral sardine (www.sernapesca.cl).

Approximately 40,000 artisanal fishermen operate in the PFCS, including shore harvesters, divers and ship owners, with a fleet of approximately 4000 registered vessels (www.sernapesca.cl). The largest fraction of this fleet belongs to the Los Lagos Region, which has been operating regularly in practically the entire area north of the Taitao Peninsula. This occurs in the context of the Benthic Fisheries Management Plan of the Contiguous Zone of the Los Lagos and Aysén Regions

(Fig. 1), whose development was grounded in the historical territorial, economic and cultural continuity existing between the regions [28, 33]. Progressive changes in productivity, exploitation areas and target species have had little effect on the structure of coastal population centers, which have a latitudinal decrease in the number and size of urban centers (Fig. 1a). As a result, the remote coastal areas of the Aysén and Magallanes regions remain almost unpopulated, although subject to the presence of a floating population of fishermen and salmon industry workers whose number exceeds 3000 people in Aysén alone.

Research, monitoring, management and control of PFCS fisheries is difficult due to the operational difficulties and the high operational costs in this vast and disconnected territory. The scientific information available on most of its ecosystems, biological communities, and populations is extremely limited. Centralized fishery controls and management measures tend to be violated, with probably significant, although not quantified, levels of underreported and illegal fishing. As a result, there are no available assessments of the ecosystem, community or population consequences of the removal of significant volumes of macroalgae, forage

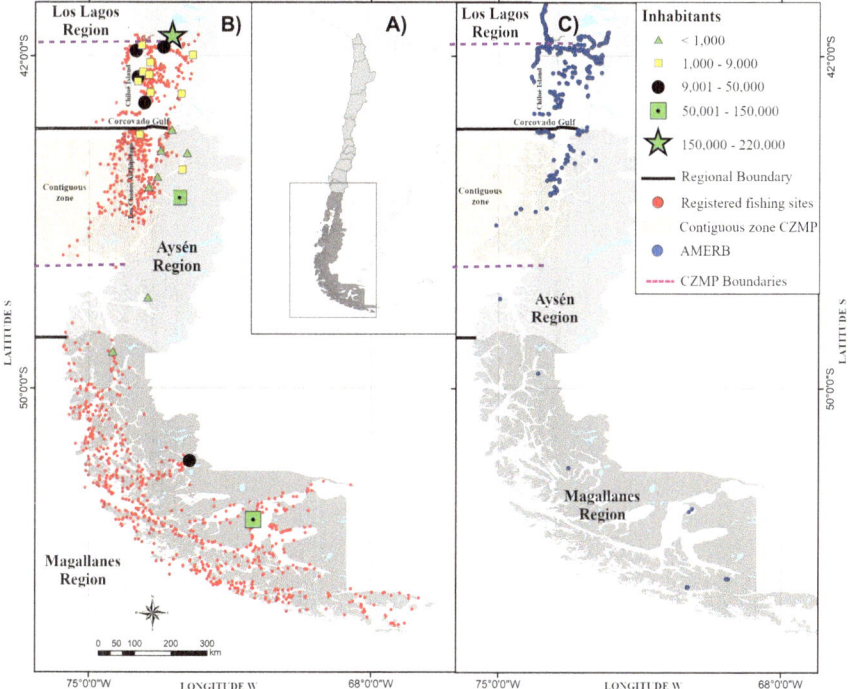

Fig. 1 Population centers, recorded sources of benthic resource extraction and Management and Exploitation Areas for Benthic Resources (in Spanish AMERBs) in the Patagonian fjord and channel system [2]

species, scavengers, herbivores and top predators that have occurred in this system since the mid-nineteenth century.

2   Scope and Objectives

This chapter is organized in four sections. In the first, we provide an overview of the recent development of benthic, demersal, and pelagic fisheries in the PFCS. We then present a summary of the main current and potential impacts attributable to these fisheries on the exploited species themselves and on other components of the ecosystems in this area. Next, we describe the main fisheries management efforts carried out for ecosystems and populations. Finally, as a corollary to the information presented, we provide a set of recommendations that we believe are fundamental to mitigate the impacts of fishing activities and promote the conservation of PFCS ecosystems.

3   Results

3.1   Development of Chile's Patagonian Fjord and Channel System's Primary Fisheries

3.1.1   Benthic Fisheries

Landings in the Los Lagos and Aysén Regions in 2019 of 192,891 t of benthic species in the PFCS were concentrated in sea urchin, *Loxechinus albus*, algae (e.g. *Gracilaria chilensis*, *Gigartina skotsbergii* and *Sarcothalia crispata*) and mollusks (e.g. *Venus antiqua*, *Concholepas concholepas*). In the Magallanes Region, urchins, algae and crustaceans such as the king crab (*Lithodes santolla*) make up the vast majority of the catch (Fig. 2). Landings of sea urchin, the main benthic fishery of the PFCS, were 31,455 t in 2019, equivalent to 95% of the national landings and close to 50% of the world landings. Since the end of the twentieth century Chile has been the main world producer of this species [15].

Although the levels of underreported and illegal harvest by the benthic fleet have not been evaluated in the PFCS, the information available for other areas of the country indicates that they could be very high. For example, illegal fishing for loco could represent 60–100% of the landings reported annually by the National Fisheries Service [3, 9, 17, 40]. Illegal and underreported fishing are more likely in the southern PFCS, due to the local geography. There is practically no monitoring or recording of loco harvest in the Aysén Region due to the ban established in 2001 on extracting loco in all open access areas (outside of AMERBs) located between Arica and Aysén, and the restrictions on capture and sale related to contamination with marine biotoxins since the late 1990s in this region [25]. Given the presence of the benthic fleet in the Aysén and Los Lagos Regions, in addition to the fleet that provides services to aquaculture (which includes divers), it is estimated that the actual catch of loco significantly exceeds the modest official landing

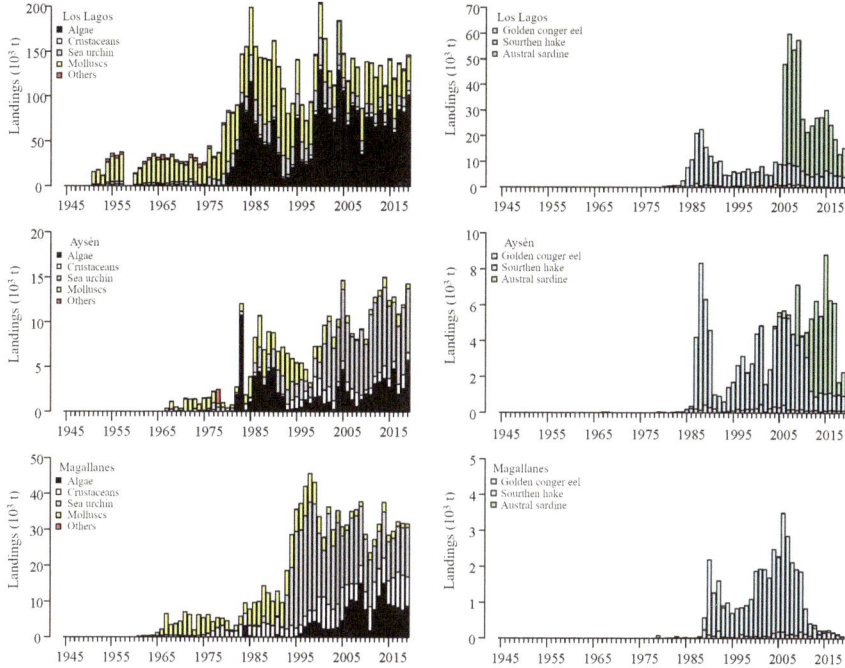

Fig. 2 Artisanal landings of the main benthic resources (left) and fish (right) in the Patagonian fjord and channel system, Chile. Sea urchin (*Loxechinus albus*), golden conger eel (*Genypterus blacodes*), southern hake (*Merluccius australis*), austral sardine (*Sprattus fuegensis*). *Source* Department of Fisheries and Hunting (1945–1959) and (1960–1977) [14]; National Fishery Service [52] (in Spanish Servicio Nacional de Pesca, SERNAPESCA

figures of approximately 7 t per year which were reported over the last ten years in the Aysén Region (www.sernapesca.cl). Illegal harvest is also a significant though unquantified concern for sea urchin, crab and king crab, due to the presence of 400–600 divers that provide services to the 220 aquaculture farms that operate in this region.

The number of benthic species exploited and recorded in the PFCS fishery increased by approximately five times between the 1950s and 2000, reaching 44 species in 2019. In the Los Lagos Region, 42 species were extracted, almost twice as many as in Aysén (27 species) and Magallanes (25 species). This greater diversity in Los Lagos is associated with a larger number of areas of fishing origins or natural "banks" (Fig. 1b, [26], whose exploitation follows an unplanned pattern of temporal and spatial rotation of exploited areas and resource species [38].

The current status and dynamics of the populations of benthic species exploited in the PFCS are poorly known, to the point that thus far only one of these fisheries is subject to regulation based on biological criteria and evaluated quantitatively. Only the urchin fishery has sufficiently developed stock assessment models for its eventual management [10, 46]. This gap in management stems from the limited availability of data on catches and fishing effort, the absence of direct abundance assessments and the unknown scale of underreported and illegal harvest. There is also limited knowledge of key aspects of the structure of exploited and unexploited PFCS populations [26], including the identification and distribution of population units, which are assumed to be distributed according to source and sink habitats [44] and interconnected through larval dispersal following a meta-population scheme [37].

An additional concern of growing importance for the sustainability of these fisheries lies in the potential negative effects of other activities that take place along the coastline, especially in the Chiloé ecoregion. Among these impacts are the increasing discharges of sediments, nutrients, and chemicals associated with aquaculture [7], agriculture, urban development and deforestation.

3.1.2 Fish Fisheries

The fisheries operating in the PFCS are of relatively recent origin (1980 onward). The southern hake *Merluccius australis* fishery stands out as a primary target species, having almost completely replaced the traditional blennie (*Eleginops maclovinus*) and silverside (*Odontesthes regia*) fisheries, which were mostly destined for local consumption. Along with these changes in target species, important changes have been observed in the number and socioeconomic profile of artisanal fishermen and in the resource species management system (Law No. 19,892 and its amendments). All this has led to an important governance crisis, aggravated by a context in which all artisanal fisheries subject to stock assessments are in the overexploited category [56].

3.1.3 Demersal Fisheries

The demersal fisheries operating in the PFCS have developed rapidly since 1984, reaching maximum annual landings of over 30,000 t between 1986 and 1988 (Fig. 2). This period was part of the social, economic, and environmental phenomenon known as "hake fever", characterized by the massive incorporation of new artisanal fishermen, generally from other fisheries, trades and/or geographic areas, primarily in pursuit of southern hake. These fishermen generated a spontaneous and transitory process of settlement of approximately 15 areas located around the Moraleda and costal channels, whose floating population reached over 5,000 people [28].

Artisanal fishing boats in Los Lagos Región, northPatagonia

Given the progressive drop in yields and catch quotas for southern hake evident at the end of the twentieth century, many of the fishers returned to their places of origin, and most of the camps were abandoned. However, two of these, Puerto Gala and Puerto Gaviota, came to be recognized by the State as permanent localities, although their populations were reduced to only 80 and 67 inhabitants, respectively, according to the 2017 census. Although all PFCS demersal fisheries are artisanal (boats <18 m), their fishing power (number of hooks) in the mid- and late 1980s reached levels similar to those of the industrial fleet harvesting this species in the adjacent ocean. Thus, the sum of both efforts has had an evident impact on the abundance of southern hake, whose official landings within the PFCS have fallen from a historical maximum of 30,200 t in 1983 to approximately 12,700 t in 2008–2010, and approximately 6300 t annually over the last seven years (Fig. 2). This last reduction, however, is not the result of lower fishing yields but rather of the enactment of Law 20,657, which in 2013 modified the General Fisheries and Aquaculture Act (Law No. 18,892 of 1989) and resulted in a reduction of close to 40% in annual catch quotas. This last drop in official landings seems to be compensated to a large extent by a significant increase in the underreported and illegal harvest, which could exceed 50% of the officially reported catch (E.J. Niklitschek, unpublished data; also see Oyanedel et al. [41], for the common hake, *Merluccius gayi*, in central Chile).

3.1.4   Pelagic Fisheries

The main pelagic fisheries currently exploited in the PFCS are the purse seine fishery for austral sardine, *Sprattus fuegensis,* and the illegal fishery for feral free-living salmonids. Commercial exploitation of the austral sardine began in the 1980s, with the purpose of providing bait for demersal fisheries. In the mid-1980s this species also began to be caught for fishmeal production, reaching reported landings of approximately 60,000 t between 2007 and 2009 (Fig. 2). Exploitation for fishmeal was initially concentrated in the Los Lagos Region, particularly in the Gulf of Ancud, but expanded to the Aysén Region in 2012, where it is of relatively less importace than in Los Lagos Region. The exploitation of free-living salmonid fisheries in the PFCS is the consequence of more than a century of repeated introductions of these species in Chile and Argentina [4] and more than three decades of massive escapes from farming centers [1]. Feral and escaped species of the genera *Oncorhynchus* and *Salmo* now support illegal and undocumented pelagic fisheries but are of growing sociopolitical and ecological importance [50].

With regard to the sustainability of these fisheries, the population of austral sardine showed clear signs of depletion as of 2010, at which time landings fell to less than 30,000 t per year, and the fleet was incapable of catching the annual quotas allowed by the national fishing authority. The real magnitude of underreported and illegal catch in the austral sardine fishery for fishmeal and for bait are unknown. The salmonid fishery, in contrast, depends on massive escapes of salmonids from aquaculture centers rather than on the abundance of already feral populations. Thus, it represents a fishery of opportunity, and contributes to mitigating the impacts of such escapes.

3.2   Effects of Fisheries on Patagonian Marine Ecosystems

3.2.1   Benthic Fisheries

While knowledge of the richness and biodiversity of benthic species in the PFCS has slowly advanced (e.g. [5, 18, 19, 61]), there are no systematic and sustained evaluations of the effects of fisheries, aquaculture and other activities on benthic species, their associated communities, or in general on the ecosystems of the PFCS. Given the information available for this and other exploited systems, it is to be expected that the removal of predators such as crab or king crab, as well as ecosystem engineering species [24] which form dense patches or banks and provide additional structure and shelter to the substrate (e.g. algae, mussels, sea urchins), tends to reduce local diversity (Fig. 3, [11, 19, 32]). Evidence of this has already been detected in the PFCS for exploited urchin banks, where an inverse relationship between exploitation indicators (i.e. size truncation) and associated community diversity has been observed [12].

The ecological effects of the reduced availability of locos, urchins, clams and other invertebrates on birds and aquatic mammals that use feeding and/or breeding areas within the PFCS are not fully known (Fig. 3). Of particular interest are the effects on threatened species such as coscoroba swan (*Coscoroba*

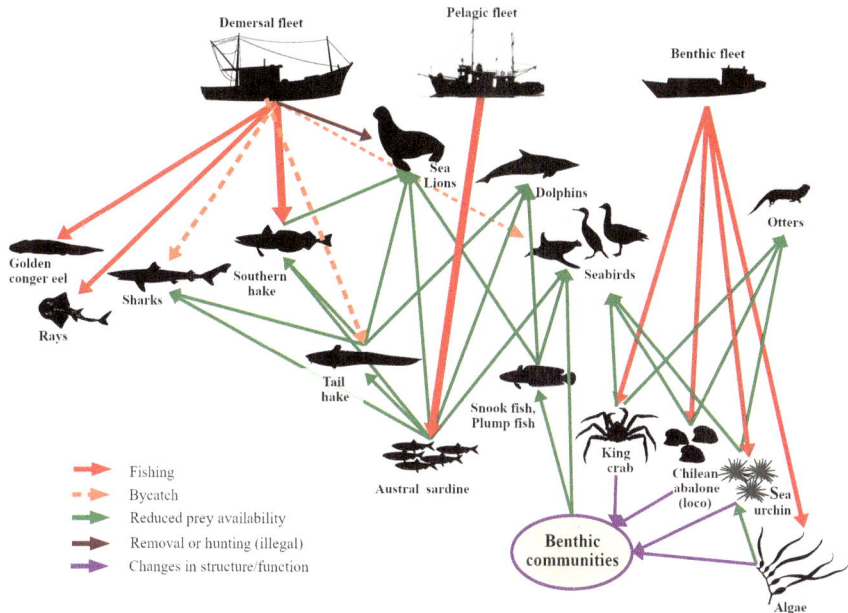

Fig. 3 Schematic and simplified representation of some potential direct and indirect effects on target and non-target fishery species present in the Patagonian fjords and channel system's main fisheries

coscoroba), widgeon *(Phalacrocorax gaimardi)*, and otters *(Lontra felina and provocax)* [13, 22, 53].

3.2.2 Demersal Fisheries

Although there has been little assessment of the direct and indirect impacts of demersal fisheries on PFCS populations (Fig. 3), the progressive reduction in the distribution and abundance of non-target demersal species subject to bycatch and often discarded, is evident and recognized by the fishermen themselves, particularly in the case of the Patagonian toothfish *(Macruronus magellanicus)* (E.J. Niklitschek, unpublished data). Even without formal evidence, similar effects are expected on elasmobranch (such as shark and ray) populations present in this system [49], which overlap spatially and vertically with demersal fisheries, and whose distribution, abundance, conservation status and ecological roles are virtually unknown in the PFCS. This bycatch of Elasmobranchii adds to the artisanal fleet's seasonal targeting of flitting skate *(Zearaja chilensis)* and thornback ray *(Dipturus trachyderma)*.

The PFCS also plays an essential role as a nursery and feeding area for different species of demersal fish, which have recognized ecological and commercial importance in coastal areas of the Pacific and Atlantic oceans [6, 59]. Three species stand out: southern hake, golden conger eel *(Genyperus blacodes)* and Patagonian toothfish. The southern hake and golden conger eel are top predators at depths beyond the regular reach of the birds and mammals that dominate this system (>200 m); the southern hake, which is essentially piscivorous, exerts strong trophic pressure on meso-pelagic species such as the Patagonian toothfish. The golden conger eel, with more territorial habits and a more diverse diet, exerts the same effect on invertebrates and benthic-demersal fish [35]. Reaching the highest abundances within this group, the Patagonian toothfish plays a fundamental role in the vertical transfer of energy as a consumer of crustaceans (e.g. Euphausiidae and *Munida* spp.) and small pelagic and meso-pelagic fish, and as the main prey of the southern hake [35]. Given the trophic role indicated above, it is expected that the substantial reduction in the abundances of exploited demersal fishes in the PFCS has generated, and is generating, important ecosystem effects [20], which have not been quantified and could affect the structure and function of the entire ecosystem and its communities, as has been observed in other highly exploited gadid fisheries [8]. It is also highly likely that the reduction of these fish has directly impacted piscivorous mammals such as the common sea lion (*Otaria flavescens*), whose feeding dives can reach 200 m depth and whose consumption of demersal fish may represent >67% of their diet [51].

Finally, given the importance of the PFCS as the main breeding area for southern hake and feeding grounds for juvenile Patagonian toothfish, it is likely that local fishing, added to aquaculture and other human activities, are affecting the recruitment of species such as those already mentioned, whose role also seems to be key to conserving the structure and function of oceanic communities and ecosystems located beyond the PFCS and may even reach those located on the Argentine shelf [6]. These are critical environmental and fisheries issues that should be research priorities in PFCS ecosystems in the future.

3.2.3 Pelagic Fisheries

Similar to benthic and demersal fisheries, the impacts of overfishing of austral sardine on the PFCS ecosystem have scarcely been studied [35]. The austral sardine dominates and probably controls the pelagic subsystem of the PFCS, which has highly productive cycles of phytoplankton [30, 31], copepods and euphausiids [23], that are the main prey for this species [29]. The austral sardine is a fundamental prey for different stages and species of fish, birds and marine mammals [35, 45, 62].

As a result of the interactions noted above, it is highly probable that overexploitation and the consequent reduction in the abundance of austral sardine are already directly affecting the growth and survival of juvenile and adult fish, birds, and mammals whose diet depends heavily on this species (e.g. Magellan penguins; Wilson et al. [62]). There are also indirect effects derived from the reduced transfer of energy to higher trophic levels [35]. Given the population reduction of the austral sardine and other planktophagous organisms (overexploited) such as clams (*Venus* spp.) and blue mussels (*Mytilus chilensis*), the natural grazing pressure on the zooplankton of the PFCS has been reduced. The qualitative and quantitative balance between this reduced natural grazing pressure on zooplankton, the effects of urban, agricultural and aquaculture nutrient discharges on primary productivity and the increased grazing pressure exerted by artificially cultured filter feeders, mainly mussels, remains unknown.

The ecosystem consequences of the feral and escaped salmonids are somewhat more evident than in the case of the austral sardine. However, the abundance of these populations, which given the irregularity of escapes is presumed to be highly variable is unknown, as is the magnitude of the fishery, given that current regulations prevent the declaration of catches since these are considered illegal. However, the diet and trophic role played by these salmonids in the PFCS is known, albeit only superficially [36]. Notwithstanding the limited existing information, it is reasonable to assume that they have a potentially negative and substantial impact on the ecosystem and that this fishery has contributed significantly to their control. Paradoxically, while this function should be promoted or subsidized, it remains prohibited under an obsolete regulatory framework that seeks to promote sport fishing and protect salmonid aquaculture.

3.3 Management and Conservation

3.3.1 Fishery Management

The fishery management applied thus far in the PFCS has followed a monospecific approach, nominally oriented to achieve the Maximum Sustainable Yield (MSY; Schaefer [48]) of each exploited stock. This management approach has used a vertical and hierarchical governance model, where the MSY management objective has been defined by law (Law No. 21.134), giving little room for effective participation of stakeholders or the pursuit of other socially relevant objectives such as income stability or social sustainability.

The vertical system of governance that has predominated in the management of the country's fisheries and the PFCS, combined with the weaknesses of the evaluation, monitoring, and control systems for exploited species, has led to an important legitimacy crisis in fisheries, such as those for the sea urchin in the early 2000s and the southern hake during the 2010s. In the former case the crisis resulted in a social and territorial conflict that, after several years, led to a change in the governance paradigm and the creation of the first participatory management plan for benthic resources implemented in Chile [33]. In the case of southern hake, a significant number of fishermen began to distrust the management system and chose to exclude themselves from the legal consultation processes. Another large number of fishermen are inadvertently marginalized by the system for registration of fishermen and allocation of access rights. As a result, a growing fraction of fishermen are now openly violating the current rules on access, reporting, and extraction quotas.

In an attempt to reverse at least partially the situation described above, the State has been promoting the creation of Management Committees and the participatory generation of management plans, as established in Law 21.134. However, the effectiveness of these committees and management plans seems to be limited by several elements, among which we highlight: (i) the still vertical and centralized character of their management and funding; (ii) the narrow decision-making space assigned to them by law, including the imposition of the MSY as a management objective; and (iii) the feedback loop of distrust that has formed between government, users, and scientists.

3.3.2   Allocation of Territorial Use Rights

An important attempt to reverse the hierarchical paradigm mentioned above and move toward co-management was the allocation of territorial rights through the system of AMERBs. However, this national system has had to face the geographic and operational difficulties of the relatively transhumant exploitation system that dominates the benthic fisheries of the PFCS, which is very different from that of the central and northern areas of the country [16].

AMERBs nominally contribute approximately 2% of the benthic landings in the regions of Los Lagos (158 operational areas, 18,000 ha) and Aysén (41 operational areas, 6700 ha), while no operational AMERBs have been recorded in Magallanes since 2001. There are also AMERBs that have been declared but not yet exploited in Los Lagos (140 areas, 15,000 ha), Aysén (36 areas, 9000 ha) and Magallanes (9 areas, 1300 ha). The exploitation of AMERBs targeting loco (*Concholepas concholepas*), macha (*Mesodesma donacium*) and algae has predominated in the Los Lagos Region, while in the Aysén Region they have been progressively concentrating on sea urchin (www.sernapesca.cl). The heterogeneous productivity and the distance between some of these areas and the population centers where the fishermen, who manage them, live have become important challenges for the profitability and operation of the AMERBs in the PFCS. Although we do not have quantitative records, we estimate that an important fraction of the landings reported by these AMERBs is in reality extracted in open-access areas (C. Molinet, personal information).

3.3.3   Conservation: Planning Instruments and Territorial Protection

The first State effort explicitly aimed at identifying marine areas for conservation and preservation within the PFCS was the zoning of the coastline promoted as part of the National Policy for the Use of the Coastline, whose first trial was carried out in the Aysén Region [27]. This effort made it possible to advance the participatory and multi-sector proposal of large areas of preferential use for conservation. However, the uncertainty associated with the lack of baseline information available for the definition of these areas and the weak hierarchy of the zoning with respect to laws and regulations have prevented its effective and direct application. Nevertheless, progress has been made in the recognition of this zoning in legal instruments such as the General Fisheries and Aquaculture Act and in the revision of areas defined as Appropriate Areas for Aquaculture under that law.

As a second conservation-oriented territorial instrument, progress has slowly begun to be made in the creation of Multiple Use Marine and Coastal Protected Area (in Spanish AMCP-MU) and marine parks, including the Francisco Coloane AMCP-MU and Marine Park (672 km^2) and the Seno Almirantazgo AMCP-MU

(764 km²) in the Magallanes Region, the Tortel AMCP-MU (6.702 km²) and the Pitipalena-Añihué AMCP-MU (239 km²) in the Aysén Region (also see Tecklin et al. [58]). While citizen support and participation have been sought in the definition and protection of these areas, there is yet no public agency in charge of their administration, which has limited their effective implementation and the achievement of their preservation, conservation, and production objectives. The expected creation of a Biodiversity and Protected Areas Service could resolve this institutional weakness of the Chilean State [22, 57, 58].

In spite of the efforts already described, recent revisions of the limits of national parks and reserves of the National System Protected Areas (in Spanish SNASPE) have shown that since their creation these zones have included an important extension of adjacent marine areas. Securing the effective protection of these areas requires resolving various legal questions, both about the maritime projection of the SNASPE parks and reserves and the implications of this protection for marine activities such as fishing and aquaculture [58].

It is very important to recognize that neither the zoning of the coastline, the creation of marine reserves or parks, nor the protection of marine areas adjacent to the SNASPE have obeyed marine conservation plans, objectives, or goals defined nationally or regionally. Except for the strategic plan and the national definition of priority sites and species, there are no public or private master plans to guide this task under the Convention on Biological Diversity 2011–2020 (www.biodivers idad.mma.gov.cl) [58]. There are only a few scientific exercises that have applied tools for the identification of priority conservation areas [21, 22, 39, 60].

4 Recommendations

Given the lack of knowledge on the impacts of fisheries on PFCS ecosystems and conservation, we recommend the following:

- Advance toward the use of new models of governance for Patagonian fisheries that are more participatory and inclusive, and increase the confidence, acceptance, and respect of the actors for the mechanisms of monitoring, evaluation, reporting and resource management. One tool that can contribute to this is the participatory design and quantitative evaluation of management strategies in the Management Committees, administered by the Undersecretary of Fisheries, including the design of non-demographic performance indicators and empirical decision rules, appropriate for the data-poor fisheries that predominate in the area.
- Strengthen fisheries monitoring programs and integrate the different monitoring efforts to obtain an ecosystem vision that includes both the status of fishery populations of interest and also that of other vulnerable populations, biodiversity, and ecosystem services [21].

- Move toward multispecies and/or ecosystem management models aimed at conserving ecosystem structure and function within acceptable limits, considering, for example, the trophic demand of top predators. Particular attention should be given to the requirements of these species during their reproductive periods, considering the possible effects of climate change.
- Design and implement a regional plan to improve management, make protection effective and increase the number and surface area of marine protected areas in the PFCS, in order to form a network aimed at protecting (i) highly vulnerable ecosystems and/or communities such as those defined by cold water corals; (ii) a significant fraction of the diversity of benthic habitats and communities existing in the PFCS; and (iii) essential habitats used by mammals, birds and fish that sustain their reproductive processes. These should have management and monitoring plans, whose objectives and management are consistent and complementary to the objectives and management of other existing territorial units (e.g. AMERBs and genetic reserves) and with existing productive development plans such as the current fishery management plans.
- Facilitate informed and timely territorial decision-making, including the definition of an effective system of protected areas. It is essential to make a multi-year and systematic investment effort, similar to the census of marine life, aimed at increasing and updating the available knowledge on the diversity and spatial distribution of biotopes, species, communities and essential habitats.

Acknowledgements The authors are grateful for the funding from the State of Chile that has allowed the development of the knowledge presented in this synthesis, as well as to our collaborators at the Fisheries Research Program. We thank the researchers of the Instituto de Fomento Pesquero for facilitating the use of the information they collected. Thanks also to Dr. Juan Carlos Castilla for his encouragement and motivation to complete this chapter. An anonymous reviewer contributed to enriching this work with their comments and corrections.

References

1. Arismendi, I., Penaluna, B. E., Dunham, J. B., de Leaniz, C. G., Soto, D., Fleming, I. A., & León- Muñoz, J. (2014). Differential invasion success of salmonids in southern Chile: Patterns and hypotheses. *Reviews in Fish Biology and Fisheries, 24*(3), 919–941.
2. Ariz, L., Grego, E., Figueroa, L., Romero, P., Wilson, A. Cortés, C., Palta, E., Aguilera, A., & González, P. (2017). *Programa de seguimiento de pesquerías bajo régimen Áreas de Manejo, 2017. Convenio de Desempeño 2016.* Valparaíso, Chile: Instituto de Fomento Pesquero. Retrieved from: http://ifop-primotc.hosted.exlibrisgroup.com/primo_library/libweb/act ion/search.do?vid=56IFOP
3. Bandin, R. M., & Quiñones, R. A. (2014). Impacto de la captura ilegal en pesquerías artesanales bentónicas bajo el régimen de co-manejo: El caso de Isla Mocha, Chile. *Latin American Journal of Aquatic Research, 42,* 547–579.
4. Basulto, S. (2003). *El largo viaje de los salmones. Una crónica olvidada. Propagación y cultivos de especies acuáticas en Chile.* Santiago, Chile: Editorial Maval.
5. Betti, F., Bavestrello, G., Bo, M., Enrichetti, F., Loi, A., Wanderlingh, A., Pérez-Santos, I., & Daneri, G. (2017). Benthic biodiversity and ecological gradients in the seno Magdalena (Puyuhuapi Fjord, Chile). *Estuarine, Coastal and Shelf Science, 198* (Part A), 269–278.

6. Brickle, P., Schuchert, P. C., Arkhipkin, A. I., Reid, M. R., & Randhawa, H. S. (2016). Otolith trace elemental analyses of South American austral hake, *Merluccius australis* (Hutton, 1872) indicates complex salinity structuring on their spawning/larval grounds. *PLoS One, 11*(1), e0145479.

7. Buschmann, A. H., Niklitschek, E. J., & Pereda, S. (2023). *Aquaculture and its impacts on the conservation of Chilean Patagonia*. Springer.

8. Bundy, A., Heymans, J. J., Morissette, L., & Savenkoff, C. (2009). Seals, cod and forage fish: A comparative exploration of variations in the theme of stock collapse and ecosystem change in four Northwest Atlantic ecosystems. *Progress in Oceanography, 81*(1–4), 188–206.

9. Castilla, J. C., Espinosa, J., Yamashiro, C., Melo, O., & Gelcich S. (2016). Telecoupling between catch, farming, and international trade for the gastropods *Concholepas concholepas* (Loco) and *Haliotis spp.* (Abalone). *Journal of Shellfish Research, 35*(2), 499–506.

10. Canales, C., Cavieres, J., Barahona, N., Araya, P., Techeira, C., Molinet, C., & Venegas, A. (2014). *Análisis de los cambios de abundancia de la población de erizo (Loxechinus albus) en la X y XI regiones*. Valparaíso, Chile: Instituto de Fomento Pesquero (IFOP). Retrieved from: http://ifop-primotc.hosted.exlibrisgroup.com/primo_library/libweb/action/search.do?vid=56IFOP

11. Coleman, F. C., & Williams, S. L. (2002). Overexploiting marine ecosystem engineers: Potential consequences for biodiversity. *Trends in Ecology & Evolution, 17*(1), 40–44.

12. Contreras, C., Niklitschek, E., Molinet, C., Díaz, P., & Díaz, M. (2019). Fishery-induced reductions in density and size truncation of sea urchin *Loxechinus albus* affects diversity and species composition in benthic communities. *Estuarine, Coastal and Shelf Science, 219*, 409–419.

13. Córdova, O., & Rau, J. R. (2016). Interacción entre la pesca artesanal y el depredador de alto nivel trófico *Lontra felina* en Chile. *Revista de Biología Marina y Oceanografía, 51*(3), 621–627. Departamento de Pesca y Caza (1945–1959). *Estadística de pesca y caza. Ministerio de Agricultura*. Santiago, Chile. Biblioteca Instituto de Fomento Pesquero, Valparaíso, Chile.

14. Departamento de Pesca y Caza (1960–1977). *Estadística de pesca y caza*. Ministerio de Agricultura. Santiago, Chile. Retrieved from: http://www.sernapesca.cl/informes/estadisticas.

15. Food and Agriculture Organization (2019). *FAO fisheries & aquaculture–fishery statistical collection—Global production*. Retrieved from: http://www.fao.org/fishery/statistics/global-production/en.

16. Gelcich, S., Fernández, M., Godoy, N., Canepa, A., Prado, L., & Castilla, J. C. (2012). Territorial user rights for fisheries as ancillary instruments for marine coastal conservation in Chile. *Conservation Biology, 26*(6), 1005–1015.

17. González, J., Stotz, W., Garrido, J., Orensanz, J. M., Parma, A. M., Tapia, C., & Zuleta, A. (2006). The Chilean turf system: How is it performing in the case of the loco fishery? *Bulletin of Marine Science, 78*, 499–527.

18. Häussermann, V., & Försterra, G. (2009). *Marine benthic fauna of Chilean Patagonia. Illustrated identification guide* (1st ed.). Nature in Focus

19. Häussermann, V., Försterra, G., & Laudien, J. (2023). *Hard bottom macrobenthos of Chilean Patagonia: emphasis on conservation of subtitoral invertebrate and algal forests*. Springer.

20. Heithaus, M. R., Frid, A., Wirsing, A. J., & Worm, B. (2008). Predicting ecological consequences of marine top predator declines. *Trends in Ecology and Evolution, 23*(4), 202–210.

21. Hucke-Gaete, R., Lo Moro, P., & Ruiz, J. (2010). *Conservando el mar de Chiloé, Palena y Guaitecas*. Síntesis del estudio *investigación para el desarrollo de área marina costera protegida Chiloé, Palena y Guaitecas*. Valdivia, Chile: Imprenta América.

22. Hucke-Gaete, R., Viddi, F. A., & Simeone, A. (2023). *Marine mammals and seabirds of Chilean Patagonia: focal species for the conservation of marine ecosystems*. Springer.

23. Iriarte, J. L., González, H. E., & Nahuelhual, L. (2010). Patagonian fjord ecosystems in southern Chile as a highly vulnerable region: Problems and needs. *Ambio, 39*(7), 463–466.

24. Jones, C. G., Lawton, J. H., & Shachak, M. (1994). Organisms as ecosystem engineers. *Oikos, 69*, 373–386.

25. Lembeye, G., Marcos, N., Sfeir, A., Molinet, C., & Jara, F. (1998). *Seguimiento de la toxicidad en recursos pesqueros de importancia comercial en la X y XI Región.* Informe final FIP-IT 97–49. Universidad Austral de Chile. Retrieved from: http://www.subpesca.cl/fipa/613/w3-article-89629.html

26. Molinet, C., Barahona, N., Yannicelli, B., González, J., Arévalo, A., & Rosales, S. (2011). Statistical and empirical identification of multispecies harvesting zones to improve monitoring, assessment, and management of benthic fisheries in southern Chile. *Bulletin of Marine Science, 87*(3), 351–375.

27. Molinet, C., Niklitschek, E. J., Coper, S., Díaz, M., Díaz, P., Fuentealba, M., & Marticorena, F. (2014). Challenges for coastal zoning and sustainable development in the northern Patagonian fjords (Aysén, Chile). *Latin American Journal of Aquatic Research, 42*(1), 18–29.

28. Molinet, C., Solari, M. E., Díaz, M., Marticorena, F., Díaz, P., Navarro, M., & Niklitschek, E. J. (2018). Fragmentos de la historia ambiental del sistema de fiordos y canales nor-patagónicos, sur de Chile: Dos siglos de explotación. *Magallania, 46*(2), 107–128.

29. Montecinos, S., Castro, L. R., & Neira, S. (2016). Stable isotope (13C and 15N) and trophic position of Patagonian sprat (*Sprattus fuegensis*) from the northern Chilean Patagonia. *Fisheries Research, 179*, 139–147.

30. Montero, P., Daneri, G., González, H. E., Iriarte, J. L., Tapia, F. J., Lizárraga, L., Sánchez, N., & Pizarro, O. (2011). Seasonal variability of primary production in a fjord ecosystem of the Chilean Patagonia: Implications for the transfer of carbon within pelagic food webs. *Continental Shelf Research, 31*(3–4), 202–215.

31. Montero, P., Daneri, G., Tapia, F., Iriarte, J. L., Crawford, D., Montero, P., Daneri, G., Tapia, F., Iriarte, J. L., & Crawford, D. (2017). Diatom blooms and primary production in a channel ecosystem of central Patagonia. *Latin American Journal of Aquatic Research, 45*(5), 999–1016.

32. Moreno, C. A. (2001). Community patterns generated by human harvesting on Chilean shores: A review. *Aquatic Conservation: Marine and Freshwater Ecosystems, 11*(1), 19–30.

33. Moreno, C. A., Barahona, N., Molinet, C., Orensanz, J. M., Parma, A. M., & Zuleta, A. (2006). From crisis to institutional sustainability in the Chilean sea urchin fishery. In: T. R. McClanahan & J. C. Castilla (Eds.), *Fisheries management: Progress towards sustainability* (pp. 43–67). Blackwell Publishing Ltd.

34. Moreno, C. A., Molinet, C., Díaz, M., Díaz, P. A., Cáceres, M. A., Añazco, B., & Niklitschek, E. (2018). Coupling biophysical processes that sustain a deep subpopulation of *Loxechinus albus* and its associated epibenthic community over a bathymetric feature. *Estuarine, Coastal and Shelf Science*, 23–33.

35. Neira, S., Arancibia, H., Barros, M., Castro, L., Cubillos, L., Niklitschek, E., & Alarcón, R. (2014). *Rol ecosistémico de sardina austral e impacto de su explotación sobre la sustentabilidad de otras especies de interés.* Informe final proyecto FIP 2012–15. Concepción, Chile: Universidad de Concepción. Retrieved from: http://www.subpesca.cl/fipa/613/articles-89322_informe_final.pdf

36. Niklitschek, E. J., Soto, D., Lafon, A., Molinet, C., & Toledo, P. (2013). Southward expansion of the Chilean salmon industry in the Patagonian fjords: Main environmental challenges. *Reviews in Aquaculture, 5*, 172–195.

37. Orensanz, J. M., & Jamieson, G.S. (1998). The assessment and management of spatially structured stocks: an overview of the North Pacific symposium on invertebrate stocks assessment and management. In: G. S. Jamieson y A. Campbell (Eds.), *Canadian Special Publication fisheries and Aquatic Sciences, Proceedings of the North Pacific Symposium on Invertebrate Stocks Assessment and Management No. 125* (pp. 441–459). NRC Research Press.

38. Orensanz, J. M., Parma, A. M., Jerez, G., Barahona, N., Montecinos, M., & Elias, I. (2005). What are the key elements for the sustainability of "S-Fisheries"? Insights South America. *Bulletin of Marine Science, 76*(2), 527–556.

39. Outeiro, L., Häussermann, V., Viddi, F., Hucke-Gaete, R., Försterra, G., Oyarzo, H., Kosiel, K., & Villasante, S. (2015). Using ecosystem services mapping for marine spatial planning in southern Chile under scenario assessment. *Ecosystem Services, 16*, 341–353.

40. Oyanedel, R., Keim, A., Castilla, J. C., & Gelcich, S. (2018). Illegal fishing and territorial user rights in Chile. *Conservation Biology, 32*(3), 619–627.
41. Oyanedel, R., Gelcich, S., & Milner-Gulland, E. J. (2020). Motivations for (non-) compliance with conservation rules by small-scale resource users. *Conservation Letters,* e12725.
42. Pickard, G. L. (1971). Some physical oceanographic features of inlets of Chile. *Fisheries Research Board of Canada, 28,* 1077–1106.
43. Pickard, G. L., & Stanton, B. R. (1980). Pacific fjords: A review of their water characteristics. In: H. J., Freeland, D. M., Farmer, & C. D. Levings (Eds.), *Fjord oceanography* (pp. 1–51). Springer.
44. Pulliam, H. R. (1988). Sources, sinks, and population regulation. *The American Naturalist, 132*(5), 652–661.
45. Riccialdelli, L., Newsome, S. D., Fogel, M. L., & Fernández, D. A. (2017). Trophic interactions and food web structure of a subantarctic marine food web in the Beagle channel: Bahía Lapataia, Argentina. *Polar Biology, 40*(4), 807–821.
46. Roa-Ureta, R. H., Molinet, C., Barahona, N., & Araya, P. (2015). Hierarchical statistical framework to combine generalized depletion models and biomass dynamic models in the stock assessment of the Chilean sea urchin (*Loxechinus albus*) fishery. *Fisheries Research, 171,* 59–67.
47. Rovira, J., & Herreros, J. (2016). *Clasificación de ecosistemas marinos chilenos de la zona económica exclusiva.* Departamento de Planificación y Políticas en Biodiversidad, División Recursos Naturales, Ministerio del Medio Ambiente. Retrieved from: https://mma.gob.cl/wp-content/uploads/2018/03/Clasificacion-ecosistemas-marinos-de-Chile.pdf
48. Schaefer, M. B. (1954). Some aspects of the dynamics of populations important to the management of the commercial marine fisheries. *Inter-American Tropical Tuna Commission, 53*(1–2), 253–279.
49. Schiønning, M.K., Ruiz-Jarabo, I., Concha, F., Straube, N., Försterra, G., & Thomasberger, A. (2018). A call for a baseline study of elasmobranchs in the Comau fjord, northern Patagonia, Chile. In *XXXVIII Congreso de Ciencias Del Mar* (p. 544). Sociedad Chilena de Ciencias del Mar. Retrieved from: https://www.schcm.cl/web/images/congresos/XXXVIIICongresodeCienciasdelMar2018.pdf
50. Sepúlveda, M., Arismendi, I., Soto, D., Jara, F., & Farías, F. (2013). Escaped farmed salmon and trout in Chile: Incidence, impacts, and the need for an ecosystem view. *Aquaculture Environment Interactions, 4*(3), 273–328.
51. Sepúlveda, M., Pavés, G., Santos-Carvallo, M., Balbontín, C., Pequeño, G., & Newsome, S. D. (2017). Spatial, temporal, age, and sex related variation in the diet of South American sea lions in southern Chile. *Marine Mammal Science, 33*(2), 480–495.
52. Servicio Nacional de Pesca. (1978–2019). *Anuarios Estadísticos de Pesca.* Chile. http://www.serna-pesca.cl/informes/estadisticas
53. Sielfeld, W., & Castilla, J. C. (1999). Estado de conservación y conocimiento de las nutrias en Chile. *Estudios Oceanológicos, 18,* 69–79.
54. Silva, N., Calvete, C., & Sievers, H. A. (1997). Características oceanográficas físicas y químicas de canales australes chilenos entre Puerto Montt y Laguna San Rafael (Crucero Cimar-Fiordo 1). *Ciencia y Tecnología del Mar, 20,* 23–106.
55. Strub, P. T., Mesías, J. M., Montecino, V., Rutllant, J., & Salinas, S. (1998). *Coastal ocean circulation off western South America.* In: A. R. Robinson & K. H. Brink (Eds.), *The sea* (Vol. 11, pp. 273–313). Wiley.
56. Subsecretaría de Pesca y Acuicultura. (2019). *Estado de situación de las principales pesquerías chilenas, año 2018.* Subsecretaría de Pesca y Acuicultura. Retrieved from: http://www.subpesca.cl/portal/618/w3-article-103742.html
57. Tacón, A., Tecklin, D., Farías, A., Peña, M.P., & García, M. (2023). *Terrestrial protected areas in Chilean Patagonia: characterization, historical evolution, and management.* Springer.
58. Tecklin, D., Farías, A., Peña, M. P., Gelvez, X. Castilla, J. C., Sepúlveda, M., Viddi, F. A., & Hucke-Gaete, R. (2023). *Coastal-marine protection in Chilean Patagonia: historical progress, current situation, and challenges.* Springer.

59. Toledo, P., Darnaude, A. M., Niklitschek, E. J., Ojeda, V., Voué, R., Leiva, F. P., Labbone, M., & Canales-Aguirre, C. B. (2019). Partial migration and early growth of southern hake *Merluccius australis*: A journey between estuarine and oceanic habitats off northwest Patagonia. *ICES Journal of Marine Science, 76*(4), 1094–1106.
60. Vila, A. R., Falabella, V., Gálvez, M., Farías, A., Droguett, D., & Saavedra, B. (2016). Identifying high-value areas to strengthen marine conservation in the channels and fjords of the southern Chile ecoregion. *Oryx, 50*(2), 308–316.
61. Viviani, C. A. (1979). Ecogeografía del litoral chileno. *Studies on Neotropical Fauna & Environment, 14*, 65–123.
62. Wilson, R. P., Jackson, S., & Straten, M. T. (2007). Rates of food consumption in free-living Magellanic penguins *Spheniscus magellanicus. Marine Ornithology, 35*, 109–111.

Carlos Molinet Marine Biologist and Ph.D. in Sciences Universidad Austral de Chile. Researcher in Patagonian coastal management, Los Lagos and Aysén regions. Full Professor at Universidad Austral de Chile, Campus Puerto Montt.

Edwin J. Niklitschek Marine Biologist, Universidad Austral de Chile. Ph.D. Marine Estuarine Environmental Sciences, University of Maryland, College Park, USA. Full Professor, Centro i-mar, Universidad de Los Lagos, Centro i~mar. Puerto Montt, Chile.

Aquaculture and Its Impacts on the Conservation of Chilean Patagonia

12

Alejandro H. Buschmann, Edwin J. Niklitschek, and Sandra V. Pereda

Abstract

Aquaculture in the Patagonian fjords and channels has experienced sustained growth during the last four decades, despite the lack of knowledge on the magnitude and intensity of its impacts. Based on a review of the latest literature, this chapter summarizes its main negative risks and delivers four recommendations aimed at reducing the risk of severe impacts on these ecosystems in a changing context. We highlight the urgent need to prevent regional and ecosystem problems. Relevant regulatory changes are required in accordance with the intensity and distribution of industrial activity. They include the implementation of an adequate monitoring system and the creation of an integrated system to effectively manage marine protected areas. Following a precautionary principle, in the short term we recommend halting the productive and territorial expansion of aquaculture, initiating essential changes to the current legislation on salmon escapes, limiting the use of chemotherapeutics and increasing public and private investment in the research and development of monitoring and mitigation technologies. The research priority should be the development of reliable assessments of carrying capacity and effective mechanisms to protect communities and species potentially threatened by this activity.

Keywords

Patagonia · Chile · Salmon farming · Blue mussel aquaculture · Carrying capacity · Environmental impacts · Legislation · Precautionary principle

A. H. Buschmann (✉) · E. J. Niklitschek · S. V. Pereda
i-mar Research Center, Universidad de Los Lagos, Puerto Montt, Chile
e-mail: abuschma@ulagos.cl

A. H. Buschmann
Center for Biotechnology and Bioengineering (CeBiB), Universidad de Los Lagos, Puerto Montt, Chile

J. C. Castilla et al. (eds.), *Conservation in Chilean Patagonia*, Integrated Science 19, https://doi.org/10.1007/978-3-031-39408-9_12

1   Introduction

As the world's population and economy continue to grow, so does the pressure on the oceans for food and services such as recreation, transportation and waste disposal. There is general consensus that the oceans are in crisis; examples include the increase in "dead" coastal areas [14], the collapse of important fisheries [29], the extensive damage that has affected diverse habitats and ecological communities, and the increasing losses of biodiversity associated with these areas [19]. Despite their remote location, Patagonian fjords and channels have been severely affected by numerous anthropogenic activities since the mid-nineteenth century [28], and in recent decades salmonid aquaculture has become one of the most important sources of anthropogenic pressure [2, 31, 35].

Aquaculture has been the fastest growing food production activity globally in recent years [21]. Most of this activity is concentrated in Asian countries, where extensive production of algae predominates, and secondarily of herbivorous and omnivorous fish and invertebrates such as carp and tilapia [11]. About one-third of aquaculture production occurs in marine environments [17], in contrast, in countries such as Chile intensive farming of marine and carnivorous organisms predominates, primarily salmonids. The footprint of salmonid production on the environment [30] is increased by a greater demand for energy and exogenous inputs (e.g. protein and oils) to the ecosystem of Patagonian fjords and channels where the fattening phase of this productive activity is carried out and where therefore the greatest biomass of its crops is concentrated.

2   Scope and Objectives

While aquaculture in the Patagonian fjords and channels has been experiencing sustained growth over the last four decades, we still do not fully understand the magnitude and intensity of its impacts. This chapter reviews and summarizes the main negative impacts of aquaculture in the Patagonian fjords and channels, organized in five sections: (i) an overview of the current development of aquaculture in Chilean Patagonia; (ii) an overview of the general environmental impacts; (iii) the magnitude of local impacts (in concession sites); (iv) the magnitude of regional impacts (impacts beyond concession boundaries); (v) aquaculture control and mitigation strategies under a scenario of global change. We provide four recommendations aimed at reducing the risks of severe impacts on the Patagonian fjords and channels in the context of climate change, highlighting the urgent need to prevent regional and ecosystem impacts on them.

3 Methods

The available scientific literature on the main negative impacts of aquaculture in Chilean Patagonia between Reloncaví Sound and the Diego Ramírez islands (41° 42′ S 73° 02′ W; 56° 29′ S 68° 44′ W) was reviewed. The authors' experience in these topics, both in Patagonia and elsewhere in Chile, is added to this research. This work is based on studies carried out over the last 20 years, incorporating some recent publications on the subject.

4 The Impacts of Aquaculture in Patagonian Fjords and Channels

4.1 Aquaculture Development in Chile

More than 1,200,000 tons (t) of aquaculture products were harvested in Chile in 2017, generating revenues of over US$ 4.91 billion FOB ("Free on Board"), equivalent to 14.3% of the country's non-copper exports. Salmonids accounted for 880,000 t (US$ 4.63 billion) of this total; the vast majority was Atlantic salmon (*Salmo salar*) (Fig. 1). Salmon farming has grown almost uninterruptedly since 1978, positioning the country since 1991 as the world's second largest producer. However, there has been a certain slowdown in its growth rate, which averaged 5% per year between 2007 and 2017. In contrast, in the same period mussel aquaculture showed an average growth of 8.6%, reaching a harvest of 338,000 t of *Mytilus chilensis* mussel in 2017 (Fig. 1).

Unlike the salmon and mussel industries, and despite Law 20.925 which encourages its cultivation, alga aquaculture has had little development in Chile. The national harvest of pelillo, the only commercially cultivated seaweed in the country, has declined by an average of 7% per year during the last decade (Fig. 1). Although aquaculture production statistics suggest an incipient diversification process in the country, with small production of various marine fish such as corvina, turbot and yellow-tail amberjack, as well as oysters, abalones, and other species of bivalves and microalgae, it is evident that the productive focus continues to be on salmonids and mussels (Fig. 1).

All Chilean salmonid and mussel production has been developed in sheltered and semi-sheltered areas of the Patagonian fjord and channel system. There has been an evident process of territorial expansion of salmon farming from the Los Lagos Region (40°–43° 38′ S) to the south (Fig. 2), triggered by the serious sanitary and productive crisis in the sector derived from the appearance of the ISA virus toward the end of the 2000s [31]. Salmonid production in the Aysén Region (43° 38′– 49° S) has risen sharply, becoming the main production area in the country by 2015 ([37]; in Spanish Servicio Nacional de Pesca, SERNAPESCA).

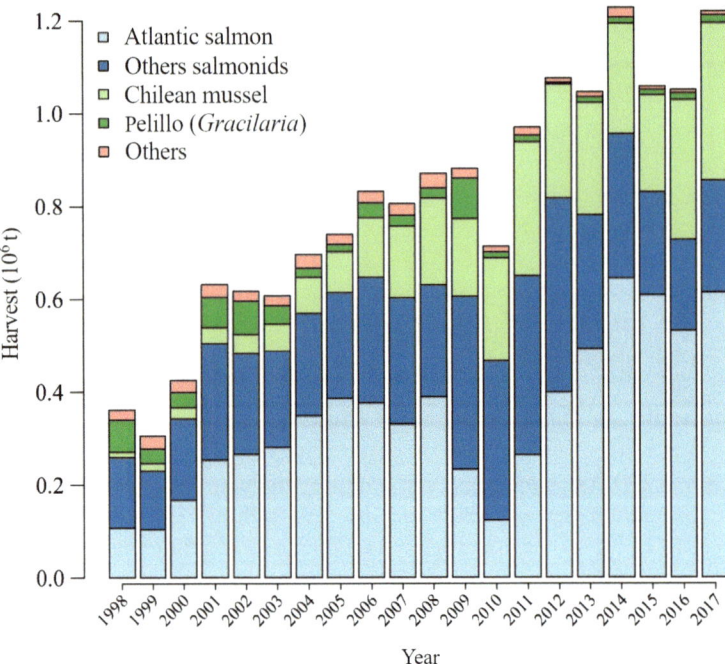

Fig. 1 Aquaculture production of Atlantic salmon, other salmonids, the blue mussel (*Mytilus chilensis*) the red seaweed pelillo (*Gracilaria chilensis*) and other species in Chile from 1998–2017 [16]. Although the figures represent production for the entire country, production outside the Patagonian fjord and channel system is marginal

The Magallanes Region (49°–56° S) shows sustained growth in its contribution to national production starting in 2010, reaching a 13% share of the national harvest in 2017 [37].

4.2 Overview of the Environmental Impacts of Aquaculture in Chilean Patagonia

Most scientific, governmental and public concern about the environmental impacts of aquaculture in Chilean Patagonia has been focused on the salmon industry (Table 1). More than 1,150,000 t of artificial food are distributed annually among the more than 320 farming centers operating in the Patagonian fjords and channels (Fig. 2), using net pens open to the environment. This feed, based on fishmeal, fish oil and various vegetable products, generates 80,400 tons of solid waste (mainly feces and residues of uneaten food) that precipitate in the environment under the cages, producing evident local impacts on the seabed [41]. In addition to these wastes, a minimum of 23,900 t of ammonium and other metabolites are released directly into Patagonian ecosystems, whose capacity to assimilate such

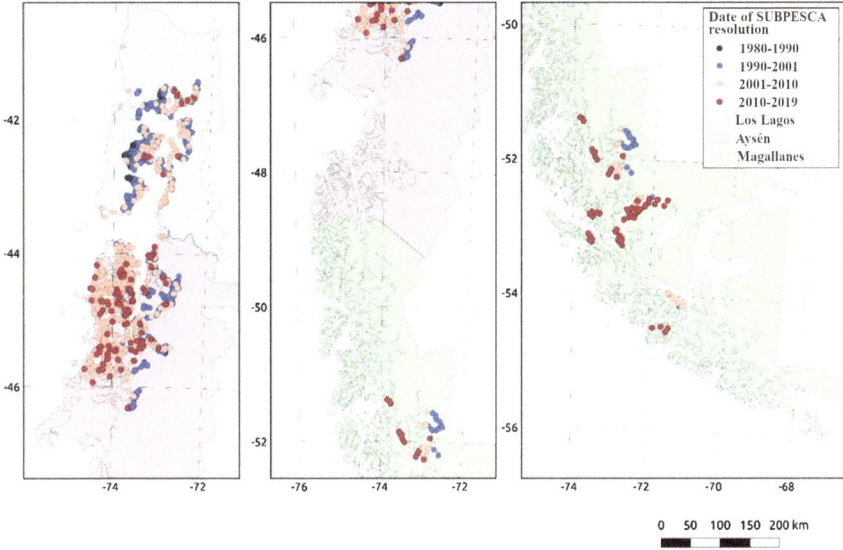

Fig. 2 Evolution of the spatial distribution of authorized aquaculture concessions of salmonids in the Patagonian fjord and channel system (1980–2019), according to the date of approval by the Undersecretariat of Fisheries and Aquaculture (SUBPESCA)

discharges remains unknown [4, 35]. Of growing concern is the escape of hundreds of thousands of salmon, which during the last two decades has averaged 487,000 individual year^{-1}, with a recorded maximum of more than 1,700,000 individuals escaping in 2013 [38]. As discussed below, these escaped fish have the potential to prey on a very large fraction of the native fauna [31], and also to generate permanent feral populations, Chinook salmon have already invaded all of Patagonia's watersheds [12]. These escapes also lead to a flow of pathogens and parasites whose health consequences on native fauna are not well known [4]. The salmon industry uses several chemotherapeutics (e.g. antiparasitics, antibiotics, antifungal agents) whose environmental effects and risks on the human population have caused public alarm [10, 47].

The environmental aspects associated with blue mussel aquaculture have received less attention than those of salmon farming, both in terms of the regulatory framework, which is basically designed for the salmon industry, and in terms of scientific research and the perception of society, which views it as a smaller industry and therefore one with less environmental impact. Although its environmental effects have not been systematically studied in Chile, the available evidence indicates that these effects should be of lesser magnitude than those generated by the salmon industry [39]. This is due to the fact that mussel farming does not require artificial feed, thus avoiding problems derived from the incorporation of inorganic matter and exogenous nutrients into the ecosystem and limiting its direct impacts to the cultivation area itself. Notwithstanding this, mussel species channel

Table 1 Main environmental impacts of salmon farming and associated research requirements. Modified and expanded from [35]

	Impacts	Research requirements	Impact scale
1	Impacts of organic matter input and increased oxygen demand on coastal ecosystems	Assess the cumulative impacts of feed and fecal residues on the structure and function of benthic and pelagic systems affected by aquaculture	Local
		Develop appropriate carrying capacity models to keep aquaculture impacts within acceptable limits	Local
	Impacts of nutrient discharges on phytoplankton composition and abundance	Study the effects of aquaculture on the generation of harmful algal blooms (HABs) and coastal eutrophication processes	Regional
	Impact of antibiotic use on biodiversity, microbial resistance and human health	Deepen studies on the impacts of antibiotic use in coastal systems and the potential risks to human health	Local
5	Transfer of diseases between natural and cultured populations	Study the transfer of pathogenic and parasitic vectors between cultivated and wild species, and propose ecosystem management models to reduce risks to biodiversity	Regional
	Impact of antimicrobial use on coastal food webs and consequences on biodiversity	Deepen studies of the impacts of the use of antiparasitics on aspects such as biodiversity, abundance of planktonic grazers and effects on larvae of commercially important benthic organisms	Regional
	Impact of compounds associated with antifouling paints on benthic communities	Study the effects of the use of antifouling compounds in aquaculture systems, especially their accumulation in sediments under culture systems	Local
	Impact of disinfectant compounds	Study the potential effects of the use of disinfectants on biodiversity	Local
	Impact of salmon escapes	Study the impact of escaped salmon on competitor species, prey, and their effects on coastal trophic webs, estuaries and freshwater bodies	Regional
	Impact of aquaculture on mammals, birds and native fishes	Study the effect of aquaculture structures on mortality of marine mammals, birds and sharks, as well as effects on their diet	Regional

(continued)

Table 1 (continued)

Impacts	Research requirements	Impact scale
Effect of organic and inorganic wastes on freshwater	Study the impacts of aquaculture on freshwater bodies in different regions of Chilean Patagonia	Regional
Effects of solid waste on bottom and beach cleanliness in aquaculture areas	Monitor the deposition of solid waste, especially plastics, and its consequences on biodiversity	Regional

* The demonstrated impacts are rather local (close to the cultivation site), but their potential effects at larger spatial scales have not received attention, so their environmental consequences at a regional scale have yet to be evaluated

and concentrate local organic matter, part of which is precipitated as feces and pseudofeces, organically enriching the sediments under the cultivation lines [27, 32]. Added to the above is the forced detachment of fouling organisms, an unregulated activity that also contributes to the accumulation of organic matter under cultivation systems [26]. In addition to increasing the sedimentation of organic matter in these culture sites and depending on both the scale of production and local circulation patterns, mussel farming can decrease the availability of phytoplankton for other filter feeders, including fish such as southern sardines, other bivalves, and zooplankton, affecting the rest of the marine food chain. These activities generally do not use chemicals or pharmaceuticals to control pathogens, which means lower levels of environmental impact. In terms of interaction with birds and mammals, there is concern about illegal hunting of seabirds such as the steamer duck (*Tachyeres* spp.) and physical obstruction of feeding and breeding areas for small cetaceans [24].

The cultivation of algae generally has a lower environmental impact than other types of crops because it does not introduce elements exogenous to the system being exploited [11]. However, pelillo, the main agar-producing alga cultivated in Chile, has been observed to modify the substrate. This has complex repercussions, modifying the abundance of herbivores and predators and generating substrates for the settlement and recruitment of mussels and epiphytic algae (see Fig. 6 in [3]). Shellfish and salmon farming share the problem of incorporation of plastics into the marine environment and their deposition on the surrounding beaches [22]. Although there is little information available on the subject, the potential use of alga cultivation to recover part of the nitrogen, phosphorus, and dissolved inorganic carbon emissions from marine animal aquaculture will be described later in the chapter.

4.3 Magnitude of Local Aquaculture Impacts

An important portion of the environmental impact of aquaculture affects the immediate surroundings of the concessions of the production sites (0–100 m); we describe these as local impacts, while other impacts reach scales of tens of kilometers which we call regional impacts [50]. Among the local-scale impacts (Table 1), that is, in and near concessions (tens of meters), we found the accumulation of organic and chemical residues, where a significant increase in carbon, phosphorus and nitrogen was observed. This has led to a significant decrease in macroinvertebrate biodiversity, from average species richness values of 7.8 (with maxima above

20) to average values of 3.5 [41]. In addition to these changes in macroinfauna, there are biogeochemical changes in sediments [7], decreased bacterial diversity [23] and accumulation of heavy metals and drugs, such as copper and antibiotics, respectively [6].

In addition to organic matter, the immediate surroundings of the cage ponds show evidence of antibiotic residues in sediments and living organisms, which have been found up to 7 km from the nearest salmon farming center [6]. Added to this are the potential risks of massive and recurrent application of antiparasite dips for farmed salmonids [47]. While antibiotic residues can affect the structure and function of microbial communities and pose potential risks to human health, antiparasitics can affect different primary plankton consumers, including larvae of commercially important arthropod populations such as crabs and spider crabs [47]. The potential for the accumulation of antiparasitics along the trophic web has been demonstrated in other regions of the world [49]. All this points to the fact that the effects of chemical use at different stages of the salmonid production cycle can go beyond the local environment alone, but more scientific research is required to confirm this.

Local impacts are considered to be generally severe and relatively permanent; nevertheless, they are mainly confined to the *ca.* 275 km² of aquaculture concessions in the Patagonian fjords and channels (as of December 2019). While this area represents a small fraction of the entire Patagonian system, the key question that remains unanswered is what specific sites are being impacted and what have been the additive effects of the suite of concessions on vulnerable habitats, species and communities such as cold-water coral reefs [25]. This is particularly relevant given the limited knowledge of the biological diversity of the exploited area and the limited number of marine protected areas (MPAs) within the Patagonian fjord and channel system. It is worth noting that in Chilean Patagonia there are 11 MPAs with coverage equivalent to 11,000 km², representing 6% of the Patagonian coastal zone [44]. With the exception of the Kawésqar National Reserve, designated in 2019 for the protection of the inland waters of the former Alacalufes Forest Reserve, the protection of marine protected areas adjacent to existing terrestrial protected areas remains legally unclear and there are unresolved policy discussions [44].

4.4 Magnitude of the Regional Impacts of Aquaculture

Regional-scale impacts are those that have an effect several kilometers from the concession area. These impacts include nutrient enrichment, introduction of exotic and pathogenic species, fish escapes, garbage deposition and incidental death of seabirds and marine mammals (Table 1). It has been estimated that despite significant improvement in feed conversion rates, uneaten feed, fecal production and excretion of inorganic metabolites (e.g. ammonium) produce 35–78 t of nitrogen and 6–13 t of phosphorus per 1000 t of salmonids produced, which are disposed of in the environment [31]. More than 70% of this waste (28 t, Fig. 3) is released directly into the water column as ammonium and urea, becoming available for immediate use by primary producers [2, 34, 46], which can support algal growth in a radius of up to 1.0 km [1]. The remaining nitrogen and most of the phosphorus precipitate to the bottom, affecting the habitat and the biogeochemical cycles of the bottom sediment, and can be dissolved and/or suspended back into the water column and become available to primary producers. The production of 855,000 t of salmon (2017 harvest, Fig. 1) implies the discharge of at least 23,865 t of nitrogen in the form of ammonium into the Patagonian fjord and channel system (Table 2), an amount that exceeds by more than 140 times the 170 t of nitrogen that entered the Pacific Ocean from the controversial disposal of 5,000 t of salmon off the Chiloé coast in 2016 [8]. This comparison becomes even more relevant if one considers that aquaculture discharges occur within a mosaic of semi-closed oceanographic systems which present large differences in volume, circulation patterns and productive load, which means that dilution is not always sufficient to minimize their environmental consequences.

Despite the large variability between production areas and the high magnitude of nutrient inputs, inorganic nitrogen is rapidly diluted to undetectable levels, even

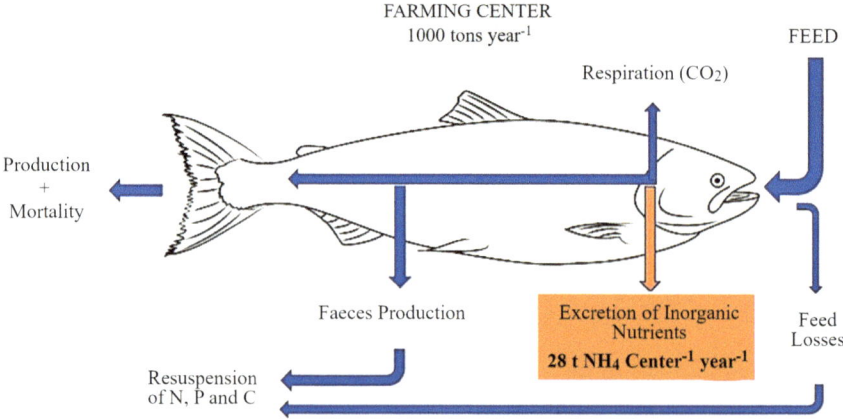

Fig. 3 Salmon farm present in the Región de Los Lagos in southern Chil

Table 2 Production value of the ammonium load (NH_4) produced by salmonid farming in the three Patagonian salmonid-producing regions in Chile in 2017. Estimated values for production of 1000 tons according to Olsen and Olsen [33] (see Fig. 3)

Region	Salmonid production (t)	NH_4 load (t region^{-1} year^{-1})
Los Lagos	351,535	9,843
Aysén	390,988	10,947
Magallanes	109,832	3,075
Total	852,354	23,865

in the vicinity of farming systems [46]. However, there is evidence that coastal eutrophication processes could occur in the vicinity of salmon farming sites, even without detectable increases in ammonium in the water column [5]. This evidence is related to the higher growth rates and nitrogen concentrations in macroalga tissues collected around salmon farming sites compared to non-salmon control sites [1, 46]. These effects would be expected to be greater in summer, when nitrogen limits primary productivity [1, 46]. Experimental evidence with experimental mesocosms suggests that effluent (water flowing in streams from aquaculture sites) from salmonids can increase the productivity of different phytoplankton species and reverse the abundance ratio of diatoms and dinoflagellates, causing the latter to predominate [4].

In summary, considering that in many production areas the salmon fattening centers are only three kilometers away from each other and that the mussel aquaculture centers that are located between them also contribute inorganic nitrogen, there may be a continuous enrichment that could trigger massive eutrophication phenomena and important changes in the abundance and/or specific composition of macroalgae and microalgae, as suggested by Buschmann et al. [4]. These aspects should be investigated with greater depth and speed, given the sustained growth of production levels and therefore their nutrient discharges.

Salmon escapes have a significant potential for predation on native fish and pelagic invertebrates. Given the current level of production, annual escapes of approximately 1.28 million individuals are predicted, which could consume approximately 15,700 t of invertebrates and native fish per year, including important forage species such as the channel prawn (*Munida* spp), silverside (*Odontesthes regia*) and the austral sardine (*Sprattus fuegensis*) [31]. Although the trophic and ecosystem impacts of feral and escaped salmonids on native fish and invertebrate populations have not been sufficiently studied [31], it is evident that the trophic impact of escaped salmonids affects not only their prey but also their counterparts, including other fish, and especially birds and marine mammals [31]. There is also a growing risk that, given the recurrent entry into the ecosystem of potentially reproductive escaped individuals, self-sustaining wild populations of farmed species may become established. This has already occurred with the Chinook salmon (*Oncorhynchus tsawytscha*), which has invaded all of Chilean Patagonia, and seems to be happening with the Coho salmon (*Oncorhynchus kisutch*) in Tierra del Fuego.

Farming facility interactions with marine mammals and seabirds are another area of growing concern [24, 36]. While avoiding consumption losses and/or damage to cages by sea lions *Otaria flavescens,* cormorants *Phalacrocorax* spp. and other species has become an important issue for the industry, there is growing concern about the levels of accidental death and/or illegal hunting of these animals on fish farms. Thus, it is of utmost importance to advance in the evaluation and systematic and mandatory monitoring of the levels of incidental mortality, the effectiveness of prevention measures and systems, and the implications of this mortality for the conservation of the affected species. In addition, the effects of the physical obstruction of breeding and feeding areas of small coastal cetaceans such as the Chilean dolphin (*Cephalorhynchus eutropia*) have not yet been evaluated [48].

Some additional aspects, thus far little addressed and quantified, are the discharge of inorganic waste (plastics) from the farm sites [22, 24], and the indirect effects of diseases and parasites partially transferred from farmed fish to wild organisms. It is necessary to address the interactions between productive systems such as salmon, mussels and algae that share the ecosystem, we do not know the complexities that may exist when they are co-cultured in the same body of water.

The manner and magnitude in which each of the environmental changes described above has affected ecologically relevant populations and communities in Chilean Patagonia remain poorly understood. Moreover, the complex interactions between the different sources of disturbance and the affected communities in the context of climate change are unknown [35].

4.5 Control and Mitigation Strategies Under Global Change Scenarios

The main axes of vulnerability to climate change in Patagonian ecosystems exploited by aquaculture are related to increased risks of harmful algal blooms, increased incidence of diseases associated with specific changes in temperature and salinity, and the multidimensional effects of the decrease in dissolved oxygen in the water [40]. The scientific literature has pointed out that as a consequence of temperature increases, ocean acidification and transient phenomena (e.g. heat waves) associated with climate change, the temperature tolerance responses and growth potential of various species used by aquaculture today will have a greater probability of having negative effects on aquaculture production in the future [17]. Changes in precipitation and sea and air temperature in Chile will affect freshwater inputs and the stratification, circulation and retention patterns of fjords and channels, with probably opposite directions in the Lagos and Aysén regions with respect to the Magallanes region [20]. There is insufficient information to predict how these changes will affect habitats, vulnerable species, and ecologically relevant processes such as primary production. However, it expected that the reduced fluvial input will reduce silica input to the fjord and channel system and thus affect diatom production and the associated trophic webs. This would negatively impact mussel aquaculture activities and significantly increase the risk of harmful dinoflagellate blooms. Detecting and responding in time to these and other changes requires immediate precautionary measures and a redesign of the coverage, frequency and characteristics of existing environmental and red tide monitoring programs in Chilean Patagonia. These monitoring programs should include marine protected areas.

Chile currently has an environmental regulatory framework for aquaculture that focuses almost exclusively on its local effects through the monitoring of existing environmental conditions under each concession [2], but ignores the cumulative, additive and/or synergistic effects that the collection of farms exerts on the region or ecosystem. Something similar occurs with environmental certification systems that are based on the promotion and certification of good practices in individual farms but neglect the regional or ecosystem problems. Thus, caution is warranted regarding the effectiveness of these certification systems in achieving global or regional conservation objectives, despite their growing importance for the salmon industry [45].

Considering the current regulatory limitations, the potential magnitude of aquaculture impacts on Patagonian ecosystems, the great uncertainty associated with their quantification and the possible aggravating effects of climate change, we consider it urgent to apply the precautionary principle and place an immediate limit on any increase in cultured biomass, nutrient discharges and areas used by salmonid and mussel aquaculture in Patagonia. The above measure should be maintained until the following conditions are met: carrying capacities (i.e. under scenarios of different levels of oxygen consumption or nutrient release) are understood and allow the establishment of objective limits for the environmentally acceptable biomass loads in the exploited ecosystems, identification and protection of the most vulnerable ecosystems including through marine protected areas; and a national monitoring system based on ecosystem indicators such as those proposed by Soto et al. [40] has been implemented which is capable of providing early warnings for the affected areas.

The proposed moratorium on the increase in the current levels of production and discharge of nutrients and chemicals should be immediate but progressively relaxed as the country develops a new regulatory system that limits the total load of cultivated biomass or the total discharge of nutrients permissible in each ecosystem, neighborhood or other management unit. All such measures are aimed at preventing unacceptable changes or risks at different scales of time and space and based on an adequate and sufficient knowledge of the relationships between the level of activity and the potential magnitude of the possible negative effects such as harmful algal blooms [35], hypoxia, escapes, incidental death of birds and mammals [31] and transmission of antibiotic resistance to human populations [9]. Under a regulatory scheme based on carrying capacity it would be possible to increase the permissible levels of productive load to the extent that there are effective technologies or measures to reduce and/or recover nutrient emissions and/or mitigate other impacts.

Mitigation measures have been proposed to reduce the vulnerability of currently exploited areas, redistribute production among areas or communities and develop a more diversified aquaculture matrix [40]. While this redistribution should not be understood as expansion to new areas, the diversification of the production matrix could be oriented to the mitigation of impacts through alga farming for nutrient recovery and oxygen production. This strategy has already been implemented in other countries such as China, where the cultivation of *Gracilaria* spp. has been demonstrated to improve water quality and decrease the quantity of nutrients and the production of phytoplankton, including species that cause harmful algal blooms [51]. It has been shown for more than two decades that it is possible to use algae such as pelillo (*Gracilaria chilensis*) and the seaweed huiro (*Macrocystis pyrifera*) to remove part of the nitrogenous inputs generated by salmon farming [7], and perhaps also have a positive effect on carbon sequestration [13, 18]. Cultivation of mussels and other filter feeders has also been proposed as an indirect way to remove nutrients of anthropogenic origin, although this option requires careful evaluation of costs and benefits, given the potential environmental effects of mussel aquaculture [42]. However, the great potential of integrated crops, the existence of legal restrictions on multispecies cultivation, the absence of regulations that oblige the industry to remove nitrogen discharged into the environment and the lack of economic incentives to move in this direction have kept this strategy merely theoretical. These co-cultivation systems could be even more relevant to maintain sustainable production under a scenario of ocean warming and acidification, as proposed by Strand et al. [43] and Fernández et al. [15].

5   Recommendations

Salmon farming has local and immediate environmental impacts, which affect biodiversity and sediments under production facilities [23, 41]. However, there are important larger spatial and temporal effects that have not been adequately quantified or addressed by current regulations (Table 1). Nor have the environmental effects of mussel and seaweed farming been quantified or addressed normatively. Under a precautionary approach, the partial and limited knowledge of the real and complex interactions of aquaculture on the coastal ecosystems of Chilean Patagonia is a great challenge that should not be a justification for inaction, especially given the rapid expansion of this activity and the threats of global change.

This chapter provides four priority recommendations to advance the conservation of Chilean Patagonia in relation to aquaculture activities:

- Halt the current production levels of salmon farming until: (i) the overall capacity of this system to receive higher nutrient discharges has been evaluated; (ii) there are estimates of carrying capacity by productive area and (iii) there is an effective system for protecting the diversity of communities and species present in this system.
- Some concrete and complementary actions to achieve this objective would be: (i) establish global or regional quotas for the production and/or discharge of nutrients (e.g. nitrogen); (ii) limit the use of products such as antibiotics and antimicrobials by defining annual quantities and developing new control mechanisms, including strategies to reduce potential health risks to local communities; and (iii) freeze the granting of new concessions for this purpose, particularly in the Magallanes Region.
- Legislate and implement as soon as possible an environmental liability system that: (i) regulates and penalizes economically the environmental damage caused by salmon escapes, stimulates preventive measures and technologies and the effective recapture of the largest possible number of escaped salmon, either directly or through third parties; and (ii) internalizes the environmental costs of

nutrient discharges and stimulates the development of mitigation measures such as integrated farming with algae and/or filter feeders.

- Install a new biological, environmental, and productive monitoring system, with open databases that are transparent and validated by a panel of experts that allow the study, evaluation of changes, processes and impacts and the generation of early warning systems at the ecosystem level. This system can include the current environmental system, but it must go far beyond it for the concessions.
- Promote research and development of technologies and productive strategies that reduce and mitigate the environmental impacts of aquaculture, considering the current and future challenges that adaptation to global change will demand.

Acknowledgements The authors thank the editors of this book for their invitation to participate. AHB and SVP gratefully acknowledge funding from ANID-FONDECYT project no. 1180647), the ANID-Chile Basal Program that funds the Center for Biotechnology and Bioengineering (CeBiB; FB-0001) and the Priority Research Areas Program (API 1), Universidad de Los Lagos.

References

1. Abreu, H., Varela, D. A., Henríquez, L., Villarroel, A., Yarish, C., I Sousa-Pinto, I., & Buschmann, A. H. (2009). Traditional vs. integrated multi-trophic aquaculture of Gracilaria chilensis. C. J. Bird, J. McLachlan y E. C. Oliveira: Productivity and physiological performance. *Aquaculture, 293*, 211–220.
2. Buschmann, A. H., Cabello, F., Young, K., Carvajal, J., Varela, D. A., & Henríquez, L. (2009). Salmon aquaculture and coastal ecosystem health in Chile: analysis of regulations, environmental impacts and bioremediation systems. *Coastal and Ocean Management, 52*, 243–249.
3. Buschmann, A. H., Correa, J. A., Westermeier, R., Hernández-González, M. C., & Norambuena, R. (2001). Red algal farming in Chile: A review. *Aquaculture, 194*, 203–220.
4. Buschmann, A. H., Riquelme, V. A., Hernández-González, M. C., Varela, D., Jiménez, J. E., Henríquez, L. A., Vergara, P. A., Guíñez, R., & Filún, L. (2006). A review of the impacts of salmon farming on marine coastal ecosystems in the southeast Pacific. *ICES Journal of Marine Science, 63*, 1338–1345.
5. Buschmann, A. H., Stead, R. A., Hernández-González, M. C., Pereda, S. V., Paredes, J. E., & Maldonado, M. A. (2013). Un análisis crítico sobre el uso de macroalgas como base de una acuicultura sustentable. *Revista Chilena de Historia Natural, 86*, 251–264.
6. Buschmann, A. H., Tomova, A., López, A., Maldonado, M. A., Henríquez, L. A., Ivanova, L., & Cabello, F. (2012). Salmon aquaculture and antimicrobial resistance in the marine environment. *PLoS ONE, 7*, e42724.
7. Buschmann, A. H., Hernández-González, M. C., Aranda, C., Chopin, T., Neori, A., Halling, C., & Troell, M. (2008). Mariculture waste management. In S. Jorgensen & B. Fath (Eds.), *Ecological engineering* (Vol. 3, pp. 2211–2217); *Encyclopedia of ecology* (vol. 5). Elsevier.
8. Buschmann, A. H., Farías, L., Tapia, F., Varela, D., & Vásquez, M. (2016). *Informe final, Comisión Marea Roja*. Universidad de Concepción, Universidad de Los Lagos y Pontificia Universidad Católica de Chile. Retrieved from: http://www.academiadeciencias.cl/wp-content/uploads/2017/04/InfoFinal_ComisionMareaRoja_21Nov2016.pdf
9. Cabello, F. C., Godfrey, H. P., Buschmann, A. H., & Dölz, H. J. (2016). Aquaculture as yet another environmental gateway to the development and globalization of antimicrobial resistance. *The Lancet Infectious Diseases, 16*, e127–e133.
10. Cabello, F. C., Godfrey, H. P., Tomova, A., Ivanova, L., Dölz, H., Millanao, A., & Buschmann, A. H. (2013). Antimicrobial use in aquaculture re-examined: its relevance to antimicrobial resistance and to animal and human health. *Environmental Microbiology, 15*, 1917–1942.

11. Chopin, T., Robinson, S. M. C., Troell, M., Neori, A., Buschmann, A. H., & Fang, J. (2008). Multitrophic integration for sustainable marine aquaculture. In S. E. Jørgensen & B. D. Fath (Eds.), *Ecological engineering* (Vol. 3); *Encyclopedia of ecology* (pp. 2463–2475). Elsevier

12. Correa, C., & Gross, M. (2008). Chinook salmon invade southern South America. *Biological Invasions, 10*, 615–639.

13. Duarte, C. M., Wu, J., Xiao, X., Bruhn, A., & Krause-Jensen, D. (2017). Can seaweed farming play a role in climate change mitigation and adaptation? *Frontiers in Marine Science, 4*, 100.

14. Díaz, R. J., & Rosenberg, R. (2008). Spreading dead zones and consequences for marine ecosystems. *Science, 321*, 926–929.

15. Fernández, P. A., Leal, P. P., & Henríquez, L. A. (2019). Co-culture in marine farms: macroalgae can act as chemical refuge for shell-forming molluscs under an ocean acidification scenario. *Phycologia, 58*, 542–551.

16. Food and Agriculture Organization. (2019). *The state of food and agriculture*. Retrieved from: http://www.fao.org/3/CA6030EN/CA6030EN.pdf

17. Froehlich, H. E., Gentry, R. R., & Halpern, B. S. (2018). Global change in marine aquaculture production potential under climate change. *Nature Ecology & Evolution, 2*, 1745–1750.

18. Froelich, H. E., Afflerbach, J. C., Frazier, M., & Halpern, B. S. (2019). Blue growth potential to mitigate climate change through seaweed offsetting. *Current Biology, 29*, 3087–3093.

19. Gall, S. C., & Thompson, R. C. (2015). The impact of debris on marine life. *Mar Pollution Bulletin, 92*, 170–179.

20. Garreaud, R. (2018). Record-breaking climate anomalies lead to severe drought and environmental disruption in western Patagonia in 2016. *Climate Research, 74*, 217–229.

21. Henriksson, P. J. G., Rico, A., Troell, M., Klinger, D., Buschmann, A. H., Saksida, S., Chadag, M. V., & Zhang, W. (2018). Unpacking factors influencing antimicrobial use in global aquaculture and their implication for management: a review from a systems perspective. *Sustainability Science, 15*, 1105–1120.

22. Hinojosa, I. A., & Thiel, M. (2009). Floating marine debris in fjords, gulfs, and channels of southern Chile. *Marine Pollution Bulletin, 58*, 341–350.

23. Hornick, K., & Buschmann, A. H. (2018). Insights into diversity and metabolic function of bacterial communities in sediments from Chilean salmon aquaculture sites. *Annals of Microbiology, 68*, 63–77.

24. Hucke-Gaete, R., Viddi, F. A., & Simeone, A. (2023). *Marine mammals and seabirds of Chilean Patagonia: Focal species for the conservation of marine ecosystems*. Springer.

25. Häussermann, V., Försterra, G., & Laudien, J. (2023). *Hard bottom macrobenthos of Chilean Patagonia: Emphasis on conservation of sublittoral invertebrate and algal forests*. Springer.

26. Lacoste, E., & Gaertner-Mazouni, N. (2014). Biofouling impact on production and ecosystem functioning: A review of bivalve aquaculture. *Reviews in Aquaculture, 7*, 187–196.

27. Lee, Y. G., Jeong, D. U., Lee, J. S., Choi, Y. H., & Lee, M. O. (2016). Effects of hypoxia caused by mussel farming on benthic foraminifera in semi-closed Gamak-bay, South Korea. *Marine Pollution Bulletin, 109*, 566–581.

28. Molinet, C., Solari, M. E., Díaz, M., Marticorena, F., Díaz, P., Navarro, M., & Niklitschek, E. J. (2018). Fragmentos de la historia ambiental del sistema de fiordos y canales nor-patagónicos, sur de Chile: dos siglos de explotación. *Magallania, 46*, 107–128.

29. Myers, R. A., & Worm, B. (2003). Rapid world depletion of predatory fish communities. *Nature, 423*, 280–283.

30. Naylor, R., Goldburg, R. J., Primaveras, J. H., Kautsky, N., Beveridge, M. C. M., Clay, J., Folke, C., Lubchenco, J., Mooney, H., & Troell, M. (2000). Effect of aquaculture on world fish supplies. *Nature, 405*, 1017–1024.

31. Niklitschek, E., Soto, D., Lafon, A., Molinet, C., & Toledo, P. (2013). Southward expansion of the Chilean salmon industry in the Patagonian fjords: main environmental challenges. *Reviews in Aquaculture, 4*, 1–24.

32. Nizzoli, D., Welsh, D. T., & Viaroli, P. (2011). Seasonal nitrogen and phosphorus dynamics during benthic clam and suspended mussel cultivation. *Marine Pollution Bulletin, 62*, 1276–1287.

33. Olsen, Y., & Olsen, L. M. (2008). Environmental impact of aquaculture on planktonic ecosystems. In K. Tsukamoto, T. Kawamura, T. Takeuchi, T. D. Beard, Jr & M. J. Kaiser (Eds.), *Fisheries for Global Welfare and Environment; 5th World Fisheries Congress* (pp. 181–196). Terrapub.
34. Pitta, P., Tsapakis, M., Apostolaki, E. T., Tsagaraki, T., Holmer, M., & Karakassis, I. (2009). Ghost nutrients from fish farms are transferred up the food web by phytoplankton grazers. *Marine Ecology Progress Series, 374*, 1–6.
35. Quiñones, R., Fuentes, M., Montes, R. M., Soto, D., & León-Muñoz, J. (2019). Environmental issues in Chilean salmon farming: A review. *Reviews in Aquaculture, 11*, 375–402.
36. Sepúlveda, M., & Oliva, D. (2005). Interactions between South American sea lions *Otaria flavescens* (Shaw) and salmon farms in southern Chile. *Aquaculture Research, 36*, 1062–1068.
37. Servicio Nacional de Pesca. (2019). *Anuarios estadísticos de pesca y acuicultura (1960–2017)*. Servicio Nacional de Pesca. Valparaíso, Chile. Retrieved from: http://www.sernapesca.cl/inf ormes/estadísticas
38. Servicio Nacional de Pesca. (2020). *Sistema integral de información y atención ciudadana. Servicio Nacional de Pesca*. Valparaíso, Chile. Retrieved from: http://www.sernapesca.cl/sistema-integral-de-informacion-y-atencion-ciudadana-siac
39. Shumway, S. E. (2011) *Shellfish aquaculture and the environment*. Wiley-Blackwell.
40. Soto, D., León-Muñoz, J., Dresder, J., Luengo, C., Tapia, F. J., & Garreaud, R. (2019). Salmon farming vulnerability to climate change in southern Chile: Understanding the biophysical, socioeconomic and governance links. *Reviews in Aquaculture, 11*, 354–374.
41. Soto, D., & Norambuena, F. (2004). Evaluation of salmon farming effects on marine systems in the inner seas of southern Chile: A large-scale mensurative experiment. *Journal of Applied Ichthyology, 20*, 493–501.
42. Stadmark, J., & Conley, D. J. (2011). Mussel farming as a nutrient reduction measure in the Baltic sea: Consideration of nutrient biogeochemical cycles. *Marine Pollution Bulletin, 62*, 1385–1388.
43. Strand, Ø., Jansen, H. M., Jiang, Z., & y Robinson, S. M. C. (2019). Perspectives on bivalves providing regulating services in integrated multi-trophic aquaculture. In A. Smaal, J. Ferreira, J. Grant, J. Petersen & Ø. Strand (Eds.), *Goods and services of marine bivalves* (pp. 209–230). Springer, Cham.
44. Tecklin, D., Farías, A., Peña, M. P., Gélvez, X., Castilla, J. C., Sepúlveda, M., Viddi, F. A., & Hucke-Gaete, R. (2023). *Coastal-marine protection in Chilean Patagonia: Historical progress, current situation, and challenges*. Springer.
45. Tlusty, M. F., & Tausig, H. (2014). Reviewing GAA-BAP shrimp farm data to determine whether certification lessens environmental impacts. *Reviews in Aquaculture, 7*, 107–116.
46. Troell, M., Halling, C., Nilsson, A., Buschmann, A. H., Kautsky, N., & Kautsky, L. (1997). Integrated open sea cultivation of *Gracilaria chilensis* (Gracilariales, Rhodophyta) and salmons for reduced environmental impact and increased economic output. *Aquaculture, 156*, 45–62.
47. Urbina, M. A., Cumillaf, J. P., Paschke, K., & Gebauer, P. (2019). Effects of pharmaceuticals used to treat salmon lice on non-target species: Evidence from a systematic review. *Science of Total Environment, 649*, 1124–1136.
48. Viddi, F. A., Harcourt, R. G., & Hucke-Gaete, R. (2016). Identifying key habitats for the conservation of Chilean dolphins in the fjords of southern Chile. *Aquatic Conservation: Marine Freshwater Ecosystems, 26*, 506–516.
49. Wang, D., Han, B., Li, S., Cao, Y., Du, X., & Lu, T. (2019). Environmental fate of the anti-parasitic ivermectin in an aquatic micro-ecological system after a single oral administration. *PeerJ, 7*, e7805.
50. Weitzman, J., Steeves, L., Bradford, J., & Figueira, R. (2019). Far-field and near-field effects of marine aquaculture. In C. Sheppard (Ed.), *World seas and environmental evaluation. Volume III, Environmental issues and environmental impacts* (pp. 197–220). Academic Press.

51. Yang, Y., Chai, Z., Wang, Q., Chen, W., He, Z., & Jiang, S. (2015). Cultivation of seaweed *Gracilaria* in Chinese coastal waters and its contribution to environmental improvements. *Algal Research, 9*, 236–244.

Alejandro H. Buschmann Marine Biologist, Universidad de Concepción. Doctor in Biological Sciences, mention Ecology, Pontificia Universidad Católica de Chile. Professor. Centro i~mar Universidad de Los Lagos, Puerto Montt, Chile. Member of the Chilean Academy of Sciences.

Edwin J. Niklitschek Marine Biologist, Universidad Austral de Chile. Ph.D. Marine Estuarine Environmental Sciences, University of Maryland, College Park, USA. Full Professor, Centro i~mar, Universidad de Los Lagos, Centro i~mar. Puerto Montt, Chile.

Sandra V. Pereda Marine Biologist, Bachelor in Marine Biology, Universidad Austral de Chile. Associate Researcher in the Algae Group, Centro i~mar, Universidad de Los Lagos, Puerto Montt, Chile.

Part V

Marine–Terrestrial Interface Ecosystems

Ecological Connections Across the Marine-Terrestrial Interface in Chilean Patagonia

13

Ricardo Rozzi, Sebastián Rosenfeld, Juan J. Armesto, Andrés Mansilla, Mariela Núñez-Ávila, and Francisca Massardo

Abstract

Chilean Patagonia encompasses the two southernmost terrestrial ecoregions of the temperate forest biome of South America (North-Patagonian and Sub-Antarctic Magellanic) and the two western marine ecoregions of the Magallanes Province (Chiloense, and Channels and Fjords of Southern Chile). These ecoregions are immersed in a complex mosaic of terrestrial (with marked altitudinal gradients), freshwater (including wetlands, rivers, lakes, and lagoons) and marine ecosystems (with myriad islands, channels, and fjords). With more than

R. Rozzi (✉) · S. Rosenfeld · A. Mansilla · F. Massardo
Cape Horn International Center (CHIC) and Parque Etnobotánico Omora, Universidad de Magallanes, Puerto Williams, Chile
e-mail: Ricardo.Rozzi@unt.edu

R. Rozzi
Departament of Philosophy and Religion and Department of Biological Sciences, University of North Texas, Denton, TX, USA

S. Rosenfeld
Facultad de Ciencias, Millennium Institute Biodiversity of Antarctic and Subantarctic Ecosystems (BASE), Universidad de Chile, Santiago, Chile

S. Rosenfeld · A. Mansilla
Laboratorio de Macroalgas Subantárticas y Antárticas, Universidad de Magallanes, Punta Arenas, Chile

S. Rosenfeld · J. J. Armesto · M. Núñez-Ávila
Estación Biológica Senda Darwin, Cruce El Quilar, Ancud, Chile

J. J. Armesto
Departamento de Ecología, Pontificia Universidad Católica de Chile, Alameda 340, Santiago, Chile

Facultad de Ciencias Naturales y Oceanográficas, Universidad de Concepción, Concepción, Chile

J. J. Armesto · M. Núñez-Ávila
Instituto de Ecología y Biodiversidad, Universidad de Concepción, Concepción, Chile

323
J. C. Castilla et al. (eds.), *Conservation in Chilean Patagonia*, Integrated Science 19, https://doi.org/10.1007/978-3-031-39408-9_13

100,000 km of coastline, most environments in the region exhibit strong land-sea interdependency in energy and nutrient flows. The goals of the chapter are to: (i) describe the main ecological features of the marine-terrestrial interface in the channels, fjords, and archipelagoes; (ii) identify major anthropogenic impacts on marine-terrestrial connectivity; (iii) describe the most important matter and energy flows across aquatic and terrestrial ecosystem; (iv) discuss the conservation status of species that are dependent on this interface; (v) identify those public protected areas that have extensive areas of marine-terrestrial interface. The major nutrient exchanges in the marine-terrestrial interface include carbon and nitrogen-rich sediment flows transported to the ocean by the rivers and streams, and abundant debris of siliceous rocks from land to ocean carried by rivers draining glaciers and ice fields. The most important vectors of biological transport of materials between the ocean and land are large marine mammals and seabirds. This includes historical records of whale landings that mobilize nutrients from ocean bottoms to the coastal zones and large populations of seabirds that nest in the archipelagos. Major threats to the marine-terrestrial interface include the massive populations of naturalized salmon that circulate in the fjords, streams, and channels. Salmon proliferation has altered the nutrient transport from the ocean to the continental rivers. Three species of exotic mammals have increased in numbers and impact at the interface between oceans, land, and freshwater systems—the beaver (*Castor canadensis*), the North American mink (*Neovison vison*), and the muskrat (*Ondatra zibethicus*). In contrast to traditional views on conservation and management that segregated land–ocean interfaces, our analysis in this chapter suggests that in order to understand ecosystem functioning in Chilean Patagonia as well as to establish comprehensive conservation programs, it will be essential to address the interrelationships of biophysical processes at the marine-terrestrial interface.

Keywords

Patagonia • Chile • Ecoregions • Sea-land interfaces • Matter and energy fluxes • Biogeochemistry • Protected areas • Conservation

1 Introduction

The interface between the land and the oceans, i.e. the narrow band at the continental margins of the planet, is subject to strong anthropogenic pressures, including high concentration of inhabitants in certain regions, the high impact of coastal recreation and tourism, and extractive exploitation of coastal and marine resources [30]. Therefore, coastlines are key scenarios for understanding the impacts of global change [17]. One of these key scenarios is the edge of South American exposed to the Pacific Ocean between Reloncaví Sound and the Diego Ramírez Islands (41° 42′ S 73° 02′ W; 56° 29′ S, 68° 44′ W), i.e. Chilean (western) Patagonia. In contrast to the temperate and subpolar latitudes of the Northern Hemisphere, the Patagonian archipelagos, fjords and channels have remained free of large-scale human impact until recent times [77].

The continental margin of the archipelagic region of Chilean Patagonia was shaped by glacial advances and retreats during the Pleistocene [96]. This has produced a zone of steep coastal fjords and inland seas with deep bottoms generated by glacial erosion, together with tectonic effects that caused the subsidence of the continental territory of the longitudinal valley south of 41° S. Flows of matter and energy are concentrated in this inland sea, and in the channels and fjords, where they reach the highest marine productivity values in the region [61]. Due to the close land-marine link in Chilean Patagonia, and the profusion of river courses, coastal ecosystem dynamics are influenced by continental ecological processes, including large tidal changes and entrainment of sediments and organic matter from rivers to the sea [61], which is especially relevant in areas of coastal marshes and wetlands.

Twenty thousand years before present (BP), during the last glacial maximum, large ice fields covered the continent between central Chiloé and Cape Horn [72, 107]. The *Nothofagus* deciduous forests persisted in fragments in areas close to the ice, east and west of the mountain ranges. Evergreen forests found refuge further north, in coastal areas northwest of Chiloé and the continental coast of the Los Lagos Region. At the end of the last glacial period (10–12,000 years BP), post-glacial warming made the expansion of forests possible throughout the southern territory, forming the north Patagonian forests from the south of Chiloé to the Gulf of Penas, and the Subantarctic forests between the latter and Cape Horn [9]. Although the fjord and channel system are ecologically linked to the mainland [61], conservation and management approaches have treated them separately. Here we propose to integrate the mosaic of marine-freshwater-terrestrial ecosystems into conservation policies that consider vital processes of exchange of matter and energy that occur at the marine-terrestrial interface in Chilean Patagonia.

2   Scope and Objectives

This chapter summarizes current knowledge on the biophysical connections between terrestrial, freshwater, and marine ecosystems in western Patagonia. It also proposes ways to develop integrated conservation of biological and cultural diversity in marine-freshwater-terrestrial ecosystem mosaics, evaluating the contribution of current and potential protected areas (PAs).

The specific objectives are to: (i) describe the main ecological and biophysical characteristics of the marine-terrestrial interface in the channels, fjords, and archipelagos, including their reciprocal influence on productivity; (ii) characterize the direct and indirect anthropogenic impacts on marine-terrestrial connectivity, such as sediment and nutrient fluxes, freshwater discharges, and others; (iii) identify and describe the flows associated with the marine-terrestrial interface in this region, considering the modulating effects of climate change; (iv) describe the conservation status of species that depend on the marine-terrestrial interface and coastal habitats, identifying key species for conservation planning; (v) identify the largest PAs in the marine-terrestrial interface to focus the development of conservation plans.

3   Methods

To conduct this analysis, a body of gray literature available at Chilean academic and government institutions was examined, and published scientific references were reviewed using the ISI Web of Knowledge core collection[1] database. Publications with a focus on the ecosystems of western Patagonia were selected. Google Scholar[2] and sources available in regional institutions were also reviewed.

3.1   Study Area and Biogeographic Location

Chilean Patagonia and its island, channel, fjord, and peninsula contours add up to more than 100,000 km of coastal environments, ranging from sheltered inland areas strongly influenced by rivers and glaciers to areas exposed to waves from the Pacific and Atlantic oceans (Tecklin et al., 2021). According to Pisano [65], four orographic regions are distinguished: Archipelagic, Cordilleran, Sub-Andean Eastern, and Coastal Plains. For the terrestrial realm, the biogeographic classifications proposed for Chilean Patagonia include the Neotropical phytogeographic region and the temperate forest biome of South America [8, 58]. Rozzi et al. [77] include the ecoregions of the Valdivian temperate forests north of 47° S and the Magellanic Subpolar forests (sensu [55]) between 47° and 56° S as the Magellanic Subantarctic Ecoregion (sensu [78, 76]).

For the marine realm we follow the classification of Spalding et al. [97]. Rovira and Herreros [75] add one more ecoregion, but essentially retain the major distinctions concordant with Spalding et al. [97]. Under this scheme the archipelagic region of southwestern South America is included in the Temperate South American Kingdom or Domain, which includes the Magallanes province in the far south.

In its western portion, this province encompasses two marine ecoregions in the latitudinal range *ca.* 41–58° S of Chilean Patagonia [97]: (i) ecoregion 188 (Chiloense); (ii) ecoregion 187 (Channels and Fjords of southern Chile). The Chiloense ecoregion (41° 30' S–46° 30' S; 277,646 km^2) extends from the Coronados Gulf and the Chacao Channel in the north to the Taitao Peninsula. The Channels and Fjords of Southern Chile ecoregion (46° 30'–58° S; 849,252 km^2) extends from the Taitao Peninsula to the Diego Ramirez Islands and Drake Passage. The Chiloé ecoregion has an intricate network of channels, fjords, and archipelagos with 10,705 km of coastline, forming the so-called Chiloé Inland Sea

---

[1] See: http://apps.webofknowledge.com.
[2] See: http://www.scholar.google.ca/intl/en/scholar/about.html.

between the island of Chiloé and the mainland [99]. In this labyrinth of channels there is a tidal range of up to 8 m, abundant freshwater inflows from rivers and copious rainfall (2–3 thousand mm annually). In contrast, towards the high seas there is an exposed oceanic system associated with westerlies, with a strong marine current with one branch that flows northward, generating the Humboldt Current, and another that flows southward, generating the Cape Horn Current [15, 16].

The fjord systems of the Channels and Fjords ecoregion of southern Chile include glaciers that descend from the Andes to the Pacific Ocean. The complex coastal geography includes myriad estuaries, small rivers, high cliffs, and marshes [99]. The network of fjords, bays, bays, channels, archipelagos, estuaries, gulfs, basins, inlets, and straits has very diverse topographic and oceanographic characteristics [102]. Heterogeneity at the land-sea interface is amplified by the influence exerted by the ice fields of the Darwin Range in Tierra del Fuego (running east–west) and the Andes on the continent (running north–south), especially in the vicinity of the Southern Ice Field [72].

These orographic and climatic characteristics have a great influence on the distribution and abundance of marine biota, which have multiple interconnections with terrestrial ecosystems. In the far south, the marine and terrestrial ecoregions are under the influence of the southeastern Pacific, southern, and southwestern Atlantic oceans. The main inter-oceanic connections are the Strait of Magellan, the Beagle Channel, and, to a greater extent, the Drake Passage.

The classifications used as a basis for defining terrestrial and marine ecoregions have a high degree of coincidence in their latitudinal limits. The location of the land boundary between the Valdivian (41–46° 30′ S) and Magellan Subantarctic (46° 30′–56° S) ecoregions coincides with that of the boundary between the marine ecoregions of Chiloé and the Channels and Fjords of southern Chile. This boundary, located at 46° 30′ S, is generated by the presence of the Gulf of Penas and the northern and southern Patagonian Ice Fields, which establish a biogeographic barrier for terrestrial species due to the presence of ice masses and a harsh climate that impacts a 110 km strip of coastline. The forests present a more complex structure north of this barrier, and are multistratified, and richer in species; this gulf is the southern limit of the distribution of woody Bambusaceae, Podocarpaceae and Gesneriaceae, among other taxa typical of the temperate and northern Patagonian forest [106]. South of this barrier, the greatest diversity of plant species is concentrated in bryophytes [76], in a mosaic of forest, tundra complex and high Andean ecosystems [65, 82]. In the marine area, the Gulf of Penas coincides with the division of the oceanic drift current into the Humboldt (northward) and Cape Horn (southward) ocean currents. This point also marks the confluence of the South American, Nazca, and Antarctic plates [15], where important nutrient upwelling occur.

4   Results and Discussion

4.1   Unique Characteristics of the Archipelagic Region of Chilean Patagonia

Chilean Patagonia exhibits marked contrasts with its Northern Hemisphere latitudinal counterparts [46]. We identified singularities in four areas: (i) geographic; (ii) climatic; (iii) biogeographic; (iv) physicochemical.

4.1.1   Geographic Singularities

The 40°–60° S latitudinal band has two notable characteristics. First, the land/ocean ratio is 2% versus 98%, in contrast to the 40°–60° N band where the land/ocean ratio is 54% versus 46% [77]. This difference generates an opposition between the climatic systems of the temperate and subtropical regions of the hemispheres, with oceanic influence prevailing in the Southern Hemisphere. Second, the great latitudinal extension of forest ecosystems in South America, which surpasses other Southern Hemisphere forests. Between the Taitao Peninsula (47° S) and Horn Island (56° S) the Subantarctic forests extend almost 10° further south than the forests of Stewart Island (47° S), the southernmost forests in New Zealand (see map in Rozzi et al. [77]). Therefore, the Magellanic Subantarctic ecoregion lacks a geographic equivalent at that latitude worldwide.

4.1.2   Climatic Singularities

The climate of the temperate-subantarctic terrestrial ecosystems is modulated by the vast expanse of ocean that produces little seasonal temperature fluctuation, with moderate winters and mild to warm summers. The greater amount of land in the Subarctic latitudes of the Northern Hemisphere generates winters with average temperatures below 0 °C and very warm summers. To illustrate this interhemispheric contrast, the 2009–2012 annual thermal amplitude recorded by the microclimatic station of the Chilean Network of Long-Term Socio-Ecological Research Network (LTSER-Chile) in Omora Park, in the Magellanic Subantarctic ecoregion, was 8.9 °C (Fig. 1). In contrast, the annual thermal amplitude for the same period recorded at the Bonanza Creek LTER site in the Subarctic region of Alaska was 36.6 °C; that is, four times greater than that of Omora Park. The absolute minimum temperature at Omora Park was −7.5 °C in August 2011, while at Bonanza Creek it reached −32 °C in January 2012 [85]. This marked thermal contrast underscores the bioclimatic uniqueness of the western Patagonian region and its potential effects on the conditions of seas and coastal environments (Fig. 1). These climatic contrasts become more relevant in the face of global climate change [13, 51].

4.1.3   Biogeographic Singularities

Unlike the Southern Hemisphere, the Northern Hemisphere presents biogeographic continuity through large boreal continental masses. The land connection between North America and Eurasia until the end of the last glaciation via the Bering Land

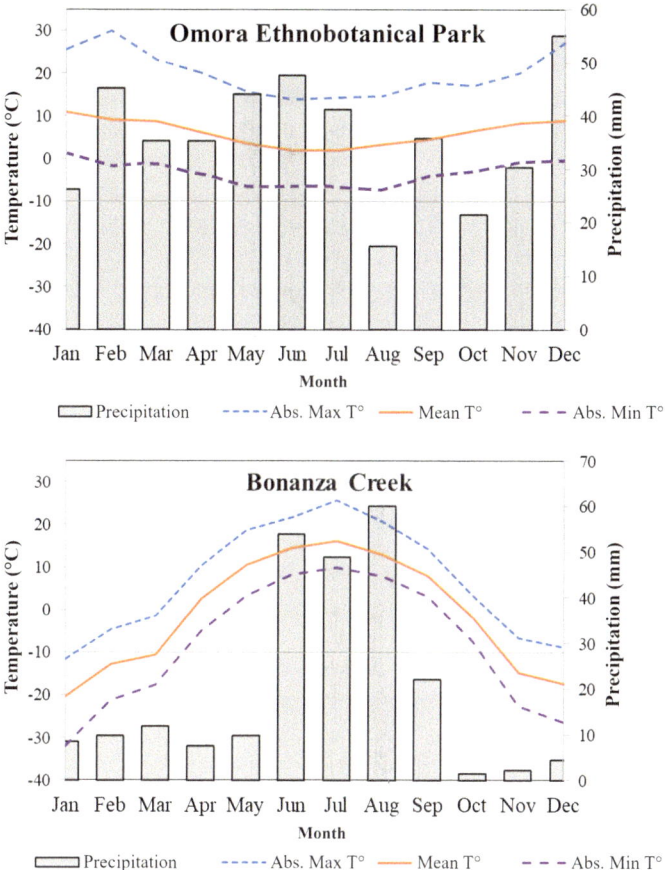

Fig. 1 Monthly mean precipitation (bars), monthly mean temperature (orange line), absolute maximum (dashed blue line) and absolute minimum (dashed purple line) temperatures recorded at the microclimate station at Omora Ethnobotanical Park (54° 56′ S, 67° 38′ W), Red Chilena de Investigación Socio-Ecológica a Largo Plazo (in English LTSER- Chile) (top) and at the microclimate station at Bonanza Creek, Alaska (64° 42′ N, 148° 08′ W), Long Term Ecological Research Network (LTER-USA) (bottom). Period January 2009-december, 2012; modified from Rozzi et al. [85]

Bridge facilitated the transit of plants, animals, and human settlement in recent times. North America represented a biogeographic barrier in the marine realm, that was partial or total in different geological epochs, preventing the contact of biota between the Atlantic and Pacific Oceans. Its disjunctive character and the climatic conditions of the tropical ocean were consolidated with the closure of the Isthmus of Panama [48].

In contrast, oceanic biotas in the Southern Hemisphere have been connected since Antarctica separated from South America with the formation of the Drake Passage (30–40 million years BP), creating the Antarctic Circumpolar Current

(ACC) that flows from the west, connecting the Pacific, Atlantic, Indian, and Southern, and Antarctic oceans [3, 47, 86]. Thus, marine biotas have maintained genetic fluxes between the edges of South America and Antarctica [69].

Biogeographic connections between the terrestrial biotas of temperate and Subantarctic regions of the Southern Hemisphere date back more than 40 million years, when the continental masses of Australia-New Zealand and South America were connected through Antarctica [90]. An iconic example of the Gondwanic connections between New Zealand, Tasmania, and southern South America is their forest ecosystems, dominated by tree species of the genus *Nothofagus* [106] and its freshwater ecosystems with galaxid fish assemblages [71, 111].

These biotas have evolved in isolation on each continent during the last 30 million years, giving rise to multiple disjunct lineages of plants and animals. Species endemism of the woody plants of the temperate forest biome of South America is 80–90%, comparable only to the oceanic islands [8], whereas bryophytes reach more than 50% species endemism in Cape Horn and the Diego Ramirez Islands [31].

4.2 Biophysical Singularities and Indirect Anthropogenic Impacts

Chilean Patagonia is one of the last regions of the planet that conserves extensive areas without major anthropogenic transformations [55]. In contrast, human activity in the rest of the planet has directly modified soil cover, altered hydrological circulation, and doubled the circulation of inorganic nitrogen in ecosystems [109]. The marine and terrestrial, temperate, and Subantarctic ecosystems located south of $40°$ S in Chile are exceptional in the world, because they are located in one of the least transformed areas of the planet and are relatively free of pollutants of industrial origin, in addition to the limited human population [7, 77]. Atmospheric aerosols and gases derived exclusively from processes such as evaporation and emission of gases from the ocean, which occur in ocean–atmosphere contact zones far from the coast, are transported to the coasts by westerly winds from the Pacific Ocean. As a consequence, rain and fog reach the continent in most of the western Patagonian region, so the islands and fjords are essentially free of industrial pollution [35, 112].

Inorganic salts (nitrates, ammonium, and sulfates) are present at trace levels or absent in the precipitation that sustains plants in coastal ecosystems in Chilean Patagonia [35]. In contrast, the terrestrial and aquatic environments of mid- and high latitudes in the Northern Hemisphere receive several kilograms per year of anthropogenic nitrogen in the form of nitrate and ammonium (dissolved inorganic nitrogen, DIN) via the atmosphere, along with other volatile products of industrial origin (fertilizers, smokestack emissions, combustion engine emissions, etc.). These substances of indirect anthropogenic influence dissolved in rain alter microbial systems, biogeochemical processes and the productivity of the sea and land.

For comparative purposes, it has been proposed that the temperate and Subantarctic terrestrial and freshwater ecosystems of Chilean Patagonia, subject to low direct and indirect human impacts, could represent a pre-industrial global "baseline" [35], which would represent the state of ecosystems before the Anthropocene [79].

Knowledge of austral ecosystems will contribute to the understanding and measurement of anthropogenic impacts on atmospheric biogeochemical cycles and the structure of terrestrial and aquatic environments since the beginning of the industrial era [10, 35, 77].

4.3 Matter and Energy Fluxes at the Land-Sea Interface

Land-sea flow in Chilean Patagonia can be linked to physical or abiotic vectors, such as wind and water, and biological vectors (e.g. birds that feed at sea and nest on land, transporting essential nutrients to riparian environments). This section examines the primary land-sea fluxes and then the converse, distinguishing their main vectors.

4.3.1 Processes that Transport Nutrients from Terrestrial and Freshwater Ecosystems to Marine Ecosystems

Within the Chiloé and Fjords and Channels Ecoregions of southern Chile, rivers are one of the largest freshwater flows, including the Puelo (approximately 600 m^3s^{-1} discharge, H. González, personal communication), the Baker and Pascua rivers and the large river systems of the ice fields. Further south, another important freshwater input comes from surface runoff and groundwater flow fed by annual rainfall, exceeding 6 m in some sectors. High precipitation decreases drastically towards the south and east of the Fjords and Channels ecoregion. The highest annual precipitation has been recorded in the northwest of this ecoregion, on the west coast of the Wellington and Guarello islands (48°–50° S), and the lowest has been recorded in the southeast, at the eastern mouth of the Strait of Magellan [6].

Hydrological dynamics influence the flux of elements to the ocean [41, 71]. There are important intra- and inter-annual variations in freshwater flow in the Aysén fjords from numerous rivers dependent on rainfall and snowfall regimes [41]. Consequently, marine primary productivity varies with the season and the dynamics of each fjord. The vertical stratification of fjords is characterized by deep waters with higher salinity and nutrients, and a surface layer of continental origin with low nutrient content and low salinity (nitrate and orthophosphate) [32]. A representative example is the coast of the Moraleda Channel (44° S), which receives a large amount of freshwater from glacial melt throughout the year and has a low-salinity surface layer, enriched in silt but deficient in nutrients. Primary productivity in this ecosystem is low, and phytoplankton communities are dominated by diatoms. In contrast, the ecosystems located north of the Moraleda Channel receive less fine sediment from glaciers and more terrestrial input from forests and wetlands [32]. These sediments reduce the penetration of PAR (Photosynthetic Active Radiation) light and limit local primary production. For example, in the Puyuhuapi Channel Subantarctic water dominates during spring and there is little glacial influence. The phytoplankton composition is a highly productive system of colonial diatoms with high nutrient and energy requirements [32].

Glacier melting accelerated by global warming is one of the processes that can alter the state of the marine-terrestrial system in the region [72, 104]. A combined effect of low temperature and low alkalinity water resulting from increased glacial melting has been observed in some fjords of Chilean Patagonia. This increases the local acidification of the water column, a corrosive condition for calcium carbonate ($CaCO_3$) of aquatic microorganisms [104]. Local acidification of channels and fjords could also affect marine communities of plankton and benthos near glaciers [103]. Larvae of the gastropod *Concholepas concholepas* and juveniles of the mytilid *Perumytilus purpuratus* affected by acidification, have altered rates of food intake. Therefore, for some marine mollusks acidification alters $CaCO_3$ absorption and the formation of their calcareous shell and produces alterations in their life cycle. Higher freshwater flux due to warming also increases turbidity in the water column, reducing primary productivity [32].

In fjords bordering forest ecosystems, freshwater is associated with the flushing of the water column of organic material from rivers and coastal forests [105]. It is estimated that water from the glacial regime contributes substantial amounts of dissolved silica and a low content of nitrate and phosphate to the surface layer of the water column. Regarding particulate organic matter, terrestrial ecosystems contribute (via rivers and glacier melting areas) around 68–86% of the carbon found in fjord ecosystems. The relevance of this allochthonous source of organic matter for fjord biota is indicated by the rate of terrestrial carbon uptake by copepods, which is equivalent to 20–50% of their body weight [105]. Therefore, the terrigenous carbon contributed to the coastal ecosystem in these fjords is particularly relevant in periods of scarce available food. Consequently, coastal productivity is linked to nutrient input derived from inland ecosystems, mainly fluvial entrainment of detritus from forest ecosystems [64, 71]. Marine upwelling in circuits associated with fjords and channels supplies nutrients to diverse assemblages of primary producers, algae, and phytoplankton [49]. The maximum fluvial discharges in the study area are recorded at 42°, 46° and 50° S (Table 1).

Up to 50% of particulate carbon in estuaries and fjords comes from terrestrial ecosystems [92]. The main nutrient exchanges at these land-sea interfaces include fluxes of organic carbon- and nitrogen-rich sediments transported to the sea by river channels [61], as well as a large input of silica from land to sea in rivers near glacier masses. Aerosol transport, associated mainly with fog influxes from the sea to the land [112], can reach tens of kilometers into the interior of the continent. Figure 2 and Table 1 show data on element fluxes and production processes in the fjord and channel zone of southern South America.

Variation in the light extinction coefficient (Kpar) through the water column is a contributing factor to the variability in chlorophyll "a" and primary productivity in the Chiloé Inland Sea (41°–43° S; [23]) and in the southernmost fjords (47°–50° S; [67]). Kpar values appear to indicate that phytoplankton primary productivity

Table 1 Fluxes from the terrestrial to the marine ecosystem and properties related to marine productivity, with emphasis on Chilean Patagonia

Flow of measured elements or rates	Magnitude of land-sea flow or measured rate	Area measured	References
Dissolved organic carbon (DOC)	8200 mg m^{-3}	40°–54° S	[64]
Concentration of total chlorophyll in the water column	57±51 mg m^{-3}	Up to 50 m depth, fjord area	[68]
Chlorophyll concentration in the water column	2.6–3.1 mg m^{-3}	Marine surface, fjord area (45S)	[40]
Freshwater discharge	2470 m^3 s^{-1}	42° S	[25]
Freshwater discharge	3480 m^3 s^{-1}	46° S	Id
Freshwater discharge	3344 m^3 s^{-1}	50° S	Id
Particulate organic carbon	76 mg m^{-3}	53°–54° S (summer)	[26]
CaCO$_3$ deposited in sediments	43 mg C m^{-2} d^{-1}	Continental slope	[49]

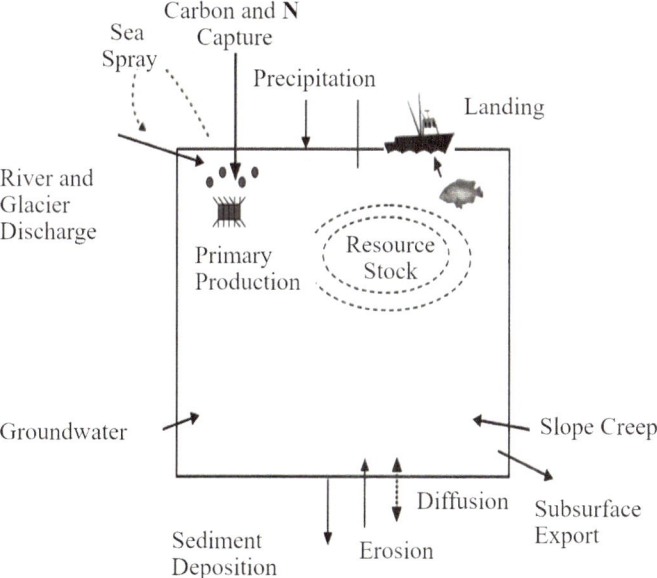

Fig. 2 Major biogeochemical flux of dissolved and particulate material in the ocean-continental margin interface. The sparse human population in many areas in the Subantarctic zone reduces anthropogenic contaminant exports. Modified from Liu et al. [49]

is light limited below 15 m depth, due to the large amount of sediment discharged by runoff from land [67].

The rivers that originate in the Coast Range and flow into the west coast of the island of Chiloé, a relatively unpopulated area protected by a national park, carry chemically pure water, i.e. similar to rainwater [35], with a high concentration of marine aerosols and a scarcity of compounds such as nitrogen and ammonium retained by microorganisms and growing trees. The old-growth forests in this area, rich in organic matter accumulated in the soils over decades and centuries, export organic nutrients such as carbon and dissolved organic nitrogen hydrologically. This characteristic distinguishes many temperate forests in southern South America from those in the Northern Hemisphere, which export high concentrations of ammonium and nitrate [35, 64] of anthropogenic origin that are not retained in soils [1].

Dissolved organic nitrogen (DON) is transported massively from coastal forests to rivers and marine estuaries [35], accounting for most of the nitrogen export by rivers in areas with little human impact in Chilean Patagonia (Fig. 3a). DON is also associated with dissolved organic carbon input, feeding rivers, groundwater and coastal seas. Both compounds coming from the organic soil layer of the forests reaches the coasts naturally, stimulating productivity [112]. The hydrological flux of organic matter in different states of decomposition (humus) are dominated by molecules of diverse chemical nature, which complete their decomposition in estuaries [112] in sectors far from pollution. Terrestrial organic matter transported by rivers also includes structures mobilized downstream, such as logs and leaf litter. The fate of these organic compounds and their relationship to marine productivity, especially in areas of fjords and islands without human intervention, is poorly understood. The use of isotopes indicates that an important part of the carbon and nutrients used by aquatic organisms in lakes and coastal seas derive, to a large extent, from the terrestrial environment (Rosenfeld et al., in preparation).

A direct contribution to the ocean from the leaves and trunks of the evergreen riparian forests of *Nothofagus betuloides* which grow on the rocky walls of the fjords, has been reported in the archipelagic region of Magallanes (Rosenfeld et al., in preparation). During high tides, the overhanging branches of these trees are submerged and become substrate for the establishment of mosses, various species of macroalgae, and other marine organisms. In fact, a species of marine mollusk of the genus *Bankia* develops in specific habitats provided by the tree trunks that fall into the sea along the coasts [113]. Another example is the marine urchin, *Pseudochinus magellanicus*, which in coastal areas is covered with mollusk remains, leaf litter or other detritus that protect it from incident radiation [63]. A study of 281 individuals of *P. magellanicus* on Navarino Island found that urchins use shells of *Nacella*, *Mytilus*, *Crepipatella*, leaves of *N. betuloides* and *N. pumilio,* and pieces of wood (Ojeda and Rosenfeld, personal communication). Thus marine-terrestrial interactions are not only linked to chemical cycles, but also to land-based material that provides habitats for algal and invertebrate species.

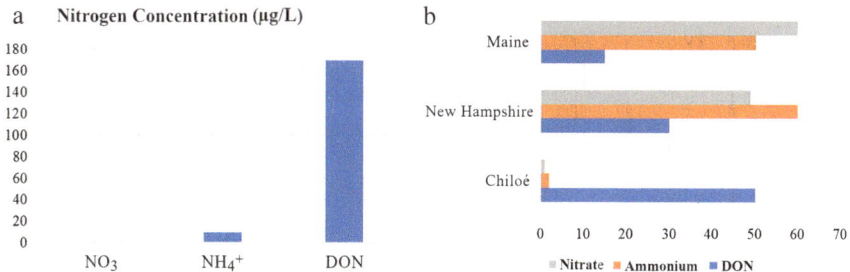

Fig. 3 **a** Concentrations of the main dissolved forms of nitrogen (nitrate, ammonium and dissolved organic nitrogen, NO_3, NH_4 and DON) in rivers descending to the sea from watersheds covered by native forest of Chiloé National Park (Modified from Hedin et al. [35]). **b** Average concentration of different forms of nitrogen in fog from Chiloé Island (n = 16), in mountainous areas of New Hampshire, USA (n = 12), and in marine fog samples from the coast of Maine, USA (n = 6). Note the differences in concentration of the three forms of dissolved nitrogen (ammonium, nitrate, and DON: modified from Weathers et al. [112])

4.3.2 Nutrient Inputs from Marine Ecosystems to Terrestrial and Freshwater Ecosystems

The Chilean Patagonian coast is exposed to westerly winds that transport clouds laden with moisture, but scarce inorganic nutrients (ammonium, nitrate, and sulfates) from the surface of the Pacific Ocean [35]. Analysis of fog water shows that clouds formed over the ocean have a high proportion of DON [112], however, inorganic nitrogen concentration (ammonium plus nitrate) in rainwater and clouds is extremely low compared to those in industrialized regions of North America and Europe (Fig. 3b). Organic nitrogen concentrations in Chiloé rainwater are up to 13 times higher than those in other remote regions of the world ([112], Fig. 3b) and the concentration of inorganic nitrogen in clouds is higher than in rain. Therefore, organic and inorganic nitrogen in clouds and haze, in addition to fixation by free microorganisms, are a significant source of nutrients for terrestrial ecosystems in southern South America, which are strongly nitrogen-limited [70], and possible sources of this element of marine origin are relevant for forest ecosystems.

According to Weathers et al. [112], the organic nitrogen present in the clouds and rainwater of the Chiloé and Subantarctic regions of Magallanes originates in organic matter from terrestrial or marine organisms, gaseous emissions from marine sources, terrestrial biomass, or from fires, even in remote regions. A stoichiometric analysis of the C:N and C:P ratios of rainwater does not suggest an origin from phytoplankton or marine bacteria, or from fires or pollen present in the atmosphere [112]. Due to the direction of the winds and considering the proportion of nitrogen that reaches terrestrial ecosystems via aerosols and fog, the organic and inorganic nitrogen from rainfall in the forests of Chiloé would have an oceanic source. This contribution is biologically significant in non-industrialized areas where the concentration of nitrate and ammonium in rainfall is extremely low (Fig. 3a, [35]). Accordingly, many terrestrial ecosystems in areas far from anthropogenic pollutants are supplied, in part, by oceanic sources.

Large mammals and seabirds are important biotic vectors between marine and terrestrial ecosystems. Along the coasts of channels and fjords there are historical and recent records of large whale strandings that can mobilize nutrients from the ocean floor to coastal areas [24, 37]. In the archipelagic zone, marine mammals and birds transport large amounts of nutrients from the sea to islands and other environments. For example, bird species that form breeding colonies can modify coastal flora and environments. Globally, large mammals such as whales can transport as much as $2.8 \times e^7$ kg y^{-1} from the sea to land [24].

Biotic vectors include naturalized salmon species that now inhabit channels and fjords, which due to their enormous proliferation have altered the transport of nutrients from the ocean to mainland rivers [14, 24, 56]. For example, during the mature stage, chinook salmon migrate from the ocean to spawn and die in rivers. This species does not feed in rivers, but when they die, the fish release nutrients into the freshwater ecosystems,they can also transport marine parasites into the region's rivers [18]. Salmon farming has both a direct biotic impact and an indirect one through the social impact of the explosive development of aquaculture since the 1990s, transforming the old tradition of artisanal fishing [100].

Estero Amalia, Bernardo O'Higgins National Park, Magallanes and Chilean Antarctica Region. Photograph by Nicolás Piwonka

Another biotic vector associated with human activity is the collection of macroalgae: red seaweed (*Gigartina skottsbergii*), black seaweed (*Sarcothalia crispate*), and huiro (*Macrocystis pyrifera*), which have been used since pre-Columbian times as agricultural fertilizer and animal feed, mobilizing nutrients from the coasts to terrestrial ecosystems. Today, innovations are being made in

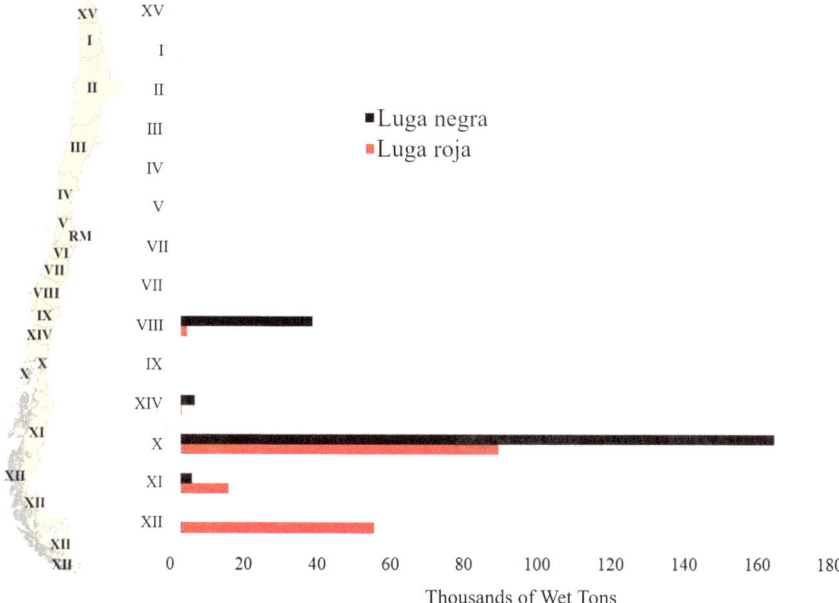

Fig. 4 Cumulative landings for each administrative region of Chile of the red alga species *G. skottsbergii* and *S. crispata* for the period 2010–2016 (*Source* National Service for Fishery and Acuiculture; in Spanish Servicio Nacional de Pesca y Acuicultura, 2018)

ways that integrate traditional and scientific knowledge to develop bio-fertilizers and alga management and cultivation practices [50]. The red seaweed harvest is concentrated in the latitudinal range of the three archipelago regions of Chilean Patagonia [50] (Fig. 4). Between 2010 and 2016, 98.5% of the seaweed biomass extracted in the country was concentrated in the three southern regions: Los Lagos (53.5%), Aysén (10.1%) and Magallanes (34.9%). The remaining 1.5% was extracted in the Bío-Bío Region (Fig. 4). The Los Lagos region concentrated artisanal black sea bass fishing activity, with 80% of the biomass harvested. Both seaweeds grow slowly and suffer a progressive decrease in their populations and biomass, and in the morphological and reproductive attributes of their fronds. It is urgent to estimate accurately the sustainability of fishing practices and harvesting effort for seaweeds.

4.4 Threats and Potential Keystone Species for Conservation in the Marine-Terrestrial Interface of Chilean Patagonia

Chilean Patagonia has been subject to rapid socio-environmental changes that bring new threats to biodiversity conservation. These changes could also open up opportunities for increasing the compatibility of local development and conservation.

4.4.1 Recent Threats to Ecosystem Integrity and Biodiversity at the Marine-Terrestrial Interface

The rapid increase in transportation connections and the expansion of tourism to the most remote sectors of the region represent two major threats to biodiversity, as well as opportunities to design sustainable forms of development [33]. The current road system seeks to connect development centers in the Patagonian archipelago region and includes building new roads through the region's primary forests [77]. New accesses will be opened from the Baker River delta (47° S) to Puerto Natales (52° S) [52], and through Tierra del Fuego Island to the Cabo de Hornos Biosphere Reserve [11]. At the same time, the diminished presence of the Chilean Navy in some marine areas and the opening of Subantarctic channels to commercial shipping and other private development projects imply increasing environmental and social pressures [11]. For example, the exponential growth of the cruise ship tourism industry in areas formerly protected by the Navy involves landings on uninhabited islands and PAs that lack basic information, infrastructure, and park rangers. Unregulated tourism is a threat to the most isolated places in this remote region [43].

Other threats are associated with modes of development that may affect the environmental, economic, and social sustainability of the area. Seven hydroelectric dams have been proposed for construction on the remote Cuervo and Baker rivers [108] the latter being the largest river in the temperate forest biome of South America—with the construction of 5,000 towers for a transmission line over 2,400 km long and 120 m wide, fragmenting and degrading ancient forest ecosystems in one of the largest continuous forest corridors in the world [108].

Forest monocultures and invasive exotic species are another threat to region's biodiversity [77]. *Pinus contorta* and *Eucalyptus* spp. plantations have recently expanded in the Los Lagos and Aysén Regions, replacing the heterogeneity of native ecosystems, and facilitating invasions into degraded native grasslands, steppes, and forests. This expansion produces loss of native species at multiple levels of organization within the ecosystem: soil microorganisms, invertebrates, plants, and vertebrates [29] Invasive plants such as *Ulex europaeus, Eucalyptus* spp., and *Cytisus scoparius* continue to expand [7]. The demand for increasing volumes of woodchips from subsidized plantations of eucalyptus (3 million ha in Chile) by the paper industry has encouraged the expansion of monocultures with high water consumption and negative impacts on hydrological cycles [7].

The rapid growth of the salmon farming industry in coastal-marine ecosystems, with an increasing number of net cages anchored directly to the seabed, has disturbed coastal ecosystems and fjord landscapes (40°–54° S). Salmon farming has multiple ecological and social impacts, including marine pollution by antibiotics, eutrophication of marine and lake waters, introduction of a voracious predator, viral infections, and displacement of traditional fishing communities from their ancestral territories [14, 56].

The impact of three exotic mammals has increased in terrestrial, freshwater, and coastal-marine ecosystems of southern South America: beaver (*Castor canadensis*), North American mink (*Neovison vison*), and muskrat (*Ondatra zibethicus*) [21]. In island ecosystems such as the Wollaston Archipelago, invasive carnivores are a major cause of vertebrate extinctions, particularly birds that lack native predators [87]. Since 2000, mink have increased in population numbers and presence in localities in Chilean Patagonia, reaching the southern tip of the region in the Cape Horn Biosphere Reserve in 2001 [19, 80]. Their impact on native fauna has been estimated through diet studies and population censuses of ground-nesting birds. In the Cape Horn Biosphere Reserve, mink diet includes similar proportions (number of food items) of native and exotic mammals, birds, and fish [20, 85–89]. Mink represents a critical threat to the biodiversity of terrestrial, freshwater, and coastal-marine ecosystems, including functionally key avifauna at the marine-terrestrial interface [21].

4.4.2 Threatened Species as Biotic Vectors Relevant to Nutrient and Energy Fluxes at the Marine-Terrestrial Interface

The islands of Chilean Patagonia provide habitat for native populations of mustelids (otters), pinniped colonies (e.g. sea lions and elephant seals) and breeding colonies of seabirds (e.g. cormorants, penguins, albatrosses). These vertebrate groups play a key ecological role in the transport of marine nutrients to terrestrial ecosystems [37, 66]. Ten species of birds and mammals that are essential for nutrient flow from sea to land (Table 2) present conservation problems derived from habitat disturbance (e.g. salmon farming), pollution, hunting (otter and pinniped fur trade), and rapid expansion of exotic species such as mink throughout the archipelagic region of Patagonia [38, 88, 89]. Among the species that contribute marine nutrients to terrestrial ecosystems are albatrosses, especially black-browed and gray-headed albatrosses, and four penguin species with abundant breeding colonies in Chilean Patagonia: Humboldt penguin, Magellanic penguin, macaroni penguin, and yellow-plumed penguin. All of these species transport large amounts of nutrients (e.g. N, P, K, Mg) from the sea to the land, modifying the vegetation of the islands [79].

At least 58 bird species affect the marine-terrestrial interface in Chilean Patagonia [66]. On Navarino Island, 65% of these species are resident and only 20% are migratory. The abundance and biomass of birds is especially high at river mouths throughout the region, so this habitat should be considered with special attention in conservation programs.

4.5 Protected Areas with Marine-Terrestrial Interface Relevant to Conservation in Chilean Patagonia

There are currently 40 units of the National System of State Wild Protected Areas (in Spanish Sistema Nacional de Areas Silvestres Protegidas del Estado, SNASPE) located in Chilean Patagonia, representing about 87% of the PA in the country

Table 2 Ten species that contribute nutrients at the land-marine interface and that may be emblematic for conservation on the coastline of the Chilean Patagonian archipelago, all of which are currently considered endangered

Species-distribution	Conservation	Feeding	Reproduction nesting
Humboldt penguin (*Spheniscus humboldti*). Endemic to the Humboldt Current in the Pacific Ocean [36]	Category: Vulnerable International Union for Conservation of Nature (IUCN) [39] Threats: Industrial fishing	Mainly fish	Habitat: sheltered coastal promontories and islands Habits: To nest they dig holes in guano or salty soil, or use existing crevices between rocks and natural caves
Magellanic penguin (*Spheniscus magellanicus*). Nests on the coasts and islands of Patagonia in Chile and Argentina [27, 98]	Category: Growing trend IUCN [39]. Threats: Climate change, impact of the fishing industry	Fish (sardines, silversides, anchovies), squid, krill and other crustaceans	Habitat: They nest on coasts with soft soils, also under bushes or forests near the sea, occasionally inland up to 1 km from the coast Habits: nests in colonies, in roosts used year after year
Macaroni penguin (*Eudyptes chrysolophus*). Nests on Subantarctic islands of Magallanes and other Subantarctic islands [12, 44]	Category: Vulnerable IUCN [39] Threats: Industrial fishing. Recommendation: Reinforce the protection of the Diego Ramírez Islands and its marine environments, where >33% of reproductive colonies are concentrated	It feeds mainly on squid, including krill and other crustaceans	Habitat: tall grass formations or "tussocks" to establish their nests Habits: nests in colonies, in roosts used year after year
Yellow-plumed penguin (*Eudyptes chrysocome chrysocome*). This subspecies nests in the Subantarctic islands of Magallanes (Chile), in the Falkland Islands, Isla de los Estados and other islands in southern Argentina [44]	Category: Vulnerable [39] Threats: Industrial fishing Recommendation: Reinforce the protection of the Diego Ramírez Islands and their marine environments, where 28% of the world's reproductive colonies are concentrated	Squid, as well as krill, fish, plankton, crustaceans, octopus and other mollusks	Habitat: Nesting colonies from sea level to cliffs and sometimes inland. Habits: Great rock and crevice jumpers, which facilitated their survival from nineteenth century whalers

(continued)

Table 2 (continued)

Species-distribution	Conservation	Feeding	Reproduction nesting
Black-browed Albatross (*Thalassarche melanophris*). Distributed from the Tropic of Capricorn to Antarctica, but only nests on subantarctic islands. Fourteen nesting sites have been recorded worldwide, six of them in Chile: Islas Diego Ramírez, Islas Ildefonso, Diego de Almagro and Islote Evangelistas, Islote Leonard, Islote Albatros [2, 5, 74, 110]	Category: Near threatened [39] Threats: Industrial fishing because they inhabit the neritic zone where accidental capture by longline fishing occurs Recommendation: Reinforce the protection of the Diego Ramírez islands and their marine environments as feeding areas, where 20% of the reproductive colonies worldwide are concentrated	Its most abundant prey are the channel shrimp (*Munida gregaria*) and the Fuegian sardine (*Sprattus fuegensis*)	Habitat: It builds its nests on cliffs or hillsides, preferably among *Poa flabellata* plants Habits: They nest in colonies. They are long-lived (>60 years), with low fecundity (maximum one offspring per year) and high philopatry (95% of juveniles recruit in their natal colony)
Gray-headed albatross (*Thalassarche chrysostoma*). Subantarctic islands of the Southern Ocean, including South Georgia, Diego Ramírez, Kerguelen, Crozet, Marion and Prince Edward, Campbell and Macquarie islands south of New Zealand and Chile [12, 73]	Category: Endangered [39] Threats: Industrial fishing because they inhabit the neritic zone where longline fishing bycatch occurs. Recommendation: Strengthen protection of the Diego Ramírez Islands and their marine environments, where 99% of Chile's population is concentrated	Mainly squid (*Martialia hyadesi*), channel shrimp (Munida gregaria) and Fuegian sardine (*Sprattus fuegensis*)	Habitat: It builds its nests on cliffs or hillsides, preferably among *Poa flabellata* plants Habits: nests in colonies

(continued)

Table 2 (continued)

Species-distribution	Conservation	Feeding	Reproduction nesting
Striated caracara (*Phalcoboenus australis*). Endemic to the Subantarctic islands of Magellanes in Chile and the Falkland/Malvinas Islands. It reaches its southern limit of distribution in the Diego Ramirez Islands [22, 45, 79]	Category: Near Threatened [39]. In Chile considered a bird of prey with high conservation priority. Threats: Hunting. Recommendation: Reinforce protection of critical nesting localities in Chile: Diego Ramirez Islands, Noir Islands and Cape Horn Archipelago	Eggs, chicks and dead seabirds. In colonies of penguins, albatrosses and other seabirds, it serves as a scavenger, distributing nutrients to the interior of terrestrial ecosystems	Habitat: Builds its nests on trees or cliffs. Habits: nests in pairs
Chungungo or sea otter (*Lontra felina*). Coasts of western South America between 9° S (Chimbote, Peru) and 56°S (Cape Horn, Chile) ([54, 81, 94, 95])	Category: Endangered [39] Threats: illegal hunting, competition with American mink for prey and habitat loss	Crustaceans (69.6%) in marine-coastal eco-systems, fish (30.4%) in the Lakes Region, mollusks, cephalopods, echinoderms and other invertebrates; birds and mammals from Chiloé to Cape Horn	Habitat: Breeds on land in burrows or crevices along rocky shores, in areas exposed to waves. Habits: Varying distances over which it ventures into coastal habitats to establish its breeding and rearing grounds

(continued)

Table 2 (continued)

Species-distribution	Conservation	Feeding	Reproduction nesting
Huillín or freshwater otter (*Lontra provocax*). Endemic to southern South America, it inhabits rivers and also coastal and estuarine ecosystems in the archipelagos between Reloncaví and Cape Horn [53, 91, 94, 95]	Category: Endangered [39] Threats: illegal hunting, degradation of riparian vegetation (causing reduction of crustaceans), modification of watercourses, channelization and construction of dams, presence of dogs and disturbance by livestock Recommendation: Reinforce the control of feral dogs	In freshwater environments they consume fish, but crustaceans are their main diet (50–100%). In estuarine environments the proportion of fish increases, reaching 75% presence in their feces	Habitat: Their burrows are located mainly under roots and trunks of riparian forests in areas protected from waves Habits: Males are territorial and only meet with females during mating season. The litter varies from 1–4 offspring that remain with the mother for 7–8 months
Southern fur seal (*Arctocephalus australis*). In Chilean Patagonia it is distributed from 43° S to the Diego Ramírez Islands (56° 30′ S), and to the east in the Malvinas Islands and the coasts of Argentine Patagonia, reaching Uruguay and southern Brazil [62]	Category of *Least Concern* [39]	It feeds on fish and invertebrates, cephalopods, crustaceans and gastropods and occasionally seabirds	Habitat: Breeds in coastal rocky areas, mainly on islets. Habits: Forms colonies (in Spanish loberías), where a territorial male mates with between 2 and 5 females during the mating and calving period, starting in November–December each year

Table 3 Surface area and number of units within the National Protected Areas System according to management category in the three administrative regions within Chilean Patagonia. Areas with coastline are distinguished (*). In the Los Lagos Region, only the areas protected from Reloncaví Sound to the south, including the island of Chiloé. Personal communication (2023) by Mr. Mariano de la Maza, Head State Wild Protected Areas, National Forestry Corporation (in Spanish Corporación Nacional Forestal, CONAF)

| Administrative region | National park | | National reserve | | Natural monument | | Total area (ha) | Total units |
	Surface area (ha)	No. units	Surface area (ha)	No. units	Surface area (ha)	No. units		
Los Lagos	950,422	5	84,924	3	209	2	1,035,555	10
Los Lagos*	950,422	5	34,098	1	8.6	1	984,529	7
Aysén	2,709,960	7	1,874,902	8	409	2	458,5271	17
Aysén*	2,261,930	5	1,772,475	2	228	1	4,076,053	8
Magellanes	8,228,355	7	31,914	2	311	3	7,744,839	12
Magellanes*	7,891,323	4	0	0	97	1	7,494,319	6
Total Patagonia	11,888,737	19	1,991,740	13	955	7	13,881,432	39
Total Patagonia*	11,103,675	14	1,806,573	3	334	3	12,910,582	20
National total	13,209,848	43	5,375,935	45	34,357	18	15,324,844	103

(Table 3). This large concentration of PAs in Patagonia stimulated the creation in 2018 of the "Chilean Patagonia National Parks Network", in order to plan and manage the parks in the region in an integrated manner, with special emphasis on human communities and the marine-terrestrial interface.

Twenty of the 39 units include coastline, representing 93% of the total protected area (Table 3). Of the 19 national parks in Chilean Patagonia, 14 (9% of the total area) include coastline. The national parks Cabo de Hornos (58,917 ha), Alberto de Agostini (1,460,000 ha), Bernardo O'Higgins (3,525,901 ha), Kawesqar (2,842,329 ha), Melimoyu (105,500 ha), Corcovado (400,011 ha), Laguna San Rafael (1,742,000 ha), Magdalena Island (249,712 ha) and Guamblin Island (10.625 ha) stand out for their territorial continuity and form a belt of parks with archipelagic zones exposed to the Pacific Ocean that protect habitats for endemic and threatened marine-terrestrial fauna (Tecklin et al., 2021).

Chilean Patagonia includes four biosphere reserves; partially one of them (Bosques Templados Lluviosos de los Andes Australes), and fully the other three: Cabo de Hornos et al. [57]. The Cabo de Hornos Biosphere Reserve (4,884,513 ha) included from its origin the protection of both land (1,917,238 ha) and ocean area (2,967,036 ha), constituting the first demonstrative unit of integrated sea-land management. The lessons learned here could be applied to the other reserves in Patagonia [78, 84]. Together, the core areas of these reserves include national parks and reserves with extensive coastlines, islands, islets, and rocky outcrops.

It is important to note that all six nature sanctuaries in Patagonia (315,292 ha) include coastlines. There are also 25 National Protected Assets (272,812 ha), 18 of which have a coastline, and represent 75% of the total area. The RAMSAR site Bahía Lomas, Magallanes Region, is a wetland located on the coastline of Tierra del Fuego Island.

The General Fisheries and Aquaculture Law of 1991 (No. 18,892) created the categories of Marine Park, Marine Reserve, and Benthic Resources Management and Exploitation Areas (AMERBs). Through the AMERBs, fishers' organizations can establish management areas for a renewable period of two years. The creation of the Ministry of the Environment in 2010 (Law No. 20.417, modification to Law No. 19.300; in Spanish Ministerio del Medio Ambiente, MMA) created and provided administrative authority over the category of Multiple Use Marine and Coastal Protected Area (MU-MCPA). These figures, decreed by the MMA, are under the responsibility of the National Fishing and Aquaculture Service. Eight of the 26 marine protected areas in Chile are in Chilean Patagonia (Tecklin et al., 2021). There are two marine reserves on the island of Chiloé, Pullinque (7.73 km^2) and Putemún (7.53 km^2), which do not border terrestrial PAs. The two marine parks in the Magallanes Region, Francisco Coloane (15.06 km^2) and Islas Diego Ramírez-Paso Drake (140,000 km^2), border terrestrial PAs.

The four MU-MCPA are adjacent to PAs: (i) Comau Fjord, Los Lagos Region, adjoins the Private Conservation Initiative (ICP) Huinay Biological Station, managed by the Huinay Foundation; (ii) Pitipalena-Añihue (238.6 km^2), Aysén region, adjoins the ICP Reserva Añihue (100 km^2); (iii) Francisco Coloane (653.3 km^2), Magallenes region, adjacent to Kawesquar National Park; (iv) Seno Almirantazgo (764 km^2), Magallenes region, is adjacent to Alberto de Agostini National Park (Tecklin et al., 2021).

The Chiloense and Channels and Fjords ecoregions of southern Chile were prioritized for marine conservation in Latin America in the 1990s [99]. South of the Magallanes Province, the Diego Ramirez Islands-Drake Passage Marine Park was created in 2018, protecting 140,000 km^2, most of this area lies within the Channels and Fjords ecoregion of southern Chile, with a section towards the Southern Ocean of the Drake Passage crossed by the Antarctic Circumpolar Current, beyond the southern end of this province [79]. The challenge is to implement and protect this vast marine park, neighboring the Cabo de Hornos Biosphere Reserve.

4.6 Final Reflections

Those ecosystems that integrate coastal-marine environments should have high priority for the conservation of Chilean Patagonian ecosystems. We have documented that in this region, the physical processes include nutrient transport associated with ocean–atmosphere interrelationships, from the ocean to terrestrial ecosystems and vice versa. Freshwater ecosystems, via watercourses, contribute nutrients from terrestrial ecosystems to the oceans. Biotic processes are mediated by large colonies of seabirds and mammal species, such as pinnipeds, which reproduce in island systems and constitute vectors of marine nutrients to terrestrial ecosystems [24].

Environmental institutions must consider the need to conserve the integrity of the coastline and regulate its multiple uses and activities in development plans. Supreme Decree No. 475 was issued in 1994, establishing the National Coastline Uses Policy, which proposes a zoning of the country's coastline in accordance with the realities of each region. Initially, this decree defined that the Undersecretary of the Navy, Chilean Ministry of National Defense, would be responsible to apply this policy to: (a) public beach land located within a strip of eighty meters wide, measured from the line of the highest tide of the coastline; (b) the beach; (c) the bays, gulfs, straits and inland channels; (d) the territorial sea of the Republic of Chile. In 1997, this responsibility was transferred to the Undersecretary of the Armed Forces and Regional Governments of Chile with the task of creating Regional Coastal Border Commissions to develop a cadaster and a zoning proposal [4].

Chilean Patagonia can become a pilot model at the national and international level for this approach to biodiversity conservation, which makes human activities, conservation and development at the land-sea interface compatible.

5   Recommendations

In order to reinforce management and protection measures for the marine-terrestrial interface in Chilean Patagonia, we propose the following recommendations:

- The SNASPE (administrator until today) and the future Biodiversity and Protected Areas Service of the Ministry of the Environment should be part of the Regional Commission for Uses of the Coastline. Lack of participation in these commissions means that proposals for managing the coastline may be dissociated from the neighboring terrestrial PAs. For example, concessions could be granted in the 80-m strip of fiscal beach for benthic resource extraction activities in areas of great importance for the conservation of threatened bird colonies or marine mammals. It is also recommended that interinstitutional and transdisciplinary collaborations be established and that a landscape-scale approach be used to integrate the conservation of marine and terrestrial ecosystems throughout the archipelagic region of Chilean Patagonia.
- That D.L. No. 1,939 on Fiscal Assets of the Ministry of National Assets (in Spanish Ministerio de Bienes Nacionales), in its policy of long-term concessions for private projects, regulate the management of the marine-terrestrial interface. Concessions on coastal lands should be in dialogue with local planning of each territory. Finally, the planning and administration of parks and terrestrial reserves, whether public or private, should consider coastal management and be involved in the decision-making of the Regional Commission for Uses of the Coastline.

- Faced with the antagonism between developmentalist and conservationist positions (see Rozzi and Feinsinger [83]), the United Nations Educational, Scientific and Cultural Organization created the Man and Biosphere Program, MaB), which seeks to integrate human societies and conservation areas to meet social, cultural, recreational and ecological needs [34]. This vision has been implemented through the establishment of an international network of biosphere reserves that today includes a mosaic of unique sites representing the planet's major ecosystems, protected through research, monitoring, education, conservation and sustainable development programs [42, 78]. However, in practice the implementation of biosphere reserves in Chile continues to be a challenge due to the complex management demands, resources, and multiplicity of actors and objectives that must be met [42].

- A pioneering example of this strategy is the Cabo de Hornos Biosphere Reserve, which may be fundamental for consolidating the Diego Ramirez Islands-Drake Passage Marine Park and could be a demonstration unit for integrating scientific studies, educational programs, special interest tourism, regulated artisanal and industrial fishing, and integrated marine-terrestrial management. Lessons learned in these areas could be implemented in other terrestrial and marine PAs in Chilean Patagonia. On this basis, given the current configuration of large terrestrial and marine parks and reserves in the territory and ocean area of Chilean Patagonia, with a strong presence of the marine-terrestrial interface and a deficiency of regulatory instruments for areas that require integrated management of oceanic and terrestrial ecosystems, we propose the creation of a biosphere reserve that includes all of Chilean Patagonia, from the Chiloé archipelago to the Diego Ramírez archipelago. This large biosphere reserve could establish the compatibility of economic and environmental sustainability and integrate national parks and marine parks as well as terrestrial and marine reserves as core areas, thus making possible territorial planning based on human communities and locally generated traditional and scientific knowledge.

Acknowledgements We appreciate the valuable revisions and comments made by Juan Carlos Castilla and David Tecklin. Ricardo Rozzi and Juan J. Armesto are grateful for the support of ANID, through projects AFC170008, and CHIC - ANID/BASAL FB210018.

References

1. Aber, J., McDowell, W., Nadelhoffer, K., Magill, A., Berntson, G., Kamakea, M., McNulty, S., Currie, W., Rustad, L., & Fernández, I. (1998). Nitrogen saturation in temperate forest ecosys tems: Hypotheses revisited. *BioScience, 48*, 921–934.
2. Aguayo, A., Acevedo, J., & Acuña, P. (2003). Nuevo sitio de anidamiento del albatros ceja negra, *Diomedea melanophris* Temmink 1828, en el seno Almirantazgo, Tierra del Fuego, Chile. *Anales del Instituto de la Patagonia (Chile), 31*, 91–96.
3. Allcock, A., & Strugnell, J. (2012). Southern Ocean diversity: New paradigms from molecular ecology. *Trends in Ecology & Evolution, 27*(9), 520–528.
4. Andersen-Cirera, K., & Balbontín-Gallo, C. (2021). La planificación del borde costero chileno. Una normativa deficiente. *Revista de Geografía Norte Grande, 80*, 227–247.

5. Arata, J., & Xavier, J. C. (2003). The diet of black-browed albatrosses at the Diego Ramírez islands, Chile. *Polar Biology, 26*(10), 638–647.

6. Aravena, J. C., & Luckman, B. (2009). Spatio-temporal rainfall patterns in southern South America. *International Journal of Climatology, 29*, 2106–2120.

7. Armesto, J. J., Manuscevich, D., Mora, A., Smith-Ramírez, C., Rozzi, R., Abarzúa, A. M., & Marquet, P. A. (2010). From the Holocene to the Anthropocene: A historical framework for land cover change in southwestern South America in the past 15,000 years. *Land Use Policy, 27*, 148–160.

8. Armesto, J. J., Rozzi, R., Smith-Ramírez, C., & Arroyo, M. T. K. (1998). Conservation targets in South American temperate forests. *Science, 282*, 1271–1272.

9. Armesto, J. J., Villagrán, C., & Kalin, M. T. (Eds.). (1996). *Ecología de los bosques nativos de Chile*. Editorial Universitaria.

10. Armesto, J. J., Smith-Ramírez, C., Carmona, M. R., Celis-Diez, J. L., Díaz, I. A., Gaxiola, A., Gutiérrez, A. G., Núñez-Avila, M. C., Pérez, C. A., & Rozzi, R. (2009). Old-growth temperate rainforests of South America: conservation, plant-animal interactions, and baseline biogeochemical processes. In C. Wirth, G. Gleixner & M. Heimann (Eds.), *Old-growth forests* (pp. 367–390). Springer.

11. Barros, E., & Harcha, J. (2004). The Cape Horn Biosphere Reserve Initiative: Analysis of a challenge for sustainable development in the Chilean Antarctic Province. In R. Rozzi, F. Massardo & C. B. Anderson (Eds.), *The Cape Horn Biosphere Reserve. A proposal for conservation and tourism to achieve sustainable development at the southern end of the Americas* (pp. 27–43). University of Magallanes Press.

12. Barroso, O., Crego, R. D., Mella, J., Rosenfeld, S., Contador, T., Mackenzie, R., Vásquez, R. A., & Rozzi, R. (2020). Colaboración científica con la Armada de Chile en estudios ornitológicos a largo plazo en el archipiélago Diego Ramírez: Primer monitoreo del ciclo anual del ensamble de aves en la isla Gonzalo. *Anales del Instituto de la Patagonia, 48*(3), 149–168.

13. Burrows, M. T., Schoeman, D. S., Buckley, L. B., Moore, P., Poloczanska, E. S., Brander, K. M., Brown, C., Bruno, J. F., Duarte, C. M., Halpern, B. S., & Holding, J. (2011). The pace of shifting climate in marine and terrestrial ecosystems. *Science, 334*, 652–655.

14. Buschmann, A. H., Niklitschek, E. J., & Pereda, S. (2023). *Aquaculture and its impacts on the conservation of Chilean Patagonia*. Springer.

15. Camus, P. A. (2001). Biogeografía marina de Chile. *Revista Chilena de Historia Natural, 74*, 587–617.

16. Castilla, J. C. (2014). Chile es Mar. In C. Aldunate (Ed.), *Mar de Chile* (pp. 14–40). Colección Santander, Museo de Chileno de Arte Precolombino.

17. Christian, R. R., & Mazzilli, S. (2007). Defining the coast and sentinel ecosystems for coastal observations of global change. *Hydrobiologia, 577*, 55–70.

18. Correa, C., & Gross, M. (2007). Chinook salmon invade southern South America. *Biological Invasions, 10*, 615–639.

19. Crego, R. D., Jiménez, J. E., & Rozzi, R. (2015). Expansión de la invasión del visón norteamericano (*Neovison vison*) en la Reserva de la Biósfera Cabo de Hornos, Chile. *Anales del Instituto de la Patagonia (Chile), 43*(1), 157–162.

20. Crego, R. D., Jiménez, J. E., & Rozzi, R. (2016). A synergistc trio of invasive mammals? Facilitative interactions among beavers, muskrats, and mink at the world's southernmost forests. *Biological Invasions, 18*, 1923–1938.

21. Crego, R. D., Rozzi, R., & Jiménez, J. E. (2018). Fur trade and the biotic homogenization of sub-polar ecosystems. In R. Rozzi, R. H. Jr. May, F. S. III Chapin, F. Massardo, M. Gavin, I. Klaver, A. Pauchard, M. A. Núñez & D. Simberloff (Eds.), *From biocultural homogenization to biocultural conservation* (pp. 233–244). Springer.

22. Cursach, J. A., Suazo, C. G., Schlatter, R. P., & Rau, J. R. (2012). Observaciones sobre el carancho negro *Phalcoboenus australis* (Gmelin, 1788) en la isla Gonzalo archipiélago Diego Ramírez, Chile. *Anales del Instituto de la Patagonia (Chile), 40*, 147–150.

23. Dellarossa, V. H. (1998). Producción primaria anual en sistemas de alta productividad biológica. [Ph.D. Thesis], Universidad de Concepción, Chile.

24. Doughty, C., Roman, J., Faurby, S., Wolf, A., Haque, A., Bakker, E., Malhi, Y., Dunning, Jr. J., & Svenning, J. (2016). Global nutrient transport in a world of giants. *Proceedings of the National Academy of Sciences USA*, 201502549.

25. Dávila, P. M., Figueroa, D., & Müller, E. (2002). Freshwater input into the coastal ocean and its relation with the salinity distribution off austral Chile (35–55° S) Continent. *Shelf Research, 22*, 521–534.

26. Fabiano, M., Povero, P., Danovaro, R., & Misic, C. (1999). Particulate organic matter composition in a semi-enclosed periantarctic system: The straits of Magellan. *Scientia Marina, 63*(S1), 89–98.

27. Frere, E., Gandini, P., & Lichtschein, V. (1996). Variación latitudinal en la dieta del pingüino de Magallanes (*Spheniscus magellanicus*) en la costa patagónica, Argentina. *Ornitologia Neotropical, 7*, 35–41.

28. Froehlich, H. E., Afflerbach, J. C., Frazier, M., & Halpern, B. S. (2019). Blue growth potential to mitigate climate change through seaweed offsetting. *Current Biology, 29*(18), 3087–3093.

29. García, A., Franzese, J., Policelli, N., Sasal, Y., Zenni, R., Nuñez, M. A., Taylor, K., & Pauchard, A. (2018). Non-native pines are homogenizing the ecosystems of South America. In R. Rozzi, R. H. Jr. May, F. S. Chapin, F. Massardo, M. Gavin, I. Klaver, A. Pauchard, M. A. Núñez & D. Simberloff (Eds.), *From biocultural homogenization to biocultural conservation* (pp. 245–264). Springer.

30. Glavovic, B. C., Limburg, K., Liu, K. K., Emeis, K. C., Thomas, H., Kremer, H., & Swaney, D. P. (2015). Living on the margin in the Anthropocene: Engagement arenas for sustainability research and action at the ocean–land interface. *Current Opinion in Environmental Sustainability, 14*, 232–238.

31. Goffinet, B., Rozzi, R., Lewis, L., Buck, W., & Massardo, F. (2012). *The miniature forests of cape Horn: Eco-tourism with a hand-lens* ("Los bosques en miniatura de cabo de Hornos: eco-turismo con lupa"). Bilingual English-Spanish edition. UNT. Press—Ediciones Universidad de Magallanes.

32. González, H. E., Castro, L., Daneri, G., Iriarte, J. L., Silva, N., Vargas, C. A., Giesecke, R., & Sánchez, N. (2011). Seasonal plankton variability in Chilean Patagonia fjords: carbon flow through the pelagic food web of Aysen fjord and plankton dynamics in the Moraleda channel basin. *Continental Shelf Research, 31*, 225–243.

33. Guala, C., Veloso, K., Farías, A., & Sariego, F. (2023). *Analysis of tourism development linked to protected areas in Chilean Patagonia*. Springer.

34. Guevara, S., & Laborde, J. (2008). The landscape approach: Designing new reserves for protection of biological and cultural diversity in Latin America. *Environmental Ethics, 30*(3), 251–262.

35. Hedin, L. O., Armesto, J. J., & Johnson, A. H. (1995). Patterns of nutrient from unpolluted, old-growth temperate forest: Evaluation of biogeochemical theory. *Ecology, 76*, 493–509.

36. Herling, C., Culik, B. M., & Hennicke, J. C. (2005). Diet of the Humboldt penguin (*Spheniscus humboldti*) in northern and southern Chile. *Marine Biology, 147*(1), 13–25.

37. Hucke-Gaete, R., Viddi, F. A., & Simeone, A. (2023). *Marine mammals and seabirds of Chilean Patagonia: focal species for the conservation of marine ecosystems*. Springer.

38. Ibarra, J. T., Fasola, L., Macdonald, D. W., Rozzi, R., & Bonacic, C. (2009). Invasive American mink in wetlands of the Cape Horn Biosphere Reserve, Southern Chile: What are they eating? *Oryx, 43*(1), 87–90.

39. International Union for Conservation of Nature. (2018). *The red list of threatened species 2018*. Retrieved from: https://www.iucnredlist.org/

40. Iriarte, J. L., Kusch, A., Osses, J., & Ruiz, M. (2001). Phytoplankton biomass in the sub-Antarctic area of the strait of Magellan (53 S), Chile during spring-summer 1997/1998. *Polar Biology, 24*(3), 154–162.

41. Iriarte, J. L., Pantoja, S., & Daneri, G. (2014). Oceanographic processes in Chilean fjords of Patagonia: from small to large-scale studies. *Progress in Oceanography, 129*, 1–7.

42. Karez, C. S., Hernández-Faccio, J. M., Schüttler, E., Rozzi, R., García, M., Meza, A. Y., & Clüsener-Godt, M. (2016). Learning experiences about intangible heritage conservation for

sustainability in biosphere reserves. Special Issue on "Intangible Cultural Heritage". *Material Culture Review,* 82–83.

43. Kirk, C., Rozzi, R., & Gelcich, S. (2018). El turismo como una herramienta para la conservación del elefante marino del sur (*Mirounga leonina*) y sus hábitats en Tierra del Fuego, Reserva de la Biósfera Cabo de Hornos, Chile. *Magallania (Chile), 46*(1), 65–78.

44. Kirkwood, R., Lawton, K., Moreno, C., Valencia, J., Schlatter, R., & Robertson, G. (2007). Estimates of southern rockhopper and macaroni penguin numbers at the Ildefonso and Diego Ramírez archipelagos, Chile, using quadrat and distance-sampling techniques. *Waterbirds, 30*(2), 259–267.

45. Kusch, A., Marín, M., Oheler, D., & Drieschman, S. (2007). Notas sobre la avifauna de isla Noir (54° 28′ S-73° 00′ W). *Anales del Instituto de la Patagonia (Chile), 35*(2), 61–66.

46. Lawford, R. G., Alaback, P. B., & Fuentes, E. (Eds.). (1996). *High-latitude rainforests and associated ecosystems of the west coast of the Americas: Climate, hydrology, ecology, and conservation.* Springer.

47. Lawver, L., & Gahagan, L. (2003). Evolution of Cenozoic seaways in the circum-antarctic region. *Palaeogeography, Palaeoclimatology, Palaeoecology, 198,* 1–27.

48. Lessios, H. A. (2008). The great American schism: divergence of marine organisms after the rise of the central American isthmus. *Annual Reviews of Ecology and Evolution Systematics, 39,* 63–91.

49. Liu, H., Jiang, Z., Cao, Y., & Wang, Y. (2010). Sedimentary characteristics and hydrocarbon accumulation of glutenite in the fourth member of Eogene Shahejie formation in Shengtuo area of Bohai bay basin, east China. *Energy Exploration & Exploitation, 28*(4), 223–237.

50. Mansilla, A. (2013). *Catálogo de macroalgas y moluscos asociados a praderas naturales de Gigartina skottsbergii de la Región de Magallanes.* Punta Arenas, Chile: Ediciones Universidad de Magallanes.

51. Marquet, P. A., Buschmann, A. H., Corcoran, D., Díaz, P. A., Fuentes-Castillo, T., Garreaud, R., Pliscoff, P., & Salazar, A. (2023). *Global change and acceleration of anthropic pressures on Patagonian ecosystems.* Springer.

52. Martinic, M. (2004). *Archipiélago Patagónico, la última frontera.* Punta Arenas, Chile: Ediciones de la Universidad de Magallanes, La Prensa Austral.

53. Medina-Vogel, G., & González-Lagos, C. (2008). Habitat use and diet of endangered southern river otter *Lontra provocax* in a predominantly palustrine wetland in Chile. *Wildlife Biology, 14*(2), 211–220.

54. Medina-Vogel, G., Bartheld, J. L., Pacheco, R. A., & Rodríguez, C. D. (2006). Population assessment and habitat use by marine otter Lontra felina in southern Chile. *Wildlife Biology, 12*(2), 191–199.

55. Mittermeier, R. A., Mittermeier, C. G., Brooks, T. M., Pilgrim, J. D., Konstant, W. R., da Fonseca, G. A. B., & Kormos, C. (2003). Wilderness and biodiversity conservation. *Proceedings of the National Academy of Sciences USA, 100,* 10309–10313.

56. Molinet, C., & Niklitschek, E. J. (2023). *Fisheries and marine conservation in Chilean Patagonia.* Springer.

57. Moreira-Muñoz, A., Carvajal, F., Elórtegui, S., & Rozzi, R. (2020). The Chilean biosphere reserves network as a model for sustainability? Challenges towards regenerative development, education, biocultural ethics and eco-social peace. In M. G., Reed & M. F. Price (Eds.), *UNESCO Biosphere Reserves: Supporting biocultural diversity, sustainability and society. Earths studies in natural resource management* (pp. 61–75). Routledge.

58. Olson, D. M., Dinerstein, E., Wikramanayake, E. D., Burgess, N. D., Powell, G. V., Underwood, E. C., & Loucks, C. J. (2001). Terrestrial ecoregions of the world: A new map of life on Earth. A new global map of terrestrial ecoregions provides an innovative tool for conserving bio-diversity. *BioScience, 51,* 933–938.

59. Oyarzún, C. E., & Huber, A. (2003). Exportación de nitrógeno en cuencas boscosas y agrícolas en el sur de Chile. *Gayana Bot, 60*(1), 63–68.

60. Oyarzún, C. E., & Hervé-Fernández, P. (2015). Ecohidrology and nutrient fluxes in forest ecosystems of southern Chile. Chapter 13. In J. A. Blanco (Ed.), *Biodiversity in ecosystems— Linking structure and function* (pp. 335–352). Universidad Pública De Navarra.

61. Pantoja, S., Iriarte, J. L., & Daneri, G. (2011). Oceanography of the Chilean Patagonia. *Continental Shelf Research, 31*, 149–153.

62. Pavés, H., & Schlatter, R. (2008). Temporada reproductiva del lobo fino austral, *Arctocephalus australis* (Zimmerman, 1783) en la isla Guafo, Chiloé, Chile. *Revista Chilena de Historia Natural, 81*, 137–149.

63. Pawson, D. (2009). Echinoidea - Erizos de Mar. In V. Häussermann & G. Försterra (Eds.), *Fauna marina bentónica de la Patagonia chilena* (pp. 850–858). Nature in Focus.

64. Perakis, S. S., & Hedin, L. O. (2002). Nitrogen loss from unpolluted South American forests mainly via dissolved organic compounds. *Nature, 415*, 416–419.

65. Pisano, E. (1977). Fitogeografía de Fuego-Patagonia chilena. I.-Comunidades vegetales entre las latitudes 52 y 56° S. *Anales del Instituto de la Patagonia, 8*, 121–250.

66. Pizarro, J. C., Anderson, C. B., & Rozzi, R. (2012). Birds as marine-terrestrial linkages in sub- polar archipelagic systems: Avian community composition, function, and seasonal dynamics in the Cape Horn Biosphere Reserve (54–55 S), Chile. *Polar Biology, 35*(1), 39–51.

67. Pizarro, G., Iriarte, J. L., Montecino, V., Blanco, J. L., & Guzmán, L. (2000). Distribución de la biomasa fitoplanctónica y productividad primaria máxima de fiordos y canales australes (47°–50° S) en octubre 1996. *Ciencia y Tecnología del Mar, 23*, 25–47.

68. Pizarro, G., Astoreca, R., Montecino, V., Paredes, M. A., Alarcón, G, Uribe, P., & Guzmán, L. (2005). Patrones espaciales de la abundancia de clorofila, su relación con la productividad primaria, y la estructura de tamaños del fitoplancton en julio y noviembre de 2001 en la Región de Aysén (43°–46° S). *Ciencia y Tecnología del Mar 28*(2), 27–42.

69. Poulin, E., González-Wevar, C. A., Díaz, A., Gérard, K., & Hüne, M. (2014). Divergence between Antarctic and South American marine invertebrates: What molecular biology tells us about Scotia Arc geodynamics and the intensification of the Antarctic Circumpolar Current. *Global and Planetary Change, 123*, 392–399.

70. Pérez, C. A., Hedin, L. O., & Armesto, J. J. (1998). Nitrogen mineralization in two unpolluted old- growth forests of contrasting biodiversity and dynamics. *Ecosystems, 1*, 361–373.

71. Reid, B., Astorga, A., Madriz, I., Correa, C., & Contador, T. (2023). *A conservation assessment of freshwater ecosystems in southwestern Patagonia*. Springer.

72. Rivera, A., Aravena, J. C., Urra, A., & Reid, B. (2023). *Chilean Patagonian glaciers and environmental change*. Springer.

73. Robertson, G., Moreno, C. A., Lawton, K., Arata, J., Valencia, J., & Kirkwood, R. (2007). An estimate of the population sizes of black-browed (*Thalassarche melanophrys*) and grey-headed (*T. chrysostoma*) albatrosses breeding in the Diego Ramírez archipelago, Chile. *Emu, 107*, 239–244.

74. Robertson, G., Wienecke, B., Suazo, C. G., Lawton, K., Arata, J. A., & Moreno, C. (2017). Continued increase in the number of black-browed albatrosses (*Thalassarche melanophris*) at Diego Ramírez, Chile. *Polar Biology, 40*(5), 1035–1042.

75. Rovira, A., & Herreros, J. (2016). *Clasificación de ecosistemas marinos chilenos de la Zona Económica Exclusiva*. Departamento de Planificación y Políticas en Biodiversidad, División de Recursos Naturales y Biodiversidad, Ministerio del Medio Ambiente de Chile. Retrieved from: https://mma.gob.cl/wp-content/uploads/2018/03/Clasificacion-ecosistemas-marinos-de-Chile.pdf

76. Rozzi, R., Armesto, J., Goffinet, B., Buck, W., Massardo, F., Silander, J., Kalin-Arroyo, M., Russell, S., Anderson, C. B., Cavieres, L., & Callicott, J. B. (2008). Changing lenses to assess bio-diversity: Patterns of species richness in sub-Antarctic plants and implications for global conservation. *Frontiers in Ecology and the Environment, 6*, 131–137.

77. Rozzi, R., Armesto, J. J., Gutiérrez, J., Massardo, F., Likens, G., Anderson, C. B., Poole, A., Moses, K., Hargrove, G., Mansilla, A., Kennedy, J. H., Willson, M., Jax, K., Jones, C., Callicott, J. B., & Kalin, M. T. (2012). Integrating ecology and environmental ethics: Earth stewardship in the southern end of the Americas. *BioScience, 62*(3), 226–236.

78. Rozzi, R., Massardo, F., Anderson, C., Heidinger, K., & Silander, J., Jr. (2006). Ten principles for biocultural conservation at the southern tip of the Americas: The approach of the Omora Ethnobotanical Park. *Ecology and Society, 11*(1), 43.

79. Rozzi, R., Massardo, F., Mansilla, A., Squeo, F. A., Barros, E., Contador, T., Frangopulos, M., Poulin, E., Rosenfeld, S., Goffinet, B., González-Weaver, C., MacKenzie, R., Crego, R. D., Viddi, F., Naretto, J., Gallardo, M. R., Jiménez, J. E., Marambio, J., Pérez, C., et al. (2017). *Parque marino Cabo de Hornos - Diego Ramírez. Informe técnico para la propuesta de creación.* Ediciones Universidad de Magallanes.

80. Rozzi, R., & Sherriffs, M. (2003). El visón (*Mustela vison*, Schereber) un nuevo mamífero exótico para la isla Navarino. *Anales del Instituto de la Patagonia (Chile), 31*, 97–104.

81. Rozzi, R., & Torres-Mura, J. R. (1990). Observaciones del chungungo (*Lutra felina*) al sur de la isla grande de Chiloé, antecedentes para su conservación. *Medio Ambiente, 11*, 24–28.

82. Rozzi, R. (2017). La cumbre austral de América. In C. Aldunate, B. Lira, H. Rodríguez, R. Rozzi, & L. Santa Cruz (Eds.), *Cabo de Hornos* (pp. 24–59). Colección Santander, Museo de Chileno de Arte Precolombino.

83. Rozzi, R., & Feinsinger, P. (2001). Desafíos para la conservación biológica en Latinoamérica. In R. Primack, R. Rozzi, P. Feinsinger, R. Dirzo & F. Massardo (Eds.), *Fundamentos de conservación biológica: Perspectivas latinoamericanas* (pp. 661–688). Fondo de Cultura Económica.

84. Rozzi, R., Massardo, F., Berghoefer, A., Anderson, C., Mansilla, A., Mansilla, M., Plana, J., Berghoefer, U., Barros, E., & Araya, P. (2006). *The Cape Horn Biosphere Reserve.* Punta Arenas, Chile: Ediciones Universidad de Magallanes.

85. Rozzi, R., Jiménez, J. E., Massardo, F., Torres-Mura, J. C., & Rijal, R. (2014). The Omora Park Long-Term Ornithological Research Program: Study sites and methods. In R. Rozzi & J. E. Jiménez (Eds.), *Magellanic Subantarctic ornithology: First decade of forest bird studies at the Omora Ethnobotanical Park, Cape Horn Biosphere Reserve* (pp. 3–40). UNT Press— Ediciones Universidad de Magallanes.

86. Scher, H. D., Whittaker, J. M., Williams, S. E., Latimer, J. C., Kordesch, W. E., & Delaney, M. L. (2015). Onset of Antarctic Circumpolar Current 30 million years ago as Tasmanian Gateway aligned with westerlies. *Nature, 523*(7562), 580.

87. Schüttler, E., Crego, R., Saavedra, L., Silva, E. A., Rozzi, R., Soto, N., & Jiménez, J. J. (2019). New records of invasive mammals from the sub-Antarctic Cape Horn archipelago. *Polar Biology, 42*, 1093–1105.

88. Schüttler, E., Cárcamo, J., & Rozzi, R. (2008). Diet of the American mink *Mustela vison* and its potential impact on the native fauna of Navarino island, Cape Horn Biosphere Reserve, Chile. *Revista Chilena de Historia Natural, 81*, 599–613.

89. Schüttler, E., Klenke, J., McGehee, S., Rozzi, R., & Jax, K. (2009). Vulnerability of ground-nesting waterbirds to predation by invasive American mink in the Cape Horn Biosphere Reserve, Chile. *Biological Conservation, 142*, 1450–1460.

90. Segovia, R. A., & Armesto, J. J. (2015). The Gondwanan legacy in South American biogeography. *Journal of Biogeography, 42*, 209–217.

91. Sepúlveda, M. A., Bartheld, J. L., Monsalve, R., Gómez, V., & Medina-Vogel, G. (2007). Habitat use and spatial behavior of the endangered southern river otter (*Lontra provocax*) in riparian habitats of Chile: Conservation implications. *Biological Conservation, 140*(3–4), 329–338.

92. Sepúlveda, J. C., Pantoja, S., Hughen, K., Lange, C. B., González, F., Muñoz, P., Rebolledo, L. V., Castro, R. P., Contreras, S., Ávila, A. A., Rossel, P. E., Lorca, G., Salamanca, M., & Silva, N. B. (2005). Fluctuations in export productivity over the last century from sediments of a southern Chilean fjord (44° S). *Geology, 65*(3), 587–600.

93. Servicio Nacional de Pesca, Chile. (2018). *Anuario estadístico de pesca 2010–2016.* Servicio Nacional de Pesca. Ministerio de Economía, Fomento y Turismo. Retrieved from: http://www.sernapesca.cl/informes/estadisticas

94. Sielfeld, W., & Castilla, J. C. (1999). Estado de conservación y conocimiento de las nutrias en Chile. *Estudios Oceanológicos, 18*, 69–79.

95. Sielfeld, W. K. (1992). Abundancias relativas de *Lutra felina* (Molina, 1782) y *L. provocax* (Thomas, 1908) en el litoral de Chile austral. *Investigaciones en Ciencia y Tecnología Serie: Ciencias del Mar, 2,* 3–11.

96. Silva, N., & Calvete, C. (2002). Características oceanográficas físicas y químicas de canales australes chilenos entre el golfo de Penas y el estrecho de Magallanes (Crucero CIMAR-FIORDOS 2). *Ciencia y Tecnología del Mar, 25,* 23–88.

97. Spalding, M. D., Fox, H. E., Allen, G. R., Davidson, N., Ferdaña, Z. A., Finlayson, M., Halpern, B. S., Jorge, M. A., Lombana, A., Lourie, S. A., Martin, K. D., Mcmanus, E., Molnar, J., Recchia, C. A., & Robertson, J. (2007). Marine ecoregions of the world: a bioregionalization of coastal and shelf areas. *BioScience, 57,* 573–583.

98. Stokes, D. L., & Boersma, P. D. (1991). Effects of substrate on the distribution of magellanic penguin (*Spheniscus magellanicus*) burrows. *The Auk, 108,* 923–933.

99. Sullivan-Sealey, K., & Bustamante, G. (1999). *Setting geographic priorities for marine conservation in Latin America and the Caribbean.* The Nature Conservancy.

100. Tecklin, D. (2015). La apropiación de la costa chilena: Ecología política de los derechos privados en torno al mayor recurso público del país. In M. Prieto, B. Bustos, J. Barton (Eds.), *Ecología política en Chile: Naturaleza, propiedad, conocimiento y poder* (pp. 121–142). Editorial Universitaria.

101. Tecklin, D., Farías, A., Peña, M. P., Gélvez, X., Castilla, J. C., Sepúlveda, M., Viddi, F. A., & Hucke-Gaete, R. (2021). *Coastal-marine protection in Chilean Patagonia: Historical progress, current situation, and challenges.* Springer.

102. Valdenegro, A., & Silva, N. (2003). Caracterización oceanográfica física y química de la zona de canales y fiordos australes de Chile entre el estrecho de Magallanes y cabo de Hornos (CIMAR 3 Fiordos). *Ciencia y Tecnología Marina, 26*(2), 19–60.

103. Vargas, C. A., Aguilera, V., San Martin, V., Manríquez, P., Navarro, J., Duarte, C., Torres, R., Lardies, M., & Lagos, N. (2014). CO_2-driven ocean acidification disrupts the filter feeding behavior in Chilean gastropod and bivalve species from different geographic localities. *Estuaries and Coasts, 38,* 1163–1177.

104. Vargas, C. A., Cuevas, A., Silva, N., González, H., De Pol-Holz, R., & Narváes, D. (2017). Influence of glacier melting and river discharges on the nutrient distribution and DIC recycling in the southern Chilean Patagonia. *Journal of Geophysical Research. Biogeosciences, 123,* 256–270.

105. Vargas, C. A., Martínez, R. A., San Martín, V., Aguayo, M., Silva, N., & Torres, R. (2011). Allochthonous subsidies of organic matter across a lake–river–fjord landscape in the Chilean Patagonia: Implications for marine zooplankton in inner fjord areas. *Continental Shelf Research, 31,* 187–201.

106. Veblen, T. T., Hill, R. S., & Read, J. (Eds.). (1996). *The ecology and biogeography of Nothofagus forests.* Yale University Press.

107. Villagrán, C. (2018). Biogeografía de los bosques subtropical-templados del sur de Sudamérica. *Hipótesis históricas. Magallania, 46,* 27–48.

108. Vince, G. (2010). Dams for Patagonia. *Science, 329,* 382–385.

109. Vitousek, P. M., Mooney, H. A., Lubchenco, J., & Melillo, J. M. (1997). Human domination of Earth's ecosystems. *Science, 277*(5325), 494–499.

110. Wakefield, E., Phillips, R. A., Trathan, P. N., Arata, J., Gales, R., Huin, N., Robertson, G., Waugh, S. M., Weimerskirch, H., & Matthiopoulos, J. (2001). Habitat preference, accessibility, and competition limit the global distribution of breeding black-browed albatrosses. *Ecological Monographs, 81*(1), 141–167.

111. Waters, J. M., Dijkstra, L. H., & Wallis, G. P. (2000). Biogeography of a southern hemisphere freshwater fish: How important is marine dispersal? *Molecular Ecology, 49,* 1815–1821.

112. Weathers, K. C., Lovett, G. M., Likens, G. E., & Caraco, N. F. (2000). Cloudwater inputs of nitrogen to forest ecosystems in southern Chile: Forms, fluxes, and sources. *Ecosystems, 3*(6), 590–595.

113. Zelaya, D. G. (2009). Gastropoda - Gasterópodos. In V. Häussermann, G. Försterra (Eds.), *Fauna marina bentónica de la Patagonia chilena* (pp. 461–504). Nature in Focus.

Ricardo Rozzi Ecologist and philosopher. Director of the Subantarctic Biocultural Conservation Program. Professor, Universidad de Magallanes, Chile and University of North Texas, USA. Director and principal researcher of the Cape Horn International Center (CHIC). Adjunct scientist, Cary Institute of Ecosystem Studies, Millbrook, New York, USA.

Sebastian Rosenfeld Researcher at Universidad de Magallanes, Chile and Cape Horn International Center (CHIC). PhD student of the EBE program, Universidad de Chile. His line of research is focused on taxonomy, systematics and ecology of high latitude marine mollusks.

Juan J. Armesto Ph.D., Botany and Plant Physiology, Rutgers University, New Jersey, USA. Full Professor, Pontificia Universidad Católica de Chile. Researcher at the Institute of Ecology and Biodiversity (IEB). Adjunct scientist, Cary Institute of Ecosystem Studies, Millbrook, New York, USA.

Andrés Mansilla Phycologist, full professor at the Universidad de Magallanes, Chile and Cape Horn International Center (CHIC), Subantarctic Biocultural Conservation Program. Laboratory of Antarctic and sub-Antarctic macroalgae (LEMAS).

Mariela Núñez-Ávila Biologist, Pontificia Universidad Católica de Chile. Master of Science, Universidad de Chile. Ph.D., in Forestry Sciences, Universidad Austral de Chile. Researcher IEB. Director of Senda Darwin Biological Station, Ancud, Chiloé, Chile.

Francisca Massardo Agricultural Engineer, University of Chile. Plant Physiologist. President, Cape Horn Subantarctic Center Foundation. Director, Omora Ethnobotanical Park, Centro Universitario Puerto Williams. Professor, Universidad de Magallanes, Chile, and Principal Researcher, Cape Horn International Center (CHIC).

Part VI

Freshwater and Cryosphere Ecosystems

A Conservation Assessment of Freshwater Ecosystems in Southwestern Patagonia

14

Brian Reid, Anna Astorga Roine, Isaí Madriz, Cristián Correa, and Tamara Contador

Abstract

Freshwater ecosystems support the highest biological diversity per unit area on the planet, despite occupying much less area than terrestrial and marine systems. Freshwater organisms are also among the most threatened worldwide. There is an urgent need to identify and conserve remaining pristine and/or vulnerable regions, and Patagonia is an excellent candidate for conservation of a significant area of relatively unaltered ecosystems. The goal of this chapter is to synthesize and evaluate the most relevant information for the conservation of westward-draining Patagonian freshwater ecosystems (41–55° S), describing the general habitat, aquatic biodiversity, ecosystem function and ecosystem services. A significant portion of Chile's freshwater resources are concentrated within this zone—ice fields, some of the world's largest, deepest, and clearest lakes, together with major rivers draining into one of the world's most extensive coastal-marine systems. Although species richness is not high, a significant portion of taxa are unique (genus and family endemism), especially fish, amphibians, and crustaceans. Impacts and threats to Patagonia's freshwater ecosystems are still limited, hence freshwater conservation efforts

B. Reid (✉) · A. A. Roine
Centro de Investigaciones en Ecosistemas de La Patagonia (CIEP), Coyhaique, Chile
e-mail: brian.reid@ciep.cl

B. Reid · A. A. Roine · T. Contador
Centro Internacional Cabo de Hornos (CHIC), Universidad de Magallanes, Puerto Williams, Chile

I. Madriz
Fulbright–National Geographic Fellowship, Coyhaique, Chile

C. Correa
Facultad de Ciencias Forestales y Recursos Naturales, Instituto de Conservación, Biodiversidad y Territorio (ICBTe), Universidad Austral de Chile, Valdivia, Chile

Centro de Humedales Río Cruces (CEHUM), Universidad Austral de Chile, Valdivia, Chile

© Pontificia Universidad Católica de Chile 2023
J. C. Castilla et al. (eds.), *Conservation in Chilean Patagonia*, Integrated Science 19,
https://doi.org/10.1007/978-3-031-39408-9_14

357

in the region are promising. Despite the overall limited habitat degradation, many key taxa, especially fish and amphibians, are in danger of extinction. Invasive salmonids native to the Northern Hemisphere are considered both the most significant impact and threat to native freshwater species in the region. We propose seven general recommendations for freshwater conservation prioritization and planning, including a systematic revision of biodiversity information, dedicated personnel for sustained biological inventories, establishment of legal conservation mechanisms for river corridors, and a geographic survey of potential freshwater ecological refuges from both biological invasions and climate change. Longer-term conservation recommendations also include the development of plans for restoration and species recovery, and citizen-science observations and atlas-based inventories of groups of conservation interest. Finally, we emphasize the importance of supporting regional and national level experts on a range of taxonomic groups, both in terms of a professional knowledge base and also as the technical foundation for public/social investment in conservation efforts.

Keywords

Southwestern Patagonia • Chile • Reference ecosystems • Fluvial networks • Regional endemism • Ecosystem services • Aquatic invasive species • Wild and scenic river • Biodiversity

1   Introduction

Until the end of the last century, Patagonia tended to evoke images of the Argentine pampas, while a vast area west of the Andes between Puerto Montt and the Strait of Magellan, with fjords and canals, extensive temperate rainforests, lakes, rivers, and ice fields, was largely ignored. The name Chilean (western) Patagonia is recent, and this is perhaps a fitting preamble to the state of knowledge of the freshwater ecosystems of Chilean Patagonia, which remains incipient. This lack of knowledge is paradoxical, considering the ecological and social importance of these systems that we are now beginning to appreciate.

Southwestern Patagonia is home to the largest rivers west of the Andes, some of the largest lakes on the continent and among the deepest in the world, as well as the most extensive temperate ice fields on the planet. Most of Chile's fresh water is located in Patagonia. Other unique features include active volcanoes, volcanic soils, limited effect of global atmospheric contaminants, convergence of biogeographic provinces including Gondwanic elements, and low species richness but high endemism. Therefore, these ecological systems may have a globally unique composition and function.

Nonetheless, Patagonian freshwater ecosystems are poorly represented in global inventories [2, 73, 92]. There are serious deficiencies in the basic description of Patagonian freshwater ecosystems with respect to their hydrology and ecosystem function. There are only sporadic biodiversity records and distribution ranges tend to be underestimates. The gaps in knowledge correspond to a shortage of researchers who work in this Patagonian area. For example, for an analogous region, the book "Freshwaters of New Zealand" [66] had the contribution of 64 specialists, many of whom are recognized worldwide. In Chilean Patagonia (Fig. 1), a territory three times the size of New Zealand, there are only a handful of researchers working on freshwater systems.

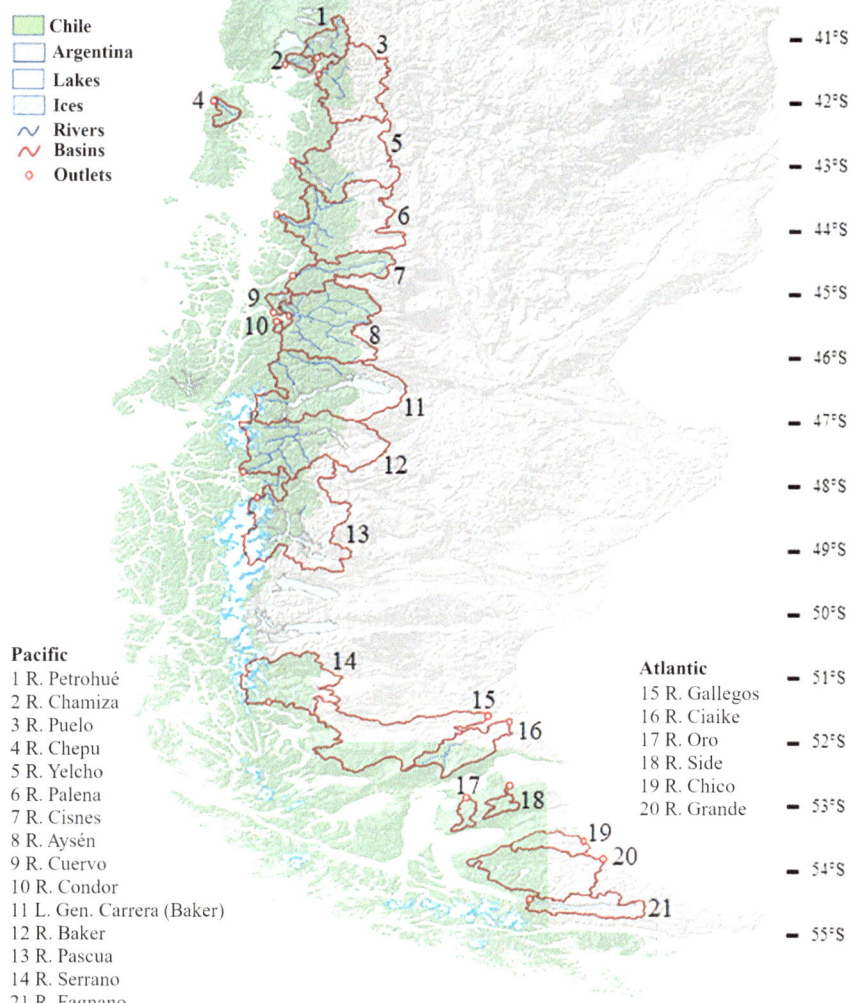

Fig. 1 Physical map of southern Patagonia showing the main Pacific and Atlantic watersheds referred to in this chapter. The study area, referred to as southwestern Patagonia, includes all watersheds (only the major basins are delineated) draining into the Pacific Ocean or its channels and fjords. Note the high incidence of trans-Andean and binational watersheds typical of the study area

In spite of the small area occupied by freshwater ecosystems, they harbor a biodiversity and proportion of species of conservation concern an order of magnitude greater than terrestrial or marine environments [48, 127]. Almost 30% of all at-risk taxa (designated of conservation concern) in the Los Lagos, Aysen, and Magallanes regions are freshwater species. Meanwhile, there is a great contrast between global vs. regional threats to freshwater biodiversity,only half of the emerging threats (sensu Reid et al. [111]) appear to be relevant to Patagonian ecosystems. Many of the threats noted for freshwater fish in Chile are applicable to populated areas (33°–41° S). This leads to a crucial point in terms of prioritizing and planning conservation in Chilean Patagonia—impacts and threats to aquatic ecosystems and conflicts over water resources are still relatively minor, so conservation efforts have a higher probability of success. This goes hand in hand with the observation that an important part of the diversity of amphibians, native fish, and aquatic insects in Chile is found within the geographic range of western Patagonia.

Chilean Patagonia has been recognized as an area of global importance for conservation [25, 92], and is also one of the few regions on the planet that already has >50% of its territory legally protected, the cornerstone of the half-earth concept promoted by Wilson [150] for biodiversity conservation. This chapter also emphasizes the less prominent part of Wilson's model, which focuses on the efficient and sustainable use of the other half of the that lies outside protected areas: this perspective is necessary in terms of the conservation of freshwater ecosystems, and Patagonia is no exception. This chapter is therefore complementary to Astorga et al. [13], in relation to the overarching priority of protecting headwater streams and undisturbed forested watersheds. Here we also emphasize the conservation status of Patagonian biodiversity and freshwater ecosystems located downstream, and often outside protected areas. The purpose of this chapter is to consolidate most relevant information and evaluate from different perspectives the conservation of freshwater systems in southwestern Patagonia, and to provide some strategic recommendations.

2   Scope and Objectives

We emphasize the importance of a hydrographic definition of the conservation of Patagonian freshwater systems, one based on drainage basins, rather than a political regionalization. The freshwater ecosystems of southwestern Patagonia constitute a unique area of often binational watersheds that drain to the Pacific Ocean via an intricate system of fjords and channels, from Chiloé to Cape Horn. Our study area comprises the basins with western slopes south of 41° S, which is predominantly Chilean, but also Argentinean (Fig. 1), which drain into and exert a tremendous influence on these interior seas, an area which we will refer to hereafter as southwestern Patagonia.

In this chapter we advance the idea that this territory offers excellent opportunities for the conservation of biodiversity and freshwater ecosystems at the landscape scale. Recommendations for improving the conservation of freshwater systems, include water resource policy research, planning actions, management, and education, with examples of conservation initiatives and tools used in other similar regions of the world.

3 Freshwater Ecosystems of Southwestern Patagonia

3.1 Western Patagonian Hydrography: Classification, Distribution, and National Importance

Southwestern Patagonia represents most of Chile's water resources. According to the Chilean Water Atlas (Directorate General of Water, 2016; in Spanish Dirección General de Aguas), almost all waterbody types or forms of water resources in the study area (Los Lagos, Aysen, and Magallanes Regions) place Patagonia as the area with the highest percentage contribution to the country's total water resources (Table 1). Some freshwater systems are treated in greater depth in other chapters: Rivera et al. [116] for glaciers, and Mansilla et al. [84] for wetlands.

Despite the division of freshwater ecosystems across chapters, we emphasize that glacier and wetland systems are an integral part of the water systems that feed, regulate, and interact with the water bodies discussed here. The following is a summary of the general characteristics of freshwater systems in the Patagonian study area.

3.1.1 Lakes and Ponds

Southwestern Patagonia has some of the largest lakes in Chile, and the second largest lake in South America (Lake General Carrera/Buenos Aires). These lakes are generally deep (>100 m). Because of the rugged Andean topography of this area and its glacial-tectonic origin, there are also two of the ten deepest lakes in the world (O'Higgins at 810 m and General Carrera at 580 m), although they do not appear in the global lists [73]. Binational and trans-Andean lakes, whose names

Table 1 Importance for Chile of the main water resources of southwestern Patagonia*

Indicator	Metrics	Proportion of national total (%)
Precipitation	>70 km^3/year	68
Number of lakes	184	50
Total lake area	>4600 km^2	64
Number of ponds	10,461	84
Total pond area	>3228 km^2	83
River runoff	>690 km^3/year	75
Glacial water equivalent**	>3100 km^3	98

*Source Chilean Water Atlas [46];
**Volume of water stored in solid state

differ between Chile and Argentina (Fig. 1), exclusive to western Patagonia, often appear truncated on official maps, and are hence poorly underappreciated in the public policy sector [46, 93]. Only a handful of the 184 major lakes identified in southwestern Patagonia have been subject to basic field observations [28, 29]. These lakes are thought to be oligotrophic or ultra-oligotrophic, due to the scarcity of nutrients and the depth of mixing induced by strong winds [51, 123]. Ponds, on the other hand, are 50 times more abundant than lakes [90], although a complete catalogue does not yet exist. They are usually shallow, dominated by littoral habitat, abundant aquatic vegetation, rich in organic matter and highly productive. They provide critical habitat for birds, amphibians, and invertebrates. Most ponds are connected to drainage networks, but there are also endorheic or isolated systems. Consequently, there are ponds with and without fish, the presence or absence of which may have a profound impact on the trophic web and biodiversity [102, 124]. Some of the endorheic ponds are hypersaline, and have a distinctive ecological function [57, 124, 153].

Apart from these generalizations, and in contrast with over 20 years of lake monitoring from 41° S (Lake Llanquihue) northward [46], background information on lakes in southwestern Patagonia is very scarce.

3.1.2 Rivers

Six of Chile's largest rivers (width > 300 m; flow > 600 m^3/s) are located in southwestern Patagonia: Puelo, Yelcho, Palena, Aysén, Baker, and Pascua rivers (Fig. 1, Table 2). Although the river networks in the region are several hundred thousand kilometers in length, the individual rivers are relatively short (<300 km), because the Andes range lies so close to the coast. Perhaps for this reason, the rivers of this region tend to be invisible at the continental level. For example, the World Wildlife Fund inventory [2] only lists the water bodies on the Patagonia ecoregion's eastern slope (in Argentina), which have longer and more visible trajectories but less flow.

From the headwaters in Patagonian peaks, ice fields and arid zones, Patagonian rivers often traverse mountain foothills and low-gradient coastal plains, ultimately discharging into fjords; the full range of fluvial habitats present in the temperate zone is probably represented in the Patagonian region. These contrasting landscapes in a relatively small area are perhaps what most distinguishes southwestern Patagonia river systems. At the same time, the variability, complexity, and network structure of river ecosystems pose a major challenge for the conservation of freshwater biodiversity in Chile.

Southern Patagonia dominates the total runoff in the western Andes [92]. Since the coastal landscape of southwestern Patagonia has an intricate arrangement of archipelagos, fjords, and inland waters, the coupling of river discharges to marine ecosystem function is perhaps more intense than anywhere else in the world, a feature that is just beginning to be appreciated [78, 133, 132]. This coastal zone has the highest rainfall in Chile and among the highest globally and has innumerable coastal watersheds that also contribute diffuse inputs (e. g. not recorded in Table 2). For example, the Madre de Dios Archipelago, with up to 9,000 mm rainfall per

Table 2 Attributes of the main watersheds of southwestern Patagonia

River	Region	Lat. S	Watershed (km^2)	Mean annual flow (m^3/s)	Length (km)
Chamiza	Los Lagos	41.4	346	55	40
Petrohué	Los Lagos	41.1	2,210	284	36
Puelo	Los Lagos	41.6	8,817	680	120
Yelcho	Los Lagos	43.1	10,979	735	330
Palena	Los Lagos/Aysén	43.9	12,887	820	240
Cisnes	Aysén	44.7	5,196	230	160
Aysén	Aysén	45.4	11,674	630	185
Baker	Aysén	47.4	26,726	1,010	170
Pascua	Aysén/Magallanes	48.1	14,760	690	62
Serrano	Aysén/Magallanes	51.2	7,347	170	38

* Historical average flows 1960–1985 from the Ministry of Public Works, 1987 (in Spanish Ministerio de Obras Públicas), from the lowest available fluviometric station

year, can produce the equivalent of 25% of the runoff of the Baker River basin which has tenfold greater surface area.

3.1.3 Ground Water

Groundwater reserves in saturated sediments and rock fissures located in geological formations known as aquifers, are both sources of freshwater flows and also habitat for freshwater organisms. Groundwater resources are better characterized and more exploited in central and northern Chile than in Patagonia. The aquifer inventory of the [46] does not include any aquifers in the Aysén and Magallanes regions. However, it is clear that aquifers are abundant in southwestern Patagonia, because of the extensive distribution of unconsolidated Quaternary material (fluvial and moraine deposits). Because of this, the aquifers of the region probably receive greater contribution from local precipitation, in contrast to regional flow typical of the central Andes. Even though their potential distribution is very heterogeneous, these aquifers probably constitute the second most important water reservoir in southwestern Patagonia, comparable to global patterns in freshwater distribution. Evidence of the widespread importance of ground water may be seen in the continuous flow of streams during periods without precipitation (base flow) and the upwelling of water in flood plains and headwater streams. The aquifers of western Patagonia also have social importance, shown by hundreds of groundwater rights or springs granted to individuals by the General Water Directorate.

Ground water as a habitat may be characterized by geochemistry [120], transmissivity, and biological community. In the Northern Hemisphere, highly transmissive fluvial and post-glacial sediment systems support unique communities of invertebrates that spend part or all of their lives in this dark environment [125]. Although similar geomorphology is present in western Patagonia, we are not aware of any studies on aquifer ecosystems in southern South America.

3.2   Freshwater Biodiversity

The freshwater biodiversity in southwestern Patagonia stands out for its uniqueness despite relatively low richness [70], with Gondwanan elements and endemism at the level of genus and family [134]. This region is a potential refuge for fish, amphibians, crustaceans, and insects, but there is limited knowledge of essential aspects, especially range/distribution. It is interesting to note that this territory is flanked to the north by the Valdivian forests of Chile, and to the south by the Cape Horn Biosphere Reserve [117], both considered to be global hotspots of terrestrial biodiversity.

Several recurrent problems became evident during the preparation of this synthesis of the distribution, habitat, and conservation value of freshwater species: unsystematic and incomplete biodiversity inventories; scarce information on geo-referenced observations in the literature; lack of knowledge on total ranges of species distributions or fragmentary distributions, poorly determined habitat for aquatic phases of the life cycle, limited information on migrations, population status, and representation in protected wild areas, and lack of local assessment of threats. At the end of this chapter, we propose some measures to address these deficiencies.

3.2.1   Aquatic Invertebrates

Aquatic invertebrates in southwestern Patagonia have high levels of endemism, for example, for the orders Plecoptera, Ephemeroptera, Trichoptera, Gastropoda, and Crustacea [139]. Although richness appears to be somewhat lower than that found in the Valdivian region (according to Valdovinos [139]), this general pattern may also be a result of the lack of information [34], augmented by logistical obstacles to biological sampling. Our main argument is that southwestern Patagonia conserves a significant percentage of Chilean freshwater macroinvertebrates in a relatively undisturbed, sparsely populated landscape with an important network of nature reserves. The southernmost distribution of several orders and families of aquatic insects is also found in this area [95].

Of the 47 species of dragonflies (Odonata) in Chile, 25 are documented in the study area [27, 68, 96]. There is one endemic family (Neopetalia) and one Gondwanic family (Austrapetaliidae), and seven families have endemic genera. Of the 63 species of stoneflies (Plecoptera) in Chile [143], 54 are in southwestern Patagonia, with one endemic family (Diamphipnoidae), four families with Gondwanic elements, and one threatened species (*Andiperla willinki*). Thirty-four of the 40 species of Chilean mayflies (Ephemeroptera) have also been recorded, including three families with Gondwanic elements and genus-level endemism. Among crustaceans, fairy shrimp (Anostraca) have four species and one endemic genus [41]. Other crustaceans such as amphipods (Amphipoda: *Hyallela* spp.), freshwater crabs (Anomura: *Aegla* spp.), and crayfish (Decapoda: Parastacidae) have species endemic to Patagonia, some in the threatened category (five species of *Aegla*, two of Parastacidae crayfish).

The "dragon of Patagonia" stonefly (Plecoptera: *Andiperla willinki*) belongs to an endemic genus and is also an indicator of an extreme ecosystem associated with ice fields (cryophilic species; [75, 128]). A groundwater crustacean documented only from the Simpson River (Crustacea: *Stygocaris patagonicus*, Noodt, 1963, type locality, Coyhaique) may be indicative of unique ecosystems in the region. Groundwater fauna are otherwise virtually explored in Chile [42], and there are possibly many other unknown and endemic species. Fairy shrimps are often emblematic of salt ponds and semi-permanent pools, which are vulnerable globally [22].

3.2.2 Fish

With a modest list of 19 freshwater fish species (Table 3, Fig. 2), southwestern Patagonia brings together ancient lineages and assemblages with diverse evolutionary and biogeographic origins. On one hand, the distribution of species in the study area includes the southernmost and most differentiated ichthyologic provinces (high endemism) within the Austral Neotropical sub-region: the Chilean Andean-Cuyana and Patagonian provinces [8, 82, 87]. On the other hand, there are two families of wide southern intercontinental distribution and of apparent Gondwanic origin (Galaxiidae and Parcichthyidae; [10, 23]). Most of these species (Table 3, Fig. 2a) are threatened in Chile or Argentina, and a high proportion are endemic to the southern Patagonian biogeographic provinces (84%) [49]. There are endemic genera, including *Aplochiton* (3 spp.; EN), *Hatcheria* (*H. macraei*; VU), and *Percichthys* (*P. trucha*; LC).

Patagonian fish are also ecologically and functionally diverse. There are strictly freshwater (e.g. *Galaxias platei* and *P. trucha*); diadromous (*Aplochiton* spp., *Galaxias maculatus,* and *Geotria australis*) and estuarine species (*Mugil cephalus* and *Eleginops maclovinus*). There are benthivorous (*O. hatcheri*), planktonic (*G. maculatus*), piscivorous (*A. marinus, G. platei, P. trout*) and omnivorous (*E. maclovinus*) species; although trophic niches are usually variable and change during ontogeny ([37, 101]). There are tiny (<5 cm*; B. bullocki, Ch. australe*), small (<15 cm; *G. maculatus, T. aerolatus*) and large fish (>30 cm; *G. platei, A. marinus, A. microlepidotus, O. hatcheri, P. trout*); introduced and invasive salmonids in the region are the largest and most voracious fish.

Southwestern Patagonia offers excellent opportunities for the conservation of these fish and for improving our knowledge of them. For example, species unknown for the region have recently been documented [8, 137], one of which appears to be restricted to Chilean Patagonia [4]. However, the fauna of significant areas remains unexplored, especially coastal areas, archipelagos, and headwater basins with difficult access.

Phylogeographic studies reveal the resilience of some taxa to prehistoric processes such as the uplift of mountain ranges and glaciations, followed by recolonization associated with climate shifts [152]. Contemporary patterns of genetic diversity and population connectivity, which are key to the design of conservation strategies, are almost unknown, but recent studies reveal a variable degree of

Table 3 Fish associated with inland waters of southern Chilean Patagonia. Colors correspond to conservation status within the latitudinal distribution (on the right): red (EN), orange (VU), yellow (NT), green (LC) and gray (not determined). Information adapted from the conservation sheets of the Ministry of the Environment; in Spanish Ministerio del Medio Ambiente, MMA (www.mma.gob.cl)

Scientific Name	Common name	Cat.	Habitat									Threats						
			Lake-Pelagic	Pond	Littoral	River	Stream	Flood Plain/Riparian	Wetland	Estuarine	Marine	Habitat Loss	Invasive Species	Harvest	Water Quality	Water Stress	Unknown	
Aplochiton marinus	Peladilla	EN	X	X	X	X				X	X	X						
Aplochiton taeniatus	Peladilla	EN	X	X	X	X	X			X	X	X	X	X	X			
Aplochiton zebra	Peladilla	EN	X	X	X	X	X					X	X					
Galaxias platei	Puye grande	LC	X	X	X	X	X					X						
Mordacia lapicida	Lamprea de agua dulce	EN				X	X					X	X		X			
Percichthys trucha	Perca trucha	LC	X	X	X	X						X	X		X			
Geotria australis [1]	Lamprea de bolsa	VU				X	X				X						X	
Galaxias maculatus	Puye	LC	X	X	X	X	X	X	X	X		X	X	X				
Odontesthes hatcheri	Pejerrey patagónico	NT	X	X		X	X					X				X		
Olivaichthys viedmensis [1]	Bagre aterciopelado	?	X		X	X						X	X					
Hatcheria macraei	Bagre patagónico	VU	X		X	X	X	X				X						
Galaxias globiceps	Puye	EN				X						X	X		X			
Cheirodon australe	Pocha del sur	VU		X	X			X	X				X		X			
Brachygalaxias bullocki	Puyecito	VU		X	X		X	X	X			X			X			
Basilichthys microlepidotus [4]	Pejerrey chileno	NT	X	X	X	X			X			X	X		X	X		
Trichomycterus areolatus	Bagre pintado	VU	X		X	X	X					X	X					
Odontesthes mauleanum	Cauque	VU	X		X	X	X			X				X	X	X		
Mugil cephalus	Lisa	LC			X				X	X	X				X	X		
Odontesthes brevianalis	Cauque del norte	VU		X		X			X	X	X	X	X			X		

(continued)

Table 3 (continued)

		Latitudinal Distribution (°S)[1]																			
		35	36	37	38	39	40	41	42	43	44	45	46	47	48	49	50	51	52	53	54	55
33																						
32																						
30																						
30																						
32																						
30																						
33																						
31																						
30																						
23																						
29																						

[1]The latitudes that define southwestern Patagonia are highlighted in bold

[2]*Basilichthys australis* was replaced by *B. microlepidotus* [147]

[3]During the editing of this chapter, Riva-Rossi et al. [115] revalidated Geotria macrostoma for Argentina, but individuals of this species have also been found in the Magallanes Region (Correa, pers. obs.)

[4]The MMA database incorrectly assigns *Diplomystes camposensis* (EN) to Region XI based on a report of *Diplomystes* sp. [97], which is currently assigned to the genus *Olivaichthys* [9]

isolation and genetic diversity among populations, associated with fluvial connections and disconnections [97], diadromy [43] and presence of invasive salmonids [140, 144]. In the current climate change scenario [85], it is important to safeguard the natural potential for genetic change that could affect the long-term survival of Patagonian fish species [21].

Native fish embody the main biodiversity crisis of freshwater systems, and perhaps of all ecosystems in the region. By virtue of low anthropogenic pressure, southwestern Patagonia is already to some extent a haven for the conservation of a unique ichthyofauna. However, proactive interventions and a better conservation policy are needed. Paradoxically, the main cause of this crisis, the invasive

Fig. 2 Distribution of inland water fish by habitat type and conservation category (**a**) and threat category (**b**). The photos on the right are from Lake Cochrane in the Baker basin (area of high biodiversity of native fish, see text); (**c**) large puye *Galaxias platei*; (**d**) catfish *Hatcheria macraei*. Conservation status categories: Endangered (EN), Vulnerable (VU), Near Threatened (NT), Least Concern (LC)

salmonids (Fig. 2b; see Sect. 3.4.) enjoy a level of cultural appreciation and legal protection that hinders the conservation of native fishes.

3.2.3 Amphibians

There are at least 21 species of amphibians in the area covered in this chapter (Table 4), accounting for more than half of those known in the country, many with family and genus endemism. There are four species with extreme endemism, which have been found only in one locality. Lack of knowledge of the geographic range of most amphibians is the greatest impediment to establishing their conservation status; the scarce local information is an obstacle to assessing the viability of populations and their metapopulation dynamics. Aquatic habitat use is often undescribed (Fig. 3b) compared to terrestrial habitat occupied by the adult phase (Fig. 3a).

The aquatic habitats most frequently mentioned for adult phase are streams and occasionally ponds, however, eggs and tadpoles are not mentioned in most cases. Threats to amphibians are associated with their low population size and distribution, habitat loss (i.e. deforestation), introduction of exotic species (Fig. 3c), and emerging diseases such as chytridiomycosis (*Batrachochytrium dendrobatidis*), which is causing a devastating pandemic for the world's amphibians [17]. Many

Table 4 Amphibians associated with inland waters of southern Chilean Patagonia. Consolidated information from the Ministry of the Environment's conservation sheets (www.mma.gob.cl), Chilean conservation status and nomenclature updated based on Correa [38]. Note: official Chilean national biodiversity information is not updated (*e.g.* at least four potential taxa are not included here: *Alsodes coppingeri, A. verrucosus, Ateleognathus salai and Batrachyla fitzroya*)

Scientific name (Species)	Common name	Cat.	Habitat adult phase										Habitat				
			Evergreen Forest	Deciduous Forest	Grasslands	Peatlands	River	Stream	Lake	Pond	Temporary Pools	Wetland	Unknown	Stream	Flooded Zones	Marsh	Lake
Nannophryne variegata	Patagonian toad	LC	x			x					x		x				
Batrachyla antartandica	Marbled wood frog	LC	x			x							x	x		x	
Chaltenobatrachus grandisonae	Puerto Edén frog	DD	x		x								x			x	
Alsodes kaweshkari	Kawésqar spiny chested frog	DD	x			x							x				
Eupsophus calcaratus	Chiloé Island ground frog	LC	x										x				
Alsodes australis	Southern spiny chested frog	NT	x	x									x	x			
Batrachyla nibaldoi	Nibaldo's wood frog	NT	x							x	x		x		x		
Hylorina sylvatica	Emerald forest frog	LC	x	x					x	x	x		x				x
Pleurodema bufoninum	Gray four-eyed frog	NT		x	x							x	x		x		
Batrachyla leptopus	Gray wood frog	LC	x										x	x			
Pleurodema thaul	Chilean four-eyed frog	NT					x	x		x			x				
Rhinoderma darwini	Darwin's frog	EN	x					x									
Rhinella spinulosa	Warty toad	LC		x	x			x					x				
Batrachyla taeniata	Banded wood frog	NT								x	x	x	x				
Eupsophus emiliopugini	Emilio's ground frog	LC	x												x		
Eupsophus roseus	Rosy ground frog	VU	x					x					x	x			
Alsodes gargola	Tonchek spiny-chested frog	EN		x									x	x			
Atelognathus ceii	Río Negro frog	DD		x	x								x				
Alsodes monticola	Island spiny-chested frog	NT	x	x		x		x			x		x				
Rhinella rubropunctata	Red-spotted toad	VU	x	x									x				
Calyptocephalella gayi	Helmeted water toad	VU						x		x		x	x	x		x	

(continued)

Table 4 (continued)

Tadpole				Threats							35	36	37	38	39	40	41	42	43	44	45	46	47	48	49	50	51	52	53	54	55
Pond	Temporary Pools	Littoral	Other/Unknown	Limited Range/Population	Habitat Loss	Invasive Species	Harvest	Disease	Water Quality	Water Stress																					
				x																											
	x			x																											
				x																											
				x																											
				x	x																										
				x																											
	x	x		x		x																									
x	x	x		x		x																									
				x																											
				x					x		21																				
		x		x	x		x																								
											33																				
x	x	x		x					x		32																				
				x	x	x																									
				x	x				x																						
				x		x																									
x				x																											
				x																											
				x	x				x																						
x				x	x	x			x	x	29																				

[1]Latitude that define southwestern Patagonia are highlighted in bold

amphibian populations probably depend on freshwater systems without fish, especially without introduced salmonids [74, 102, 129]. This suggests the potential value of isolated naturally fishless freshwater systems for freshwater conservation.

3.2.4 Birds

There are 23 species of aquatic birds in conservation categories, although observations are often scarce in the conservation platform of the Ministry of the Environment, and global databases such as "eBird" often have sporadic coverages that concentrates on the coast of Aysén and touristic areas such as Puerto Natales (e.g. Garay et al. [58]). The specific aquatic habitat associations are not well described, and there is little information on migratory patterns [31, 62] or

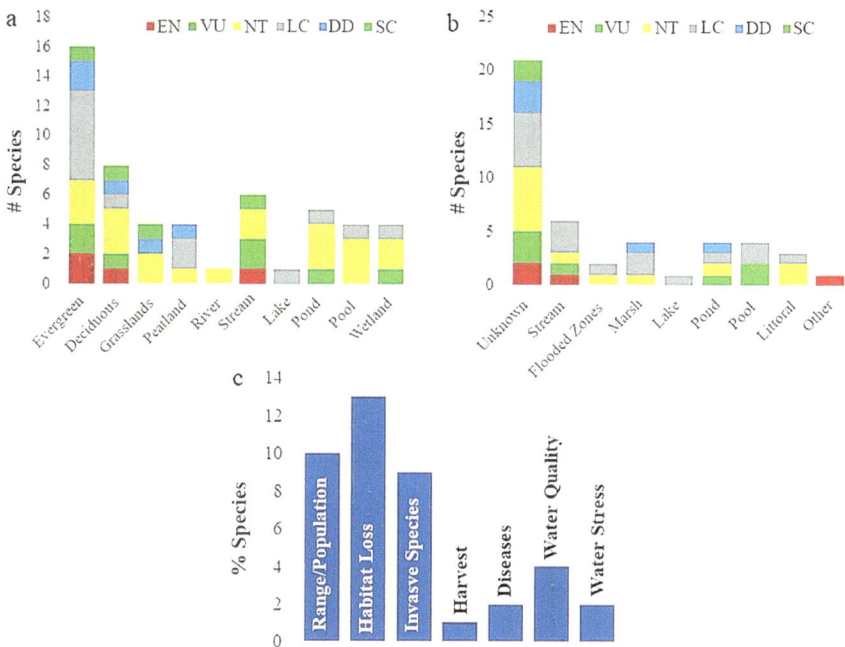

Fig. 3 Distribution of amphibians associated with inland waters by habitat type and conservation category of the adult stage (**a**), tadpole stage (**b**) and threat category (**c**). Conservation status categories as in Fig. 2

nesting of colonial birds (e.g. yeco, heron, bandurria). There are three principal threats to aquatic birds in Patagonia: (i) predation by exotic species (the most mentioned are mink and salmonids; [54, 64, 72]), (ii) loss of their food source (e.g. piscivorous species, [76]), (iii) habitat loss [83].

The species most at risk of extinction, about which there is somewhat more knowledge, is the Ruddy-headed goose *Chloephaga rubidiceps* (considered endangered), with losses of >90% of its population in the last century. It was recently estimated that there is a 50% extinction risk within a decade and/or three generations [39]. Although there is relatively comprehensive and coordinated management between Chile and Argentina, predation by introduced species demands permanent active intervention. The work of Cossa et al. [39] may be the best potential example of a recovery plan, with a comprehensive treatment of life cycle, habitat, threats, management, while noting information gaps. Another aquatic bird of conservation importance is the Magellan Plover *Pluvianellus socialis*, the only migratory and endemic species of the southern zone, but in this case without a designated conservation status or sufficient information or observations for a general conservation assessment.

3.2.5   Mammals

The huillín (*Lontra provocax*), or river otter, inhabits inland waters with abundant riparian vegetation, from approximately 38°–56° S. During the period 1910–1954 huillín suffered from heavy hunting pressure. There has been limited evidence of recovery, despite large areas of habitat that could be potentially suitable for the huillín, based on the availability of macro-crustaceans in rivers [32]. Genetic studies have confirmed low diversity (genetic bottleneck) in inland freshwater sub-populations, compared to coastal populations south of the Taitao Peninsula [145]. A reduction of up to 50% of the current river and lake population sizes is predicted over the next 30 years for huillín [121]. It is important for the conservation of the huillín that the management and protection of riparian environments be included as part of conservation measures for environmental protection in the region.

3.3   Ecosystem Services of Southwestern Patagonian Freshwater Systems

The benefits that an ecosystem provides to human society, in terms of provisioning, supporting/biodiversity, regulating, or cultural, are collectively referred to as ecosystem services (ES). Since the Millennium Ecosystem Assessment [91], the economic and sociocultural valuation of ES has become prominent in territorial planning and national environmental governance, including Chile. However, despite the great importance of freshwater ES, less than 10% of studies in Chile focus on fresh waters or water resources [14–16] and there is no synthesis of ES for Patagonia (although see the terrestrial classification of Martínez-Harms and Gajardo [86]). Valuation of ES in the planning stage of conservation strategies is especially relevant in populated areas, where multiple pressures and interests, and complex balance of provision and demand for ES is intensified. Conversely, it is noted that the main limitation of an ES framework applied to southwestern Patagonia is that there is limited local demand for ES in remote and unpopulated areas, which could affect the sustainability of this approach compared to other conservation measures.

Appreciation of ES is useful in cases where there are complex relationships between multiple ES or conflicts of ES valuation. For example, wild populations of introduced salmonid fish is perceived by many as beneficial, as a food source in rural areas and recreational fishing opportunity that translates into local economic dividends meanwhile others may view salmonids as the most significant threat to native ecosystem integrity [65, 146]. This sort of dichotomy in an ESs context may trigger rather than resolve conflicts, because actions that favor some ES may diminish others. However, it may be possible to find intermediate solutions, such as managing the density of introduced salmonids, which may favor coexistence with native species, simultaneously improving the quality of recreational fishing (robust fish) and the conservation of native biodiversity [36].

Freshwater, terrestrial and marine systems and their ES are interconnected, from the runoff from small headwater basins affected by livestock grazing or forestry practices, to the provision of potable water in lowlands, and transport of materials/nutrients or terrestrial sourced contaminants, to productive marine areas and fjords. Conservation of headwater streams and watersheds without intervention [13] can be justified by ES values related to water quality, carbon sequestration, or biodiversity. While dense forest cover favors carbon sequestration and habitat provision, water storage and supply depend on the geographic context, populations, and productive uses downstream. The hydrological effect of mature forest in terms of water balance vs. vegetation development is, however, complex, conflictive, and often inconclusive (e.g. [53, 55, [61]). Responsible application of ES as an environmental management tool generally involves finding agreements that maximize material, immaterial, and regulatory benefits among different sectors of society, while specifically adding a directional upstream–downstream freshwater context.

These examples illustrate the need to incorporate non-economic criteria in the valuation of ES. It has been recognized recently that certain reductionist implementations of ES translate only into material (e.g. monetary) values, which has motivated the adoption of more holistic and inclusive methods in the United Nations Intergovernmental Platform on Biodiversity and Ecosystem Services (IPBES) [52]. An initial strategy would be to strengthen scientific understanding of the ES functioning of Patagonian ecosystems, and to identify stakeholders and socioecological systems that benefit from the diverse ES of the region (e.g. [117]).

3.4   Impacts, Stressors, and Threats to Freshwater Ecosystems

Patagonia is an anomaly with regard to the threats to the diversity of freshwater systems. Perhaps only five of the 12 emerging global threats for freshwater systems [111] are relevant to the region: climate change, invasive species, disease, harmful algal blooms (HAB), and hydrological alteration (dams). There is a general lack of information on other threats in the region, such as emerging contaminants, nanomaterials, microplastics, light and noise impacts, salinization, calcium reduction, and accumulative stressors, but we assume that these are minor threats compared to those in other parts of the world. Climate change is among the most disruptive, especially in terms of a threat multiplier. The effect on terrestrial ecosystems in Chilean Patagonia [85] may be distinct from the effects on freshwater systems. Changes in water balance affect river flows and water temperature, water levels in lakes, and connectivity between water bodies. An increase in the frequency and duration of minimum flows is expected, such as all-time minimum recorded flows observed during the 2015–2016 ENSO episode [59]. High extreme events are also expected, due to changes in rainfall intensity and patterns in the rain/snow transition. However hydrologic trends in the region have been subject to very limited analysis, despite the existence of data for many large rivers along the climatic gradient in southwestern Patagonia [59].

Another generalization with respect to climate change is that the effect on freshwater organisms may be more acute than in other ecosystems, mainly due to four factors: (i) warming of water bodies may be disproportionate to increases in air temperature [105]; (ii) stenothermic species (which cannot regulate body temperature through their metabolism) have little capacity to withstand an increase in temperature [33], (iii) dispersal options are more constrained; (iv) many anthropogenic impacts are concentrated near water bodies [151].

Although the thermal niche of most Patagonian species is unknown, some native species (*G. platei, P. trout* and *O. hatcheri*) tolerate heat better than salmonids and could benefit from global warming [19, 21]). Regional and global studies have not yet demonstrated a temperature increase in Chile's southern lakes [105, 108]. Possible explanations for this anomaly include the large volume and thermal mass of these lakes, the compensatory effect of accelerated melting of ice and snow and strong winds that deepen the mixing layer.

Invasive species present the most direct and imminent threat and impact to the freshwater ecosystems of Chilean Patagonia. Invasive salmonids such as rainbow trout (*Oncorhynchus mykiss*), chinook salmon (*O. tshawytscha*) and brown trout (*Salmo trutta*) represent major challenges for the conservation of native species in Patagonia [63, 106]. They are also considered to be among the most harmful invasive species worldwide [26].

Ironically, much of the knowledge of freshwater systems in southwestern Patagonia is based on studies of invasive species: salmonids [36, 101, 103]; floodplain plants such as willow and lupine [80, 90, 124]; American beaver as an ecosystem engineer [5–7], and the invasive diatom *Didymosphenia geminate* [24, 113, 111]. Hence, studies of species composition, trophic structure, and functioning of these unique ecosystems have almost always been overwhelmed by the impact of invasive species that are difficult to control and almost impossible to eradicate. There is an urgent need to identify, study, and protect hydrologically isolated areas that can still provide ecological refugia, such as islands and areas upstream of hydrologic barriers such as waterfalls [36, 65, 112].

Exotic pathogens are another type of potentially harmful invasive species. Of greatest current concern is the chytrid fungus *B. dendrobatidis* (Bd), which causes the panzootic (*cf.* pandemic) chytridiomycosis of amphibians, in turn is responsible for the largest recorded reduction in biodiversity worldwide attributable to a single pathogen [118]. Chytridiomycosis is considered an emerging epidemic in Chile, it has been documented in a dozen localities between 41°–46° S [17],twelve of the 14 amphibian species evaluated in Chilean Patagonia were Bd+[17]. Another future threat that could affect humans is giardiasis, caused by a parasite associated with invasive beavers (beaver fever, [50]).

The increase in HABs and their adverse effects on marine ecosystems is discussed by Marquet et al. [85]. A global increase in cyanobacterial blooms in freshwater bodies is expected, due to the combined effect of climate change, land use change, and species introductions [119]. Toxins from new and potentially invasive cyanobacteria species have appeared in Chile in recent years [98]. There are currently no studies or monitoring of this phenomenon in Chilean Patagonia, although there are anecdotal observations of unusual blooms of unknown algae during the El Niño period in 2015–16 (Servicio Nacional de Pesca y Acuicultura, Aysén, personal communication).

Hydroelectric power plants have a great impact on aquatic systems globally, altering the flow regime, reducing sediment and nutrient inputs [126], and acting as barriers to the migration of aquatic species. Water regime alteration due to dams is a gray area of knowledge and uncertain threat in southwestern Patagonia. Proposals for mega hydroelectric projects in Chilean Patagonia have been met by strong social opposition, hydroelectric plants are currently scarce and of low impact. The dam in Los Alerces National Park (Argentina), with a flow regime that alternates with high frequency between minimum and maximum flow, affects the flow of the Futaleufú River in Chile. In this case, the operation of a hydroelectric power plant can have a damaging effect downstream far beyond the footprint of the reservoir.

An impact that is still a mild stressor in southwestern Patagonia should be added to the global list: atmospheric pollution of nitrogen, sulfur, heavy metals, and persistent organic compounds (POPs; [3, 18]). Mercury and POPs have been recorded in Patagonian fish [11, 94]. Although Patagonia is currently one of the areas of the world with the least atmospheric pollution [44, 117], increases in atmospheric nitrogen deposition are predicted, the impact of which can be transformative for low productivity freshwater systems [51, 107].

While emerging threats (sensu Reid et al. [111]) are not as evident in southwestern Patagonia, local impacts, stressors, and threats are evident, with potentially significant cumulative effect over the longer term. Physical habitat alteration, unsustainable water use, pollution, and altered trophic status of water bodies are impacts and threats in Patagonia, as in other regions of the world.

The drastic transformation of the Patagonian landscape during the period of contemporary colonization and fires is discussed in Astorga et al. [13]. Downstream, where land use and impacts have historically been more intense, water bodies have probably borne the brunt of eroded soils, affecting the aquatic food web simultaneously by changing light (primary production) and organic carbon input (allochthonous energy sources). There is a lack of local documentation of these effects, although they are known in freshwater systems worldwide.

Riparian vegetation strips are frequently absent due to extensive cattle and sheep ranching. Valleys and slopes converted to grasslands have a higher risk of sediment entrainment into streams and large rivers. Slopes converted to plantations of exotic species to control erosion have been shown to have very negative effects on the water cycle [76–79], although the proportion of the landscape affected by these plantations in the region is small compared to central regions of Chile. Direct impacts include channel modifications for river defenses and aggregate extraction.

These impacts are often located near urban areas or the Southern Highway (in Spanish Carretera Austral). The extraction of sand and gravel in certain reaches used for recreational fishing is currently generating acute conflicts in Coyhaique. Unregulated encroachment on flood plains in urban areas (B. Reid, personal observation) may generate future flood control problems. However, compared to other areas of Chile, current hydroelectric project planning [135, 136] shows a low level of intervention in the morphology of the main rivers of the Patagonian region: Puelo, Yelcho, Palena, Cisnes, Aysén, Baker, and Pascua.

Conflicts over water use and rights are not yet common in Chilean Patagonia as compared to central Chile [46]. Nor is the impact of pollution on watercourses a serious problem [141], except in urban areas, reaches affected by fish farm effluent, and drainage from mine tailings (B. Reid, unpublished data).

Mining exploration has increased in recent years; however, although specific projects have not yet materialized. Organic contaminants are probably of low intensity, considering the current low-intensity state of agriculture and low urban population densities.

Patagonia is at a crucial point regarding freshwater conservation high biodiversity and reference ecosystems coincides with levels of impacts, stressors, and threats that are still slight. Most of the local threats and impacts recorded in fish conservation datasheets (Sect. 3, Fig. 2b) are based on regions with greater urbanization and economic development (33°–41° S) and are not as applicable to the geographic area considered in this chapter. Meanwhile, a significant portion of native amphibians and fish, and the vast majority of aquatic insect species, have ranges that extend to the less intervened areas of southwestern Patagonia.

Local management options are limited with respect to global stressors such as climate change and acute impacts such as invasive species. However, there are many local opportunities exist in terms of habitat protection, best practices, restoration, and mitigation of land use impacts and stressors. Taken together, conservation planning at both scales is relevant to the principle of safe operating space [119]: local actions to offset global stressors should be prioritized.

3.5 Parameters and Tools for the Conservation of Freshwater Ecosystems in Southwestern Patagonia

The conservation of freshwater ecosystems presents special challenges related to connectivity, water flow directionality, snowmelt regime dynamics, and hydrological disturbance. Below, we highlight some of these challenges for species, communities, and ecosystems, and discuss some knowledge gaps and shortcomings, as well as management of the National System of State Wild Protected Areas (SNASPE in Spanish), and watershed management.

3.5.1 Distribution and Conservation of Freshwater Species

There are some key differences in the conservation of freshwater systems with respect to terrestrial systems. In the former, habitat delineation for species in conservation categories can be difficult due to small-scale spatial complexity, hierarchical organization, and directionality in movement and dispersal patterns ([67, 131]). This is even more complicated for the diadromous fish that move between fresh and sea water (e.g. family Galaxiidae, [88]), aquatic insects with aerial/terrestrial dispersal [131], and amphibians whose breeding and rearing sometimes occur in small, isolated water bodies that may be difficult to characterize or delineate. Connectivity between populations is restricted to drainage networks for fluvial species, with corresponding physical barriers such as waterfalls upstream or lakes/estuaries downstream. Knowledge about metapopulations is often essential for conservation purposes. Species-based freshwater conservation approaches require a solid knowledge base of connectivity and integrity of populations at the watershed scale [47, 112, 131]. River connectivity is also related to genetic isolation among populations. Conservation actions should focus on evolutionary blocks or lineages that make up species (evolutionarily significant units, [149]). Several studies on freshwater and diadromous fish [97, 140, 152], amphibians [105], and freshwater crustaceans [150] have revealed significant evolutionary units in southwestern Patagonia, consistent with the degree of population isolation and historical and present fluvial connectivity. Future efforts should target the conservation of more refined evolutionary units than species.

Niche-based models have been used where there is a lack of information on the distribution of native and invasive inland water species (Energy Center, 2016). This tool could be useful as a first filter to assess the future risk of invasive species [30]. However, the bioclimatic variables and land cover attributes used to develop niche models must distinguish between local (i.e. immediate vicinity of a water body) and watershed-level effects, especially where there are strong bioclimatic gradients. Although there are many alternatives for modeling the distribution of aquatic species, the limiting factor in Chilean Patagonia is the lack of direct observations to calibrate and validate these models.

3.5.2 Classification of Communities and Ecosystems

The typology of surface waters is important to regulate, monitor, and manage aquatic ecosystems in a systematic and/or representative manner. The proposed classification of freshwater ecosystems for Chile is based on fish ecoregions crossed with abiotic geomorphological, hydrological, physical, and chemical criteria [56]. However, this classification is especially uncertain for southwestern Patagonia, given the lack of observations of these same parameters of aquatic ecosystems. While we recognize the operational need for a classification, its refinement with field data is indispensable.

Reference systems are complementary to classification systems, also being key to conservation and restoration. This concept includes reference watersheds [47]. Astorga et al. [13] provided a systematic approach to mapping pristine headwater and low-order watersheds in southwestern Patagonia. Most downstream freshwater systems, such as medium and large lakes and rivers, lack a similar analysis of reference condition. For example, Lake Yulton, west of Puerto Aysén, may be among the few large lakes in the world without introduced fish [36], although there is currently no formal recognition of this value.

3.5.3 Site-Based Conservation

Conservation initiatives in parks and reserves need to integrate an aquatic perspective, complementary to the current terrestrial approach. The SNASPE tends to protect high elevation areas such as mountain ranges; however, the most productive freshwater ecosystems are often in low-lying areas excluded from reserves (Fig. 4). Mechanisms must be found to strengthen downstream conservation. The

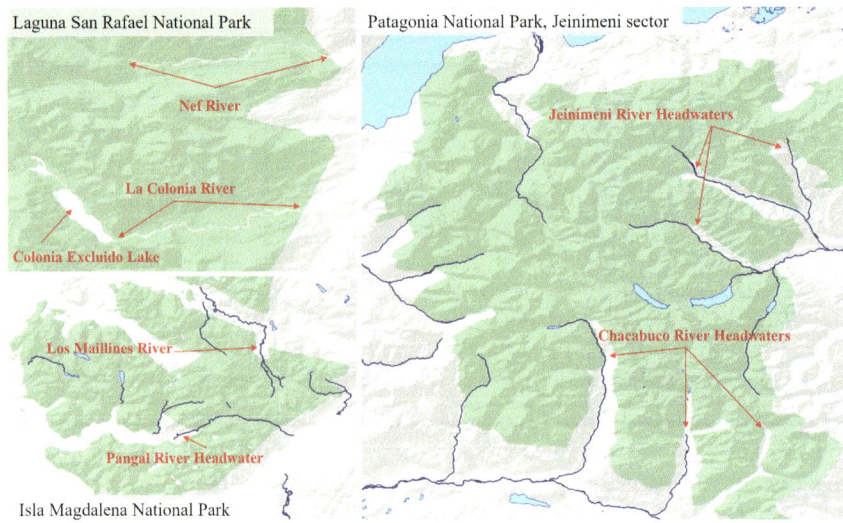

Fig. 4 Gaps in river conservation within the current system of protected areas. Three examples are presented: Laguna San Rafael National Park, Magdalena Island National Park, and Jeinimeni sector in Patagonia National Park. Note that SNASPE boundaries (in green) are imprecise and sometimes do not show private land. (e.g. most of the Nef River and Colonia River valleys are private)

designation of wild and scenic rivers (WSR) [60] was implemented in the USA in the 1960s to offset public incentives for river interventions for hydroelectricity and irrigation, among others. Simultaneously, the WSR designation is complementary to protected areas, aimed especially at river corridors, and may be designated by presidential decree, with the sponsorship of a public agency or local community initiatives.

Zone of a rapid on the Futaleufú River, Los Lagos Region. Photograph by Jorge Gerstle

3.5.4 Water Resources

Conservation aspects unique to freshwater ecosystems include water quality, quantity, and hydrologic regime. The Water Code, administered by the General Water Directorate, establishes the national framework regulating water use, water quality standards (Chilean Standard 1333, 409), the National Water Quality Monitoring Network, and the Lake Monitoring Network [46, 141]. Lake monitoring has not been officially implemented in southwestern Patagonia, despite the superabundance of lakes and being a distinct zone in the categories of Fuster (2015), as discussed above. Secondary standards for water quality have also not been developed, except for the Serrano River (Decree 75, 2010). Water quality is generally exceptional, corresponding with lower population density and limited industry in southern Chile [141].

Minimum ecological flows defined in the Water Code (Article 129 bis 1, Law 20.017) are an environmental management tool established to guarantee a minimum flow for maintenance of aquatic ecosystems. Restrictions on the granting of water use rights may preserve up to 20% of the average annual flow as ecological flow, or up to 40% in exceptional cases [114]. However, the application is not systematic and does not adhere to ecological criteria, and most importantly presumes stationarity (e.g. does not consider climate change trends). On the surface, the rivers of western Patagonia with limited existing conflicts over water rights and uses, might offer an exceptional opportunity for establishing ecological flows. However, river flows undergo changes that are difficult to predict in a climate change scenario, which suggests the need for a more conservative criterion. An evaluation of the Murta River sub-basin (Baker River basin; [45]) revealed that a potential ecological flow of 20% or even 40% would still be well below the minimum flows recorded in 25 years of monitoring and would permit the disruption of the natural hydrological regime. The problem is conceptually similar to mitigating the impact of dams and hydroelectric plants by emulating natural flows, aquatic species and their habitats responding to parameters other than averages, such as peak flows, and frequency and duration of floods [109]. While there are currently limited alterations to natural river flow in southwestern Patagonia, the evident impact of hydropeaking caused by an Argentine dam on the Futaleufú River could be minimized by applying similar strategy based on analysis of flow regime.

Another more promising legal instrument given the near reference flow conditions of Chilean Patagonia is the Water Reserve (in Spanish Reserva de Caudal) decrees (Article 147 bis, paragraph 3 of Law No. 20,017 of 2005). This measure allows the setting aside of unoccupied or otherwise unused water rights. The limitation of this tool is that it is used case by case, relies on presidential decree, is reversable and without binding long-term commitments. Conversely, southern Patagonia has the advantage of potential opportunities for application in regions of low anthropogenic intervention, and where unallocated flows are still considerable. In Chilean Patagonia Water Reserves have been applied to the Petrohué, Cochamó, Murta, Figueroa, and Del Oro rivers.

3.5.5   Integrated Watershed Management

Watershed conservation and management is based on a key distinction, which is that the conservation unit does not consist of patches (e.g. vegetation), but rather links terrestrial and aquatic ecosystems through a drainage network. There are few examples in Chile that can serve as a model. One of the main challenges is that differences between stakeholders and the complexity of value conflicts that can be overwhelming, especially for larger watersheds. Thus, for the moment, in southwestern Patagonia it seems more feasible to focus on smaller scales (e.g. rural drinking water systems) or to consider the individual elements that make up good watershed conservation practices—watershed conservation, water management, and water quality of intact headwaters, riparian buffer zones, floodplain conservation, and conservation planning/management of areas of high biodiversity [1, 47].

The importance of conserving the headwaters, the most degradation-sensitive part of drainage networks, is recognized globally [1, 47]. The initiative to map intact watersheds in Patagonia as a basis for freshwater conservation and watershed management is an innovative beginning [13]. Unlike other regions of Chile, in southern Patagonia the headwaters that supply public water provisioning systems are often located in private properties. This presents an opportunity to encourage the proactive development of best practices. For example, riparian vegetation corridors are recommended as a buffer measure in areas dominated by agriculture and forest management. The integrity of the buffer depends on its continuity, while its width may vary [105]. Riparian buffer corridors are included in forest management plans in Chile, however, they are poorly implemented. Headwaters are often in public lands in mountainous regions, but exceptions exist [12]. Downstream, the concept of corridors expands to floodplains, among the world's most threatened productive ecosystems [130], where large floods are received and buffered, sediment accumulates, and water is stored. In some countries, a public agency analogous to Chile's National Emergency Office (in Spanish ONEMI), are responsible for regulating activities such as extraction of sand and gravel and urbanization in this zone. In Chile these areas are administered by the Ministry of National Assets (in Spanish MBN, Decreto Supremo 609), but with limited assessment and oversight.

4   Conclusions and Recommendations

A freshwater friendly policy: challenges and recommendations

The extraordinary abundance of water resources in southwestern Patagonia contrasts with the scarcity of knowledge about their corresponding biological diversity, threats and impacts that affect them, and the lack of tools for their sustainable use and conservation. Fortunately, there is still time to move towards a future where freshwater ecosystems are recognized as an essential part of our well-being. This path requires cultural, social, and political changes that are difficult, but possible to achieve in the long term. With this goal in mind, we offer seven recommendations. Some are initiatives that can be implemented in the short term, with limited resources, while others are aimed at the medium and long term because of their greater complexity and need for political and legislative support. These first steps will help in the characterization, assessment, and protection of key freshwater ecosystems:

- **Systematization of biodiversity information**. There are huge gaps in the knowledge of freshwater biodiversity and its values, functions and services and current threats. The existing knowledge is mostly separated in different sources or formats, for example scientific versus gray literature, expert or cultural knowledge. We propose synthesis and systematization of biodiversity observations (not just freshwater) followed by the comprehensive and systematic generation of new observations, to facilitate the identification of priority areas, and the design and implementation of monitoring and conservation plans. This progress involves two complementary components: investment in human capital dedicated to the collection of information on biodiversity in the area, and a public platform dedicated to information management. A substantial part of the observations on global biodiversity comes from grey literature, museum collections, and unpublished work of experts. The dedicated work of experts with refined observational skills, taxonomic expertise, and knowledge of natural history would have the added benefit that these experts are also often skilled communicators of biodiversity values. Regarding the second component, we draw attention to the scarcity of geo-referenced information on freshwater species in the national conservation status classification system. Centralized information management is essential and complementary to the work of specialists in taxonomic groups. Finally, it is important to consider that such platforms require human resources for their maintenance, and incentives or mechanisms to solicit contributions from experts.

- **River connectivity inventory.** Considering the impact of invasive species on native species in western Patagonia, some key questions arise: where are the refuges for native species? What are the physical barriers to dispersal and invasion? We propose an inventory of barriers, principally natural (waterfalls, steep slopes) and a few anthropogenic (dams) within the river networks of the region, to identify potential refuges for aquatic biodiversity. The mapping we propose may be done based on a digital elevation model, complemented with on-site validation. These maps would facilitate the development of conservation and restoration plans for native fish and systems without fish (as refugia for amphibians and other aquatic fauna; [142]).

- **Recovery plans for the species in greatest danger.** A number of freshwater species in southwestern Patagonia are in danger of extinction, including fish of the genus *Aplochiton* (peladilla), the huillín or river otter, and several species of amphibians whose risk is defined by extremely limited known distribution. Without exception, knowledge of the life history and distribution of these species is fragmentary. Proactive strategies (i.e. recovery plans) for the conservation of native species are just beginning to implemented in Chile, and there are no such initiatives for at-risk freshwater species. Meanwhile, public agencies and existing policies have traditionally favored the propagation and care of introduced salmonids to the detriment of the protection of local biodiversity, including native fish [20]. The lack of knowledge and the use of inappropriate management practices is a dangerous combination that can compromise the future of our natural heritage. Our recommendation is to generate management and conservation plans for threatened species (e.g. [110]) with objectives determined by their population size and distribution, and to involve local stakeholders in monitoring, and adaptive management.

- **Wild and scenic rivers (WSR) model.** The WSR model is complementary to the system of public reserves for river sections that are not within the limits of the SNASPE (Fig. 4). This possibility is being discussed preliminarily in Chile, but there are currently no political or legal mechanisms for this region. Regional energy planning [135, 136] and the Reserved Flow concept could serve as base line for WSR proposals.

- **Ecological Flow versus National Water Reserve.** In a territory where a substantial part of the flows of large rivers remains in the hands of the Chilean State or effectively unused despite allocation, the legally binding safeguard of Ecological Flows of up to 20–40% of the average annual flow seems arbitrary and insufficient (see above, Sect. 3.5.4). The existing Water Reserve decree or Reserved Flows represents an imperfect option, but the best currently available to guarantee more secure and comprehensive protection, especially combined with Ecological Flow. An analysis based on the Murta River [45], in which the procedure for determining a reserved flow is based on the probability of exceedance, conserving up to 80% of the average flow, would represent an exceptional yet feasible precedent for Chile, considering the opportunities of the region and the needs for ecosystems (i.e. it is complementary to WSR).

- **Binational collaboration in the management of natural resources in Patagonia.** Due to the importance of binational watersheds (Chile and Argentina), there is a need to facilitate binational communication and management of water resources in southwestern Patagonia. The first obstacle is political, as any interaction of public services is funneled to the national level. Second, Chile-Argentina binational research in this area is almost non-existent, perhaps because funding for international collaborations typically promotes opportunities with other countries much further away geographically.

- **Environmental standards for territorial planning and development.** Potential opportunities exist for mitigating impacts on freshwater systems, such as the Environmental impact Assessment System, regional development plans, and other public subsidies for agricultural or forest production. A major advance would be to have well-defined performance standards for road projects near watercourses. Another example is the incentive programs for agriculture (machinery, fertilizers, etc.) that may result in misapplication and contamination of watercourses over the long term. If such incentives were implemented together with conditional permitting such as buffer zones (vegetation strips) or fencing barriers to protect watercourses, they would be potentially effective in ensuring water quality conservation at the landscape scale.

References

1. Abell, R., Allan, J., & Lehner, B. (2007). Unlocking the potential of protected areas for freshwaters. *Biological Conservation, 34*, 48–63.
2. Abell, R., Thieme, M. L., Revenga, C., Bryer, M., Kottelat, M., Bogutskaya, N., Coad, B., Mandrak, N., Balderas, S. C., Bussing, W., Stiassny, M. L. J., Skelton, P., Allen, G. R., Unmack, P., Naseka, A., Ng, R., Sindorf, N., Robertson, J., Armijo, E., Petry, P., et al. (2008). Freshwater ecoregions of the world: A new map of biogeographic units for freshwater biodiversity conservation. *BioScience, 58*, 403–414.
3. Aguayo, P., Gonzáles, C., Barra, R., Becerra, J., & Martínez, M. (2014). Herbicides induce change in metabolic and genetic diversity of bacterial communities from a cold oligotrophic lake. *World Journal of Microbioogy and Biotechnology, 30*, 1101–1110.
4. Alò, D., Correa, C., Samaniego, H., Krabbenhoft, C., & Turner, T. (2017). Otolith microchemistry identifies diadromous populations of Patagonian river fishes. *PeerJ, 7*, e6149.
5. Anderson, C., Lencinas, M., Wallem, P., Valenzuela, A., Simanonock, M., & Pastur, G. (2014). Engineering by an invasive species alters landscape-level ecosystem function but does not affect biodiversity in freshwater systems. *Diversity and Distributions, 20*, 214–222.
6. Anderson, C., Pastur, G., Lencinas, M., Wallman, P., Moorman, M., & Rosemond, A. (2009). Do introduced North American beavers *Castor canadensis* engineer differently in southern South America? An overview with implications for restoration. *Mammal Reviews 39*, 33–52; Anderson, C., & Rosemond, A. (2007). Ecosystem engineering by invasive exotic beavers

reduces in-stream diversity and enhances ecosystem function in cape Horn, Chile. *Oecologia, 154*, 141–153.

7. Anderson, C., & Rosemond, A. (2010). Beaver invasion alters terrestrial subsidies to subantarctic stream food webs. *Hydrobiology, 652*, 349–361.

8. Arratia, G. (1983). Preferencias de hábitat de peces siluriformes de aguas continentales de Chile (Fam. Diplomystidae y Trichomycteridae). *Studies on Neotropical Fauna and Environment, 18*, 217–237.

9. Arratia, G., & Quezada-Romegialli, C. (2017). Understanding morphological variability in a taxonomic context in Chilean diplomystids (Teleostei: Siluriformes), including the description of a new species. *PeerJ, 5*, e2991.

10. Arratia, G., & Quezada-Romegialli, C. (2019). The South American and Australian percichthyids and perciliids. What is new about them? *Neotropical Ichthyology, 17*, e180102.

11. Arribere, M., Dieguez, M., Guevarra, S., Queimaliños, C., Fajon, V., Reissig, M., & Horvat, M. (2010). Mercury in an ultraoligotrophic north Patagonian Andean lake (Argentina): Concentration patterns in different components of the water column. *Journal of Environmental Science, 22*, 1171–1178.

12. Astorga, A., Moreno, P., & Reid, B. (2018). Watersheds and trees fall together: An analysis of intact forested watersheds in southern Patagonia (41–56 S). *Forests, 9*, 385.

13. Astorga, A., Moreno, P., Rojas, P., & Reid, B. (2023). *Conserving the origin of rivers: Intact forested watersheds in western Patagonia.* Springer.

14. Bachmann, P. (2009). *Comparación de la exportación de nitrógeno desde un ecosistema forestal versus un ecosistema pastoril, a través de la aplicación de modelos de simulación.* [Master Thesis]. Universidad de Chile.

15. Bachmann, P. (2013). *Ecosystem services modeling as a tool for ecosystem assessment and support for decision making processes in Aysén region, Chile (northern Patagonia).* [Master Thesis]. Christian-Albrechts-Universität.

16. Bachmann, P., Barrera, F., & Tironi, A. (2014). *Recopilación y sistemización de información relativa a estudios de evaluación, mapeo y valorización de servicios ecosistémicos en Chile.* Informe final Ministerio Medioambiente por Cienciambiental Consultores S.A. Retrieved from: http://bibliotecadigital.ciren.cl/handle/123456789/26106

17. Bacigalupe, L., Vásquez, I., Estay, S., Valenzuela-Sánchez, A., Alvarado-Rybak, M., Peñafiel-Ricuarte, A., Cunningham, A., & Soto-Azat, C. (2019). The amphibian-killing fungus in a biodiversity hotspot: Identifying and validating high-risk areas and refugia. *Ecosphere, 10.* e02724.

18. Barra, R., & Quiroz, R. (2011). Mountain ecosystems as a temporal sink for persistent organic pollutants. In Richards, K. (Ed.), *Mountain ecosystems: dynamics, management and conservation* (pp 79–92). Nova Science.

19. Barrantes, M. E., Lattuca, M. E., Vanella, F. A., & Fernández, D. A. (2017). Thermal ecology of Galaxias platei (Pisces, Galaxiidae) in South Patagonia: Perspectives under a climate change scenario. *Hydrobiologia, 802*, 255–267. https://doi.org/10.1007/s10750-017-3275-3.

20. Basulto, S. (2003). *El largo viaje de los salmones: una crónica olvidada, propagación y cultivo de especies acuáticas en Chile.* Editorial Maval Ltda.

21. Becker, L., Crichigno, S., & Cussac, V. (2018). Climate change impacts on freshwater fishes: A Patagonian perspective. *Hydrobiologia, 816*, 21–38.

22. Belk, D. (1998). Global status and trends in ephemeral pool invertebrate conservation: implications for californian fairy shrimp. In C. W. Witham, E. T. Bauder, D. Belk, W. R. Ferren Jr. & R. Ornduff (Eds.), *Ecology, Conservation, and Management of Vernal Pool Ecosystems—Proceedings from a 1996 Conference* (pp. 147–150). California Native Plant Society.

23. Burridge, C. P., McDowall, R. M., Craw, D., Wilson, M. V. H., & Waters, J. M. (2012). Marine dispersal as a pre-requisite for Gondwanan vicariance among elements of the galaxiid fish fauna. *Journal of Biogeography, 39*(2), 306–321. https://doi.org/10.1111/j.1365-2699.2011.02600.x

24. Bus, P., Cerda, J., Sala, S., & Reid, B. (2014). Mink (*Neovison vison*) as a natural vector in the dispersal of the invasive diatom *Didymosphenia geminata. Diatom Research, 29*(3), 259–266.

25. Callicott, J., Rozzi, R., Delgado, L., Monticino, M., Acevedo, M., & Harcome, P. (2007). Bio-complexity and conservation of biodiversity hotspots: Three case studies from the Americas. *Philosophical Transactions of the Royal Society, B, 362*, 321–333.
26. Cambray, J. (2003). Impact on indigenous species biodiversity caused by the globalization of alien recreational freshwater fisheries. *Hydrobiologia, 500*, 217–230.
27. Camousseight, A., & Vera, A. (2007). Estado del conocimiento de los Odonata (Insecta) de Chile. *Boletín del Museo Nacional de Historia Natural, Chile, 56*, 119–132.
28. Campos, H. (1996). *Estudios limnológicos de los lagos Elizalde y Riesco. Informe final.* Universidad Austral de Chile.
29. Campos, H. (1999). *Diagnóstico limnólogico de los principales lagos de la comuna de Coyhaique. Informe final.* Universidad Austral de Chile.
30. Campos, M., de Andrade, A., Kunzmann, B., Galvao, D., Silva, F., Cardoso, A., Carvalho, M., & Mota, H. (2014). Modeling of the potential distribution of *Limnoperna fortunei* (Dunker, 1857) on a global scale. *Aquatic Invasions, 9*, 253–265.
31. Canevari, P. (1996). The austral goose (*Chloephaga* spp.) of southern Argentina and Chile: A review of its current status. *Gibier Faune Savage, Game Wildlife, 13*, 355–366.
32. Cassini, M., & Sepúlveda, M. (2006). El huillín *Lontra provocax*: Investigaciones sobre una nutria patagónica en peligro de extinción. *Serie Fauna Neotropical, 1*, 162.
33. Comte, L., & Olden, J. (2017). Climatic vulnerability of the world's freshwater and marine fishes. *Nature Climate Change, 7*, 718–722.
34. Contador, T., Kennedy, J., & Rozzi, R. (2012). The conservation status of southern South American aquatic insects in the literature. *Biodiversity Conservation, 21*, 2095–2107.
35. Correa, C., & Gross, M. (2007). Chinook salmon invade southern South America. *Biological Invasions, 10*, 615–639.
36. Correa, C., & Hendry, A. P. (2012). Invasive salmonids and lake order interact in the decline of puye grande *Galaxias platei* in western Patagonia lakes. *Ecological Applications, 22*, 828–842.
37. Correa, C., Bravo, A. P., & Hendry, A. P. (2012) Reciprocal trophic niche shifts in native and invasive fish: Salmonids and galaxiids in Patagonian lakes. *Freshwater Biology, 57*(9), 1769–1781.
38. Correa, C. (2022). Lista viva de las especies de anfibios de Chile. Publicaciones RECH: www.herpetologiadechile.cl
39. Cossa, N., Fasola, L., Roesler, I., & Reboreda, J. (2016). Ruddy-headed goose *Chloephaga rubidiceps*: Former plague and present protected species on the edge of extinction. *Bird Conservation International, 27*, 269–281.
40. Cussac, V., Habit, E., Ciancio, J., Battinim, M., Rossi, C., Barriga, J., Baigun, C., & Crichigno, S. (2016). Freshwater fishes of Patagonia: Conservation and fisheries. *Journal of Fish Biology, 89*(1), 369–370.
41. De los Ríos-Escalante, P. (2010). Crustacean zooplankton communities in Chilean inland waters. *Crustaceana Monographs, 12*, 109 Pp. Brill.
42. De los Ríos-Escalante, P., Parra, L., Peralta, M., Perez-Schultheiss, J., & Rudolph, E. (2016). A checklist of subterranean water crustaceans of Chile (South America). *Proceedings of the Biological Society of Washington, 129*, 114–128.
43. Delgado, M., Górski, K., Habit, E., & Ruzzante, D. (2019). The effects of diadromy and its loss on genomic divergence: The case of amphidromous *Galaxias maculatus* populations. *Molecular Ecology, 28*, 5217–5231.
44. Dentener, F., Devet, J., Lamarque, J., Bey, I., Eickhout, B., Fiore, A., Hauglustaine, D., Horowitz, L., Krol, M., Kulshrestha, U., Lawrence, M., Gay-Lacaux, C., Rast, S., Shindell, D., Stevenson, D., Van Noije, T., Atherton, C., Bell, N., Bergman, D., Butler, T., Cofala, J., Collins, B., Doherty, Ellingsen, K., Galloway, J., Gauss, M., Montanaro, V., Müller, J., Pitari, G., Rodríguez, J., Sanderson, M., Solmon, F., Strahan, S., Schultz, M., Sudo, K., Szopa, S., & Wild, O. (2006). Nitrogen and sulfur deposition on regional and global scales: a multimodel evaluation. *Global Biogeochemical Cycles 20*: GB002672.

45. Dirección General de Aguas. (2009). *Informe técnico No 5. Reserva del río Murta para la conservación ambiental y el desarrollo local de la cuenca.* S.D.T No 208, Ministerio Obras Públicas, Gobierno de Chile. Retrieved from de: https://snia.mop.gob.cl/

46. Dirección General de Aguas. (2016). *Atlas del Agua de Chile.* Ministerio Obras Públicas. Retrieved from: http://bibliotecadigital.ciren.cl/handle/123456789/26705

47. Doppelt, B., Scurlock, M., Frissel, C., & Karr, J. (1993). *Entering the watershed.* Island Press.

48. Dudgeon, D., Arthington, A., Gessner, M., Kawabata, Z. I., Knowler, D., Leveque, C., Naiman, R., Prieur-Richard, A. H., Soto, D., Stiassny, M., & Sullivan, C. (2006). Freshwater biodiversity: Importance, threats, status and conservation challenges. *Biological Reviews, 81,* 163–182.

49. Dyer, B. (2000). Systematic review and biogeography of the freshwater fishes of Chile. *Estudios Oceanológicos (Chile), 19,* 77–98.

50. Dykes, A., Juranek, D., Lorenz, R., Sinclair, S., Jakubowski, W., & Davies, R. (1980). Municipal waterborne giardiasis: An epidemiologic investigation. *Annals of Internal Medicine, 92,* 165–170.

51. Díaz, M., Pedrozo, F., Reynolds, C., & Temporetti, P. (2007). Chemical composition and the nitrogen-regulated trophic state of Patagonian lakes. *Limnologica, 37,* 17–27.

52. Díaz, S., Pascual, U., Stenseke, M., Martín-López, B., Watson, R. T., Molnár, Z., Hill, R., Chan, Baste, K. M. A., Brauman, I. A., Polasky, K. A., Church, S., Lonsdale, A., Larigauderie, M., Leadley, A., Oudenhoven, P. W., van Plaat, A. P. E., van der Schröter, F., Lavorel, M., Aumeeruddy-Thomas, S., Bukvareva, Y., Davies, E., Demissew, K., Erpul, S., Failler, G., Guerra, P., Hewitt, C. A., Keune, C. L., H., Lindley, S., & Shirayama, Y. (2018). Assessing nature's contributions to people. *Science 359,* 270–272.

53. Farley, K., Jobbagy, E., & Jackson, R. (2005). Effects of afforestation on water yield: A global synthesis with implications for policy. *Global Change Biology, 11,* 1565–1576.

54. Fasola, L., & Valenzuela, A. (2014). Invasive carnivores in Patagonia: Defining priorities for their management using the American mink (*Neovison vison*) as a case study. *Ecología Austral, 24,* 173–182.

55. Frene, C., Dorner, J., Zuñiga, F., Cuevas, J., Alfaro, F., & Armesto, J. (2020). Eco-hydrological functions in forested catchments of southern Chile. *Ecosystems, 23,* 307–323.

56. Fuster, R., Escobar, C., Lillo, G., & de la Fuente, A. (2015). Construction of a typology system for rivers in Chile based on the European Water Framework Directive (WFD). *Environmental Earth Sciences, 73,* 5255–5268.

57. Gajardo, G., & Redón, S. (2019). Hypersaline lagoons from Chile, the southern edge of the world. In A. J. Manning (Ed.), *Lagoon environments around the world—A scientific perspective.* IntechOpen. Retrieved from: https://www.intechopen.com/books/lagoon-environments-around-the-world-a-scientific-perspective

58. Garay, G., Johnson, W., & Franklin, W. (1991). Relative abundance of aquatic birds and their use of wetlands in the Patagonia of southern Chile. *Revista Chilena de Historia Natural, 64,* 127–137.

59. Garreaud, R. (2018). Record-breaking climate anomalies lead to severe drought and environmental disruption in western Patagonia in 2016. *Climate Research, 74,* 217–229.

60. Gillian, D., & Brown, T. (1997). *Instream flow protection.* Island Press.

61. Goeking, S., & Tarboten, D. (2020). Forests and water yield: Synthesis of disturbance effects on streamflow and snowpack in western coniferous forests. *Journal of Forestry, 118,* 172–192.

62. Goldfeder, S., & Blanco, D. (2006). The conservation status of migratory waterbirds in Argentina: towards a national strategy. In G. Boere, C. Galbraith & D. Stroud (Eds.), *Waterbirds around the world* (pp. 189–194). The Stationery Office.

63. Habit, E., & Cussac, V. (2016). Conservation of the freshwater fauna of Patagonia: An alert to the urgent need for integrative management and sustainable development. *Journal of Fish Biology, 89,* 369–370.

64. Habit, E., González, J., Ortiz-Sandoval, J., Elgueta, A., Sobenes, C., Habit, E., González, J., Ortiz- Sandoval, J., Elgueta, A., & Sobenes, C. (2015). Efectos de la invasión de salmónidos en ríos y lagos de Chile. *Revista Ecosistemas, 24*(1), 43–51.

65. Habit, E. M., Piedra, P., Ruzzante, D. E., Walde, S. J., Belk, M. C., Cussac, V. E., González, J., & Colin, N. (2010). Changes in the distribution of native fishes in response to introduced species and other anthropogenic effects. *Global Ecology and Biogeography, 19*, 697–710.
66. Harding, J., Mosley, P., Pearson, C., & Sorrell, B. (Eds.). (2004). *Freshwaters of New Zealand.* Caxton Press.
67. Hawkins, C. P., Kershner, J. L., Bisson, P. A., Bryant, M. D., Decker, L. M., Gregory, S. V., McCullough, D. A., Overton, C. K., Reeves, G. H., Steedman, R. J., & Young, M. K. (1993). A hierarchical approach to classifying stream habitat features. *Fisheries, 18*, 3–11.
68. Heckman, C. (2006). *Encyclopedia of South American aquatic insects: Odonata-Anisoptera.* Springer.
69. Higgins, J., Lammert, M., Bryer, M., DePhilip, M., & Grossman, D. (1998). *Freshwater conservation in the Great Lakes basin: Development and application of an aquatic community classification framework.* Report to the George Gund Foundation: The Nature Conservancy, Great Lakes Program.
70. Hoekstra, J. M., Molnar, J. L., Jennings, M., Revenga, C., Spalding, M. D., Boucher, T. M., Robertson, J. C., Heibel, T. J., & Ellison, K. (2010). *The atlas of global conservation: Changes, challenges, and opportunities to make a difference.* University of California Press.
71. Hopkinton, C., & Valino, J. (1995). The relationship among man's activities in watersheds and estuaries: A model of runoff effects on patterns of estuarine community metabolism. *Estuaries, 18*, 598–621.
72. Ibarra, J., Fasola, L., Macdonald, D., Rozzi, R., & Bonadac, C. (2009). Invasive American mink *Mustela vison* in wetlands of the Cape Horn Biosphere Reserve, southern Chile: What are they eating? *Oryx, 43*(1), 87–90.
73. Kalff, J. (2002). *Limnology: Inland water ecosystems.* Prentice
74. Kats, L., & Ferrer, R. (2003). Alien predators and amphibian declines: Review of two decades of science and the transition to conservation. *Diversity and Distributions, 9*, 99–110.
75. Koshima, S. (1985). Patagonian glaciers as insect habitats. In K. Nakajima (Ed.), *Glaciological studies in Patagonia northern icefield 1982–1984* (pp. 94–99). Data center for Glacier Research, Japanese Society of Snow and Ice.
76. Lancelotti, J., Marinone, M., & Roesler, I. (2017). Rainbow trout effects on zooplankton in the reproductive area of the critically endangered hooded grebe. *Aquatic Conservation: Marine and Freshwater Ecosystems, 27*, 128–136.
77. Lewerentz, A., Eggers, G., Householder, E., Reid, B., Garofano-Gómez, V., & Braun, A. (2019). Functional assessment of invasive *Salix fragilis* L. in north-western Patagonian flood plains: A comparative approach. *Acta Oecologica, 95*, 36–44.
78. León-Muñoz, J., Urbina, M., Garreaud, R., & Iriarte, J. (2018). Hydroclimatic conditions trigger record harmful algal bloom in western Patagonia (summer 2016). *Scientific Reports, 8*, 1–10.
79. Little, C., Cuevas, J., Lara, A., Pino, M., & Schoenholtz, S. (2015). Buffer effects of streamside native forests on water provision in watersheds dominated by exotic forest plantations. *Ecohydrology, 8*, 1205–1217.
80. Little, C., & Lara, A. (2010). Restauración ecológica para aumentar la provisión de agua como un servicio ecosistémico en cuencas forestales del centro-sur de Chile. *Bosque, 31*(3), 175–178.
81. Little, C., Lara, A., McFee, J., & Urrutia, R. (2009). Revealing the impact of forest exotic plantations on water yield in large scale watersheds in south-central Chile. *Journal of Hydrology, 374*, 162–170.
82. López, H., Menni, R., Donato, M., & Miquelarena, A. (2008). Biogeographical revision of Argentina (Andean and Neotropical regions): An analysis using freshwater fishes. *Journal of Biogeography, 35*, 1564–1579.
83. Madsen, J., Tombre, I., & Eide, N. (2009). Effects of disturbance on geese in Svalbard: Implications for regulating increasing tourism. *Polar Research, 28*(3), 376–389.

84. Mansilla, C. A., Domínguez, E., Mackenzie, R., Hoyos-Santillán, J., Henríquez, J. M., Aravena, J. C., & Villa-Martínez, R. (2023). *Peatlands in Chilean Patagonia: Distribution, biodiversity, ecosystem services, and conservation.* Springer.

85. Marquet, P. A., Buschmann, A. H., Corcoran, D., Díaz, P. A., Fuentes-Castillo, T., Garreaud, R., Pliscoff, P., & Salazar, A. (2023). *Global change and acceleration of anthropic pressures on Patagonian ecosystems.* Springer.

86. Martínez-Harms, M., & Gajardo, R. (2008). Ecosystem value in the western Patagonia protected areas. *Journal of Nature Conservation, 16,* 72–87.

87. Matthews, W. J. (1998). *Patterns in freshwater fish ecology.* Springer.

88. McDowell, R. (2006). Crying wolf, crying foul, or crying shame: Alien salmonids and a biodiversity crisis in the southern cool-temperate galaxoid fishes? *Review of Fish Biology and Fisheries, 16,* 233–422.

89. Meier, C., Reid, B., & Sandoval, O. (2013). Effects of the invasive plant *Lupinus polyphyllus* on vertical accretion of fine sediment and nutrient availability in bars of the gravel-bed Paloma River. *Limnologica, 43,* 381–387.

90. Messager, M., Lehner, B., Grill, G., Nedeva, I., & Schmitt, O. (2016). Estimating the volume and age of water stored in global lakes using a geo-statistical approach. *Nature Communications, 7,* 13603.

91. Millennium Ecosystem Assessment. (2005). *Ecosystems and human well-being: synthesis,* Island Press. Retrieved from: http://www.millenniumassessment.org/en/Synthesis.aspx

92. Milliman, J., & Farnsworth, K. (2011). *River discharge to the coastal ocean.* Cambridge; Mittermeier, R., Mittermeier, C., Brooks, T., Pilgrim, J., Konstant, W., da Fonseca, G., & Kormos,C. (2003). Wilderness and biodiversity conservation. *Proceedings of the National Academy of Science, U.S.A., 100,* 10309–10313.

93. Ministerio de Obras Públicas, Dirección General de Aguas (1987). *Balance Hídrico de Chile.* Retrieved from: http://bibliotecadigital.ciren.cl/handle/123456789/12442

94. Montory, M., Habit, E., Fernandez, P., Grimalt, J., & Barra, R. (2010). PCBs and PBDEs in wild chinook salmon (*Oncorhymchus tshawytscha*) in the northern Patagonia, Chile. *Chemosphere, 78,* 1193–1199.

95. Moorman, M., Anderson, C., Gutiérrez, A., Charlin, R., & Rozzi, R. (2006). Watershed conservation and aquatic benthic macroinvertebrate diversity in the Alberto D'Angostini National Park, Tierra del Fuego, Chile. *Annals of the Institute of Patagonia (Chile), 34,* 41–58.

96. Muzon, J., Pessacq, P., & Lozano, F. (2014). The Odonata (Insecta) of Patagonia: A synopsis of their current statuswith illustrated keys for their identification. *Zootaxa, 4,* 346–388.

97. Muñoz-Ramírez, C., Unmack, P., Habit, E., Johnson, J., Cussac, V., & Victoriano, P. (2014). Phylogeography of the ancient catfish family Diplomystidae: Biogeographic, systematic and conservation implications. *Molecular Phylogenetics and Evolution, 73,* 146–160.

98. Nimptsch, J., Woelfl, S., Osorio, S., Valenzuela, J., Moreira, C., Ramos, V., Castelo-Branco, R., Leao, P., & Vasconcelos, V. (2016). First record of toxins associated with cyanobacterial blooms in oligotrophic north Patagonian lakes of Chile-a genomic approach. *International Review of Hydrobiology, 111,* 57–68.

99. Nowak, P., Norman, J., & Mulla, D. (2006). *Watershed-scale tools.* University of Wisconsin.

100. Núñez, J., Wood, N., Rabanal, F., Fontanella, F., & Sites, J. (2011). Amphibian phylogeography in the antipodes: Refugia and postglacial colonization explain mitochondrial haplotype distribution in the Patagonian frog *Eupsophus calcaratus* (Cycloramphidae). *Molecular Phylogenetics and Evolution, 58,* 343–352.

101. Ortiz-Sandoval, J., Gorski, K., González-Diaz, A., & Habit, E. (2015). Trophic scaling of Percichthys trucha (Percichthyidae) in monospecific and multispecific lakes in western Patagonia. *Limnologica, 53,* 50–59.

102. Ortubay, S., Cussac, V. E., Battini, M., Barriga, J. P., Aigo, J., Alonso, M., Macchi, P. J., Reissig, M., Yoshioka, J., & Fox, S. (2006). Is the decline of birds and amphibians in a steppe lake of northern Patagonia a consequence of limnological changes following fish introduction? *Aquatic Conservation: Marine and Freshwater Ecosystems, 16,* 93–105.

103. Ortíz-Sandoval, J., Gorski, K., Sobenes, K., Gonzalez, J., Manosalve, A., Elgueta, A., & Habit, E. (2017). Invasive trout affect trophic ecology of *Galaxias platei* in Patagonian lakes. *Hydrobiologia, 790*, 201–212.

104. Oyanedel, A., Valdovinos, C., Sandoval, N., Moya, C., Kiessling, G., Salvo, K., & Olmos, V. (2011). The southernmost freshwater anomurans of the world: Geographic distribution and new records of Patagonian aeglids (Decapoda: Aeglidae). *Journal of Crustacean Biology, 31*(3), 396–400.

105. O'Reilly, C., Sharma, S., Gray, D., Hampton, S., Read, J., Rowley, R., Schneider, P., Lenters, J., McIntyre, P., Kraemer, B., Weyhenmeyer, G., Straile, D., Dong, B., Adrian, R., Allan, M., Anneville, O., Arvola, L., Austin, J., Bailey, J., Baron, J., Brookes, J., Zhang, G., et al. (2015). Rapid and highly variable warming of lake surfacewaters around the globe. *Geophysical Research Letters, 42*, GL066235.

106. Pascual, M. A., Cussack, V., Dyer, B., Soto, D., Vigliano, P., Ortubay, S., & Macchi, P. (2007). Freshwater fishes in Patagonia in the 21st century after a hundred years of human settlement, species introductions, and environmental change. *Aquatic Ecosystem Health, 10*, 212–227.

107. Perakis, S., & Hedin, L. (2002). Nitrogen loss from unpolluted South American forests mainly via dissolved organic compounds. *Nature, 416*, 416–420.

108. Pizarro, J., Vergara, P., Cerda, S., & Briones, D. (2016). Cooling and eutrophication of southern Chilean lakes. *Science of the Total Environment, 541*, 683–691.

109. Poff, N., Allan, D., Bain, M., Karr, J., Prestegaard, K., Richter, B., Sparks, R., & Stromberg, J. (1997). The natural flow regime. *BioScience, 47*, 769–784.

110. Raadick, T. (1995). A research recovery plan for the barred galaxias in southeastern Australia. Department of Conservation and Natural Resources, Victoria. *Fauna and Flora Technical Report, 141*, 1–24.

111. Reid, A., Carlson, A., Creed, I., Eliason, E., Gell, P., Johnson, P., Kidd, K., MacCormick, T., Olden, J., Ormerod, S., Smol, J., Taylor, W., Tockner, K., Vermaire, J., Dudgeon, D., & Cooke, S. (2018). Emerging threats and persistent conservation challenges for freshwater biodiversity. *Biological Reviewsof the Cambridge Philosophical Society, 94*(3), 849–873.

112. Reid, B., Hernández, K., Frangopolis, M., Bauer, G., Lorca, M., Kilroy, C., & Spaulding, S. (2012). The invasion of the freshwater diatom *Didymosphenia geminata* in Patagonia: Prospects, strategies, and implications for biosecurity of invasive microorganisms in continental waters. *Conservation Letters, 5*, 432–440.

113. Reid, B., & Torres, R. (2013). *Didymosphenia geminata* invasion in South America: Ecosystem impacts and potential biogeochemical state change in Patagonian rivers. *Acta Oecologica, 54*, 101–109.

114. Riestra, F. (2017). Environmental flow policy. In G. Donoso (Ed.), *Water policy in Chile* (pp. 103–116). Springer.

115. Riva-Rossi, C., Barrasso, D. A., Baker, C., Quiroga, A. P., Baigún, C., & Basso, N. G. (2020). Revalidation of the Argentinian pouched lamprey Geotria macrostoma (Burmeister, 1868) with molecular and morphological evidence. *PLoS ONE, 15*(5), e0233792. https://doi.org/10.1371/journal.pone.0233792

116. Rivera, A., Aravena, J. C., Urra, A., & Reid, B. (2023). *Chilean Patagonian glaciers and environmental change.* Springer.

117. Rozzi, R., Armesto, J., Gutiérrez, J., Massardo, F., Likens, G., Anderson, C., Poole, A., Moses, K., Hargrove, E., Mansilla, A., Kennedy, J., Wilson, M., Jax, K., Jones, C., Callicott, J., & Arroyo, M. (2012). Integrating ecology and environmental ethics: Earth stewardship in the southern end of the Americas. *BioScience, 62*, 226–236.

118. Scheele, B., Pasmans, F., Skerratt, L., Berger, L., Martel, A., Beukema, W., Acevedo, A., Burrows, P., Carvalho, T., Catenazzi, A., de la Riva, I., Fisher, M., Flechas, S., Foster, C., Frías-Álvarez, P., Garner, T., Gratwicke, B., Guayasamin, J., Petersen, M., Canessa, S., et al. (2019). Amphibian fungal panzootic causes catastrophic and ongoing loss of biodiversity. *Science, 363*, 1459–1463.

119. Scheffer, M., Barrett, S., Carpenter, S., Folke, C., Green, A., Holmgren, M., Hughes, T., Kosten, S., van de Leemput, I., Nepstad, D., van Ness, E., Peeters, E., & Walker, B. (2015). Creating a safe operating space for iconic ecosystems. *Science, 347*, 1317–1319.

120. Schoeller, H. (1967). Qualitative evaluation of ground water resources. In H. Schoeller (Ed.), *Methods and techniques of groundwater investigation and development* (pp. 44–52). Water Resource Series No. 33. UNESCO

121. Sepúlveda, M., Valenzuela, A., Pozzi, C., Medina-Vogel, G., & Chehébar, C. (2015). *Lontra provocax*. The IUCN red list of threatened species 2015: E. T12305a21938042. Retrieved fom: https://www.researchgate.net/publication/291161582_Lontra_provocax

122. Serra, M., Albariño, R., & Villanueva, V. (2012). Invasive *Salix fragilis* alters benthic invertebrate communities and litter decomposition in northern Patagonian streams. *Hydrobiologia, 701*, 173–188.

123. Soto, D. (2002). Oligotrophic patterns in southern Chilean lakes: The relevance of nutrients and mixing depth. *Revista Chilena de Historia Natural, 75*, 377–393.

124. Soto, D., Campos, H., Steffen, W., Parra, O., & Zuñiga, L. (1994). Limnology of the Torres del Paine lake district (Chilean Patagonia): A case of pristine N-limited lakes and ponds. *Archiv fur Hidrobiologie, 99*, 181–197.

125. Stanford, J. A., & Ward, J. V. (1988). The hyporheic habitat of river ecosystems. *Nature, 335*, 64–66.

126. Stanford, J., & Ward, J. (2001). Revisiting the serial discontinuity concept. *Regulated Rivers Research and Management, 17*, 303–310.

127. Strayer, D., & Dudgeon, D. (2010). Freshwater biodiversity conservation: Recent progress and future challenges. *Journal of the North American Benthological Society, 29*, 344–358.

128. Takeuchi, N., & Koshima, S. (2004). A snow algal community on a Patagonian Glacier, Tyndall Glacier in the southern Patagonia Icefield. *Arctic, Antarctic, and Alpine Research, 6*(1), 91–98.

129. Tiberti, R., & Hardenberg, A. (2012). Impact of introduced fish on common frog (*Rana temporaria*) close to its altitudinal limit in alpine lakes. *Amphibia-Reptilia, 33*, 303–307.

130. Tockner, K., & Stanford, J. A. (2002). Riverine flood plains: Present state and future trends. *Environmental Conservation, 29*(3), 308–330.

131. Tonkin, J., Altermatt, F., Finn, D., Heino, J., Olden, J., Pauls, S., & Lytle, D. (2017). The role of dispersal in river network metacommunities: Patterns, process and pathways. *Freshwater Biology, 63*, 141–163.

132. Torres, R., Reid, B., Frangópulos, M., Alarcón, E., Márqueza, M., Häussermann, V., Förster, G., Pizarro, G., Iriarte, J., & González, H. (2020). Freshwater runoff effects on the production of biogenic silicate and chlorophyll-a in western Patagonia archipelago (50–51° S). *Estuarine Coastal and Continental Shelf Research, 241*, 106597.

133. Torres, R., Silva, N., Reid, B., & Frangopulos, M. (2014). Silicic acid enrichment of subantarctic surface water from continental inputs along the Patagonian archipelago interior sea (41°–56° S). *Progress in Oceanography, 129A*, 50–61.

134. United Nations Environment Program. (1998). *Freshwater biodiversity: A preliminary global assessment*. WCMC Biodiversity Series No. 8. Retrieved from: https://www.unep-wcmc.org/resources-and-data/freshwater-biodiversity--a-preliminary-global-assessment

135. Universidad de Chile. (2016). *Análisis de las condicionantes para el desarrollo hidroeléctrico en las cuencas del Maule, Biobío, Toltén, Valdivia, Bueno, Puelo y Yelcho, desde el potencial de generación a las dinámicas socio-ambientales*. Informe al Ministerio de Energía, Gob. Chile. Centro de Energía, Facultad de Ciencias Físicas y Matemáticas. Retrieved from: http://www.biblioteca.digital.gob.cl/handle/123456789/635

136. Universidad de Concepción. (2016). *Análisis de las condicionantes para el desarrollo hidroeléctrica en las cuencas de los ríos Palena, Cisnes, Aysén, Baker y Pascua, desde el potencial de generación a las dinámicos socio-ambientales*. Informe al Ministerio de Energía ID: 584105-40-LP15. Centro de Ciencias Ambientales EULA-Chile. Retrieved from: http://www.biblioteca.digital.gob.cl/handle/123456789/635

137. Unmack, P., Habit, E., & Johnson, J. (2009). Nuevos registros de *Hatcheria macraei* (Siluriformes, Trichomycteridae) en la provincia chilena. *Gayana, 73*, 102–110.
138. Valdovinos, C., Kiessling, A., Mardones, M., Moya, C., Oyanedel, A., Salvo, J., Olmos, V., & Parra, O. (2010). Distribution of macroinvertebrates (Plecoptera and Aeglidae) in fluvial ecosystems of the Chilean Patagonia: Do they show signals of the postglacial geomorphological evolution? *Revista Chilena de Historia Natural, 83*, 267–287.
139. Valdovinos, C. (2008). Invertebrados Dulceacuicolas. In J. Rovira, J. Ugalde & M. Stutzen (Eds.), *Biodiversidad de Chile: patrimonio y desafíos* (pp. 202–222). Comisión Nacional Medio Ambiente.
140. Vanhaecke, D., Leoniz, C., Gajardo, G., & Consuegra, S. (2015). Genetic signatures of historical dispersal of fish threatened by biological invasions: The case of galaxiids in South America. *Journal of Biogeography, 42*(10), 1942–1952.
141. Vega, A., Lizama, K., & Pasten, P. (2017). Water quality: Trends and challenges. In G. Donoso (Ed.), *Water policy in Chile* (pp. 25–52). Springer.
142. Ventura, M., Tiberti, R., Buchaca, T., Buñay, D., Sabas, I., & Miro, A. (2017). Why should we preserve fishless high mountain lakes? In J. Catalán, J. M. Ninot & M. M. Aniz (Eds.), *High mountain conservation in a changing world. Advances in global change research* (vol. 60, pp 181–205). Springer.
143. Vera, A., & Camousseight, A. (2006). Estado de conocimiento de los plecópteras de Chile. *Gayana, 70*, 57–64.
144. Vera-Escalona, I., Habit, E., & Ruzzante, D. (2019). Invasive species and postglacial colonization: Their effects on the genetic diversity of a Patagonian fish. *Proceedings of the Royal Society B: Biological Sciences, 286*, 20182567.
145. Vianna, J. A., Medina-Vogel, G., Chehébar, C., Sielfeld, W., Olavarría, C., & Faugeron, S. (2011). Phylogeography of the patagonian otter *Lontra provocax*: Adaptive divergence to marine habitat or signature of southern glacial refugia? *BMC evolutionary Biology, 11*, 53.
146. Vigliano, P. H., Alonso, M., & Aquaculture, M. (2007). Salmonid Introductions in Patagonia: A mixed blessing. In T. M. Bert (Ed.), *Ecological and genetic implications of aquaculture activities* (pp. 315–331). Springer.
147. Véliz, D., Catalán, L., Pardo, R., Acuña, P., Díaz, A., Poulin, E., & Vila, I. (2012). The genus *Basilichthys* (Teleostei: Atherinopsidae) revisited along its Chilean distribution range (21° to 40° S) using variation in morphology and mtDNA. *Revista Chilena de Historia Natural, 85*, 49–59.
148. Waldron, A., Mooers, A., Miller, D., Nibbelink, N., Redding, D., Kuhn, T., Roberts, J., & Gittleman, J. (2013). Targeting global conservation funding to limit immediate biodiversity declines. *Proceedings of the National Academy of Sciences, 110*, 12144–12148.
149. Waples, R. (1991). Pacific salmon, *Oncorhynchus spp.*, and the definition of "species" under the Endangered Species Act. *Marine Fisheries Review, 53*, 11–22.
150. Wilson, E. (2016). *Half-Earth*. Liveright; Xu, J., Pérez-Losada, M., Jara, C. G., & Crandall, K. A. (2009). Pleistocene glaciation leaves deep signature on the freshwater crab *Aegla alacalufi* in Chilean Patagonia. *Molecular Ecology, 18*, 904–918.
151. Woodward, G., Perkins, D., & Brown, L. (2010). Climate change and freshwater ecosystems: Impacts across multiple levels of organization. *Philosophical Transactions of the Royal Society of London Series B, 365*, 2093–2106.
152. Zemlak, T., Habit, E., Walde, S., Battini, M., Adams, E., & Ruzzante, D. (2008). Across the southern Andes on fin: Glacial refugia, drainage reversals and a secondary contact zone revealed by the phylogeographical signal of *Galaxias platei* in Patagonia. *Molecular Ecology, 70*, 5049–5071.
153. Zúñiga, O., Wilson, R., Amat, F., & Hontoria, F. (1999). Distribution and characterization of Chilean populations of the brine shrimp *Artemia* (Crustacea, Branchiopoda, Anostraca). *International Journal of Salt Lake Research, 8*, 23–40.

Brian Reid BS in Neurobiology, Cornell University, USA. Ph.D., Aquatic Ecology, Montana University, USA. Between 1994–2001 worked in conservation NGOs in the USA. Resident researcher at CIEP, Coyhaique, Aysén Region, Chile.

Anna Astorga Roine BS in Biology, Pontificia Universidad Católica de Chile. Ph.D., in Ecology, specializing in Stream Ecology, Oulu University, Finland. Researcher at Centro de Investigaciones en Ecosistemas de la Patagonia, Coyhaique, Región de Aysén, Chile.

Rubén Isaí Madriz Entomologist with specialization in freshwater Dipera. Ph.D., University of Iowa, USA. His research focus on the systematics and life cycle of insect groups with gondwanic lineages.

Christian Correa Evolutionary ecologist, specialist in native and introduced fishes. Ph.D., McGill University, Canada. Interdisciplinary work in conservation in Chilean Patagonia, including community ecology, reproductive biology and genetics.

Tamara Contador Ph.D., in Biology, University of North Texas, USA. Researcher at Cape Horn International Center, Institute of Biodiversity and Antarctic Ecosystems, Nucleus of Austral Invasive Salmonids, and Universidad de Magallanes, Chile.

Chilean Patagonian Glaciers and Environmental Change

15

Andrés Rivera, Juan Carlos Aravena, Alejandra Urra, and Brian Reid

Abstract

Patagonian glaciers (41°–56°S) have experienced strong volume losses and retreats during recent decades in response to the climatic changes affecting this part of Chile, contributing significantly to global sea level rise. These changes have had an impact on the region's ecosystems, due to processes such as the expansion of fjords and lakes, altered hydrology and geology risks, higher sediment loads contributed to rivers, and changes in the altitude and composition of nearby vegetation. These factors affecting the ecosystem services provided by glaciers, such as runoff and flood regulation, slope stability, biodiversity, and cultural services they generate as one of the few remaining pristine components of the Earth. The recent changes in glacier volume make them highly vulnerable to the adverse effects of ongoing climate change, a condition that affects other Subantarctic natural systems of Chile. We emphasize the need to enhance the systematic monitoring of glacier volume and surface extent in Patagonia.

Keywords

Patagonia • Chile • Glaciers • Geological risks • Ecosystem services

A. Rivera (✉)
Department of Geography, Universidad de Chile, Santiago, Chile
e-mail: arivera@uchile.cl

J. C. Aravena
Centro de Investigación Gaia Antártica (CIGA), Universidad de Magallanes, Punta Arenas, Chile

A. Urra
Rio Cruces Wetland Centre, Universidad Austral de Chile, Valdivia, Chile

B. Reid
Centro de Investigación en Ecosistemas de La Patagonia (CIEP), Coyhaique, Chile

© Pontificia Universidad Católica de Chile 2023
J. C. Castilla et al. (eds.), *Conservation in Chilean Patagonia*, Integrated Science 19,
https://doi.org/10.1007/978-3-031-39408-9_15

1   Introduction

Glaciers are perennial ice masses with firn and snow, originating on the land surface by the recrystallization of snow or other forms of solid precipitation, and showing evidence of present or past flows. This definition includes four elementary characteristics of a glacier: (i) composition (water in a solid state, mainly ice); (ii) origin (solid precipitation on land that recrystallizes to form ice); (iii) temporality (long-lived); (iv) dynamics (flow). This implies that glaciers are climatically open systems, and therefore their volumetric and dynamic changes are susceptible to modification. Each of these characteristics has increasing levels of complexity, for example, in terms of what makes up a glacier, as it may contain water in a liquid state, especially in summer when melting can be very high; it may also contain a large amount and diversity of supra-, intra- and sub-glacial rock material, including cryoconites, coarse sediments, detritus and large (erratic) clasts. In many cases glaciers have their own ecosystem, so they can be considered as biomes [1], containing different types of extremophile organisms including microalgae and insects, such as the "Patagonian dragon" stonefly (*Andiperla winklii*), of which very little is known [2].

Therefore, a glacier is not only the ice that exists in the mountain range; it is a complex natural ecosystem that is associated with a contributing basin that generates rivers with meltwater (surface and/or underground), and also produces icebergs when it terminates in a lake or fjord. The environmental importance and effects of glaciers and their changes are not restricted to their basins, but cascade downstream, even impacting coastal marine environments where the water from melting snow and ice flow into them [3].

These multiple spatial and temporal cascading relationships mean that glaciers provide a range of ecosystem services relevant to human well-being [4], including: (i) water provision (e.g. drinking water, irrigation, power generation); (ii) sediment and nutrient inputs (e.g. fertilization of rivers, lakes, and coastal areas associated with aquaculture resources); (iii) flow regulation (*hydrological regime in dry periods); (iv) flood mitigation* (regulation of flash floods, glacial lake outburst floods or GLOFs); (v) slope conservation (e.g. slope stability and its geological hazards); (vi) biodiversity conservation (e.g. maintenance of wildlife corridors and glacier lake outburst floods, high Andean vegetation, microbial components of glacial surfaces); and (vii) cultural services (e.g. tourism, sports, national identity, sense of belonging). These ecosystem services [5] have decreased over time due to deglaciation, which has affected most of the planet's mountain glaciers since the mid-nineteenth century, a period that has been called the Anthropocene [6].

This chapter will analyze the glaciers located in Chilean Patagonia, understood as the region of the southern Andes located from 41°S to the southern tip of the continent, which historically has also been called western Patagonia. In this region there are glaciers on the western slope of the Andes and some on the eastern slope, especially in the Southern Patagonia Icefield (SPI), the largest ice mass in South America [7].

Fig. 1 Location and
identification of the main
glaciers in Patagonia,
together with the great
Patagonian ice fields

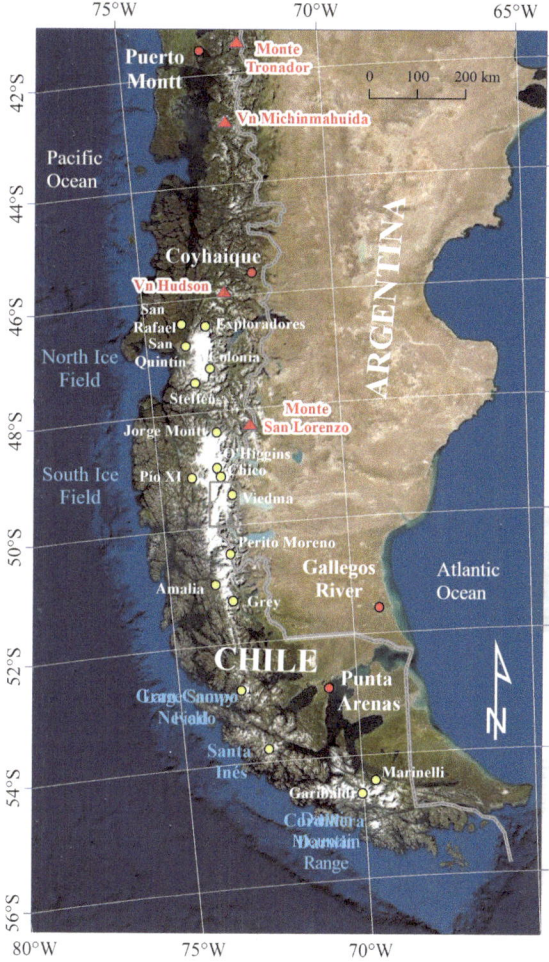

2   Scope and Objectives

This chapter describes the current state of glaciers in Chilean Patagonia (Fig. 1),
their ecosystem services, and the possible environmental consequences of the
current process of glacial retreat affecting this territory of Chile.

3   Materials and Methods

To conduct this analysis, published scientific literature and grey literature in the
ISI Web of Knowledge core collection database (http://apps.webofknowledge.com)
were reviewed. Publications with a focus on glaciers in western Patagonia were

selected. Google Scholar and sources available in regional institutions were also reviewed.

4   Results

4.1   The Glaciers of Chilean Patagonia

Glacier fluctuations during the Quaternary were the main agent shaping the present landscape of the entire Patagonia region. During the last glaciations, the glaciers formed a large ice cap that covered practically the entire western slope of the continent south of the Island of Chiloé, reaching the oceanic shelf (except for some possible local ecological refuges, as will be seen later). On the eastern slope the glaciers extended towards the Patagonian plains with different magnitudes, forming large lobes that contributed meltwater to the Atlantic Ocean. The onset of deglaciation about 19,000 years before present (BP) [8] transformed the western slope of Patagonia into the Aysén and Magallanes archipelagos, divided by hundreds of fjords and channels, while on the eastern slope the main ice lobes transformed into the large piedmont lakes, General Carrera/Buenos Aires or O'Higgins/San Martín, whose waters began to drain into the Pacific Ocean along deep valleys that dissected the remnants of the mountain range. South of the SPIF, the depth of Quaternary erosion was so significant that the Pacific fjords joined the basins abandoned by the glacial lobes, forming the Otway and Skyring sounds, or joined the Atlantic Ocean, forming the Strait of Magellan.

The current remnants of this great glaciation are thousands of glaciers located in the upper parts of the mountains, many of which are volcanoes associated with the Liquiñe Ofqui mega fault [9]. However, the main glacier masses are concentrated in the North and South Patagonian ice fields, which are semi-continuous masses of about 4,000 and 12,300 km^2 area, respectively (Dusaillant et al. 2019). The present long-term deglaciation, with numerous Holocene fluctuations [10], has been accentuated since the Little Ice Age, the last period that showed advances in the vast majority of the planet's glaciers, and which ended in Patagonia approximately 150 years BP [11].

Many glaciers have experienced retreat and thinning in recent decades in Chile, synchronously with volume losses throughout the Andes [12]. These changes relate importantly to increasing atmospheric temperatures, especially the warming experienced by the high Andes in Chile south to 46 °S [13], and by the precipitation decrease observed in several regions of the country, including eastern Patagonia, to about 49°S [14]. However, from the SPIF to the Darwin Range precipitation trends appear to reverse [15] and no definite tendencies are detected in atmospheric temperatures at higher altitudes (no data, only models), although an increase of 0.73 °C was recorded in the city of Punta Arenas between 1960 and 2010 [16].

The combined effect of both processes is that the elevation of the snow line has risen, especially in those glaciers outside the ice fields, negatively impacting snow accumulation and also causing a phase change in precipitation, shifting from solid to liquid in the lower areas of the glaciers, even in winter [17]. Detected a contrasting picture between 2000 and 2015 in the Northern Patagonia Icefield (NPI) and SPI (Fig. 1), with positive trends in snow accumulation on the western slopes and negative trends on the eastern slope. This would imply that the increased snow accumulation is partly offsetting the effects of regional atmospheric warming on glacial retreat, which could even lead to stability or even re-advancement of some glacier tongues, as in the few exceptions observed in the Andes [11].

There are 18,954 glaciers in Chilean Patagonia, with a total area of 22,463 km^2, equivalent to almost 95% of Chile's ice area [18]. It has been estimated that a specific rate of 0.78 ± 0.25 m of water equivalent were lost per year between 2000 and 2018 in this southern region [12], a figure only surpassed by Alaska and other Arctic regions. This amount has been increasing for several decades, contributing significantly to global sea level rise [19].

Due to the large number and diversity of glaciers in Patagonia, an important part of the country's glaciological research is concentrated in this southern area, particularly in the Northern and Southern Ice Fields, where the number and size of proglacial lakes has also expanded due to the retreat and thinning of the ice, many of which have experienced (or could suffer) sudden emptying [20]. Important lahars or rapid debris flows have been generated in this region as a consequence of ice melting due to volcanic activity, such as that detected in the Hudson volcano [9].

Clove Peninsula, Alberto de Agostini National Park, Magallanes and Chilean Antarctica Region. Photograph by Nicolás Piwonka

Very diverse glacier behavior has been detected in the NPI, the largest temperate ice mass in the Southern Hemisphere outside Antarctica; some have experienced strong retreat, while acceleration of ice flow has been observed in several cases in recent decades [21]. These retreats have been accompanied by significant volume loss [12], although there are few exceptions to this trend, the most important being the Pío XI glacier (also called Brüggen), which has been in a maximum neoglacial position since the early 1990s, when it began to overtake centennial trees present on its margins [22].

These different behaviors are attributed to the fact that most NPI glaciers produce icebergs in fjords or lakes, where glacial behavior is correlated with the depth of these bodies of water. When a glacier ends in a fjord or deep lake, its front can be unstable and respond dynamically with strong retreat once it loses its equilibrium with medium-term climatic conditions [23].

The strong losses in area in recent decades also appear to be accelerating in the vast majority of NPI and SPI glaciers. This implies that glaciers remain significantly out of balance with respect to current climate conditions. Recent modelling indicates that there are positive mass balances in the last decades, so that the dynamic responses of the ice (thinning and acceleration of ice flow) are in equilibrium with the current climate conditions. Thinning and acceleration of ice flow by longitudinal extension due to the higher rate of iceberg production at the terminal front are the main factors explaining mass loss in the area, followed by strong melting in the lower parts or ablation zones [21]. South of the Strait of Magellan there has also been contrasting behavior in recent decades, with strong retreats in some glaciers, (e.g. Marinelli Glacier) and advances in others, such as the Garibaldi Glacier [24]. Patagonian glaciers located outside the ice fields have also shown strong retreats and mass losses, such as in Mount San Lorenzo [25] and continental Chiloé [26].

Despite the important advances in glaciological knowledge in Chile, we are just discovering some of the basic characteristics of glaciers, such in ice thicknesses. [27] determined that the existing ice of the NPI and SPI together has a total volume of 4,756 km^3, almost 40 times more than the glacier volume in the Alps. There is also insufficient knowledge about the ecosystem impacts of the deglaciation process, potential effects are addressed in the following sections.

4.2 Environmental Effects of Glacial Changes in Patagonia

Areas freed from ice since the Little Ice Age have given way to hundreds of proglacial lakes, many of which have been dammed by thrust moraines formed when glaciers over-excavated pre-existing valleys. In other cases, these lakes have formed on the margins of glaciers, being dammed by ice (Fig. 2). It can be argued that not all meltwater has been lost to the sea, since an important part has been stored in lakes and wetlands, which have subsoil with unconsolidated Quaternary material. Since proglacial lakes are not stable in many cases, the risk of sudden

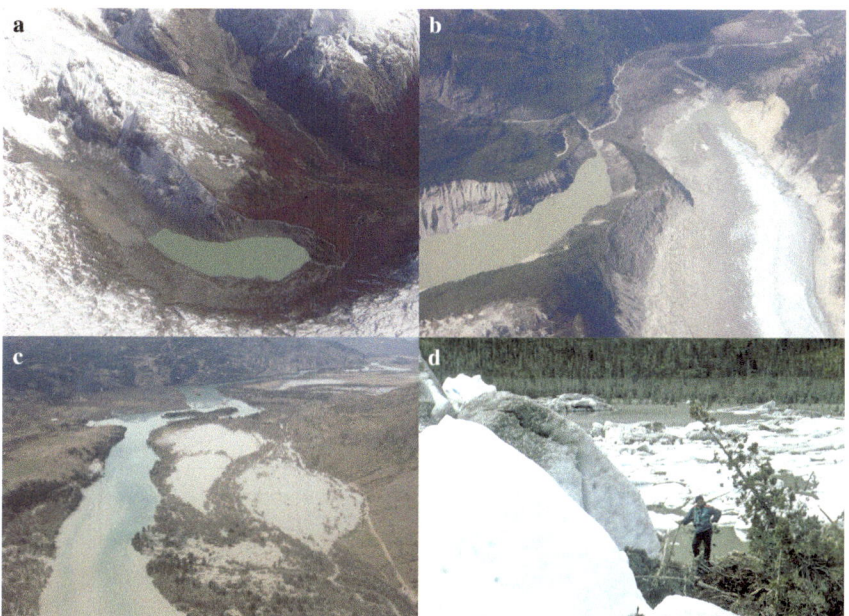

Fig. 2 Chilean Patagonia: examples of proglacial lakes and environmental effects of their changes. **a** Lake dammed by thrust moraines that are uncovered/covered by vegetation on its inner/outer margins. **b** On the left a lake formed by the retreat of the Hyades Glacier and on the right another lake in formation by the retreat of the Soler glacier of the NPI. **c** Flooding of the Baker River valley due to the emptying of Lake Cachet 2 in 2008. **d** Proglacial lake dammed by the advance of the Pío XI Glacier of the SPI in 1993. Photos: Andrés Rivera

emptying is high, as has occurred repeatedly in Lake Cachet 2, dammed by the NPI Colonia Glacier [20].

Another consequence of glacial retreat is that the areas abandoned by the ice have significant geological instability. The glaciers carry and push moraine material that accumulates on the margins and fronts, being partly retained by the ice. These areas may collapse as the buried ice disappears, covering the glaciers or generating rock avalanches such as the one that occurred. on the margin of the Chico Glacier of the SPI (Fig. 3a). In other cases, lateral terraces are formed while buttressed by ice, which can later crack and are susceptible to landslides when they lose support. Cracks may also form in valley walls indicating the position which a glacier reached in the past (trimline), often without vegetation following retreat and thinning, such as those that occur in the terrain that was occupied by the Upsala glacier until about 1945 (Fig. 3b).

An increase in hydrogeological risks has also been observed due to the instability of the fluvioglacial plains formed downstream of the thrust moraines, which are eroded by glacial streams (Fig. 3c), especially during periods of flooding and/ or high flows produced by rapid melting of ice during heat waves or heavy rainfall. These phenomena have become more recurrent in recent years [28]. Another

Fig. 3 Examples of hydrological and geological risks associated with glaciers in Patagonia. *Photos* Andrés Rivera (**a**, **c** and **d**), Esteban Lannutti **b**

example are mudflows resulting from heavy rainfall events, such as those that saturated the slopes of the Burritos River valley, causing a process of slope failure that fell partly on a mountain lake and partly on a glacier in 2017 (Fig. 3d) leading to the devastation Villa Santa Lucía.

One of the most prominent effects of glacial changes is geomorphological, particularly in lakes and fjords, many of which have expanded (Fig. 4a and b), generating significant hydrographic and lacustrine changes (Fig. 4c). In some exceptional cases the opposite has occurred, as in the Pío XI of the SPI, which has advanced and formed a thrust moraine due to the high rate of accumulated sedimentation at the head of Eyre Fjord (Fig. 4c).

Finally, there is the glacio-volcanic interaction in Patagonia, where volcanic activity is frequent near glaciers, causing landslides, mudflows, deformations and sudden melting of ice and formation of lahars. Deglaciation has generally been associated with increased volcanic activity due to the effect of ice discharge on volcanic edifices [29], and it cannot be ruled out that this is happening today at Reclus Volcano (Fig. 4a).

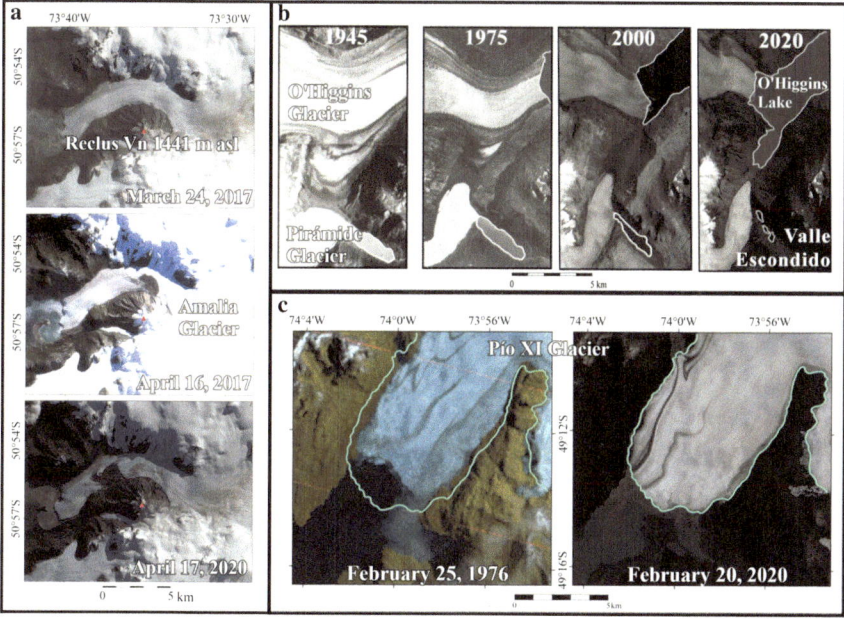

Fig. 4 Chilean Patagonia: examples of glacial changes in the SPI and their effects on the environments. **a** North slope landslide of Reclus Volcano in April, 2017 when part of the tongue of Amalia Glacier was covered. **b** Changes in flow direction of Pirámide Glacier in response to the retreat of O'Higgins Glacier. **c** Generation of a frontal moraine on the Pío XI Glacier in response to its almost uninterrupted advance since 1976. The green line in the 1976 image indicates the extension of the glacier in 2020

4.3   Effects of Glacial Fluctuations on Vegetation and Wildlife

The long-term fluctuations of Patagonian glaciers described in the previous sections of this chapter have generated important environmental modifications in periglacial ecosystems. For example, some effects on vegetation have produced changes in the altitude of high Andean plant community formations, in the underlying forest formations and wetlands, as well as variation in their floristic composition [3, 30], along with changes in the dynamics of natural and anthropogenic disturbances, such as variations in the recurrence of fires and insect attacks.

These types of environmental variation generated by glacial fluctuations also determine the increase or decrease in the surface areas covered by high Andean and forest vegetation formations that share a common limit in the altitudinal gradient. Sustained periods of decrease in the altitude of the zero-isotherm due to lower temperatures and increased ice masses have led to an expansion of the area of high Andean vegetation and a consequent compression of the area occupied by forest formations, with decreases in the altitude of the upper tree line [31].

Glacial advances have conditioned the upper limit of the forests, further compressing the area available for vegetation. The opposite has occurred in warm periods, where the area that expands is that occupied by forests, with an increase in the altitude of tree line and the zero isotherm, compressing the area covered by high Andean vegetation [32]. Changes such as these have repercussions at various levels. The floristic composition of the aforementioned communities can vary from changes in the abundance of taxa to modifications due to the exit/entry of species, which can be expressed in modifications in the diversity of the communities [33]. More extreme changes may involve complete replacement of plant formations with variation in the proportion of arboreal versus non-arboreal vegetation, or of species assemblages with an affinity to grow in conditions of higher or lower humidity [34]. Environmental variations also have an impact on the natural disturbance regime to which plant formations are subjected. For example, the occurrence of fires decreases in cold and humid phases and increases in dry and warm phases. A record that provides evidence of this effect for the last 3,000 years in Chilean Patagonia [34] shows a very good fit of dry and warm phases with the increase in charcoal particle counts associated with the increase in natural and/or anthropogenic fires. All these aspects demonstrate the importance of these periglacial ecosystems subject to glacial fluctuations and their importance in the study of macroecological patterns [28] and evolutionary and successional processes of plant communities.

Environmental fluctuations associated with glacial advance or retreat have determined the formation or interruption of biological corridors associated with periglacial environments of great importance in the connectivity of animal populations, with a significant effect on their conservation, evolution, and genetic diversity, for example the populations of huemul (*Hippocamelus bisulcus*) that inhabit the periglacial areas of the SPI [35]. These populations are relicts of the last ice age, this species is endemic to the southern Andes of South America, adapted to cold conditions and high slopes and irregular terrain, including rocky terraces and cliff edges. The species forms small, isolated groups, which scientists have interpreted as an adaptation to periglacial mountain environments subjected to recurrent and extensive glaciations and the subsequent fragmentation of their habitat, and the restriction and reduction of available resources. These conditions of climate and vegetation oscillation, and recurrence of glacial periods during the Pleistocene meant the creation of geographic barriers that currently result in disjunct animal populations, with refuges of high genetic diversity such as Wellington Island (49°S), west of the SPI. Glaciers and their long-term fluctuations have also determined environmental refuges in semi-permanent water bodies for fish such as catfish (*Hatcheria*), and macroinvertebrates such as *Plecoptera* and *Aeglidae* [36].

4.4   Biogeochemical Effects of Patagonian Glaciers

The fjord systems of Chilean Patagonia are fed by freshwater discharges from continental runoff, rivers, and glacier tributaries. These systems have low salinity, low density, strong stratification and are low in nitrate (NO_3^-) and orthophosphate (PO_43-), but high in silicic acid derived from the erosion of rocky substrates [37]. The fjords are also fed by Subantarctic oceanic waters which have high nutrient concentration ($NO_3^- > 12$ μM and $PO_{43}^- > 1$ μM) but a low concentration of silicic acid [38].

These sub-Antarctic waters interact strongly with freshwater input from large river discharge and glacier melt, causing marked horizontal and vertical salinity gradients in the coastal marine environments of the Chilean Patagonian region. The horizontal input of freshwater runoff in these cases has an important effect on the spatio-temporal patterns of phytoplankton, including biomass distribution, mainly because freshwater has effects on phytoplankton productivity in both surface ocean waters and fjords [39].

The impact that increased freshwater input generated by glacial melt due to climate change could have on Chilean Patagonia is important for the following reasons: (i) this water creates a stratification in the water column that favors phytoplankton growth, but at the same time can reduce the supply of nutrients in the vertical mixing of the water column [38], (ii) glacial runoff and ice floes detached from glacial fronts that end in fjords are likely to act as fertilizers, as they are rich in organic matter in solution derived from terrestrial ecosystems, and have other nutrients that come from substrates such as iron, phosphorus, nitrogen, and silica [3, 40], (iii the high concentration of suspended particles in glacial runoff can have a double effect, adding more nutrients downstream [41], but they can also create light-limiting conditions for plankton within the fjords [37]. These impacts are likely to intensify in the Patagonian fjords as freshwater flows increase in the face of higher increased temperature scenarios.

In any case, the effects of glacial melt on ecosystems (positive such as fertilization or negative such as low transparency/salinity) may undergo profound changes over time depending on the characteristics (e.g. geological conditions) of each large watershed.

5   Conclusions and Recommendations

Since the end of the period known as the Little Ice Age, which began about 150 years BP, most of the glaciers in Chilean Patagonia have been losing mass at an accelerated rate, affecting the regional environment with increased geological, hydrogeological, and glacio-volcanic risks. The ongoing deglaciation is also affecting the vegetation surrounding the glaciers, modifying their altitudinal zoning and floristic composition. The flow and chemical composition of rivers have also been modified [42]. The lakes and fjords that receive glacial meltwater flows have also experienced variations in their nutrient concentrations and productivity [43]. All these changes are a warning that cannot be ignored, especially when all scenarios of future climate change indicate that impacts could be exacerbated, possibly crossing important ecological thresholds [28].

These impacts are much more marked in smaller glaciers and those located at lower altitudes, which are more vulnerable to disappear, affecting the surrounding natural and human communities. Numerous proposals have emerged to address these impacts, which aim to overcome the gaps and deficiencies in the existing information in Patagonia [11]. These include the need to determine more precisely the total volume of glacier ice, to validate future change models with more and better data, and to determine more precisely the associated natural risks. A multidisciplinary approach is required to carry out these tasks, which promotes cooperation for the establishment of systematic long-term glacier monitoring programs, and to this end we recommend:

- The installation of networks of measuring instruments close to glaciers (especially at their source and terminus), spatially arranged in high altitude areas and on rocky outcrops surrounded by ice (nunataks). These networks of monitoring stations should record multiple parameters, including oceanographic, limnological, hydrological, and climatological variables. It would be desirable for the instruments to have the capability to transmit data in real time, and their sensors should meet international standards to ensure quality data capture. These networks should be subject to periodic maintenance and their data should be available for unrestricted public access and use.
- The intensification and extension of campaigns to measure glaciers and surrounding areas with geophysical prospecting methods such as radar, sonar, gravimeters, and LiDAR (Laser Imaging Detection and Ranging), all of which would allow a more precise characterization of glaciers. The main unknowns of Patagonian glaciers are ice thicknesses, mass balances, flow, and sediment input.

Acknowledgements This work is part of FONDECT project 1171832. Francisca Bown collaborated in the discussion and recommendations of this chapter.

References

1. Anesio, A. M., & Laybourn-Parry, J. (2012). Glaciers and ice sheets as a biome. *Trends in Ecology and Evolution, 27*(4), 219–225
2. Kohshima, S., Yoshimura, Y., & Takeuchi, N. (2002). Glacier ecosystem and biological ice-core analysis. In: F. Sepúlveda, G., Casassa, and R. Sinclair (Eds.), *The Patagonian icefields: a unique natural laboratory*, pp. 1–8. New York: Kluwer/Plenum
3. Rozzi, R., Rosenfeld, S., Armesto J. J., Mansilla, A., Núñez-Ávila, M., & Massardo, F. (2023). *Ecological Connections Across the Marine-Terrestrial Interface in Chilean Patagonia.* Springer
4. Milner, A., Khamis, K., Battin, T., Brittain, J., Barrand, N. E., Füreder, L., Cauvy-Fraunie, S., Gislason, G., Jacobsen, D., Hannah, D., Hodson, A., Hood, E., Lencioni, V., Olafsson, J., Robinsonn, C., Tranter, M., & Brown, L. (2017). Glacier shrinkage driving global changes in downstream systems. *Proceedings of the National Academy of Sciences USA, 114*(37), 9770–9778
5. Millennium Ecosystem Assessment (2003). *Ecosystems and human well-being: a framework for assessmenet.* Washington D. C., USA: Island Press

6. Steffen, W., Rockström, J., Richardson, K., Lenton, T. M., Folke, C., Liverman, D., Summer-hayes, C., Barnosky, A. D., Cornell, S. E., Crucifix, M., Donges, J. F., Fetzer, I., Lade, S. J., Scheffer, M., Winkelmann, R., & Schellnhuber, H. J. (2018). Trajectories of the earth system in the Anthropocene. *Proceedings of the National Academy of Sciences USA, 115*(33), 8252–8259

7. Rivera, A., Bown, F., Napoleoni, N., Muñoz, C., & Vuille, M. (2016). *Balance de masa glaciar*. Valdivia, Chile: Ediciones CECs

8. Mendelova, M., Hein, A. S., McCulloch R., & Davies B. (2017). The last glacial maximum and deglaciation in Central Patagonia, 44°–49° S. *Geographical Research Letters, 43*(2), 719–750

9. Rivera, A., & Bown, F. (2013). Recent glacier variations on active ice capped volcanoes in the southern volcanic zone (37° 46°S), Chilean Andes. *Journal of South American Earth Sciences, 45*, 345–356

10. Kaplan, M., Schaefer, J. M., Strelin, J. M., Denton, G. H., Anderson, R. F., Vandergoes, M. J., Finkel, R., Schwartz, R., Travis, S., Garcia, J. L., Martini, M., & Nielsen, S. H. H. (2016). Patagonian and southern south Atlantic view of Holocene climate. *Quaternary Science Reviews, 141*, 1–14

11. Masiokas, M., Rabatel, A., Rivera, A., Ruiz, L., Pitte, P., Ceballos, J. L., Barcaza, G., Soruco, A., Bown, F., Berthier, E., Dussaillant, I., & MacDonell, S. (2020). A review of the current state and recent changes of the Andean cryosphere. *Frontiers in Earth science, 8*, 99

12. Dussaillant, I., Berthier, E., Brun, F., Masiokas, M., Hugonnet, R., Favier, V., Rabatel, A., Pitte, P., & Ruiz, L. (2019). Two decades of glacier mass loss along the Andes. *Nature Geoscience, 12*, 802–808

13. Falvey, M., & Garreaud, R. (2009). Regional cooling in a warming world: recent temperature trends in the southeast Pacific and along the west coast of subtropical South America (1979–2006), *Journal of Geophysical Research, 114(D4)*

14. Boisier, J., Alvarez-Garreton, C., Cordero, R., Damiani, A., Gallardo, L., Garreaud, R., Lambert, F., Ramallo, C., Rojas, M., & Rondanelli, R. (2018). Anthropogenic drying in central-southern Chile evidenced by long-term observations and climate model simulations. *Elementa Science of the Anthropocene, 6*, 74

15. Garreaud, R., Lopez, P., Minvielle, M., & Rojas, M. (2013). Large-scale control on the Patagonian climate. *Journal of Climate, 26*(1), 215–230

16. Carrasco, J. (2013). Decadal changes in the near-surface air temperature in the western side of the Antarctic peninsula. *Atmospheric and Climate Sciences, 3*, 275–281

17. Bravo, C., Bozkurt, D., González-Reyes, A., Quincey, D., Ross, A., Farías-Barahona, D., & Rojas, M. (2019). Assessing snow accumulation patterns and changes on the Patagonian icefields. *Frontiers in Environmental Science, 7*(1), 30

18. Barcaza, G., Nussbaumer, S., Tapia, G., Valdés, J., García, J. L., Videla, Y., Albornoz, A., & Arias, V. (2017). Glacier inventory and recent glacier variations in the Andes of Chile. *South America. Annals of Glaciology, 58*(75), 166–180

19. Zemp, M., Huss, M., Thibbert, E., & Cogley, J. G. (2019). Global glacier mass changes and their contributions to sea-level rise from 1961 to 2016. *Nature, 568*, 382–386

20. Wilson, R., Glasser, N., Reynolds, J., Harrison, S., Iribarren, P., Schaefer, M., & Shannon, S. (2018). Glacial lakes of the central and Patagonian Andes. *Global and Planetary Change, 162*, 275–291

21. Bown, F., Rivera, A., Petlicki, M., Bravo, C., Oberreuter, J., & Moffat, C. (2019). Recent ice dynamics and mass balance of Jorge Montt glacier, southern Patagonia icefield. *Journal of Glaciology, 65*(253), 732–744

22. Rivera, A. (2018). Glaciar Pío XI: La excepción a la tendencia de desglaciación en Patagonia. *Revista Geográfica de Chile Terra Australis, 54*, 1–12

23. Rivera, A., Koppes, M., Bravo, C., & Aravena, J. C. (2012). Little Ice Age advance and retreat of Jorge Montt glacier, Chilean Patagonia. *Climate of the Past, 8*, 403–414

24. Masiokas, M., Rivera, A., Espizúa, L., Villalba, R., Delgado, D., & Aravena, J. C. (2009). Glacier fluctuations in extratropical South America during the past 1000 years. *Palaeogeography, Palaeoclimatology, Palaeoecology, 281*, 242–268

25. Falaschi, D., Lenzano, M., Villalba, R., Bolch, T., Rivera, A., Repetto, L. V., & A. (2019). Six decades (1958–2018) of geodetic glacier mass balance in monte San Lorenzo, Patagonian Andes. *Frontiers in Earth Science-Cryospheric Sciences, 7*, 326

26. Paul, F., & Molg, N. (2014). Hasty retreat of glaciers in northern Patagonia from 1985 to 2011. *Journal of Glaciology, 60*(224), 1033–1043

27. Millán, R., Rignot, E., Rivera, A., Martineau, V., Mouginot, J., Zamora, R., Uribe, J., Lenzano, G., De Fleurian, B., Li, X., Gim, Y., & Kirchner, D. (2019). Ice thickness and bed elevation of the northern and southern Patagonian icefields. *Geophysical Research Letters, 46*, 6626–6635

28. Marquet, P. A., Buschmann, A. H., Corcoran, D., Díaz, P. A., Fuentes-Castillo, T., Garreaud, R., Pliscoff, P., & Salazar, A. (2023). *Global Change and Acceleration of Anthropic Pressures on Patagonian Ecosystems.* Springer

29. Watt, S., Pyle, D., & Mather, T. (2013). The volcanic response to deglaciation: Evidence from gla- ciated arcs and a reassessment of global eruption records. *Earth-Science reviews, 122*, 77–102

30. Mansilla, C. A., Domínguez, E., Mackenzie, R., Hoyos-Santillán, J., Henríquez, J. M., Aravena, J. C., & Villa-Martínez, R. (2023). *Peatlands in Chilean Patagonia: Distribution, Biodiversity, Ecosystem Services, and Conservation.* Springer

31. Körner C. (2003). *Alpine plant life: functional plant ecology of high mountain ecosystems, 2nd ed.* Berlin: Springer

32. Arroyo, M. T. K., Dudley, L. S., Pliscoff, P., Cavieres, L. A., Squeo, F. A., Marticorena, C., & Rozzi, R. (2010). A possible correlation between the altitudinal and latitudinal ranges of species in the high elevation flora of the Andes. In E. Spehn & C. Körner (Eds.), *Data mining for global trends in mountain biodiversity* (pp. 29–38). CRC Press, Taylor and Francis Group

33. Rundel, P., Arroyo, M. T. K., Cowling, R., Keeley, J., Lamon, B., & Vargas, P. (2016). Mediterrean biomes: Evolution of their vegetation, floras, and climates. *Annual Review of Ecology, Evolution and Systematics, 47*, 383–407

34. Moreno, P., Vilanova, I., Villa-Martínez, R., Garreaud, R., Rojas, M., & De Pol-Holz, R. (2014). Southern Annular Mode-like changes in southwestern Patagonia at centennial timescales over the last three millennia. *Nature Communications, 5*, 4375

35. Vila, A., Saucedo, C., Aldridge, D., Ramilo, E., & Corti, P. (2010). *South Andean huemul (Hippocamelus bisulcus, Molina 1782).* In: Barbanti Duarte, J. M. & González, S. (Eds.), *Neotropical cervidology: biology and medicine of Latin American deer,* pp. 89–100. São Paulo, Brazil: FUNEP-IUCN

36. Valdovinos, C., Kiessling, A., Mardones, M., Moya, C., Oyanedel, A., Salvo, J., Olmos, V., & Parra, O. (2010). Distribución de macroinvertebrados (Plecoptera y Aeglidae) en ecosistemas fluviales de la Patagonia chilena: ¿muestran señales biológicas de la evolución geomorfológica postglacial? *Revista Chilena de Historia Natural, 83*, 267–287

37. Aracena, C., Lange, C., Iriarte, J., Rebolledo, L., & Pantoja, S. (2011). Latitudinal patterns of export production recorded in surface sediments of the Chilean Patagonian fjords (41–55 S) as a response to water column productivity. *Continental Shelf Research, 31*(3–4), 340–355

38. González, H., Castro, L., Daneri, G., Iriarte, J., Silva, N., Vargas, C., Giesecke, R., & Sánchez, N. (2011). Seasonal plankton variability in chilean Patagonia fjords: Carbon flow through the pelagic food web of Aysen fjord and plankton dynamics in the Moraleda channel basin. *Continental Shelf Research, 31*(3–4), 225–243

39. Iriarte, J., González, E., Liu, K., Rivas, C., & Valenzuela, C. (2007). Spatial and temporal variability of chlorophyll and primary productivity in surface waters of southern Chile (41.5–43 S). *Estuarine, Coastal and Shelf Science, 74*(3), 471–480

40. Wadham, J., De'ath, R., Monteiro, F., Tranter, M., Ridgwell, A., Raiswell, R., & Tulaczyk, S. (2013). The potential role of the Antarctic ice sheet in global biogeochemical cycles. *Earth and Environmental Science Transactions of the Royal Society of Edinburgh, 104*(1), 55–67

41. Hawkings, J., Wadham, J., Tranter, M., Lawson, E., Sole, A., Cowton, T., Tedstone, A. J., Bartholomew, I., Nienow, P., Chandler, D., & Telling, J. (2015). The effect of warming climate

on nutrient and solute export from the Greenland ice sheet. *Geochemical Perspective Letters, 1*, 94–104

42. Reid, B., Astorga, A., Madriz, I., & Correa, C., and Contador, T. (2023). *A Conservation Assessment of Freshwater Ecosystems in Southwestern Patagonia.* Springer
43. Hucke-Gaete, R., Viddi, F. A., and Simeone, A. (2023). *Marine Mammals and Seabirds of Chilean Patagonia: Focal Species for the Conservation of Marine Ecosystems.* Springer

Andrés Rivera Full Professor, Department of Geography, Universidad de Chile. PhD Bristol Glaciology Centre, University of Bristol, UK.

Juan Carlos Aravena B.Sc and M.Sc in Biological Sciences, Universidad de Chile. Ph.D in Environmental Sciences, University of Western Ontario, Canada. Associate Professor and Director of the GAIA-Antarctic Research Center, Universidad de Magallanes, Chile.

Alejandra Urra Ph.D in Geography, Bristol Glaciology Centre, University of Bristol, UK. Researcher at Rio Cruces Wetland Centre, Universidad Austral de Chile, Valdivia, Chile.

Brian Reid BS in Neurobiology, Cornell University, USA. Ph.D Aquatic Ecology, Montana University, USA. Between 1994-2001 worked in conservation NGOs in the USA. Resident researcher at CIEP, Coyhaique, Aysén Region, Chile.

Part VII
Socio-environments

Conservation and Indigenous People in Chilean Patagonia

16

José Aylwin, Lorena Arce, Felipe Guerra, David Núñez, Ricardo Álvarez, Pablo Mansilla, David Alday, Leticia Caro, Cristián Chiguay, and Carolina Huenucoy

Abstract

The purpose of this chapter, which was prepared by an interdisciplinary and intercultural research team, was to analyze the tensions and challenges associated with biodiversity conservation in Chilean Patagonia as it relates to the

J. Aylwin (✉)
Globalization and Human Rights Program, Observatorio Ciudadano, Faculty of Legal and Social Sciences, Universidad Austral de Chile, Antonio Varas 428, Temuco, Chile
e-mail: jose.aylwin@gmail.com

L. Arce
Biodiversity and Alternatives to Development Program, Observatorio Ciudadano; Southern Cone Coordinator, TICCA Consortium, Territories and Areas Conserved By Indigenous People and Local Communities, Temuco, Chile

F. Guerra
Legal Area, Observatorio Ciudadano, Faculty of Legal and Social Sciences, Universidad Austral de Chile, Antonio Varas 428, Temuco, Chile

D. Núnez
NGO Poloc, Cousin 216, Providencia, Santiago, Chile

R. Álvarez
School of Archaeology, Universidad Austral de Chile, Balneario Pelluco, Los Pinos S/N, Puerto Montt, Chile

P. Mansilla
Institute of Geography, Pontificia Universidad Católica de Valparaíso, Valparaíso, Chile

D. Alday
Yagán Community, Bahía Mejillones, Chile

L. Caro
Community Grupos Familiares Nómades del Mar, Vina del Mar, Chile

C. Chiguay
Mon Fen de Yaldad Community, Quellón, Chile

© Pontificia Universidad Católica de Chile 2023
J. C. Castilla et al. (eds.), *Conservation in Chilean Patagonia*, Integrated Science 19,
https://doi.org/10.1007/978-3-031-39408-9_16

Mapuche- Williche, Kawésqar, and Yagán Indigenous peoples that have histor-ically inhabited this vast territory. We describe how Indigenous people settled and occupied this territory over time, then identify the specific forms and scope of relationships between these peoples and protected areas through seven case studies. The chapter confirms the existence of clear overlaps between protected areas and the ancestral lands of Indigenous people, as well as their current use by these peoples. It also documents the absence of free, prior, and informed consent in the establishment of these protected areas, as well as the current exclusion of Indigenous people from their governance. We verified the lack of awareness and recognition of Indigenous people's own initiatives for the pro-tection and conservation of this territory, as well as their contributions to the conservation of biodiversity. Finally, we provide recommendations based on international standards concerning Indigenous rights and emergent conservation guidelines, in order to promote a respectful, collaborative, and synergic relation-ship between public, private, and Indigenous people's conservation strategies in Patagonia.

Keywords

Patagonia • Chile • Indigenous people • Protected areas • Conservation • Human rights

1   Introduction

Nature conservation models and protected areas (PA) have their origins in the late 19th and early twentieth centuries, in the processes of consolidation of the states that created them, ignoring the native peoples, their lands and territories of customary use [42]. The first PAs created in the Americas were conceived as spaces administered by states whose objective was to preserve and strictly protect nature from any human intervention [18]. This model generated the exclusion and displacement of numerous native communities that had inhabited and preserved their ecosystems for centuries, impacting their ways of life and cultures [41]. Chile was no stranger to this trend. Efforts began to create PAs under the name of "Forest Reserves" at the end of the nineteenth century and the beginning of the twentieth century, a time of expansion of state borders towards the north and south, and most of these were established on lands of ancestral use and occupation by Indigenous people.[1]

C. Huenucoy
Kawésqar Community Resident, Puerto Edén, Chile

[1] The first Fiscal Reserve was created in Malleco in 1907. The forest reserves of Tirúa, Alto del Bío-Bío, Villarrica, Llanquihue, Petrohué, Puyehue, and Chiloé were established between 1907 and 1913. All of these are territories of ancestral occupation and use by the Mapuche people, with 600,000 hectares distributed between Concepción and Puerto Montt. In contrast, the State of Chile recognized Mapuche peoples; land rights for only 407,695 hectares divided into 2,318 titles (known as títulos de merced) in the process of settlement between 1880 and 1927 [6].

2 Scope and Objectives

The objective of this chapter is to identify the key issues in the relationship that has existed historically between the Indigenous peoples of Patagonia, inhabitants of the region, and the territories where the current PAs have been progressively established.

3 Methodology

An interdisciplinary and intercultural work team was formed to draft this chapter, composed of academics and professionals, together with representatives of the different Patagonian Indigenous people that occupy this territory. The analysis was based on bibliography, interviews with members and representatives of Indigenous communities, and available spatial information on PAs, Indigenous people and communities, and their conservation initiatives, contrasting this with industrial activities, particularly in the marine area.[2]

The study begins with the identification of the Indigenous people that historically inhabited and still inhabit this part of Chile, their history and current situation together with their practices, uses, and traditional knowledge of their ancestral lands. Then the evolution of conservation and human rights standards is presented, in particular those referring to Indigenous people and their relevance to the emergence of new conservation approaches. Subsequently, the public and private situation of various PAs in Chilean Patagonia is reviewed. The spatial overlap of these protected areas with the lands and territories of ancestral occupation of Indigenous people is examined, as well as their forms of participation in PA management and governance. It then analyzes the reception that these standards have had in Chile, in particular the applicable legislation, including the law that establishes Indigenous People's Coastal Marine Spaces (in Spanish Indigenous Peoples's Coastal Marien Spaces, ECMPO), the bill that creates the Biodiversity and Protected Areas Service (in Spanish SBAP), and the National System of State Wild Protected Areas (in Spanish Sistema Nacional de Areas Silvestres Protegidas del Estado, SNASPE).

---

[2] Only the marine and coastal spaces of Indigenous people were mapped, so other Indigenous people's conservation initiatives were not described spatially.

4   Results

4.1   Indigenous People in Chilean Patagonia

The vast terrestrial and maritime territory known today as Patagonia was inhabited by different people since the late Pleistocene (Fig. 1; [20]). The earliest dates for South America are about 15,000 years before the present (BP); they indicate a settlement route that followed the Pacific continental coastline [8]. Settlement dates of the southernmost areas of Patagonia are more than 10,000 years BP [32]. In historical times, that is, since the use of writing to describe the distribution of native peoples in the national territory, the Indigenous people inhabiting the regions coastline included, from north to south: the Williche (or Veliche), Chono, Kawésqar (or Alacalufes), and Yagán (or Yámana). In the continental steppe zone this included the Aónikenk (or Tehuelches) and in Tierra del Fuego, the Selk'nam (or Onas) and Haush (or Mánekenks). However, these identities were complex and included a diversity of sub-identities with their own mobile borders [4].

These native people generated a heterogeneous way of life, given the nature of the territory they have inhabited since ancestral times. These ways of life balanced between subsistence practices and their cosmogonic universes [1]. Their acts of provisioning, hunting, fishing, and gathering were strongly influenced by taboos, beliefs, and rituals, which operated at the moment of entering the hunting and gathering spaces, together with their daily practices. This cosmogony, and territorial occupation persist to the present day, mixed with new cultural elements (Fig. 2). Hence, one way to understand the biodiversity of Patagonia is through the languages and knowledge of its oldest populations and those that inherited and crossbred this heritage.

4.1.1   Williche and Chonos

The Williches inhabited the archipelagos of Chiloé and the continental coasts of the current provinces of Llanquihue and Palena and shared with the Chonos significant nautical mobility [26]. The wide biodiversity of edible species in the austral area, as well as the enormous variety of potatoes (*Solanum tuberosum*), are due to these peoples and their horticultural knowledge. European contact with the Williches occurred during the first millennium of the Christian era. This Indigenous people had an agro-pottery tradition,their language is Mapudungun [11].

The Chonos were nomadic fishermfolks, hunters and gatherers [35]. Their routes extended between Chiloé and the Gulf of Penas. However, the strong pressure exerted by colonial authorities and missionaries led to their cultural dismantling as an ethnic group [37]. During the eighteenth century, under pressure from the authorities and the Church, the last Chono nomads settled in Chiloé, gradually integrating into Williche society.

The archipelagos of Chiloé and Calbuco were occupied by the Spanish in 1567, establishing a regime of slavery during the sixteenth century, and of feudal estates (*encomienda*) and missionization of the entire Williche population in the following centuries [48]. At the beginning of the colonial period, as a consequence of

Fig. 1 Indigenous territories in Patagonia before the European colonization. *Source* own elaboration

slavery, forced labor, and epidemics, the Indigenous population in the Chiloé area decreased considerably. Later there was a demographic recovery that went hand in hand with the founding of small coastal villages. Around 1823, the Spanish Crown recognized part of the Indigenous property in the south of Chiloé, through the figure of royal estates (known as *potreros realengos*, [33]. In 1826, after the

Fig. 2 Lands and territory occupied and under current use by Indigenous people in Chilean Patagonia

forced incorporation of Chiloé into the Republic of Chile, and although these titles were initially recognized, from the beginning of the twentieth century their lands were alienated and transferred to third parties.

Today, most Williche communities are on the coast, with a way of life that combines small-scale agriculture with shellfish gathering and artisanal fishing, which was also adopted by the region's non-Indigenous population, giving rise to the culture known today as "cultura chilota" [40]. The sea has always been a fundamental source of food and health for these people. Beaches in particular are spaces of socialization loaded with cultural significance. The Williche communities of the Aysén coast and southern Chiloé retain cultural traits of the ancient Chono navigators, especially their spatial occupation patterns that have led them to inhabit the entire marine area between Chiloé and the southern channels (to the municipality of Cape Horn and the Chilean and Argentine Patagonia). This group has been strongly affected by extractive commercial logic [39]. They were used as labor in the fur industry and the Guaitecas cypress industry during the nineteenth century, and later in the extraction of shellfish and fin fish for the fishing industry, which is still the case today. Many families belonging to this group lives in small villages on the Aysén coast and are highly dependent on marine resources.

During the twentieth century, the descendants of this people made several requests to the State for land regularization claiming a status as settlers in the Aysén Region, which to date have been ignored or rejected. This situation has not changed since the creation of the National Corporation for Indigenous Development (in Spanish Corporación Nacional de Desarrollo Indígena, CONADI) in 1993, since this entity has not recognized the existence of Indigenous ancestral territories in the Aysén Region. This is expressed in the refusal to accept and review Indigenous land claims, as well as in the responses from the Ministry of Education to requests by several Indigenous representatives request to implement intercultural education in the region.[3]

4.1.2 Kawésqar

Further south, the Kawésqar inhabited the channels between the Gulf of Penas to the north and the west coast of Tierra del Fuego to the south more than 6,000 years ago [27], settling in transient stopovers in relatively large family groups. They obtained their subsistence mostly from the sea, transiting seasonally between the sheltered islands of the interior and the exposed coasts of the Pacific. They built boats that allowed them to move around this vast territory to capture their food and collect materials, leading a nomadic life [25].

Although the initial contacts with European navigators date back to the end of the sixteenth century, by the end of the eighteenth century, the expeditions of foreign sea lion hunters and, above all, Chilotes [19], had a negative influence on the life of the Kawésqar, who were affected by disease contagion, abuse, and aggression, and even the capture of boys and girls who were sent to Chiloé as "striplings".[4] This had serious demographic consequences, producing a drastic

---

[3] In recent years there have been at least four land claims under Law 19,253 of 1993, known as the "Indigenous peoples Law" and ILO Convention 169 on Indigenous and Tribal Peoples, which have been ignored by public institutions, thus maintaining the policy of invisibility. Among them are the Nahuelquin Delgado family, Traiguen Island, the Cabero Risco family, Luchin and Pomar islands, and José Huaiquen Gamin's family on part of Elena Island.

[4] Obliged to work in the field or in the home in exchange for food and lodging.

reduction of their population, as well as affecting the physical and psychological health of the survivors.

In the second half of the nineteenth century, during the colonization process promoted in Magallanes by the Chilean State, many Kawésqar were taken to the Salesian mission of Dawson Island and Río Grande where they died (Historic and New Deal Commission, 2003; in Spanish Comisión de Verdad Histórica y Nuevo Trato, CVHNT). At the end of the 1930s a radio station of the Chilean Air Force and the San Pedro lighthouse in charge of the Navy were installed in Puerto Eden in Kawésqar territory. The population was concentrated around these two centers, which produced a forced sedentary life with the consequent changes in their traditional ways of living. The population of Puerto Eden was estimated at less than a hundred people in the mid-twentieth century [19]. At the beginning of the 1990s the Kawésqar population numbered around 100, only 12 of whom lived in Puerto Eden, while the majority of the population lived in Punta Arenas and Puerto Natales in marginal housing and social conditions, generally subsisting on fishing and service industries [7].

Artisanal fisherfolks in El Manzano cove, Hualaihue, Los Lagos Región. Photograph by Jorge López

The Indigenous Law No. 19,253 of 1993 recognized the existence of the southern canoe peoples and established the State's duty to ensure their protection and development, providing support for health and social security, job training and subsistence (art. 72 and following). Policies have included the purchase of land for families residing in Puerto Natales.[5] In the 2017 Census, (National Institute of Statistics, 2018 (in Spanish Instituto Nacional de Estadísticas, INE) 3,448 people self-identified as Kawésqar, living mostly in the cities of Puerto Natales and Punta Arenas, in precarious social conditions. Information on the complexity of the Kawésqar world and territory and their current challenges as a people are beginning to be uncovered by recent research, including that carried out by their own members, who question the prejudices with which traditional literature has referred to them.[6]

4.1.3 Yagán

The Yagán inhabited the channels and coasts located in the southern area of Tierra del Fuego between the Beagle Channel and Cape Horn [38]. Their life and culture were very similar to those of the Kawésqar. Distributed in family groups, they led a nomadic life in the seas and channels of the territory, moving in wooden bark canoes and living by fishing, hunting, and gathering in the rich ecosystem they inhabited, developing a long canoeing tradition and a refined traditional ecological knowledge of their territory [9].

Although contact with European navigators began early, it was the Anglican missions of the nineteenth century that had the greatest impact on their way of life.[7] An important part of their population was forced to settle in the Ushuaia mission, generating serious health consequences and resulting in a critical decrease of the population. In the twentieth century, the missionaries settled in Mejillones, on Navarino Island, where in the 1950s many Yagán families settled until their transfer to Villa Ukika, on the edge of the Puerto Williams naval base [13]. By the 1920s, their population was estimated at 70 people [22].

Today, the Yagán population lives in Villa Ukika and Puerto Williams, as well as in different cities throughout the country and in Argentina. This peoples' main economic activities include fishing, construction, and for women, the sale of handicrafts [7]. The State has transferred around 4,000 hectares (ha) to the Yagán community since the enactment of the Indigenous Law, including the lands they occupied in Mejillones Bay on Navarino Island, Douglas Bay, and Dientes de Navarino Lagoon.[8]

---

[5] After 20 years of demands, the MBN announced the transfer to the Kawésqar community Ekcewe Lejes Woes -which groups a total of 15 families in the municipality of Río Verde- on a total of 400 hectares. http://www.elmostrador.cl/noticias/pais/2018/03/05/reivin dicacion-territorial- bienes-nacionales-hara-transferencia-de-mas-de-400-hectáreas-a-comunidad-kawesqar-de-magallanes/.

[6] Tonko [44], Aguilera and Tonko [1], Aguilera and Tonko [3], Aravena et al., [5].

[7] It was one of these expeditions, Fitzroy's, that in 1830 brought four Yagáns to England, among them Jemmy Button, where an attempt was made to "civilize" them.

[8] https://prensaantartica.com/2016/06/18/bienes-nacionales-transfiere-cerca-de-4-mil-hectár eas-al- pueblo-yagan/.

An Indigenous Development Area (ADI Cabo de Hornos) was established in the area in 2010, but only began operating in 2016. The Yagan population has experienced a significant recent increase. A total of 1,600 people self-identified as Yagán in 2017, most of whom live in poverty.[9] As in the case of the Kawésqar, there is now a new and growing scientific production on the Yagán territory and culture, which offers a different view of their archipelago, considering the threats that affect them, including aquaculture development, privatization of water and land, and the deconstruction of the image of an extinct people.

4.1.4 Aónikenk, Haush, and Selk'nam

The Patagonian steppes were inhabited by the Aónikenk in the continental portion and by the Haush and Selk'nam on the island of Tierra del Fuego [12, 28]. These nomadic land hunters moved in family groups in search of food and shelter, hunting animals such as guanaco, rhea and other species of birds, small mammals, and marine fauna.

The founding of Punta Arenas in the mid-nineteenth century, European colonization, and the granting of extensive land concessions for sheep farming to large foreign companies fomented after 1880 by Chile and Argentina, had serious consequences for these peoples. Far from being guaranteed their traditional lands, they were victims of displacement, poisoning by the ranchers, and genocide.[10] At the same time, the few surviving Haush were decimated by sea lion hunters. To this was added the agreement between governments, ranchers, and the Catholic Church, by virtue of which about 1,500 Selk'nam were transferred to the Salesian missions of Dawson Island in Chile and Rio Grande in Argentina in the last decades of the nineteenth century. There, as noted, most died from uprooting and disease [13].

The remaining Selk'nam population is currently settled in communities in the province of Santa Cruz in Argentina. The official literature reported the death of the last Selk'nam woman in 1974 [12], but in recent years there has been a reorganization of the descendants of this people, who live in the town of Tolhuin in Tierra del Fuego in Argentina.[11] Migrants displaced to Santiago, Chile during the twentieth century have formed the Covadonga Ona community, which was constituted in 2015, in order to work for the rescue and valuation of their cultural identity, obtaining legal personality as a private law corporation.

---

[9] This figure, as in the Kawésqar case, includes Yagán population throughout the country. (National Institute of Statistics, 2018).

[10] In the case, The Chilean State only recognized 10,000 hectares of land to chief Mulato of the Aónikenk in 1883, which led to their displacement to Argentina [7].

[11] In 1995 the Argentine State recognized the Rafaela Ishton community of the Selk'nam people. http://red23noticias.com/el-museo-de-la-plata-restituyo-restos-a-la-comunidad-selknam/.

4.2 Conservation Standards and Indigenous People

4.2.1 International Human Rights and Biological Conservation Standards

New conservation approaches have emerged in recent decades that recognize the virtuous relationship and influence of human communities, particularly Indigenous people and their cultures, on the biological diversity of the ecosystems they inhabit.[12] The recognition of these virtuous relationships, as well as of the rights of Indigenous people, has led to profound changes in classical conservation approaches, and the importance of biocultural conservation is increasingly recognized [6]. In parallel, various international instruments, including International Labor Organization (ILO) Convention 169 on Indigenous and Tribal Peoples (1989), the United Nations Declaration on the Rights of Indigenous people (UNDRIP, 2007), and the American Declaration on the Rights of Indigenous people (DADPI, 2016) have progressively recognized the rights to participation and consultation,free prior and informed consent before measures are taken that affect them; self-determination and autonomy; as well as rights over their lands, territories, and natural resources of traditional use and occupation. The ILO Convention 169 recognized the right of ownership and possession of Indigenous people over the lands they traditionally occupy, including those of nomadic peoples and shifting cultivators (art. 14.1). The UNDRIP has recognized the rights of these peoples to the natural resources in their territories, which cover the totality of the habitat they occupy or use in some way, including waters, coastal seas, and other traditional resources (art. 25 UNDRIP). These rights, in accordance with ILO Convention 169, include participation in the use, administration, and conservation of such resources (art. 15.1). Both the UNDRIP and the standards developed by the bodies of the Inter-American Human Rights System have gone further, recognizing the right of ownership of Indigenous people over their resources by reason of ancestral ownership or other traditional occupation or use (Inter-American Commission on Human Rights, 2009). This framework has been progressively taken up by the International Union for Conservation of Nature, IUCN (in Spanish Unión Internacional para la Conservación de la Naturaleza)). Since the 2003 World Parks Congress, which approved the Durban Accord and its Action Plan, the IUCN has promoted, along with respect for the rights of Indigenous people, the recognition of their contributions to conservation. It also developed guidelines that recognize the diversity of types of governance of protected areas by different actors, including state, co-managed, private, and governance by Indigenous people and local communities (Borrini-Feyerabend et al., 2013).

The IUCN has made an explicit call to conservation actors to apply the aforementioned UNDRIP, an instrument that includes the duty of States to restitute the lands, territories and natural resources that have been confiscated from these

---

[12] The interdependence between biological and cultural diversity has been conceptualized as biocultural diversity (Borrini-Feyerabend, et al., 2010).

peoples (IUCN, 2008). It coined the concept of "Indigenous conservation territories and local community conserved areas" (ICCAs) to refer to a diversity of areas that are collectively conserved by communities [6]. ICCAs have been defined as "natural and/or modified ecosystems containing significant biodiversity values, ecological benefits and cultural values voluntarily conserved by Indigenous people and local communities, both sedentary and mobile, through customary laws or other effective means" (Borrini-Feyerabend, et al., 2010). Both the IUCN and the Convention on Biological Diversity (CBD) have encouraged governments and conservation organizations to recognize and support ICCAs as examples of effective collective governance of biocultural diversity.[13]

The IUCN recently urged the development of good practice guidance in identifying, recognizing and respecting ICCAs that overlap with PAs, "before including any protected area on the IUCN Green List of Protected and Conserved Areas or before recommending inclusion in the World Heritage List."[14] It has also urged parties to "…promote the establishment of appropriate approaches, including fair and equitable access to information and meaningful participation of Indigenous communities in decision-making processes, to avoid negative impacts, especially from unsustainable externally-driven developments and other forms of land and ecosystem degradation."[15]

4.2.2   Implementation of International Standards in Chile

Terrestrial and marine PAs in Chile are regulated by numerous laws and regulations [43], many of which promote conservation approaches that do not adequately address their relationship with Indigenous people and local communities. As of 2011, the country had more than 20 laws and regulations on PAs (Ministry of Environment, 2011,in Spanish Ministerio del Medio Ambiente, MMA). The only law referring to Indigenous people is No. 19,253 of 1993 on the Protection, Promotion and Development of Indigenous people, which states that the participation of Indigenous people and local communities will be considered in Indigenous Development Areas (MMA, 2011) and in the administration of state protected areas, which must be determined in agreement with the National Forestry Corporation (in Spanish Corporación Nacional Forestal, CONAF, art. 35). In addition to these regulations, numerous international conventions ratified by Chile are currently in force (MMA, 2011), which have resulted in 13 PA categories administered by different

---

[13] Thus in 2008 IUCN urged its members to "Fully recognize the conservation importance of Indigenous Conservation Territories and other Indigenous Peoples and Community Conserved Areas (ICTs and ICCAs)—which include conserved sites, territories, landscapes, seascapes and sacred sites—that are managed and administered by Indigenous people and local communities, including mobile peoples" (IUCN, 2008: Recommendation. 4.049, 2008). (IUCN, 2008: Recommendation. 4.049, 2008).

[14] (IUCN, 2016): WCC-2016-Res-030-SP.

[15] (IUCN, 2016): WCC-2016-Res-088-SP.

bodies under various ministries [36], which do not have technical competency for the relationship between conservation and Indigenous people.[16]

The administration of protected areas in Chile is still exercised by CONAF, a private law entity created in 1973 under the Ministry of Agriculture.[17] Law No. 20,417 was enacted on January 12, 2010, as part of a reform of the country's environmental institutions, which created the MMA, the Environmental Evaluation Service, and the Superintendence of the Environment. In 2011, the government of President Sebastián Piñera sent a bill to Congress (Bulletin n°7487–12) for the creation of the National System of State Protected Areas, which would be under the supervision of the Biodiversity and Protected Areas Service. However, this initiative did not advance significantly. In June 2014, former President Michelle Bachelet sent to the Senate a new bill for the creation of SBAP and SNASPE (Bulletin No. 9,404–12), that suffered from a series of limitations from the perspective of human rights and biocultural conservation guidelines referred to above, and which in 2016 was submitted to a consultation process with Indigenous people.[18] In November, 2017, the Senate Environment Committee approved a second version of the bill, which incorporated some proposals that emerged from the consultation, such as the recognition of biodiversity conservation practices of local communities and Indigenous people (art. 50); the creation of a new category of protected area specifically for Indigenous people called "Indigenous peoples' Conservation Areas" (art. 67); and empowering the SBAP to enter into management agreements with local authorities, organizations, and Indigenous communities (art. 71). Although important, these advances are far from the international guidelines referred to, particularly in terms of land restitution and recognition of territories of traditional Indigenous occupation and governance. After its approval by the Senate, the bill passed to the Chamber of Deputies in August 2019, where it is currently being analyzed.

Another important piece of legislation for the Indigenous people of Chilean Patagonia is Law No. 20,249, which created the ECMPO, known as the Lafkenche law. This law aims to "safeguard the customary use of these spaces, in order to maintain the traditions and use of natural resources by the communities linked to the coast" (Art. 3), and is considered by coastal Indigenous communities as a tool for the recognition of marine spaces of ancestral occupation and use, as well

---

[16] The ministries most directly involved in the creation and administration of protected areas are: Ministry of Agriculture (CONAF) and MBN in terrestrial environments; the Ministry of Defense (Undersecretariat of the Armed Forces, DIRECTEMAR) and the Ministry of Economy (SER-NAPESCA) in marine environments. It should be noted that most of the PAs in Chile are owned by the State and administered by public agencies: CONAF in terrestrial environments, and SER-NAPESCA and DIRECTEMAR in marine environments.

[17] CONAF's background includes the 1931 Forestry Law and the creation of the Administration of National Parks and Forest Reserves in the 1960s.

[18] See questioning of the bill by Indigenous people and human rights organizations. Available at: https://observatorio.cl/organizaciones-cuestionan-proyecto-de-ley-de-biodiversidad-y-su-avance-sin-haber-concluido-consulta-indigena/.

as for the protection and conservation of ecosystems linked to their ways of life and cultures [49].[19] However, the implementation of this law has been slow and arbitrary. As of September 2020, only 13 of the 98 ECMPOs requested by Indigenous communities[20] have been decreed, all after long processing periods exceeding 4 years[21] [29]. There are currently more than 79 applications in process in Chilean Patagonia,only 12 have destination decrees. Indigenous communities have encountered various obstacles to the use of this law as a tool to protect marine spaces and their customary uses [29, 49]. Such obstacles are fundamentally the result of the overlap between ECMPO areas and the interests of the aquaculture and fishing industry (Fig. 3), whose adverse impacts on marine biodiversity are recognized [10, 34].[22] This is due to the fact that requests for the creation of ECMPOs by law have preference over any other request to affect these areas for other purposes, such as aquaculture concessions. In addition, the Indigenous communities have found little support from the State and conservation organizations in their ECMPO applications and have had to face these processes with their own means and by generating alliances between communities.[23] In spite of all this, ECMPO applications have had the practical effect of halting the expansion of the salmon industry into the fjords and canals of Chilean Patagonia.[24]

4.3   Case Analysis

This section reviews seven cases of terrestrial and marine protected areas, both public and private, that are closely related to Indigenous people in Chilean Patagonia. These cases were identified through spatial analysis (Fig. 4) and interviews with local experts.

---

[19] References to the sustainable use of coastal marine areas, ecosystem restoration, and the preservation of areas of high ecological value are present in the ECMPO requests made by Indigenous people, such as: the request of the Huichas Islands, that of the Association of Carelmapu Communities, and that of the Pu Wapi communities of Melinka and the Association of four communities of Puerto Aguirre in the Guaitecas archipelago.

[20] The 13 requests submitted during 2018 to the Undersecretariat of Fisheries are registered with "entry penalties" by this entity. The reason for this is unknown.

[21] Status of ECMPO Applications in Process. Available at: http://www.subpesca.cl/portal/616/w3-propertyvalue-50834.html.

[22] Several studies document such impacts, especially if we consider that the production is concentrated in a priority area for marine conservation worldwide, recognized by the Convention on Biological Diversity as an "ecologically or biologically significant area" (EBSA) See https://chm.cbd.int/database/record?documentID=204089.

[23] This is the case of organizations such as Identidad Territorial Lafkenche and the recently created Coordinadora Willi Lafken Weichan, which brings together more than 40 communities of the Chiloé and Aysén archipelagoes, who are playing a key role in these processes.

[24] A 2018 report by the Sociedad de Fomento Fabril (SOFOFA) states that 612 aquaculture concession applications are affected by ECMPO requests. Of this total, 56.9% of suspended aquaculture concession applications are located in the Los Lagos Region and 30.7% in the Magallanes and Chilean Antarctica Region (SOFOFA, 2018).

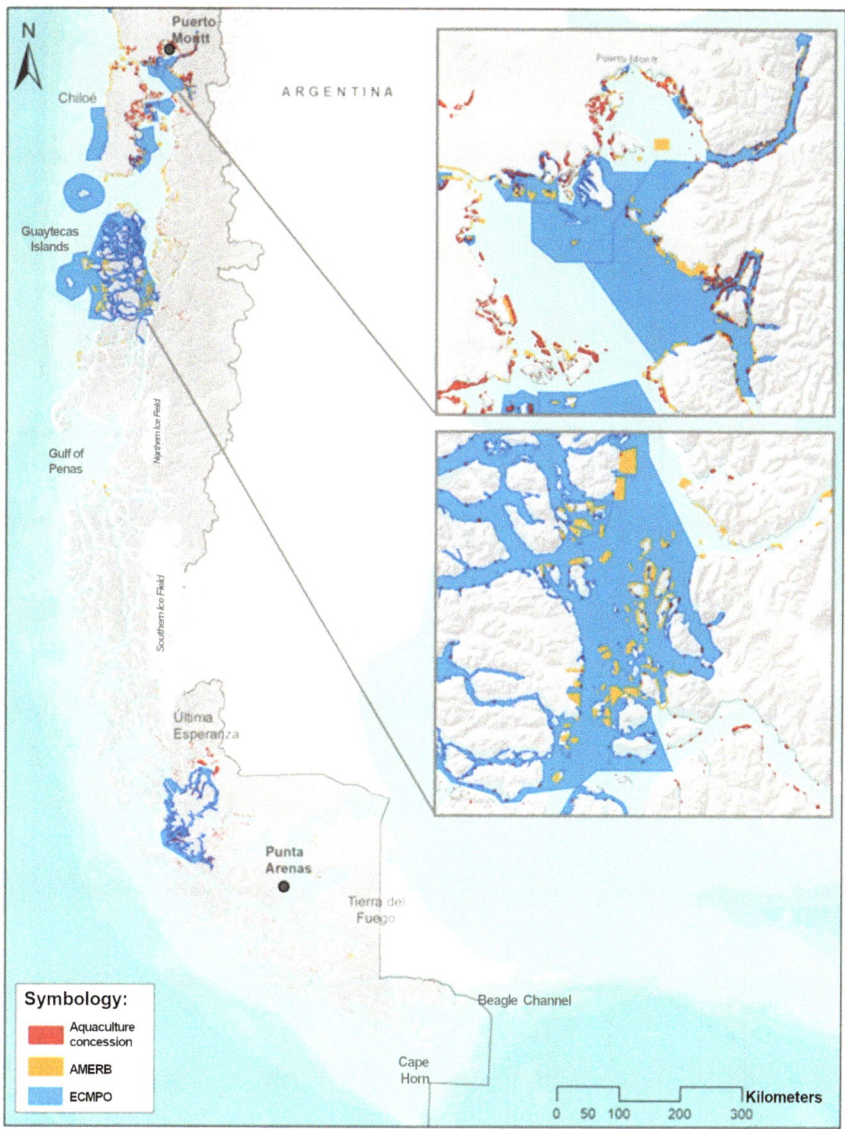

Fig. 3 Original Indigenous People's Coastal Marine Space (in Spanish Espacios Costeros ECMPOs; Salmon Concessions (in Spanish Concesiones de Salmonicultura); Management and Exploitation Areas for Benthic Resources (in Spanish Areas de Manejo y Explotación de Recursos Bentónicos, AMERB). September 2018

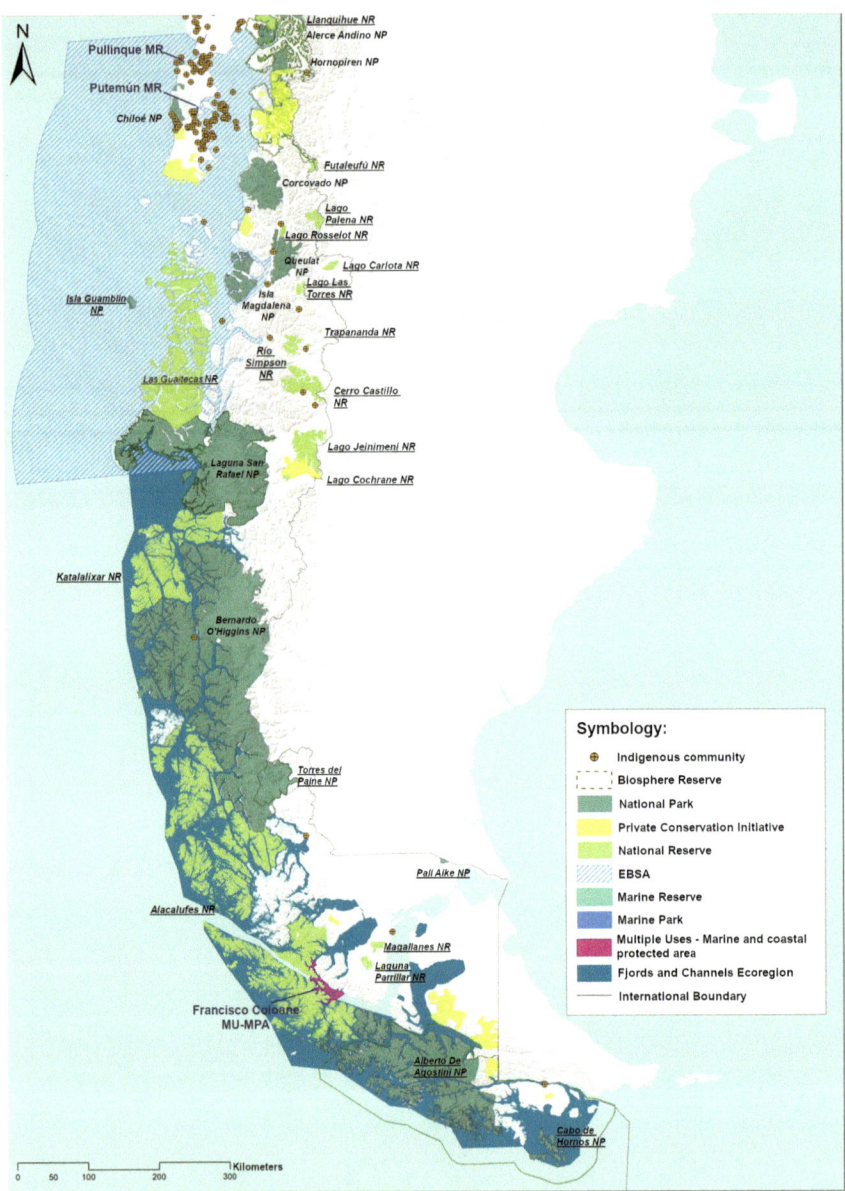

Fig. 4 Protected areas and Indigenous communities. *Source* own elaboration

Table 1 shows the overlaps between PAs and Indigenous communities. It shows that there is a total overlap between the area that today forms part of the identified PAs and the lands of current or traditional use and occupation of the related Indigenous communities in six of the seven cases. There are different types of claims by the communities involved, ranging from the restitution of lands occupied by PAs to shared governance. The last column shows the type of governance of each protected area according to IUCN guidelines: state, shared, private, Indigenous people, and local communities. It is noteworthy that according to CONAF five of the 7 cases identified have state-only governance types, one has private governance, but none has community governance and only one case, with modifications in 2018, has shared governance, which however is restricted to the terrestrial area without including the maritime area. Below, we describe and analyze each case.

4.3.1 Williche Communities of Hualaihué and Hornopirén National Park

In the municipality of Hualaihué there are at least two Williche communities that face overlapping conflicts between their ancestral territories and Hornopirén National Park, which was created in 1988 without their consent.[25] These are the Rüpü L'afken' and Mapu Peñi communities, both from the Paillan-Peranchiguay family lineage, communities that for decades have demanded the lands of the former Colimahuidan estate ranch, which surrounded Lake Cabrera. The estate was subdivided during the military dictatorship and the part corresponding to the lake was sold to a private company.[26] However, the Pillán family, historically linked to this territory, maintained its occupation. In 2007, the company installed a barrier that was removed by the community, the family was sued for this act as an invasion of private property. The family's countersuit requested free access to the lake, since it contains places of cultural significance, a reminder of the families buried by a landslide, and economic values associated with summer tourism, with facilities and thermal water rights. The right of use was recognized in the judgement; however, ownership of the land was not restored. In exchange, the government offered to give them lands of the same Colimahuidan estate ranch, not beside the lake, which were later incorporated into Hornopirén National Park, ignoring the customary property rights.

---

[25] DS. No. 884/1988 MBN.
[26] Members of the Rüpü L'afken' and Mapu Peñi communities report that CORFO was the body that sold the land, and that at the same time the State Defense Council filed a lawsuit against the sale, but the Puerto Montt court ruled that the claim was time-barred (more than 5 years). The lands of the lake inhabitants were surveyed, and a regularization process was initiated during the agrarian reform, but they were not given titles before the military coup. At some point during the dictatorship, a police picket entered Lake Cabrera and destroyed the houses, expelling the inhabitants from the area (Interview with representatives of the Rüpü L'afken' and Mapu Peñi communities of the Lake Cabrera sector, September 28, 2018).

Table 1 Protected areas (PAs) and indigenous people in Chilean Patagonia

Region, province and municipality	PA (date of creation)	Surface (ha)	Management instrument	Indigenous Community	Overlap	Type of governance
Los Lagos: Llanquihue, Palena Cochamó, Hualaihué	Hornopirén NP (1988)	Land: 66,196	Management Plan 1999 (Not in effect)	Rüpü L'afken' and Mapu Peñi (communities may have been left out of the study)	Partial	State (CONAF)
Los Lagos: Chiloé Ancud, Castro, Chonchi, Dalcahue	Chiloé NP (1982)	Land: 42,568	Management Plan 1997 (Not in effect)	Huentemó, Palihue, Chanquin and Cucao	Total	State (CONAF)
Los Lagos: Chiloé Quellón	Tantauco Park (2005)	Land: 118,000	No information	Mon Fen de Yaldad, Cocauque, Trincao, Inio, Inkopulli, Tuweo, Coldita, Piedra Blanca and Huequetrumao	Total	Private (Fundación Futuro)
Aysén: Aysén Aysén, Cisnes	Las Guaitecas FR(1938) Guamblin Island NP (1967)	Ground: 1,097,975 Land: 10,625	None None	**Puerto Aysén**: Melipichun Nitor, Guaquel Mariman; Melinka: Pu Wapi Community **Puerto Aguirre**: Pewmayen, Antunen Raín, Fotum Mapu and Aliwen	Total	State (CONAF)
Aysén/ Magallanes: Capitán Prat, Última Esperanza O'Higgins, Tortel, Natales, Torres del Paine	Bernardo O'Higgins NP 1969	Land: 3,525,901 Navy: 1,025,902	Management Guide 2000 (Not in force)	Kawésqar community Resident in Puerto Eden	Total	State (CONAF)

(continued)

Table 1 (continued)

Region, province and municipality	PA (date of creation)	Surface (ha)	Management instrument	Indigenous Community	Overlap	Type of governance
Magallanes: Magallanes, Última Esperanza Punta Arenas, Río Verde, Natales	Kawésqar NP Kawésqar NR	Land: 2,842,329.1 Maritime: 2,628,429.2	None	Communities: As Wal La Iep, Atap, Rio Primero residents, Grupos Familiares Nómades del Mar Kawésqar people in general	Total	NP: shared NR: State (CONAF)
Magallanes: Antarctic, Magallanes, Tierra del Fuego Cape Horn, Punta Arenas, Timaukel	Alberto D'Agostini NP 1965 Cape Horn NP (1945)	Land: 1,460,000 Marine: 1,158,837 Land: 58,917	None None	Bahía Mejillones Yaghan Community (Jeeinikin Usi Yagan)	Total	State (CONAF)

Source Compilation based on the National Register of Protected Areas of the Ministry of Environment, Decree creating Kawésqar NP and NR, and interviews with local experts. Available at: http://areasprot egidas.mma.gob.cl/

As part of the State's commitment to establish a National Parks Network (RPN) in Chilean Patagonia, an administrative process began of donations of land, belonging to foundations and/or companies related to ecologist Douglas Tompkins, the integration of state property into existing National Parks, and the reclassification of protected areas. Supreme Decree n°2 of the MBN was issued on January 15, 2018, which expanded Hornopirén NP, locating it in the municipalities of Cochamó and Hualaihué, in the Los Lagos Region.[27] Although the communities do not oppose the declaration of part of their territory as a National Park, they do demand to participate in its administration and governance.[28] At present, the community continues to use Lake Cabrera and request its ownership, while also demanding the administration of the area recently incorporated into the National Park.

---

[27] By means of this act, the executive incorporates public lands as part of this conservation unit, including the Hornopirén volcano cone, as well as the land called Lot 13 "a", with an approximate area of 108.2 hectares located in the Chaqueihua sector, municipality of Hualaihué.

[28] The communities made an unsuccessful attempt to work inside the park, within the framework of the Hornopirén NP Management Plan (Resolution No. 239/1999, CONAF). About 8 years ago they had reached an agreement with CONAF to carry out tourism activities inside the park, including The communities would be in charge of the infrastructure construction. However, when they needed wood at the construction site, CONAF indicated that they had to bring the wood from outside the park because they did not have authorization to extract it inside the park. Because the only way to bring materials to the mountain site was by helicopter, the initiative did not prosper.

4.3.2   Williche Communities and Chiloé National Park

Chiloé NP was created in 1982.[29] It currently covers an area of 42,567 hectares (ha) located in the municipalities of Ancud, Castro, Chonchi, and Dalcahue.[30] Williche presence in the area is ancestral,[31] and the establishment of the NP on historically Williche lands included episodes of repression and violence (Correa, 2003, unpublished: "El Parque Nacional Chiloé y las comunidades huilliches", paper presented at the seminar Taller Áreas Protegidas y Comunidades Humanas, Chiloé, August 24–26, 2003). After its creation, around 80 Williche families remained inside the park as occupants [17]. In 1995, during the government of President Eduardo Frei Ruiz Tagle, an agreement was signed between the Chanquín and Huentemó communities to exclude the land they occupied from the NP [17]. In 2000, these lands were transferred to CONADI in order to grant 2,765 ha to the Williche communities (Decree 368 of the MBN,2000). Another 1,841 ha were then added to the NP, In total, 4,727 ha were granted in individual and communal titles. This concluded in 2015 and benefited a hundred families from these communities as well as Chanquin Palihue.

The NP's Management Plan, which dates back to 1997, was only supposed to be in effect until 2007, but it has not been updated. The plan recognizes the historic settlement of the Williche communities, and one of its objectives is to "promote and encourage interaction with neighboring communities as a factor contributing to their development" (CONAF, 1997; 128). However, it did not consider the inclusion of these communities in its administration. After the titling of the communities, CONAF created a consultative council for the NP with the inclusion of the Williche communities. However, this plan is not operational. During the same period, an administrative agreement was signed for the NP visitor center that involved the communities.[32]

The Williche communities carry out tourism initiatives in the area, including work as trail guides, horseback riding, and cabin rentals.[33] However, to date there is no community participation in the governance of the NP, despite the growing influx of visitors, totaling nearly 50,000 in 2015 (CONAF, 2017), nor do they participate in the benefits that the NP generates.[34] The Williche communities maintain their claim over NP lands, as well as their participation in the governance of the NP.[35] The same communities, together with four others bordering the NP (Cucao,

---

[29] DS No. 734 of the MBN.

[30] CONAF, available at: http://www.conaf.cl/parques/parque-nacional-chiloe/.

[31] By the end of that century, the town of Cucao had a population of 120 people. Today, 450 people live there. Available at: https://www.chile365.cl/es-region-10-isla-de-chiloe-cucao.php.

[32] According to Jorge Huenuman, Lonko (traditional chief) of the Chanquín community, this agreement allowed them to receive income from visitors to the NP (Interview with Jorge Huenuman, October 15, 2018).

[33] Interview with Jorge Huenuman, Lonko of Huentemo, September 21, 2018.

[34] CONAF recently called for bids for the administration of tourism services in the NP, which was awarded to a private company until 2020.

[35] According to Jorge Huenuman, Lonko of Huentemo, "the park has more than 40,000 ha and they gave us back a small slice, and the management that our ancestors did was much more than the lines

Quilque, Chaique Cole Cole, and Montaña Chonchi) submitted in 2016 an ECMPO request for an area of 203,154.26 ha of adjacent marine area, to carry out seaweed harvesting, hunting, diving, and fishing activities.[36] The approval of this ECMPO is pending.

4.3.3 Williche Communities of Southern Chiloé and Tantauco Park

In the municipality of Quellón in the south of the large island of Chiloé the Williche communities of the territories of Weketrumao, Yaldad, Coldita, and Inío, have varying degrees of overlap with lands that currently form part of the 118,000 ha Tantauco Park, owned by Fundación Futuro (belonging to former President Sebastián Piñera) and declared a private conservation initiative in 2005. These communities claim areas that are currently within the park and recognize its entire area as a territory of ancestral use.

According to CONADI, the communities claim an area of approximately 18,000 ha corresponding to the old royal title (in Spanish Títulos de Realengo) granted to them. This title was registered for Yaldad in the Castro land registry in 1889, but ownership of the land was not recognized by the State, which gave it in concession to natural resource exploitation companies at the beginning of the twentieth century. As for the use of the rest of the territory, Lonko Cristian Chiguay of the community indicates that "[…] our people have made use of it far beyond those limits, they have made spiritual use of it, hunting, gathering *l'awen'* and medicinal plants, to the west towards the Pacific, and also from Lake Chaiguata north […]". Elías Colivoro of the Cocauque sector points out that towards the west side of the community there is no territory boundary, and that ancestral use extends to the Pacific Ocean, where until a few decades ago otters and sea lions were hunted and *quilineja*, a fiber used to make rope, was collected.

Most of the current Tantauco Park had no ancestral use or occupation except on the coast, because the communities assign this space a high spiritual value, since the forests and mountains are inhabited by the *ngen'* that sustain the balance and good health of the territory.[37] There is currently no dialogue between the Foundation that manages Tantauco Park and the Williche communities in the territory, with the exception of the Inío community, which signed an agreement for the loan of half a hectare per family. For the leaders of the Yaldad communities, the ideal would be for the communities to be able to administer the protected areas that are

---

that remained," which is why they maintain their claim to it. Moreover, the communities cannot make traditional use of the area, such as working with dead larch, or promote tourism activities such as trails, as they are prevented by the NP administration (Interview with Jorge Huenuman, Lonko of Huentemo, September 21, 2018).

[36] Available at: www.subpesca.cl/portal/616/articles-79852_recurso_1.xls.

[37] In the words of the Lonko of the Mon Fen de Yaldad community, "the people of the communities decided not to occupy those spaces, to leave them as a natural reserve, to give them more life, to not interrupt the itrofill mongen [biodiversity], to not create an imbalance of life on the island; it seems that nobody occupies those spaces, but there is a need to nourish the island, the inhabitants, and all the living beings on this island […].for us nature is very important to nourish us with the spiritual strength that we need […]. (September 2018).

part of the ancestral use territory.[38] These communities are currently working to apply for ECMPO status for the entire southern part of the large island of Chiloé, including the coast of Tantauco Park.

4.3.4 Williche Communities of Melinka, Puerto Aguirre and Protected Areas

Most of the inhabitants of the towns of Puerto Aguirre and Melinka, in the Aysén region, identify themselves as Williche,[39] and are currently organized into five Indigenous communities, one in Melinka and four in Puerto Aguirre. The territory of ancestral use and occupation of these communities includes the Chonos Archipelago and the Guaitecas Islands. A large part of this territory is protected by the Las Guaitecas Forest Reserve (FR), created in 1938[40] and the Guamblin Island NP, created in 1967.[41] The creation of the Las Guaitecas RF was carried out with the objective of protecting and regulating cypress extraction, which meant the confiscation of lands and forests of ancestral use of the Williche and Chono communities of the archipelago. However, these protected areas have not been able to avoid the intense pressure from the salmon industry to develop aquaculture in the area,[42] affecting both marine biodiversity and the main livelihoods of Indigenous and local communities [10, 23]. Given this situation, the communities of Melinka and Puerto Aguirre have submitted applications for the creation of ECMPOs, in order to safeguard their customary use rights and pursue their development priorities.

To date, the communities have not had any conflicts with CONAF because neither the Guaitecas NR nor Guamblin Island NP have specialized personnel or infrastructure in situ. Daniel Caniullan, Lonko of the Pu Wapi de Melinka community, notes that there is no CONAF presence and that the community has

---

[38] "I believe that the administration of the parks should be passed to the communities, because in all these National Parks or Reserves, even if they are private, there are ancestral uses of the inhabitants, then it is necessary that this use be recognized (...) it is important that NGOs work with the communities, because who else but the communities know how to preserve and conserve..." (Lonko Cristian Chiguay, Lonko Mon Fen de Yaldad community, September 2018).

[39] According to the 2017 Census, in the municipality of Guaitecas and Puerto Aysén 51% and 34% of the population respectively self-identify as Mapuche-Williche. Puerto Aguirre is a locality of the municipality of Aysén and its Indigenous proportion is above the communal average.

[40] D.S. No. 2612/1938 Ministry of Lands and Colonization (in Spanish Ministerio de Tierras y Colonizacion,MTC).

[41] D.S. No. 321/19767 Ministry of Agriculture (in Spanish Ministerio de Agricultura, MA).

[42] Article 158 of Law No. 18,892 on fishing and aquaculture as a general rule prohibits all extractive fishing and aquaculture activities in the portions of water—be they lake, river or maritime zones—that form part of the SNASPE. However, by exception, these extractive activities may be authorized if they are carried out in maritime areas that are part of national and forest reserves, also allowing the use of land portions that are part of these reserves to complement maritime aquaculture activities, with prior authorization from the competent bodies. An interesting pronouncement in this sense can be found in Opinion No. 38,429 of the Office of the Comptroller General of the Republic in 2013, where it ruled on the development of aquaculture activities in the interior of the Bernardo O'Higgins and Alberto de Agostini National Parks [21].

complained about the contamination produced by the salmon farms in the channels. Nelson Millatureo, an Indigenous leader from Puerto Aguirre, says the same. Both communities claim rights of use and administration of both the sea and the land. The communities of Puerto Aguirre have stated the need for co-management of these protected areas and have agreed with the conservation objectives, but have allowed the use of, and in specific cases effective occupation of some areas. The history of the Williche people of southern Chiloé and the Aysén coast is strongly linked to this territory, which is why both protected areas, Las Guaitecas RF and Guamblin Island NP, overlap with the ancestral territory of these communities.

4.3.5 Kawésqar Community Residing in Puerto Eden and Bernardo O'Higgins NP

In 1969, the Bernardo O'Higgins NP was created in Kawésqar territory, without the consent of the community, with an initial area of approximately 1,761,000 hectares and with the objective of conserving the area and protecting it from various threats, as it is "[...] land exposed to occupation, forest fires and because its flora and fauna are threatened with extinction[43]". In this extensive area included in the SNASPE, hunting and gathering of natural goods is prohibited thus depriving the Kawésqar of their main means of livelihood, affecting their ancestral practices as nomadic hunter-gatherers.

Since its creation to date, this National Park has undergone important modifications through two different decrees,[44] reaching its current area of 3,525,901 ha of land and 1,025,902 ha of sea, making it the largest state protected area in Chile and the second largest in Latin America. It contains the Southern Ice Fields and important channels of western Patagonia, extending west to the Pacific Ocean and south to Puerto Natales in the province of Última Esperanza. In the 1985 modification, it is stated "to protect by all possible means the anthropological values such as the remains of the human communities of the channels (Kawésqar), recognizing the presence and importance of this ancestral community. Then in 1989, together with the expansion of the park to its current area, some land within the NP and adjacent to the town of Puerto Eden was exempted in order to regularize the ownership situation of lands occupied in the "Villa Puerto Eden", mostly belonging to Kawésqar families, who would have a space "to meet their firewood needs" (DS 392). During the 1990s, in the face of political changes in Chile, the Kawésqar of Puerto Eden began a process of vindication and exercise of their rights, using the tools provided by Chilean and international laws. The Kawésqar community residing in Puerto Eden was recognized in 1994. One of the community's main strategies has been to form alliances with researchers and scientists to demonstrate the ancestral occupation of the territory by the Kawésqar. Several archeological studies and collaborations with botanists, biologists, and linguists have made it

---

[43] D.S. No. 264 of the MA, available at: https://www.leychile.cl/Navegar?idNorma=269844.

[44] D.S. No. 135, Ministry of Agriculture, April 24, 1985, available at: https://www.leychile. cl/Consulta/m/norma_plana?org = &idNorma = 164271; D.S. No. 392, Ministry of Agriculture, June 14, 1989, available at: https://www.leychile.cl/Navegar?idNorma=93678.

possible to produce an Ethnogeographic Guide [2], identifying channels, fjords and bays with Kawésqar names, as well as funerary, birth, and taboo sites that were used ancestrally, demonstrating the deep knowledge of this vast territory by the Kawésqar community resident in Puerto Eden.[45]

CONAF commissioned a baseline of the NP which was developed between 2009 and 2011 to provide information for the management plan and a tourism development plan, which "included the active participation throughout the project of the Kawésqar Indigenous Community Resident in Puerto Eden [5]. However, this management plan has yet to be approved and is still under review by CONAF. In parallel to this planning process, the possibility of establishing Areas Apt for Aquaculture (AAA) emerged within the framework of the coastal uses zoning process for the Magallanes Region between 2008 and 2014 in the framework of the zoning of the coastal border of the Magallanes Region, opening a controversy as to the legality of granting aquaculture concessions within the coastal zone of terrestrial National Parks. As a result of the requests for an official response on this issue by the Kawésqar Community of Puerto Edén, in 2013 the Comptroller General of the Republic confirmed that the marine areas located within the perimeter of this NP are protected and that aquaculture may not be developed in them, applying the commitments assumed by the State of Chile when it ratified the Washington Convention in 1967 [21], see also [43].[46]

In 2013, the community drafted and made public the Jetárkte Declaration, which states that its territory, called *Kawésqar-wæs*, extends from the entrance of the Gulf of Penas in the north to Diego de Almagro Island in the south, where the Bernanrdo O'Higgins NP, the Katalalixar NR, the Archipelago Madre de Dios National Protected Asset, and the northern area of the Alacalufes Reserve (now Kawésqar NP and NR), including the Southern Ice Field, are located. The Kawésqar community of Puerto Edén recently indicated that CONAF urgently needs to make official the Management Plan that was jointly elaborated over five years ago.

4.3.6 Kawésqar Communities and the Kawésqar National Park and Reserve

An agreement was signed between then President Michelle Bachelet and the Tompkins family foundations in 2017, formalizing the transfer to the Chilean State of 407,625 ha and the commitment of 949,000 ha of public lands for the creation of a network of eight NPs in Patagonia. This included the decision to expand and reclassify the Alacalufes National Reserve[47] as a National Park. Given that this

---

[45] Kawésqar Historical Photos, made by the community during 2014. Available at: https://www.iccaconsortium.org/index.php/2018/08/27/parque-nacional-bernardo-ohiggins-territorio-kaw esqar-waes-conservation-and-management-in-an-ancestral-territory/?fbclid=IwAR2ZKo2tyUK_ KuLgTHy6MEIvw13t_PEJ7_CMFaiXnL20coCcmc3uNxcPnOQ.

[46] Office of the Comptroller General of the Republic, Opinion No. 38,429 on Development of Aquaculture Activities in National Parks, Santiago, June 18, 2013.

[47] Created by Decree No. 618 of the MBN 3, 1987.

area "[…] is located in a territory that has been declared a National Park and is located in a territory that has cultural and symbolic significance for the Kawésqar people, since it corresponds to navigation areas ancestrally occupied by this people and that surround the protection area", the MBN n Indigenous people.[48] The communities that participated in the consultation accepted the proposed measure on the condition that its name be changed to Kawésqar NP. They also demanded co-administration of the future NP, including its waters, through the creation of a consultative and decision-making Kawésqar Indigenous Council.[49] The consultation process concluded with consent regarding the need to reclassify the Alacalufes RF to NP and the expansion of its area. The communities also rejected that the future Kawésqar NP would exclude interior waters and channels.[50]

In January 2018 the decree was issued to replace the Alacalufes RF with the Kawésqar NP and NR,[51] the former covering the terrestrial area of approximately 2.3 million ha and the latter on the adjacent maritime territory with an approximate area of 2.6 million ha.[52] The decision not to apply NP status to the maritime area of traditional occupation and navigation by the Kawésqar people generated frustration in the area's Indigenous communities. In response to the government's refusal to fully protect the marine territory,[53] the communities As Wal La Iep, Atap, Residentes del Río Primero, and the Grupos Familiares Nómades del Mar initiated an

---

[48] Exempt Resolution No. 1322 of June 29, 2017, of the MBN. Available at: https://www.leychile.cl/Navegar?idNorma=1105048. Page of the Consultation in Bienes Nacionales: http://www.bienes nacionales.cl/?page_id=28877.

[49] Position of the Kawésqar people regarding the consultation for the reclassification to park and expansion of the Alacalufe Reserve" (MBN, 2017). In the document sent to the Council of Ministers of Sustainability, it states, "We want governance of the park, by virtue of the right to self-determination. We want the administration of the Protected Area. As the Rapa Nui, we want to administer and decide our territory as self-government". They also demand "freedom of navigation and landing on the beaches and coasts of our ancestral territory", among other demands such as the development of ancestral activities, access to sacred sites and places of historical significance, hunting, fishing, and gathering, as well as the safeguarding of ancestral knowledge (ibid).

[50] The consultation report states: "The Kawésqar people, as a nomadic sea canoe people, culturally understand the sea and the land as a whole in relation to themselves. The sea is part of their cosmovision, of how they decode reality and interpret it. This explains their disappointment that the sea is not within the proposal, although they agree with the National Park for the protection of the environment against the advance of aquaculture and mining exploitation, as well as intensive tourism." (MBN, 2017).

[51] Decree No. 6 of January 26, 2018, MBN, published in the Official Gazette on January 30, 2019.

[52] Available at: https://www.diariooficial.interior.gob.cl/publicaciones/2019/01/30/42267/01/153 7812. pdf.

[53] The reasons for the State's refusal to include the protection of the sea in this new protected area stem from the interests of the salmon industry in the Patagonian channels. See http:// www.aqua.cl/2018/01/10/subpesca-responde-por-futuro-parque-nacional-que-impactaria-la-salmonicultura/.

ECMPO application process in order to protect the sea, continue with their ancestral activities of fishing, hunting, and gathering, and protect their ancestral right to
access and use the sea.[54]

4.3.7 Yaghan Communities and Protected Areas

Cape Horn NP and adjacent islands were created as a Virgin Region Reserve in the
Wollaston Archipelago, Navarino municipality in 1945 (D.S. 995/1945 Ministry
of Agriculture), with an approximate area of 63,093 km². In 1965, the Alberto
de Agostini NP was established on the islands located to the west of Navarino
Island, with an approximate area of 480,000 ha and with the purpose of protect-
ing the channels, fjords, fauna, and forests of the area (D.S. 80/1965 Ministry of
Agriculture). The Holanda FR and part of the Hernando de Magallanes NP were
incorporated into this NP in 1985, with an area of 1,460,000 ha. Finally, in 2013,
Yendegaia NP was created in the southern part of Tierra del Fuego adjacent to the
Beagle Channel, with 1,118 km², which includes part of the coastline of traditional
Yagán use.

The creation of these conservation units did not consider the traditional occupation and use of the area by the Yagán, reducing them to settlements and generating
human concentrations in small spaces. However, for the Yagán community this is
still a territory that belongs to them, and which they warn is obstructed by multiple regulatory restrictions that prevent their customary use (Martín Calderón,
in Serrano and Azócar (2016); Documentary Tanana *Ready to set sail*, 74 min).
The Yagán Community of Villa Ukika and its members demand a landscape that
goes beyond conservation (strict or multiple uses) and that allows them to manifest freely their economic, relational, and cultural practices (from the collection
of reeds for basketry to the capture of crab for commercial purposes, combining
both customary and non-customary uses) and to project their life plans and their
own development priorities. In this context, they consider the ECMPO Law (No.
20,249) referred to above as an option to protect the maritime coastal space of
traditional use and to maintain, in collaboration with other local actors, their own
way of life in accordance with their culture.

5   Discussion

These case studies of the relationship between PAs and Indigenous people in
Chilean Patagonia indicate that: (i) There is an evident overlap between the PAs
analyzed here and the lands and territories traditionally occupied and currently

---

[54] This process has not been without difficulties, however, given that they had to file a constitu-
tional appeal for protection against the granting of aquaculture concessions in the area requested as
ECMPO. Appeal filed on July 25, 2018, Rol: 684-2018 before the Court of Appeals of Punta Arenas, against the Undersecretary of Fisheries and Aquaculture and the Undersecretary of the Armed
Forces.

used by communities of different Indigenous people. In several cases the overlap is produced because the areas now considered PAs have been occupied by Indigenous people since ancestral times. In most cases, as pointed out in interviews with stakeholders from these communities, the overlap is total, determined both by their permanent or temporary use of the geographic spaces where the PAs are located, and by the relationship, both material and spiritual, that their members continue to have with these territories. Thus, in most cases the communities have claims for the restitution and/or use of all or part of the lands and territories of these PAs. (ii) The establishment of PAs, both public and private, with the exception of Kawésqar NP and the Cape Horn Biosphere Reserve, has been carried out without consultation with potentially affected communities, and therefore without their free, prior, and informed consent. Although this is a right established by ILO Convention 169, in force since 2009, the creation of most PAs by the State or private parties have occurred without consultation processes, which weakens their legitimacy, and in some cases, generates conflicts that have not been adequately addressed. It is of concern that in one of the cases where consultations were conducted with Indigenous people prior to the establishment of a PA—the Kawésqar NP—the communities have questioned both the form and results of the process. The same communities question the fact that the maritime space of traditional use and occupation has not been protected in the same way under the NP category, as was done with the terrestrial space. (iii) The PAs analyzed do not have up to date management plans and are administered by the State or private entities, with minimal or no participation of Indigenous people. Initiatives to include Indigenous people in the management of these areas have been limited to consultative spaces that are not maintained over time (Chiloé NP) or to the design of tourism development and management plans that have not yet been implemented (Bernardo O'Higgins NP) and lack mechanisms for their effective participation in the governance of the PAs. (iv) Nor are there initiatives that allow Indigenous people to participate directly in the economic benefits that some of these PAs generate, and benefits received are only secondary in nature (sale of handicrafts, marine tourism, minor services). Thus, in most of the cases analyzed here, there is an unsatisfied demand for shared governance of PAs by the Indigenous people and communities that inhabit these areas.

As a consequence of this exclusion, and in contradiction to international conservation guidelines applicable to Indigenous people, particularly the provisions of the Convention on Biological Diversity, PAs have not considered or integrated ancestral practices, uses, and knowledge that can contribute to biodiversity conservation. Except in the case of Bernardo O'Higgins NP and the Cape Horn Biosphere Reserve, it was not possible to identify research or documentation in this regard. The scarce existing information does not come from the entities in charge of the governance of these areas, but from the communities themselves (such as the Kawésqar community of Puerto Edén), from academia or from environmental organizations such as the Omora-UMAG Ethnobotanical Park. The restrictions in current legislation regarding the use and exploitation of natural resources in PAs have contributed to biodiversity conservation. However, they have also limited the sustainable use and practices that local communities make of these areas and their natural resources. Practices such as navigation, fishing, and gathering are limited due to general restrictions, without considering the ways of life and culture of the Indigenous people. In contrast, PAs such as Bernardo O'Higgins NP and Kawésqar NP have been affected by major threats, especially the granting of aquaculture concessions within their maritime areas, threats that have only been counteracted by the intervention of the Kawésqar communities themselves.

The PAs analyzed, particularly those belonging to the State, lack adequate governance and care, both because they do not have management plans in place and because in most cases, they do not have sufficient human and financial resources to provide adequate in situ protection. This is particularly critical in the case of the largest and most isolated areas (Las Guaitecas NR, Katalalixar NR, Bernardo O'Higgins NP, Kawésqar NP and NR, and Agostini NP), which is paradoxical considering that in all of these areas there are local communities with a clear interest in their protection and preservation.

Except for the public and private entities in charge of the governance of PAs, there is a lack of knowledge of the conservation initiatives that Indigenous people are developing in both their terrestrial and marine environments in Chilean Patagonia. A case of special interest is that of the ECMPO applications made so far by Indigenous people in Patagonia, which have purposes fundamentally linked to the sustainable management of resources and the conservation of these areas of traditional use, whose processing by the State has been extremely slow and bureaucratic. It is worth noting, however, that various conservation NGOs have begun to support the communities' requests for ECMPOs, as well as the few management plans that have so far been granted, which is indicative of the understanding that these organizations have gained regarding the conservation potential of these areas.

Linked to this, we found that there is no regulatory framework in the country that guarantees the protection of Indigenous people's rights in relation to public or private conservation initiatives that are promoted in their lands and territories. There is also no regulatory framework that recognizes Indigenous initiatives that contribute to the protection of biological diversity and their relationship with PAs, in accordance with the international guidelines on Indigenous people's rights and conservation referred to here.

The lack of a regulatory framework and a policy to process the demands of Indigenous people and communities in relation to public or private PAs located on their lands and territories of traditional and current use and occupation, ranging from the restitution of lands and territories to shared governance, generates a growing conflict that must be addressed. Together with the need for a regulatory framework that is appropriate to the aforementioned Indigenous rights and conservation guidelines, dialogue processes must be promoted that result in constructive agreements with each people or community in order to overcome this conflict.

6 Conclusions and Recommendations

Among the primary results case studies presented here, there is a clear overlap between the PAs decreed and the lands of ancestral occupation and current use by Indigenous people in Chilean Patagonia. There is a general absence of processes of consultation and free, prior, and informed consent in the conformation of PAs, as well as the exclusion of Indigenous people from their governance. There is also a lack of recognition and understanding of Indigenous people's own conservation initiatives, as well as of their practices, uses, and traditional ecological knowledge that contribute to conservation. There is an absence of legal frameworks and public policy using a rights-based approach that could promote synergies between public, private, and Indigenous stakeholders in favor of conservation strategies in Chilean Patagonia.

In order to strengthen the respectful, collaborative, and synergetic relationship between public and private strategies and those of the Indigenous people themselves that promote conservation in PAs in Chilean Patagonia, we make the following recommendations:

- The MMA, CONAF, academia, and private conservation entities, should carry out research and analysis with active Indigenous participation, to identify the nature and scope of this relationship in each PA, including the total or partial overlap between PAs and the lands and territories of ancestral occupation. These studies should identify the practices, uses, and traditional knowledge of these peoples and their contributions to conservation, as well as analyze compatible forms of governance.
- Promote dialogue between stakeholders in each PA, bearing in mind the IUCN guidelines and recommendations regarding the types of governance and the rights of Indigenous people over their lands and territories, in order to analyze together and define proposals for solutions to current and potential conflicts. These proposals may consider, among other modalities: (i) the definition of forms of shared governance; (ii) recognition of these PAs, or some of them, as Indigenous and Community Conservation Areas (ICCAs) under the governance of Indigenous people; (iii) establishment of agreements and mechanisms that regulate the uses of PAs so as to guarantee Indigenous ways of life and culture; (iv) agreements to share the benefits generated by activities in PAs; (v) total or partial restitution to Indigenous people of PAs when this is a demand based on demonstrated traditional occupation and is consistent with conservation purposes.
- Consideration of the international guidelines on Indigenous people and conservation referred to above by the competent public entities in the creation of new PAs in Chilean Patagonia. In particular, to consider the development of studies on potential spatial overlaps or other forms of impact on Indigenous people and communities, as well as the promotion of a consultation process with a view to achieving the agreement or consent of these peoples and communities.

- Prompt approval by the National Congress of the Biodiversity and Protected Areas Service bill, which establishes a regulatory framework for PAs and biodiversity in the country that considers the rights of Indigenous people and enables the protection and promotion of their conservation initiatives.
- Related to this, consideration should be given by the competent public and private entities that manage PAs in Chilean Patagonia to the potential of ECMPOs as conservation initiatives of native peoples in their coastal marine spaces, that are often adjacent to protected areas. To this end, we recommend that: (i) the political and administrative obstacles that limit the approval of ECMPOs within the deadlines established by law be documented and analyzed by academia, with the participation of the communities involved; (ii) CONADI and the Undersecretariat of Fisheries and Aquaculture provide support to the communities in the ECMPO application processes for purposes compatible with conservation; (iii) there be a reform of the review and evaluation processes for management and administration plans in order to reduce their processing periods; (iv) SUBPESCA provides support to strengthen the communities' capacities for the collective governance of the ECMPOs and the sustainable use of their resources.
- Promote, through the competent entities, a public policy of recognition and support for ICCAs in Chilean Patagonia, thus promoting compliance with the Convention on Biological Diversity's Aichi Targets for the Strategic Plan 2011-2020, particularly Targets 11, 14, and 18[55]
- Grant through the competent public entities, particularly the MMA and the future Biodiversity and Protected Areas Service, effective and financed public protection in accordance with the international guidelines outlined above for the PAs and ICCAs in Chilean Patagonia, in the face of the threats to which they are currently subject. In particular, this should address the threats from industrial fishing and aquaculture projects, mining, road building, and other projects causing environmental impact, as well as real estate projects.

Acknowledgements We thank the representatives of all the communities and Indigenous people's organizations that collaborated with their time and knowledge in the development of this research, without whose contribution it would not have been possible.

References

1. Aguilera, O., & Tonko, J. (2007). Literatura oral kawésqar: cuento del pájaro carpintero y su esposa, la mujer tiuque. *Onomázein, 2*(16).
2. Aguilera, O., & Tonko, J. (2011). *Guía etnográfica del Parque Nacional Bernardo O'Higgins.* Comunidad kawésqar de puerto Edén, Cequa: CORFO, CONAF. Retrieved from http://www.cequa.cl/cequa/libros.php?lib=44.
3. Aguilera, O., & Tonko, J. (2013). *Relatos de viaje kawésqar, nómadas canoeros de la Patagonia occidental.* Ofqui Editores.

[55] Aichi Targets, available at: https://www.cbd.int/sp/targets/.

4. Álvarez, R. (2002). *Reflexiones en torno a las identidades de las poblaciones canoeras, situadas entre los 44° y 48° de latitud sur, denominadas "chonos"*. Chile: Anales del Instituto de la Patagonia. Universidad de Magallanes. Retrieved from http://bibliotecadigital.umag.cl/han dle/20.500.11893/1540?show=full.

5. Aravena, J. C., Vela-Ruiz, G., Torres, J., Huenucoy, C., & Tonko, J. C. (2018). Parque Nacional Bernardo O'Higgins/territorio Kawesqar Waes: conservación y gestión en un territorio ancestral. *Magallania, 46*(1), 49–63.

6. Arce, L., Guerra, F., & Aylwin, J. (2016). *Cuestionando los enfoques clásicos de conservación en Chile: el aporte de los pueblos indígenas y las comunidades locales a la protección de la biodiversidad*. Temuco, Chile: Observatorio Ciudadano. Retrieved from https://observatorio. cl/cuestionando-los-enfoques-clasicos-de-la-conservacion-en-chile-el-aporte-de-los-pueblos-indigenas-y-las-comunidades-locales-a-la-proteccion-de-la-biodiversidad/.

7. Aylwin, J. (1995). *Comunidades indígenas de los canales australes*. Temuco, Chile: CONADI.

8. Braje, T. J., Dillehay, T. D., Erlandson, J. M., Klein, R. G., & Rick, T. C. (2017). Finding the first Americans. *Science, 358*(6363), 592–594.

9. Bridges, E. L. (2003). *El último confín de la tierra*. Buenos Aires, Argentina: Editorial Sudamericana. Borrini-Feyerabend, G., Dudley, N., Jaeger, T., Lassen, B., Pathak Broome, N., Phillips A., and Sandwith T. (2013). *Gobernanza de áreas protegidas: de la comprensión a la acción*. Serie Directrices para buenas prácticas en áreas protegidas No. 20, Gland, Switzerland: UICN. xvi. Retrieved from https://portals.iucn.org/library/sites/library/files/documents/ PAG-020-Es.pdf.

10. Buschmann, A. H., Niklitschek, E. J., & Pereda, S. (2023). Aquaculture and its Impacts on the Conservation of Chilean Patagonia. Springer.

11. Cárdenas, R., Montiel, D., & Hall, G. (1991). *Los Chonos y los Veliche de Chiloé*. Santiago, Chile: Editorial Olimpo.

12. Chapman, A. (1986). *Los selk'nam. La vida de los onas*. Emecé.

13. Comisión de Verdad Histórica y Nuevo Trato. (2003). *Informe de la Comisión de Verdad Histórica y Nuevo Trato con los Pueblos Indígenas*. Retrieved from http://bibliotecadigital. indh.cl/handle/123456789/268.

14. Comisión Interamericana de Derechos Humanos (2009) *Derechos de los pueblos indígenas y tribales sobre sus tierras ancestrales y recursos naturales*. Retrieved from https://www.oas. org/es/cidh/indigenas/docs/pdf/tierras-ancestrales.esp.pdf.

15. Corporación Nacional Forestal. (1997). *Plan de manejo Parque Nacional Chiloé*. Retrieved from https://www.conaf.cl/wp-content/files_mf/1382465935PNChiloe.pdf.

16. Corporación Nacional Forestal. (2017). *Bases de licitación de infraestructura del Parque Nacional Chiloé para servicios ecoturísticos*. Retrieved from www.conaf.cl/wp-content/.../lic itacion-parque_chiloe-chanquin.docx.

17. Cox, M. (2007). *Plan integral de conservación y desarrollo, zona sur de la cordillera de Piuchén, Chiloé: una propuesta de desarrollo participativo*. [Thesis]. Universidad Austral de Chile, Chile. Retrieved from http://cybertesis.uach.cl/tesis/uach/2007/ffc877p/doc/ffc877p. pdf.

18. Diegues, A. (2010). *El mito moderno de la naturaleza intocada*. Quito, Ecuador: Ediciones Abya-Yala. Retrieved from http://digitalrepository.unm.edu/cgi/viewcontent.cgi?art icle=1461&context=abya_yala.

19. Emperaire, J. (1963). *Los nómades del mar*. Universidad de Chile.

20. Falabella, F. (Ed.). (2016). *Prehistoria en Chile: Desde sus primeros habitantes hasta los Incas*. Editorial Universitaria.

21. Guerra, F. (2015). ¿Se puede intervenir en las áreas silvestres protegidas del Estado? Una aproxi- mación al contexto chileno a partir del dictamen N° 38.429 de la Contraloría General de la República. *Revista Derecho Ambiental, 7*(7), 189–201. Retrieved from https://cl.boell. org/sites/default/files/libro_fima_interior_y_tapas.pdf.

22. Gusinde, M. (1986). *Los indios de Tierra del Fuego: T. II, los Yamana*. Buenos Aires, Argentina: Centro Argentino de Etnología Americana, CNIT.

23. Häussermann, V., Försterra, G., & Laudien, J. (2023). Hard Bottom Macrobenthos of Chilean Patagonia: Emphasis on Conservation of Subltitoral Invertebrate and Algal Forests. Springer.
24. Instituto Nacional de Estadísticas. (2018). *Segunda entrega resultados definitivos censo 2017*. Retrieved from http://www.censo2017.cl/wp-content/uploads/2018/05/presentacion_de_la_segunda_en-trega_de_resultados_censo2017.pdf.
25. Lira, N., & Legoupil, D. (2014). Navegantes del sur y las regiones australes. In: Aldunate, C., (Ed.), *Mar de Chile*, pp. 102–143. Santiago, Chile: colección Santander, Museo Chileno de Arte Precolombino.
26. Lira, N. (2017). *Antiguos navegantes en los mares de Chiloé*. Capítulo III, Chiloé. Santiago, Chile: Museo de Arte Precolombino.
27. Legoupil, D., & Fontugne, M. (1997). El poblamiento marítimo en los archipiélagos de Patagonia: Núcleos antiguos y dispersión reciente. Anales del Instituto de la Patagonia. *Serie Ciencias Humanas (chile), 25*, 75–87.
28. Martinic, M. (1992). *Historia de la Región Magallánica. Volumen I*. Universidad de Magallanes.
29. Meza-Lopehandía, M. (2018). *La ley lafkenche. Análisis y perspectivas a 10 años de su entrada en vigor. Retrieved from* https://www.bcn.cl/obtienearchivo?id=repositorio/10221/254 31/1/BCN FINAL La_Ley_Lafkenche_10_anos_despues_2018.pdf.
30. Ministerio de Bienes Nacionales. (2017). *Sistematización del proceso de consulta indígena al pueblo kawésqar por la ampliación y re-clasificación de la Reserva Nacional Alacalufes. Informe final*. Retrieved from: http://www.bienesnacionales.cl/wp-content/uploads/2017/12/InformeFinal_PCI_revLTV_NBA_LPA_FINAL_05102017_F2.pdf.
31. Ministerio del Medio Ambiente. (2011). *Las áreas protegidas de Chile. Antecedentes, institucionalidad, estadísticas y desafíos*. Santiago de Chile: Ministerio del Medio Ambiente. Retrieved from: http://www.mma.gob.cl/1304/articles-50613_pdf.pdf.
32. Miotti, L., & Salemme, M. (2004). Poblamiento, movilidad y territorios entre las sociedades cazadoras-recolectoras de Patagonia. *Complutum, 15*, 177–206.
33. Molina, R., & Correa, M. (1996). *Territorios huilliches de Chiloé*. Santiago, Chile: Corporación Nacional de Desarrollo Indígena y Agencia Española de Cooperación Internacional.
34. Molinet, C., & Niklitschek, E. J. (2023). Fisheries and Marine Conservation in Chilean Patagonia. Springer.
35. Ocampo, C., & Aspillaga, E. (1984). Breves notas sobre una prospección arqueológica en los archipiélagos de las Guaitecas y los chonos. *Revista Chilena De Antropología, 4*, 155–156.
36. Praus, S., Palma, M., & Domínguez, R. (2011). *La situación jurídica de las actuales áreas protegidas de Chile*. Proyecto GEF, PNUD, MMA Creación de un Sistema Nacional Integral de Áreas Protegidas para Chile". Retrieved from: http://www.proyectogefareasprotegidas.cl/wpcontent/ uploads/2012/05/La_Situacion_Juridica.pdf.
37. Reyes, O. (2019). El poblamiento del archipiélago de los Chonos (43°-47° S). Patagonia occidental, Chile. *Arqueología, 25*(2), 277–281.
38. Rivas, P., Ocampo, C., & Aspillaga, E. (1999). Poblamiento temprano de los canales patagónicos: el núcleo ecotonal septentrional. *Anales del Instituto de la Patagonia, Serie Ciencias Históricas (Chile)*: 221–230.
39. Saavedra, G. (2016). La pesca artesanal en el sur austral de Chile. Controversias territoriales en el espacio marino-costero. *Revista Antropologías Del Sur, 5*, 65–83.
40. Skewes, J., Álvarez, R., & Navarro, M. (2012). Usos consuetudinarios, conflictos actuales y conservación en el borde costero de Chiloé insular. *Magallania (Punta Arenas), 40*(1), 109–125. Sociedad de Fomento Fabril (2018). *Análisis impacto económico y sectorial*. Ley n° 20.249 (Ley Lafkenche).
41. Stevens, S. (1997). Lessons and directions. In: S. Stevens (Ed.), *Conservation through cultural survival: Indigenous people and protected areas*. pp. 265-298. Washington D. C., USA: Island Press.
42. Tauli-Corpuz, V. (2016). *Informe de la relatora especial del Consejo de Derechos Humanos sobre los Derechos de los Pueblos Indígenas*. A/71/229. Retrieved from https://documents-dds-ny.un.org/doc/UNDOC/GEN/N16/241/12/PDF/N1624112.pdf?OpenElement.

43. Tecklin, D., Farías, A., Peña, M. P., Gélvez, X. Castilla, J. C., Sepúlveda, M., Viddi, F. A., y Hucke-Gaete, R. (2023). Coastal-Marine Protection in Chilean Patagonia: Historical Progress, Current Situation, and Challenges. Springer.
44. Tonko, J. (2007). Relatos De Viaje Kawésqar. *Onomázein, 18*(2), 11–47.
45. Unión Internacional para la Conservación de la Naturaleza. (2003). *Acuerdo de Durban.* Retrieved from: https://cmsdata.iucn.org/downloads/durbanaccordes.pdf.
46. Unión Internacional para la Conservación de la Naturaleza. (2008). *Resoluciones y recomendaciones Congreso Mundial de la Naturaleza*, Barcelona, España. Retrieved from: https://cms data.iucn.org/downloads/wcc_4th_005_spanish.pdf.
47. Unión Internacional para la Conservación de la Naturaleza. (2016). *Resoluciones, recomendaciones y otras decisiones de la UICN Congreso Mundial de la Naturaleza.* Honolulu, Hawai. Retrieved from: https://portals.iucn.org/library/sites/library/files/docume nts/WCC-6th-005-Es.pdf.
48. Urbina, R. (1983). *La periferia meridional indiana Chiloé en el siglo XVIII.* Ediciones Universitarias de Valparaíso, Universidad Católica de Valparaíso.
49. Zelada, S., & Park, J. (2013). Análisis crítico de la Ley Lafkenche (N°20.249): el complejo contexto ideológico, jurídico, administrativo y social que dificulta su aplicación. *Universum (Talca), 28*(1), 47–72.

Drivers of Change in Ecosystems of Chilean Patagonia: Current and Projected Trends

17

Laura Nahuelhual and Alejandra Carmona

Abstract

Identifying and analyzing the effects of anthropogenic drivers on ecosystems is a critical step in conservation planning. This chapter identifies and analyzes the evolution of direct and indirect drivers of change in the ecosystems of Chile's Patagonia region. We analyzed native forest degradation, mining expansion, tourism, energy generation, agriculture and livestock, and fisheries and aquaculture production as direct drivers. As indirect drivers we included demographic dynamics, economic growth, and institutional factors. Using a cluster analysis, we identified eight types of municipalities that reflect differentiated territorial characteristics and dynamics in terms of these drivers. These included municipalities characterized primarily by dynamics of urban growth, mining, tourism, aquaculture, forest exploitation, livestock development, and particularities in the ethnic composition of the territories, as well as municipalities where the drivers were concentrated in a non-distinguishable way. This diversity of situations suggests the need for differentiated conservation strategies that target the specific pressures on the different ecosystems and territories of South Patagonia.

L. Nahuelhual (✉)
Departament of Social Sciences, Universidad de los Lagos, Osorno, Chile
e-mail: laura.nahuel@gmail.com

Centro de Investigación Dinámica de Ecosistemas Marinos de Altas Latitudes (IDEAL), Valdivia, Chile

Instituto Milenio en Socioecología Costera (SECOS), Santiago, Chile

A. Carmona
Prospectiva Local Consultores, Bueras 1070, Valdivia, Chile

Centro de Educación Continua, Universidad Austral de Chile, Valdivia, Chile

J. C. Castilla et al. (eds.), *Conservation in Chilean Patagonia*, Integrated Science 19,
https://doi.org/10.1007/978-3-031-39408-9_17

445

Keywords

Patagonia • Chile • Drivers • Ecosystem change • Conservation planning •
Biodiversity

1 Introduction

According to the Millennium Ecosystem Assessment [1], (hereafter MEA), drivers
of change are anthropogenic factors that directly or indirectly cause alterations in
ecosystems. MEA [1] divided them into direct and indirect drivers; the difference
is that the former directly influence ecosystem processes, while the latter operate in
an underlying manner by altering one or more direct drivers [2, 3]. These drivers,
and actions to mitigate their impacts, have been recognized and formalized in
multilateral agreements, such as the Convention on Biological Diversity and the
United Nations 2030 Agenda for Sustainable Development. A key objective of
these agreements is to identify baselines and trends in drivers [4, 5].

Direct drivers include energy use, introduction of invasive species [6, 7],
greenhouse gas emissions and climate change, land use/land cover change, and
extraction of renewable (e.g. fisheries) and non-renewable (e.g. mining) natural
resources. Indirect drivers include population growth, climate change, land use/
land cover change, and economic growth, technological change, socio-political
factors, culture, and religion [5, 1].

In most cases, ecosystem changes occur through interactions among multiple
drivers at different spatial and temporal scales (e.g. [2]). The effect of interactions between drivers depends on a variety of context-specific factors (e.g. [8]).
Ecosystem changes also interact with drivers in complex ways [1, 5]. Altered
ecosystems create new opportunities and constraints on land use, induce institutional changes in response to degradation and resource scarcity, and result in
social effects such as changes in income inequality [9]. For example, the filling of
wetlands for urbanization brings with it the opportunity to build new housing, but
also generates significant negative impacts on ecosystem services, which has led
countries to implement incentive policies for their protection and restoration [10].
Finally, capitalism is the origin of all drivers and impacts. The second contradiction
of capitalism posits that as a political-economic system, it relies on expropriation
and exploitation on an ever-increasing scale, creating social inequalities and environmental degradation [11–13]. Indeed, recent research indicates that these drivers
are increasing, causing ecosystem destruction at unprecedented rates across natural
systems [14, 15]. Despite extensive efforts to predict the trajectories of drivers and
assess their impacts (e.g. [1, 5, 16, 17]), there has been no synthesis of the status
of current research on drivers, particularly in developing countries.

Although there is information on the temporal dynamics of variables that could
be considered drivers of change for Chilean Patagonia, there are practically no
studies linking these drivers to ecosystem changes. The drivers most analyzed
are direct drivers, including land use change. In contrast, very few studies have

explored the effects of indirect drivers, which is at least partially due to the limited availability of spatio-temporal data. Probably the most complete information on the dynamics of environmental change drivers available for Chile is contained in the country Environmental Performance Assessments (Economic Commission for Latin America and the Caribbean (ECLAC) and Organization for Economic Cooperation and Development (OECD) [18]; in Spanish ECLAC and OECD) and in the Ministry of the Environment, 2019 (in Spanish Ministerio del Medio Ambiente, MMA).

2   Scope and Objectives

This chapter focuses on analysis of the spatial and temporal dynamics of direct and indirect drivers of ecosystem change in Chilean Patagonia. The direct drivers include the degradation and loss of native forests, the expansion of mining, tourism, energy generation, agriculture and livestock, and fishing and aquaculture. Indirect drivers include demographic dynamics, economic growth, and institutional factors.

3   Methods

The region known as Chilean Patagonia is located at the southern tip of the American continent (41° 42′ S, 73° 02′ W–56° 29′ S, 68° 44°). It has 456,225 inhabitants (2.7% of the population of Chile, National Institute of Statistics, 2017, (in Spanish INE) and an area of 280,000 km² (37% of the area of continental Chile). The study area comprises 34 municipalities, 85% of which are below the average poverty rate measured by income at the municipal level (based on the Ministry of Social Development [19]; in Spanish Ministerio de Desarrollo Social, MDS) and 53% belonged to the hundred municipalities with the highest isolation index in 2011 (Under-secretary of Regional Development [20]; in Spanish Subsecretaría de Desarrollo Regional).

3.1 Identification of Drivers of Change in the Ecosystems of Chilean Patagonia

The MEA [1] framework was used for the analysis of drivers which is considered the most appropriate for social-ecological studies, as it provides a more flexible definition and analysis of drivers of change. A panel of biodiversity experts of the MMA analyzed conservation strategies for the Patagonia (https://biodiversidad.mma.gob.cl/), in order to identify the environmental problems that generate relevant pressures on biodiversity. These problems were associated with drivers and then indicators were selected for evaluation (Table 1). The indicators were chosen on the basis of the following criteria: existence of data, geographic coverage, reliability of sources, consistency with the driver, and accessibility of the data.

3.2 Definition of the Scale of Work

The municipalities were selected as administrative units of analysis; however, the rivers were studied at three scales, depending on the characteristics of the spatial information available: (i) territorial (Chilean Patagonia); (ii) administrative region; (iii) municipal. Following Tzanopoulos et al. [26], a method was used that guarantees comparability between administrative levels and the proposed indicators, so as to be able to represent the drivers of change numerically. The method proposed by these authors consists of assessing the change in two key variables at the different scales analyzed: (i) change in driver intensity: (ii) change in uniformity. Intensity measures whether the values of an indicator are over or under-represented within the scales of analysis. This was evaluated based on the relative change of the median of an indicator at the scale of region, compared to the scale of municipality. A negative value represents an over-representation of low values for the indicator, while a positive value expresses an over-representation of high values.

Uniformity is an average of similarity between the values of the indicator for different municipalities within the regions, within the study area and ranges from 0 to 1, being 1 the greatest uniformity. The Shannon uniformity index [27], derived from Shannon's diversity index, was used to measure uniformity. This index measures the homogeneity of the spatial distribution of the indicator values among the municipalities, taking into consideration the value for the region and territory [26]; also see [28].

3.3 Data Collection

An exhaustive review of existing literature and technical documents was carried out to compile a list of drivers affecting the biodiversity of the Chilean Patagonian territory (Table 1), and together with experts, indicators were identified to evaluate the trend of each driver. The review included papers published in scientific journals, databases, and government reports. Based on the information collected, a temporal database was constructed by municipality.

3.4 Trends in Drivers of Ecosystem Change

The time trend of the indicators for each selected variable was analyzed by projecting the variables over time. The variables selected to represent each indicator and their trends were spatialized with geographic information systems. For indicators that had more than one variable and whose prospection was not homogeneous in

Table 1 List of drivers and sources of available information

Driver	Indicator	Data sources
Degradation and loss of native forest	Annual change rates (CA) of native forest and by municipality CA = ln (a2/a1)*/ (t2 − t1) ln: natural logarithm a2: area of native forest at time 2 a1: area of native forest at time 1 t1: year time 1 t2: year time 2	Cadaster of Chile's vegetation resources (National Forestry Corporation, CONAF): Los Lagos Region 1998 (cadaster) and 2013 (update) [21, 22] Aysén Region 1998 (cadaster) and 2006 (update) Magallanes Region 1996 (cadaster) and 2005 (update) [21, 23–25]
Mining expansion	Annual production of metallic and non-metallic minerals by municipality between 1999 and 2018 – Number of mining projects entered into the environmental evaluation system between 1992 and 2018 by municipality	– Servicio de Evaluación Ambiental (SEA) 1999–2018 https://www.sea.gob.cl/ – Chilean Copper Commission Statistics (COCHILCO) 1999–2018 https://www.cochilco.cl/Pag inas/Estadisticas/Bases% 20of%20Data/Bases-de-Datos. aspx
Tourism expansion	Visitor rates to consolidated and emerging tourist destinations (persons/year) by municipality	Visits to tourist destinations National Tourism Service (SERNATUR) 2015–2016 http://www.subturismo.gob.cl/ documentos/statistics/
Expansion of energy generation	Energy generation (kw) by community	Statistical Yearbook of the National Energy Commission (2015–2018) https://www.cne. cl/nuestros-servicios/reportes/ informacion-y-estadisticas/
Expansion of agriculture and livestock	Annual rate of increase in agricultural land and heads of livestock by municipality	Statistics Office of Agrarian Studies and Policies (in Spanish ODEPA) 2013–2019 https://www.odepa.gob.cl/sec tor-statistics/production-statis tics

(continued)

Table 1 (continued)

Driver	Indicator	Data sources
Fish production and aquaculture	Annual fishing landings tons (t) by region	– Landing statistics National Fisheries and Aquaculture Service 2011–2018 (SERNAPESCA) http://www.sernapesca.cl/informes/statistics?qt-quicktabs_work_area=5 – Servicio de Evaluación Ambiental (SEA) 1992–2018 https://www.sea.gob.cl/
Economic growth	Regional Gross Domestic Product (GDP) ($/year)	Central Bank of Chile 2011–2018 https://www.bcentral.cl/
Demographic dynamics	Population growth rate by municipality (%) Rurality rate per municipality (%) Proportion of indigenous people by municipality (%)	National Statistics Institute (INE)

time and space, a multi-criteria analysis was carried out to define the future areas of concentration of the driver. Weights were established for the variables based on expert criteria and according to the importance indicated in the literature. Table 2 shows the type of projection made according to each indicator.

3.5 Development of a Municipal Typology of Exposure to Drivers of Change

A typology of municipalities was developed according to the magnitude of the different types of indicators, for which a cluster analysis was used to group municipalities that showed similar patterns in the magnitude of the indicators. Cluster analysis is a quantitative statistical method that uses unsupervised learning to explore, find, and classify characteristics, and to obtain information about the nature or structure of the data [29]. This study was conducted on group analysis units with similar behavior, based on combinations of indicators. The analysis used was non-hierarchical and based on the centroid method (squared Euclidean distance), which ensures that the distance between observations in the same cluster is smaller than the distance between observations belonging to different clusters [30, 31].

Table 2 Chilean Patagonia. Type of projection according to indicator

Indicator	Projection for 2030
Rates of change in the area of native forest by municipality	Multi-criteria analysis based on:
	– Projection of the rate of native forest loss
	– Forest type
	– Projection of total municipal firewood consumption over time
Number of mining projects submitted to the Environmental Impact Assessment System (SEIA) between 1992 and 2018 by municipality	Correlation between the growth of mining production and the number of projects submitted to SEIA
Visitor rates to tourist destinations (persons/year) by municipality	Linear regression of the rate of change of visitors over time
Energy generation (kw) by community	Correlation between community consumption and regional demand projection (http://datos.energiaabierta.cl/visualizations/31955/CONSU-DE-ENERG-ELECT-EN/)
Increase in agricultural land and increase in livestock (ha and head of livestock) by municipality	– Projection of the rate of change of agricultural land area by municipality
	– Projection of livestock increase. Correlation of agricultural land and livestock head projections (carrying capacity)
Fishing landings (t) by fishing port	Landings projections over time by fishing port according to past trend
Gross regional domestic product ($/year)	GDP projection over time based on past trend
Population growth rate by municipality (%)	Population projection according to past trend
Rurality rate by municipality (%)	
Proportion of indigenous people by municipality (%)	

4 Results

4.1 Degradation[1] and Loss of Native Forests

Forest degradation and conversion are among the most significant transformations of the land surface globally [33]. Its causes are heterogeneous and change over time and from one region to another [34, 35]. Degradation is one of the most

---

[1] Degradation is the result of a progressive decrease in the structure, composition, and functions on which the vigor and resilience of a forest is based. A degraded forest is one whose structure, function, species composition, or productivity have been severely modified or permanently lost as a result of detrimental human activities [32].

significant direct drivers of the loss of remnant native forests in Chile and particularly in the study area, ultimately leading to loss of forest cover [32]. Direct causes include selective logging (legal and illegal), forest fires, overgrazing, and invasive species [36]. Indirect causes have been reported including poverty, inadequate conservation and management policies, institutional weaknesses, and various economic and technological factors [37–39]. The increase in the demand for fuelwood due to population growth has been identified as the main indirect driver in recent decades [38, 40]. To the extent that the rate of fuelwood extraction is greater than the rate of forest recovery [41], there is a serious risk of conservation as long as other cheaper energy sources are not developed.

Herding sheep in the ex-Estancia Chacabuco, Patagonia National Park, Aysén Región, Photo by Jorge López

The firewood penetration rate in the study area, understood as the percentage of households using firewood for heating, is 94.8% in the Los Lagos Region, 98.2% in the Aysén Region, and 12.7% in the Magallanes Region. Consumption per household fluctuates between 13 m^3st (cubic meters in stereo) in the Los Lagos Region, 13.75 m^3st in the Aysén Region and 17.5 m^3st in the Magallanes Region.

Although firewood constitutes an important income for forest owners, its market is highly informal and therefore the precise numbers of extraction and owners who extract firewood are unknown [38, 42]. For example, it is estimated that 80%

of the firewood arriving in the city of Coyhaique comes from forests without management plans [43]. The extraction of firewood without proper management leads to what is called high grading, a practice that involves the removal of the best trees, generating an increasingly impoverished forest, ultimately leading to its loss [44]. Lenga and evergreen forest types are the most affected (National Forestry Corporation [45]; in Spanish Corporación Nacional Forestal, CONAF).

Between 1998 and 2013, a total of 23,370 ha changed from native forest to shrubland or from shrubland to grassland, affecting 0.26% of the total area of native forest in the study area.[2] This change in coverage is very evident in the Chiloé archipelago, where it represents between 70 and 90% of native forest loss, and in the mountain municipalities of Futaleufú, Hualaihué, and Palena, where it represents 23% of native forest loss. The main cause of this change is firewood extraction [38, 40, 43].

This change is also significant in some municipalities of the Magallanes Region, especially in Porvenir, Primavera, and Timaukel, (reaching 70% of native forest loss), but it is associated with other direct causes, specifically forest fires and destruction by beavers [46].

Based on past rates of native forest degradation, firewood consumption rates, and the distribution of the forest types most affected by firewood and beaver pressures, the projection indicates that the Aysén province (Aysén Region) and Tierra del Fuego province would concentrate the greatest pressures from this driver in the future. In these sectors there is a concentration of areas of standing forest susceptible to firewood harvesting and also affected by beaver damage (Tierra del Fuego Province) (Fig. 1A).

4.2 Mining Expansion

Mineral extraction in Latin America is part of an extensive history of dispossession and environmental degradation, which has literally produced sacrifice areas, and Chile is no exception [47, 48]. Although the major environmental and social impacts in Chile are mostly located in the Norte Grande, associated with large-scale copper mining, more recently there are cases of conflicts in Chilean Patagonia, of which the conflict in Riesco Island is emblematic [49]. The effects of mining on ecosystems and biodiversity have been well documented and there is extensive literature on the subject [50–52]. The effects are produced mainly by: (i) waste production and sedimentation in water bodies, acid surface and subsurface drainage, and the effects of metal and waste rock deposits; (ii) habitat alteration; (iii) indirect effects associated with work such as roads. Mining projects generally

---

[2] The values presented were calculated based on the Cadaster and Evaluation of Native Vegetation Resources of Chile and its update, which for the Los Lagos Region is dated 1998 (cadaster) and 2013 (update), for the Aysén Region 1998 (cadaster) and 2006 (update), and for the Magallanes Region 1996 (cadaster) and 2005 (update) [21–25].

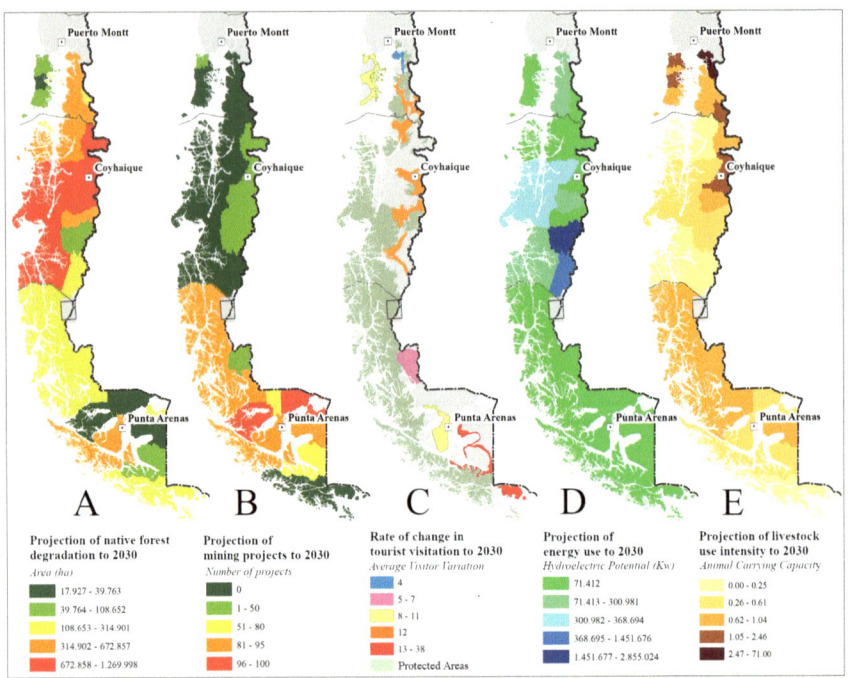

Fig. 1 Chilean Patagonia. Projection of direct drivers of ecosystem change to 2030. Panel A, 30-year projection of the driver of degradation and loss of native forest, measured in area (ha); in green smaller areas, in red larger areas. Panel B, 30-year projection of the mining driver, measured in number of projects, in green fewer projects, in red more projects. Panel C, 30-year projection of tourism development driver measured in visitor variation rate, in light blue lower variation rate, in red higher variation rate. Panel D, 30-year projection of energy use driver, measured in hydroelectric potential (kw); in green lower potential, in blue higher potential. Panel E, 30-year projection of livestock intensity driver, measured in animal carrying capacity; in yellow lower carrying capacity, in brown higher carrying capacity

affect biodiversity adversely and irreparably, and the remediation measures contemplated are not always sufficient to recover the flora and fauna present in the areas of impact [53].

Between the years 1998 and 2017, the Chilean Patagonia region produced 7% of the metallic mining production nationally, 4% of non-metallic mining, and 91% of fuel (coal, oil and natural gas) (Comisión Chilena del Cobre [54]; in Spanish COCHILCO).

4.2.1 Metal Mining

Metal mining in the study area is concentrated exclusively in the Aysén Region and mostly focuses on silver extraction (75% of total national metal ore level), zinc (25%), and to a lesser extent gold and lead. This mining takes place almost entirely in mountainous areas and at the headwaters of river basins, producing effects on water bodies, mostly associated with sedimentation and contamination with heavy metals [53].

4.2.2 Non-metal Mining

Non-metal mining is concentrated in the Magallanes Region (99.92% of total non-metal mining in the study area) and includes the extraction of calcium carbonate (55%), limestone (45%) and peat on a very low scale [55]. Peat extraction is in itself an unsustainable activity, similar to topsoil extraction, which has been studied and denounced for its serious consequences for the regional water balance and for the global carbon balance [56]. Effects in the study area such as the arrival of exotic species [57] and the water crisis in Chiloé [58] are reported.

4.2.3 Fuel Extraction

Fuel extraction from the study area takes place entirely in the Magallanes Region, primarily focused on crude oil (20.37%) and coal (79.49%) [54]. During the period 1999–2018, 100% of domestic oil, 88% of coal, and 35% of natural gas came from this region.

Coal exploitation reached 2,256,656 t in 2018, 98% of the national production. The deposits are concentrated in the Brunswick Peninsula, Riesco Island, and Puerto Natales. It is estimated that the probable reserves are 555 million t (Ministry of Energy [59]; in Spanish Ministerio de Energía, ME), whose potential development has already generated socio-environmental conflicts, such as the one that occurred with the Isla Riesco mining project [60]. Given the existing coal reserves in the study area (mostly concentrated in Skyring Sound) and depending on the feasibility of using coal as a raw material for the industrial process of hydrogen production, this driver could become especially important in the future, with consequent impacts on landscape alteration and water pollution [61].

Between 1992 and 2018, 580 mining projects were submitted for environmental assessment (93% of which were approved); only nine of them were submitted via Environmental Impact Assessment whereas the rest was submitted via Declaration of Environmental Impact,[3] without citizen participation or proposals for mitigation and/or remediation strategies. This situation is particularly serious given that 33 of the projects presented via Declaration of Environmental Impact were for hydraulic fracturing (fracking) (see Footnote 3). Most of these projects are concentrated in the municipalities of San Gregorio and Primavera, in the Magallanes Region.

Our projection indicates that new mining projects (metal, non-metal, and fuel) will be concentrated by 2030 mainly in the mountain municipalities of the Aysén and Magallanes Regions and especially in the Seno Skyring (Magallanes Region), which coincides with areas proposed for conservation in the regional biodiversity conservation strategies (Fig. 1B), which could lead to potential territorial conflicts within a decade.

---

[3] Consultation made through Internet, available at: http://www.sea.gob.cl/.

4.3 Tourism Expansion

Tourism is an important source of income generation in developing countries. However, it has also been associated with numerous adverse environmental consequences, such as vehicular congestion, air pollution, solid waste, and water scarcity, especially during months of high tourist flow [62].

There are twelve tourist destinations identified in the study area (Ministry of Public Works [63]; in Spanish Ministerio de Obras Públicas, MOP), related to the presence of tourist attractions such as parks and reserves. One third of these are consolidated destinations, 50% are emerging, and 16.6% are potential. Tourist arrivals in Chilean Patagonia reached 722,028 people in 2016, equivalent to 7.5% of tourist arrivals in the country; most were domestic tourists (62.7%) and 37.3% were foreigners, a pattern analogous to the national pattern. 49% of arrivals in 2016 occurred between the months of November and February, marking a summer seasonality pattern.[4] The average variation rate with respect to the 2015 period was 12%, highlighting the destinations of Antarctica, King George Island, Cape Horn, Puerto Eden, and Tierra del Fuego. However, the Chiloé Archipelago and Carretera Austral-Parque Nacional Queulat destinations had a negative variation. It should be noted that the population received in the month with the highest tourist concentration in the region (February) is equivalent to 48% of the total population of the study area estimated in 2017 [64]. None of the existing municipal planning (regulatory plans, waste management plans, etc.) incorporates this floating population in its basic services, which results in a low capacity to receive and manage tourists, producing impacts on ecosystems [63]. Between the years 2000 and 2018, 2,107 forest fires occurred in the study area, mostly associated with recreational activities and the transit of people, vehicles, and aircrafts, which affected an area of 65,428 ha.[5] On average, 58% of the burned vegetation was scrubland and native forest.

The trends in tourist arrivals to destinations and the projections made by the National Tourism Service for new tourist routes and destinations (emerging destinations) project an increase in tourist pressure in the tourist destinations of Cape Horn and Tierra del Fuego (municipalities of Punta Arenas, Porvenir, Timaukel, and Cape Horn) and to a lesser extent in already consolidated destinations such as the Carretera Austral (municipalities of Chaitén, Futaleufú, Palena, Cisnes, and Lago Verde) (Fig. 1C; also see [65]).

4.4 Expansion of Energy Production

A recent study in Chile shows energy consumption still coupled to GDP, low energy efficiency, and higher end-use export sectors (e.g. mining) [66], and

---

[4] Statistics on tourist lodging establishments. Undersecretary of Tourism year 2017.
[5] Forest fire statistics National Forestry Corporation.

coal use for electricity generation that has increased in absolute terms. Non-conventional renewable energies (NCRE) remain controversial in some cases, including in particular hydroelectric generation [67].

We find the following trends in the energy matrix of the study area: (i) Fire-wood, an NCRE, is the prevalent energy source for heating purposes in both the Chiloé Archipelago and the Aysén Region. It is estimated that 96% of the population uses this resource as the main source [68]. In the Magallanes Region the main heating source is natural gas (non-renewable) [59]. Firewood extraction is not only a driver of native forest degradation, but also the cause of high levels of particulate matter pollution in the municipalities that consume it [69]. (ii) The Chiloé Archipelago is connected to the central interconnected electricity grid, with eight electrical substations fed by three thermoelectric plants (Quellón I and II, Degañ, and Danisco Biomar), three hydroelectric plants (Colil, Piruquina, and Dongo) and one wind power plant (San Pedro de Dalcahue). As a result, the consumption matrix of the archipelago is highly diversified with a maximum power generation of 267,800 MW.[6] Thermoelectric power plants have local impacts affecting the quality of life of the surrounding communities, due to the local pollution they produce, which affects the increase in the levels of conflict and social rejection [70]. (iii) Aysén's electricity matrix system is based on medium-sized and isolated systems, which together have an installed capacity of 67.74 MW, with hydroelectric, diesel thermal, and wind generated sources. There are also two private mining companies, Cerro Bayo and El Toqui, which produce their own electricity, with an installed capacity of 22.38 MW, and an energy project that supplies 1.3 MW to Pesquera Tornagaleones. The region's installed capacity totals 91.42 MW. (iv) The Magallanes system has a final consumption matrix highly concentrated in natural gas (64.8% of the total). Natural gas represents 94.1% of the fuel used in the consumption matrix for electricity generation (5.6% is Diesel and 0.2% wind energy).

The National Energy Commission [71] indicates a historical energy consumption growth rate of 5.8% for the Aysén Region and 2.6% for the Magallanes Region in electricity and 1.6% for natural gas. The environmental impacts of electricity generation are usually related to the technology used and can be summarized as follows: (i) fragmentation of the landscape,(ii) alteration of lifestyles; (iii) atmospheric emissions; (iv) impact on regional fauna; (v) impact on regional flora; (vi) alteration of flows [72]. It should be noted that the greater inclusion of NCRE in the country is focused towards medium-sized electricity systems located in extreme areas, in order to replace current diesel generation [73, 74]. These regions have increased from 3% NCRE generation to reach the national average, which is around 18% [74].

The energy resources considering renewable sources are outstanding in the study area, especially hydroelectric and wind power, by which the current demand

---

[6] Consult at: http://energiaregion.cl/region/X.

could be amply covered. It should be noted that hydraulic estimates as gross theoretical potential reach 6,876 MW in the study area (all located in the Aysén Region). No less than 13.21% of these are located within national parks and conservation areas. Finally, it is important to note that for marine sources, the gross resource of wave energy is estimated at 47,170 MW, while for tides it is 220 MW. However, the technical resource, i.e. that which is feasible to harness, is estimated to be negligible for waves, and 22 MW for tides, mainly due to the large distances to centers of consumption [74]. Considering these energy resources, projections to 2030 of possible pressures on ecosystems as a result of this driver will be concentrated mainly in the municipalities of Cochrane, O'Higgins, and Aysén (Fig. 1D).

4.5 Expansion of Agriculture and Livestock Farming

The occupation of the Patagonian territory was strongly conditioned by the development of extensive livestock grazing, relying predominantly on sheep, which gradually developed into a few establishments of very large extensions and low population density, which coexisted with numerous small and medium-sized operations [75].

The soil erosion generated by this phenomenon in the past and the current droughts in critical months for fodder production are factors that condition livestock production and at the same time damage a key Patagonian ecosystem, the steppe[7] (see [76]). This is even more aggravated when 84.6% of the regional surface of Aysén shows severe to moderate desertification and 91.4% of the Magallanes Region is in the same situation [77]. Since its origins, livestock activity also came into conflict with native species, mainly foxes and pumas, as domestic animals became part of the diet of these carnivores, with the consequent economic impact for those who make a living from this activity [78].

The degradation of native forests also leads to the use of degraded forests and scrublands for livestock grazing, without limitations in stocking rates that would allow the systematic recovery of those sectors with adequate conditions for their reversion to the initial forest stages [76, 79]. This has serious effects on biodiversity and landscape conservation.

Although the area dedicated to agriculture and livestock in Chilean Patagonia has been maintained over time, with the incentive of public policies, there has been a conversion to more intensive livestock farming.[8] In 1997 there was an average of 5% improved pastures in the territory, which increased to 12% in 2007 [80, 81]. Despite the above, there has been a negative variation in the number of sheep, at a much higher rate in the municipalities of the Los Lagos Region and on

---

[7] Bulletin accessed online: https://chile.wcs.org/DesktopModules/Bring2mind/DMX/Download.aspx?EntryId=33826&PortalId=134&DownloadMethod=attachment.
[8] One of the examples is the Degraded Soil Reclamation System.

average higher than the national variation rate. There was a positive variation in bovine herds in Chilean Patagonia, specifically in the regions of Aysén and Magallanes, contrary to the national trend, which may be evidence of a reconversion of the industry, related to the commercial opening of the southern regions of Chile towards China [82].

A study on livestock carrying capacity in the study area shows that in most of the municipalities it would be advisable to reduce it, especially in the Chiloé Archipelago [83]. Our analysis shows that the variations obtained in this driver in the Aysén Region were of lesser magnitude than those obtained in the municipalities of the Los Lagos Region; indicating underutilization for some areas of the resource (where the livestock mass could be increased), with the exception of the municipalities of Coyhaique and Río Ibáñez, where overutilization of the resource is confirmed. The livestock systems of the Magallanes Region are in the range classified as medium to very good, with the exception of the municipality of Cabo de Hornos, where the degree of intensity of livestock exploitation is considered insufficient.

It is projected that the pressure on ecosystems due to this driver will continue to occur strongly in the municipalities of Hualaihué, Ancud, Castro, Chonchi, Palena, Coyhaique, and to a lesser extent in the rest of Chiloé Island, Chaitén, Futaleufú, Lago Verde, Río Ibáñez, and almost the entire Magallanes Region (Fig. 1E).

4.6 Expansion of Fishery and Aquaculture Production

It is widely recognized that fisheries are in crisis, which has four dimensions [15]: (i) ecological, as they lose productive capacity; (ii) socioeconomic, because industrial fisheries are sustained by large subsidies while simultaneously harming small-scale fisheries; (iii) intellectual, because fishery sciences have lost credibility by supporting industry interests and refusing to use accumulated ecological knowledge to support management based on precautionary principles; (iv) ethical, because the fisheries sector has discarded any notion of protecting the resources on which it depends. These crises have their greatest expression in Chile in the number of overexploited and fully exploited fisheries [84], and the sustained conflicts between artisanal and industrial fisheries (Subsecretaría de Pesca y Acuicultura [85]; in Spanish SUBPESCA).

The study area concentrates high-value fisheries, many of which have undergone enormous transformations in scale in recent decades, from small volumes destined for local markets to huge volumes destined almost entirely for export, in response to increasing global demands [86, 87]. The main species caught in the provinces of Palena and Chiloé are anchovy (*Engraulidae rigens*), clam (*Venus antiqua*), southern hake (*Merlucius australis*), jack mackerel (*Trachurus murphyi*), and pelillo (*Gracilaria chilensis*), which between 2000 and 2017 have had an average decrease in landings of 49%. In the Aysén Region the main species are red sea bass (*Gigartina skottsbergii*), urchin (*Loxechinus albus*) and southern sardine

(*Sprattus fuegensis*). Since 2000, red sea bass and southern sardine have experienced an increase in landings; in contrast, sea urchin had a 63% decline between 2000 and 2017.

The main species landed in the Magallanes Region are king crab (3,350 t, 65% of the national total in 2018) and sea urchin (12,506 t, 43% of the national total in 2018). Although both species have had increases in their landings since 2000, during the last five years the dynamics varied, exhibiting a sustained decline [85]. The stock statuses are declared fully exploited for king crab, sardine, and sea urchin and overfished for hake, clam, and jack mackerel. The overfished fisheries are over their maximum sustainable yield and risk collapse without radical changes in their management.

In addition to overfishing, the marine ecosystems and fisheries of Chilean Patagonia are exposed to other threats, including climate change and illegal fishing [6, 84, 86, 87]. The main impacts of climate change in the Patagonian seas are the increase in sea level temperature, acidification, and change in salinity due to water freshening [88, 89]. These disturbances can lead to stock displacements, increased mortality of crustaceans due to sea acidification and a general disturbance in food chains [90]. The main illegal fishing infractions reported in the study area are due to non-compliance with seasonal closures, failure to accredit the origin of fishing, and non-compliance with established quotas [84]. The effects of illegal fishing and climate change act synergistically with overfishing, intensifying the degradation of the marine system [85].

Aquaculture, and salmon farming in particular, have been in Patagonia since 1989, with an explosive development [91]. A boost to private investment through Fundación Chile achieved a rapid dynamism of the sector, reaching historic salmon harvests in 2014 of over 800,000 t per year. There is a strong concentration of aquaculture concessions in the province of Chiloé (more than 900 concessions granted,23% for salmonids, 9.6% for algae, and 66% for mollusks). In the Aysén Region there are 728 concessions granted, of which 0.6% are for mollusks and 99% for salmonids. In the Magallanes Region there are 130 concessions granted, of which 4.6% are for mollusks and 95% for salmonids.[9]

Among the most significant effects of the expansive dynamics of salmon farming in Chilean Patagonia are the following: ISA virus outbreaks in salmonids in 2007, episodes of significant increase in sea lice (*Caligus rogercresseyi*) infection since 2007, decrease in seed catch in mussel farming during the years 2009 and 2011, and a reduction of fattening stock during the years 2011 and 2012, population explosions (bloom) of microalgae harmful to salmonids in the summer of 2016 and that of the microalga which produces paralytic toxin in mytilids during the fall of 2016. All these events produced social and environmental effects that were widely reported in the press and in scientific reports [91–94]. Recent studies account for the counterproductive effects of aquaculture on regulatory and cultural marine ecosystem services, in addition to the welfare of local communities [95],

---

[9] Statistics Undersecretariat of Fisheries and Aquaculture.

despite the large generation of jobs associated with the activity. The aforementioned crises have been concentrated mainly in the province of Chiloé, resulting in the closure to the entry of new aquaculture concessions in the province and the displacement of aquaculture activity to the south. Currently, the entire study area is closed to new aquaculture concession applications; however, applications already submitted are still being processed in the Magallanes Region, and therefore an increase in aquaculture activity is expected in the region.

4.7 Demographic Dynamics

Population growth underlies all environmental problems, from global warming to habitat loss. Throughout the twentieth century and so far, this century, there has been unprecedented urban expansion worldwide, driven by population growth, economic growth, land use policies, and transportation costs [96].

Chilean urban growth has seen an almost uncontrolled physical expansion of cities, which has had profound environmental repercussions [97, 98]. Huge areas of land of high agricultural capacity or covered by remnants of natural forests, wetlands, and river and stream beds have been urbanized, seriously disturbing the natural flow of energy and matter [98, 99].

The study area has a population of 452,946 inhabitants, 76% urban and 24% rural [64]. However, there are eight eminently rural municipalities (with more than 70% of the population in this condition), five in the Magallanes Region. Chilean Patagonia has a negative average regional growth, with a percentage variation of −0.9% between 2012 and 2017. However, 24 of the 34 municipalities analyzed (70.5%) had an increase in population between census periods (2012–2017), ranging from 1.24% (municipality of Palena) to 63.60% (municipality of Torres del Paine). At least six of these municipalities are located in the Chiloé Archipelago; Castro increased by 11.28%; Chonchi by 18.18%; Curaco de Vélez by 12.52%; Dalcahue by 28.7%; Queilén by 4.81%; and Quellón by 24.70%.

It should be noted that these municipalities suffered marked growth in the previous decade, associated with the establishment of the aquaculture industry [100], which in turn was associated with phenomena of agricultural abandonment [101] and subdivision of land for second homes, which contributed to the fragmentation of the landscape [102]. Other municipalities with an increase in population in 2012–2017 are Río Verde (72.35%) and Torres del Paine (63.6%), which is caused by the growth of the international tourist destination.

It should be noted that the regions of Aysén and Magallanes have been classified in population terms as erratic zones, i.e. a group of regions that are distinguished by their volatility, made up of regions of smaller demographic size, which makes them more susceptible to more intense rates of migration, associated with temporary migratory flows determined both by public policies and by growth in industrial activities, especially mining [103].

The mean proportion of inhabitants who recognize themselves as belonging to an Indigenous people is 29.9%. The municipalities of Guaitecas (51.4%), Queilén

(51.7%), and Quinchao (50.51%) stand out, with more than 50% of their population belonging to Indigenous people. The lowest proportion of inhabitants belonging to Indigenous people is the Magallanes Region, with an average of 22.3%. The highest proportion is the Los Lagos Region, with 34.92%, and 30.9% in the Aysén Region (see [104]).

4.8 Economic Growth

Although the relationship between economic growth and environmental degradation is still debated, it is widely recognized that several critical planetary boundaries have already been crossed as a result of the current capitalist model [105, 106]. Economic growth in Chilean Patagonia depends significantly on natural resources, as in the rest of Chile. Thus, for example, the fisheries and aquaculture sector is of great importance to Patagonia's GDP (12% of total GDP in 2016).[10] In the Aysén Region it represents the main contribution to growth (28% of GDP), which is mainly due to aquaculture [107].

The Magallanes Region led the growth of the study area in 2016 and 2017 (4.7% for the 2015–2016 period), followed by Aysén (1.9%). The increase in 2016 is explained by the increase in the activity of the industrial manufacturing sector, associated with the increase in production of the fishing industry subsector; while for 2017 the increase was associated with the fuel processing subsector.[11] This growth has been indisputably associated with environmental effects derived from the expansion of economic activities. In the latest environmental performance evaluation of Chile, ECLAC and OECD [18] indicated that: "as a result of the growing economic activity, greater extraction and use of natural resources, and the development and expansion of infrastructure, the pressures on Chile's varied biological diversity are intensifying. In addition, deep income inequality exacerbates environmental conflicts and fuels mistrust." Biodiversity conservation objectives are progressively being integrated into other policy areas such as agriculture, forestry, and mining, but tangible results have yet to materialize. As a result, the country and the region face large-scale conservation challenges that involve more than just the green and blue discourses of sustainable growth.

4.9 Institutional Factors and Environmental Governance

Governance has been recognized as a driver of ecosystem change [4]. The norms or institutional frameworks are a relevant dimension of governance. In Chile these frameworks are inscribed within meta-norms such as neoliberalism, sustainability, and the "polluter pay" principle. Neoliberalism in particular is the dominant

---

[10] Considering only the regions of Aysén and Magallanes.
[11] Information extracted from the Central Bank platform.

model for political economic practice, and therefore environmental governance has been determined by the neoliberal imperative to deregulate, liberalize trade and investment, commercialize, and privatize [108–111]. This has had implications for nature conservation in Chile, from changing social values towards nature and the environment to conservation and resource management instruments (e.g. [112–115].

The literature presents Chile as an iconic example of neoliberal policies in action (e.g. [116]) and an example of continued economic dependence on nature, with an unresolved tension in its accumulation strategy, which has not been able to absorb the negative externalities in the use of nature [113]. The cause-effect relationship between governance and its effect on ecosystems is probably the most difficult to establish, and certainly requires a historical perspective. Therefore, in this chapter we will only focus on exemplifying some regulatory norms (Table 3) and their current limitations to achieve good governance.

The main conservation instrument in Chilean Patagonia is the National System of State Protected Areas (in Spanish SNASPE). Several studies show that the design and coverage of terrestrial and marine protected areas (PAs) are still inadequate to conserve key biodiversity features, that their management approaches are insufficient to address threats and pressures on biodiversity, and that they have a low degree of ecological integrity due to increasing levels of environmental degradation [117]. Many of these areas also lack up-to-date management plans, have delimitation problems, and have undergone changes in their size and boundaries. This is in addition to the fact that there are no territorial management instruments that regulate the use of the areas surrounding PAs [118].

Approximately 50% (130,225.02 km^2) of the terrestrial landscape in the study area is in some SNASPE conservation category; national parks represent 85% of this total. Private terrestrial protected areas represent 3.5% (9296.13 km^2) of the total territory. There is also marine conservation, especially in the Kawésqar National Reserve, as reported by Tecklin et al. [119] and the Seno Almirantazgo Multiple Use Coastal Marine Protected Area (in Spanish AMCP-MU) with 115,200 ha.

Despite the recognition of their importance, terrestrial and marine PAs in Chilean Patagonia are located in places with limited access, which has less effect on preventing changes in ecosystems [120], as is the case in the Aysén and Magallanes regions. There are no metrics to measure the effectiveness of conservation, and even if they did exist, resources are required for monitoring and effective protection. Chile is one of the Latin American countries that invests the least in its protected areas [121].

A similar situation occurs with private protected areas (PPAs), where it is even more complex to determine the effectiveness of conservation measures, given their diversity in terms of tenure, motivations, and spatial distribution. While the number of PPAs has grown in Chilean Patagonia, many of them lack legal, institutional, and administrative recognition, which has resulted in a lack of integration within the broader scope of conservation [122]. At least part of land acquisition for private conservation is speculative and driven by rising land prices, which is why several

Table 3 Scope, instruments, and limitations of the institutional framework in Chile, with emphasis on the Patagonia region

Scope	Main instrument	Type of instrument	Actors involved	Examples of recognized limitations
Biodiversity conservación	National Protected Area System (SNASPE)	Command and control	State	– Lack of financing – Lack of representativeness of threatened ecosystems
	Private Protected Areas (PPAs)	Market-based	Individual home owners, NGOs, foundations, companies, corporations, corporations, etc	– Lack of standards – Lack of incentives, beyond the motivation of those involved
	Real conservation rights	Market-based	State, individual owners	– Little jurisprudence
	Conservation landscapes	Planing	Individual home owners, NGOs, foundations, companies, corporations, corporations, etc	– Non-binding, only guidance
Forest management	Management plans	Command and control	State, individual owners	– Limited control capacity – Lack of environmental education and financial capacity of the owners
	Community management	Mixed: state-civil society	State, individual landowners, indigenous communities	– Insecurity of land tenure and access to forests – Weakness of community organizations – Unfavorable market access for forest products and services by communities – Shortage of capital and limited access to credit in rural sectors
	Subsidy	Command and control	State, individual owners	– Low number of bonuses – Lack of monitoring of the impact on local development and conservation

(continued)

Table 3 (continued)

Scope	Main instrument	Type of instrument	Actors involved	Examples of recognized limitations
Fisheries management	Management and Exploitation Areas for Benthic Resources (in Spanish: (AMERB)	Mixed: Co-management	State-Fishermen's associations	– Generate exclusion of non-owners of AMERBs – Fixed zoning in dynamic marine spaces
	Access Permits (Registry of Artisanal Fishers)	Command and control	Fishers and vessels	– They generate exclusion and are limited when the species are in full exploitation
	Fees	Market-based	Fishermen's associations	– Generate exclusion and are a driver of illegalities – Conflict arises between associations over quota distribution
Environment	Environmental Impact Assessment Strategic Environmental Assessment	Command and control	State, companies, communities	– Weak territorial planning – Weak citizen participation – Practical case of adequate citizen participation (PACA) – Strategic environmental assessment not required for all projects

PPAs have faced various conflicts with local and Indigenous communities [123, 124]. Other emerging forms of private conservation are the so-called "real right of conservation" and conservation landscapes, which however suffer from the same limitations as PPAs: lack of implementation effectiveness, conflicting values, and low economic efficiency [125].

Management tools have not been able to reduce or prevent pressure on natural resources due to: (i) their limited scope (e.g. subsidies under the Native Forest Law); (ii) the lack of monitoring of their implementation, as in the case of forest and fisheries management plans; (iii) the lack of resources for their control, as is the case with the Artisanal Fishers Register or fishing quotas. In some cases, these instruments lack legitimacy, leading to non-compliance, as has been documented in the case of the Artisanal Fishers Registry and illegal crab fishing in the Magallanes Region (see [37, 86]). Our review of land-use planning instruments indicates that biodiversity conservation is largely contingent on the economic development strategies of the regions [126, 127]. Environmental assessment instruments are still insufficient to compensate for the negative effects of investment projects. Much of the economic activity in the study area is subject to no more than Declarations of Environmental Impact and lacks Environmental Impact Assessments, as is the case for almost all mining exploitations and prospections, the installation of aquaculture centers and their intensification [60]. The strategic environmental assessment has not fulfilled its role as a coordinating mechanism for different territorial planning instruments, so it has not been possible to address conflicts over priority land uses and resolve contradictory activities or those that present proximity conflicts [128]. For example, this is reflected in the pressure currently suffered by the Skyring Sound, which is a Patagonian area where, as mentioned above, multiple drivers of change and incompatible uses are concentrated (e.g. mining projects presented for environmental assessment, aquaculture development, tourism development, among others).

4.9.1 Synergistic Effects of Direct and Indirect Drivers

Eight types of clusters were configured in the 34 municipalities within the study area using the methodology described (Fig. 2). It should be noted that the spatial configuration defined by this analysis is consistent with the conclusions obtained in the description of the indicators and in the analysis of uniformity and intensity [28]. The resulting zones do not correspond to defined administrative areas, but rather group municipalities that behave similarly in terms of drivers:

Type 1. Includes only the municipality of Coyhaique, which is home to 13% of the population of the study area, and is the second most populated municipality after Punta Arenas, with 86% urban population. This type of municipality has a higher growth rate than that of the country and the region. The progressive concentration of population in the city could be explained by the mobility or migration of people from municipalities of lower rank in the hierarchy of the region and other regions of the country to the city of Coyhaique. This is manifested in higher consumption of electricity (third highest electricity consumption) and firewood. The high consumption of firewood is also associated with a high annual deforestation rate (0.03%). The municipality has productive industrial and business activities related to aquaculture [91]; however, this activity does not occur in areas around the municipality. There are also metal mining concessions in the mountainous areas of the municipality.

Type 2. Includes only the municipality of Quellón, which with 6% of the population of the study area is the third most important urban center. One of the differentiating characteristics of this municipality is that 49% of its population are Indigenous people. This condition is associated with multiple activities that include seaweed extraction and the expansion of livestock activities, with a carrying capacity of 0.94 animals/ha. Firewood extraction is also an important activity, although the annual rate of forest loss is low (0.019%) compared to the municipalities that host the largest cities in the archipelago (e.g. 0.15% in Ancud). Quellón is the second municipality after Punta Arenas with the highest energy consumption per inhabitant; important non-conventional energy development projects are being developed there. The high energy consumption is attributable to aquaculture and fishing activities.

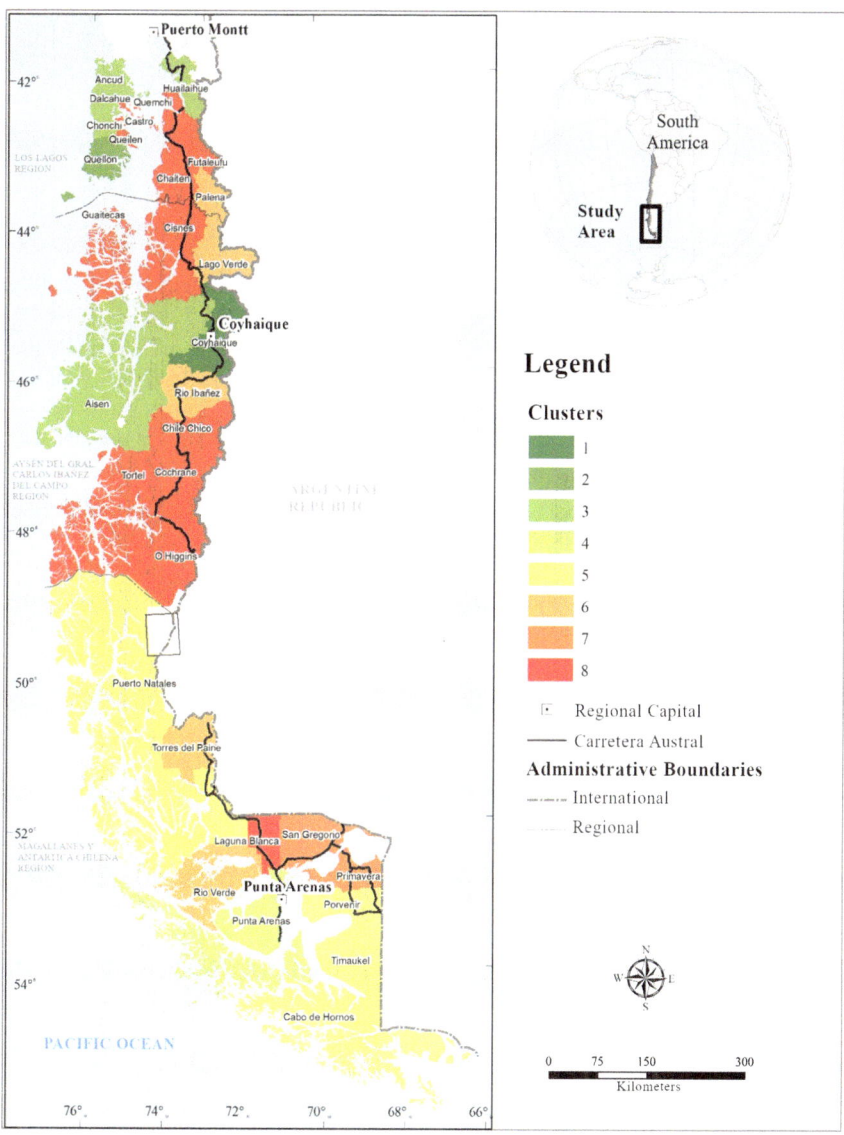

Fig. 2 Chilean Patagonia. Types of municipalities resulting from the cluster analysis. Each of the colors groups municipalities that in terms of drivers of ecosystem change behave similarly

Type 3. Comprises seven municipalities, most of which are located in the Chiloé Archipelago, except for Aysén and Hualaihué. These are municipalities with a strong cultural heritage of native peoples, mainly in terms of varied activities. These municipalities have greater livestock development with above-average stocking capacities (1.24 animals per hectare compared to an average of 0.8 in the study area). There was peat extraction pressure in these municipalities in the past, but despite the fact that this activity has declined, these ecosystems continue to suffer pressures for Sphagnum extraction [58, 55]. Their greatest pressures are mostly related to the loss of native forest at an average annual rate of 0.05% (maximum 0.15% in Ancud and minimum 0.007% in Quinchao) and to the expansion of tourism in municipalities belonging to consolidated (Chiloé Archipelago) and emerging (municipality of Hualaihué) destinations.

Type 4. Includes only the municipality of Punta Arenas, the largest urban center in the Magallanes region, which concentrates 29% of the population of the study area, with 95% living in the urban zone. Punta Arenas is divided into two large areas, continental and insular. The insular sector includes Dawson Island, the archipelago and Kawésqar National Park, where the islands of Santa Inés, Clarence, Capitán Aracena, and Desolación stand out. With difficult habitability conditions due to their geomorphology, climate, and precarious accessibility, these sites coexist with nature conservation, fishing activities, and tourism development. The urban area of Punta Arenas is where the socioeconomic activities of the municipality are based, mainly those related to industrial and tourism development. This area has one of the largest increases in the annual visitor rate within the study area (0.23%, much higher than the study area average of 0.04%) (see also [65]). This municipality is the gateway to Antarctic and sub-Antarctic ecosystems that are emerging as new tourist destinations. This municipality has the highest energy consumption, probably associated with its industrial development.

Type 5. Includes almost all of the Magallanes Region, with the exception of the municipalities of Punta Arenas, Primavera, San Gregorio, Laguna Blanca, Torres del Paine, and Río Verde. The municipalities contained in this category have marked rurality, low population and are isolated. They have an incipient tourist development, with a variation rate of 0.22% for Timaukel. It should be noted that the municipality of Puerto Natales belongs to the tourist destination Puerto Eden, classified as a potential destination, where it seeks to promote tourist circuits in its northern zone through infrastructure [63]. This group of municipalities is widely affected by mining projects (it is the group with the second largest number of projects after Type 6), in addition to presenting an incremental development of aquaculture, particularly salmon farming [91].

Type 6. Includes the tourist municipalities of Palena, Lago Verde, Río Verde, and Torres del Paine. The differentiating characteristic is that it groups emerging and consolidated tourist destinations with positive trends in visitor flow. These municipalities have had positive population growth, except Lago Verde, which is attributed to the development of new economic activities such as tourism. Large population growth rates stand out in the municipalities of Río Verde and Torres del Paine with 2012–2017 increases of 72.35% and 63.60%, respectively. However, these are municipalities with low populations, ranging from 1,711 inhabitants to 617 and with a low proportion of Indigenous people (21.4%).

Type 7. Includes the municipalities of San Gregorio and Primavera, whose differentiating characteristic is the strong effects of mining development (oil and coal). The Primavera municipality belongs to the province of Tierra del Fuego, which has been identified as a potential destination for tourism development. These municipalities have a low number of inhabitants (average 978) and their population variation is erratic in time and space. For example, for San Gregorio there was a negative 2012–2017 variation of −31% and for the municipality of Primavera a positive variation of 13.98%.

Type 8. Comprises a good part of the municipalities of the Aysén Region, whose development is focused mainly on aquaculture and fishing activities and to a lesser extent on tourism, which has shown a negative variation in visitors in recent years. Another activity that puts pressure on this territory is mining of silver, zinc, and to a lesser extent gold and lead. These municipalities generally have more inhabitants than the rest of the types that are not capital cities (types 1 and 4). They represent 4.9% of the population of the study area and have a high proportion of Indigenous people (average 30%). They also have important forest loss rates, averaging 0.016% per year, containing a considerable area and forest types used for fuelwood extraction.

5   Discussion

The ways in which ecosystems respond to anthropogenic pressures are complex, and our understanding of the effect of human activities on ecosystems is limited [129, 130]. The spatial patterns of drivers are varied, and the diversity of recent changes is largely unknown. In many places, we know little about which drivers are having the greatest impact on ecosystem condition, their cumulative effect, or how the composition of drivers is changing over time [131]. Given the availability of information, this study has focused on characterizing these drivers and their current and projected trends, as well as their spatial dynamics, information that is a valuable input for conservation planning.

The expansion trends of the drivers we observed coincide with regional (e.g. [4]) and country assessments [18, 132]. All the drivers analyzed generally exhibit a past increase (exponential) or constant trend, but no evidence of a decrease, which is in line with global trends [14, 15]. The case of governance is different as it cannot be assessed as increasing or decreasing, adequate or inadequate in a normative sense, which is why only some limitations of the institutional framework have been presented. Most of the drivers are related, with mining-energy-economic growth and population-tourism interactions being the most identifiable with the available data. The dynamics of the different drivers vary spatially, as shown by the cluster analysis. In some cases, these dynamics are circumscribed or punctual and therefore more easily attributable (e.g. mining expansion), while in other cases the effect of the drivers is more difficult to visualize. The main trends can be summarized as follows: (i) The activities that have expanded the most over time in Chilean

Patagonia are mining, aquaculture, and tourism. Along with expanding, they have become concentrated, as a result of which recent conflicts are evident (e.g. Riesco Island; areas saturated by aquaculture in Skyring Sound, Xaultegua Gulf, Puerto Natales, Beaufort Bay) and future conflicts are projected given the incompatibility of these activities with biodiversity and landscape conservation [60, 113, 133]. The saturation of areas for aquaculture generates new pressures for salmon farming in other areas, while expanding tourist routes with land and transoceanic marine routes. (ii) The activities that have remained constant or have been brought in are cattle ranching and exotic plantations, which may eventually have an impact on the recovery of the native forest or at least on slowing down its degradation and consequent loss. However, the flip side of this trend is the intensification of agriculture and livestock farming through technology, which has created significant environmental problems in other regions of the country (e.g. [134]). (iii) The territories that could be considered under most pressure are types 7 and 8, where the drivers of expansion of mining, fishing, aquaculture, and tourism are concentrated and where these activities are growing at the highest rates. As mentioned above, the expression of the interaction between drivers is highly context-specific, hence the importance of identifying their simultaneous presence and magnitude in a given space. (iv) The least pressured territories belong to type 5 (municipalities of Puerto Natales, Timaukel, and Porvenir), where the drivers identified have lower effects on ecosystems (e.g. emerging tourism), and expansion rates are the lowest. (v) Potential territories in socio-environmental conflict are places where Indigenous people are concentrated and where some of the economic drivers are beginning to make their presence felt in new territories. The Chiloé Archipelago and the Aysén Region are the most susceptible to this. (vi) While it is not possible to rigorously link direct and indirect drivers, the literature strongly supports the direct relationship between economic growth (affluence), population, and environmental impact [135]. This is most clearly observed in the relationship between mining growth and expansion and energy production and between population growth, affluence, and tourism expansion, as documented by other studies (e.g. [66]).

6   Conclusions and Recommendations

The results shed light on actions and recommendations to minimize the simultaneous impact of these drivers in Chilean Patagonia, including the following:

- To reform the environmental impact assessment regulations by adding the criterion of risk to the presence and expansion of certain drivers.
- To reformulate the current strategic environmental assessment tool, with a view to orienting it to regional biodiversity conservation strategies, incorporating evaluation of the synergistic effects of economic drivers.

- To promote legally binding territorial planning. Regional territorial planning currently provides only guidance for the territory, without any legal attribution to shape the proposed zoning. Also, territorial and coastal use planning instruments have few environmental considerations [127, 136, 137].
- To design territorial strategies that include financial strategies, considering public and private resources to increase financing for biodiversity conservation management.
- To promote the integration of biodiversity management in territorial planning instruments (e.g. establish metrics related to carrying capacity as a requirement in areas of tourist interest) and productive development (e.g. promote agroforestry in incentive instruments for farmers and ranchers).
- Finally, it is necessary to recognize the complexity of the environmental problems facing both the country and the study area. This condition requires both disciplinary collaboration and the connection of researchers with the practice and experience of territorial problems in all their magnitude. While a practice-based approach may challenge our static scientific mindset regarding who we are as environmental researchers and educators, scientific research is often insufficient to address the complexity of environmental problems [138].

Acknowledgements Our thanks to Aldo Farías for his support in the improvement of maps. This work was partially funded by FONDAP 15150003 and Fondecyt 1190207 projects.

References

1. Millennium Ecosystem Assessment. (2005). *Ecosystems and human well-being: current state and trends.* Island Press. https://www.millenniumassessment.org/documents/document.766.aspx.pdf
2. Curtis, P. G., Slay, C. M., Harris, N. L., Tyukavina, A., & Hansen, M. C. (2018). Classifying drivers of global forest loss. *Science, 361*(6407), 1108–1111.
3. Nelson, G. C., Bennett, E., Berhe, A. A., Cassman, K., DeFries, R., Dietz, T., Dobermann, A., Dobson, A., Janetos, A., Levy, M., Marco, D., Nakicenovic, N., O'Neill, B., Norgaard, R., Petschel-Held, G., Ojima, D., Pingali, P., Watson, R., and Zurek, M. (2006). Anthropogenic drivers of ecosystem change: An overview. *Ecology and Society, 11*(2).
4. Intergovernmental Science-Policy Platform on Biodiversity and Ecosystem Services. (2018). The IPBES regional assessment report on biodiversity and ecosystem services for the Americas. In J. Rice, C. S. Seixas, M. E. Zaccagnini, M. Bedoya-Gaitán, & N. Valderrama (Eds.), Secretariat of the Intergovernmental Science-Policy Platform on Biodiversity and Ecosystem Services. https://doi.org/10.5281/zenodo.3236252
5. Intergovernmental Science-Policy Platform on Biodiversity and Ecosystem Services. (2019). Summary for policymakers of the global assessment report on biodiversity and ecosystem services of the Intergovernmental Science-Policy Platform on Biodiversity and Ecosystem Services. In S. Díaz, J. Settele, E. S. Brondízio, H. T. Ngo, J. Guèze, J. Agard, A. Arneth, P. Balvanera, K. A. Brauman, S. H. M. Butchart, K. M. A. Chan, L. A. Garibaldi, K. Ichii, J. Liu, S. M. Subramanian, G. F. Midgley, P. Miloslavich, Z. Molnár, D. Obura, A. Pfaff, S. Polasky, A. Purvis, J. Razzaque, B. Reyers, R. Roy Chowdhury, Y. J. Shin, I. J. Visseren-Hamakers, K. J. Willis, & C. N. Zayas (Eds.), IPBES Secretariat.

6. Marquet, P. A., Buschmann, A. H., Corcoran, D., Díaz, P. A., Fuentes-Castillo, T., Garreaud, R., Pliscoff, P., and Salazar, A. (2023). *Global change and acceleration of anthropic pressures on Patagonian ecosystems*. Springer.

7. Rozzi, R., Rosenfeld, S., Armesto J. J., Mansilla, A., Núñez-Ávila, M., & Massardo, F. (2023). *Ecological connections across the marine-terrestrial interface in Chilean Patagonia*. Springer.

8. Kolb, M., & Galicia, L. (2018). Scenarios and story lines: Drivers of land use change in southern Mexico. *Environment, Development and Sustainability, 20*(2), 681–702.

9. Laterra, P., Nahuelhual, L., Vallejos, M., Berrouet, L., Arroyo, E. A., Enrico, L., Jiménez-Sierra, C., Mejía, K., Meli, P., Rincón-Ruiz, A., Salas, D., Spiric, J., Villegas, J., & Villegas-Palacio, C. (2019). Linking inequalities and ecosystem services in Latin America. *Ecosystem Services, 36*, 100875.

10. Peimer, A. W., Krzywicka, A. E., Cohen, D. B., Van den Bosch, K., Buxton, V. L., Stevenson, N. A., & Matthews, J. W. (2017). National-level wetland policy specificity and goals vary according to political and economic indicators. *Environmental Management, 59*(1), 141–153.

11. Clark, B., Auerbach, D., & Longo, S. B. (2018). The bottom line: Capital's production of social inequalities and environmental degradation. *Journal of Environmental Studies and Sciences, 8*, 562–569.

12. Moore, J. W. (2017). The Capitalocene, Part I: On the nature and origins of our ecological crisis. *The Journal of Peasant Studies, 44*(3), 594–630.

13. O'Connor, J. (1991). On the two contradictions of capitalism. *Capitalism Nature Socialism, 2*(3), 107–109.

14. Mazor, T., Doropoulos, C., Schwarzmueller, F., Gladish, D. W., Kumaran, N., Merkel, K., Di Marco, M., & Gagic, V. (2018). Global mismatch of policy and research on drivers of biodiversity loss. *Nature Ecology & Evolution, 2*(7), 1071–1074.

15. Pauly, D. (2019). *Vanishing fish: Shifting baselines and the future of global fisheries*. Greystone Books Ltd.

16. Newbold, T., Hudson, L. N., Arnell, A. P., Contu, S., De Palma, A., Ferrier, S., Hill, S., Hosking, A., Lysenko, I., Phillips, H., Burton, V., Chng, C., Emerson, S., Gao, D., Pask-Hale, G., Hutton, J., Jung, M., Sánchez-Ortiz, K., Simmons, B., … Purvis, A. (2016). Has land use pushed terrestrial biodiversity beyond the planetary boundary? A global assessment. *Science, 353*(6296), 288–291.

17. Segan, D. B., Murray, K. A., & Watson, J. E. (2016). A global assessment of current and future biodiversity vulnerability to habitat loss-climate change interactions. *Global Ecology and Conservation, 5*, 12–21.

18. Comisión Económica para América Latina y el Caribe y Organización para la Cooperación y el Desarrollo Económicos. (2016). *Evaluaciones del desempeño ambiental: Chile 2016*. Santiago. https://repositorio.cepal.org/bitstream/handle/11362/40308/S1600413_es.pdf

19. Ministerio de Desarrollo Social. (2015). *Informe de desarrollo social 2015*. http://www.desarrollosocialyfamilia.gob.cl/pdf/upload/IDS_INAL_FCM_3.pdf

20. Subsecretaría de Desarrollo Regional. (2012). *Estudio identificación de localidades en condiciones de aislamiento 2012*. http://www.subdere.gov.cl/sites/default/files/documentos/zonas_aisladas2.pdf

21. CONAF. (1998). Catastro y Evaluación de los Recursos Vegetacionales de Chile región de Aysén.

22. CONAF. (2013). Actualización del Catastro de los recursos vegetacionales de la región de Los Lagos, levantamiento de información a escala 1:50.000. Informe Final. Santiago, Chile.

23. CONAF. (1996). Catastro y Evaluación de los Recursos Vegetacionales de Chile región de Magallanes.

24. CONAF. (2005). Actualización del Catastro de los recursos vegetacionales de la región de Magallanes y Antártica Chilena, levantamiento de información a escala 1:50.000. Informe Final. Santiago, Chile.

25. Corporación Nacional Forestal. (2006). *Monitoreo y actualización. Catastro de uso del suelo y vegetación, Región de Magallanes y Antártica chilena*. https://www.conaf.cl/nuestros-bos ques/bosques-en-chile/catastro-vegetacional/
26. Tzanopoulos, J., Mouttet, R., Letourneau, A., Vogiatzakis, I. N., Potts, S. G., Henle, K., Math-evet, R., & Marty, P. (2013). Scale sensitivity of drivers of environmental change across Europe. *Global Environmental Change, 23*(1), 167–178.
27. Ibáñez Martí, J. J., & Alonso, A. (2000). Pedodiversity and scaling laws: Sharing Martín and Rey's opinion on the role of the Shannon index as a measure of diversity. *Geoderma, 98*, 5–9.
28. Carmona, A., & Nahuelhual, L. (2020). *Intensidad y uniformidad de los impulsores de cambio en los ecosistemas de la Patagonia*. https://doi.org/10.13140/RG.2.2.27543.39841
29. Long, J., Nelson, T., & Wulder, M. (2010). Regionalization of landscape pattern indices using multivariate cluster analysis. *Environmental Management, 46*(1), 134–142.
30. Hair, J., Anderson, R., Tatham, R., & Black, W. (2007). *Análisis multivariado* (5ª edición). Pearson Prentice Hall.
31. Reynolds, A. P., Richards, G., de la Iglesia, B., & Rayward-Smith, V. J. (2006). Cluster-ing rules: A comparison of partitioning and hierarchical clustering algorithms. *Journal of Mathematical Modelling and Algorithms, 5*(4), 475–504.
32. Vásquez-Grandón, A., Donoso, P. J., & Gerding, V. (2018). Forest degradation: When is a forest degraded? *Forests, 9*(11), 726.
33. Smil, V. (2013). *Harvesting the biosphere: What we have taken from nature*. MIT Press.
34. Lambin, E. F., Turner, B. L., Geist, H. J., Agbola, S. B., Angelsen, A., Coomes, J., Bruce, J. W., Coomes, O., Dirzo, R., Fischer, G., Folke, C., George, P., Homewood, K., Imbernon, J., Leemans, R., Li, X., Moran, E., Moltimore, M., Ramakrishnan, P., … Xu, J. (2001). The causes of land-use and land-cover change: Moving beyond the myths. *Global Environmental Change, 11*(4), 261–269.
35. Lim, C. L., Prescott, G. W., De Alban, J. D. T., Ziegler, A. D., & Webb, E. L. (2017). Untan-gling the proximate causes and underlying drivers of deforestation and forest degradation in Myanmar. *Conservation Biology, 31*(6), 1362–1372.
36. Reyes, R., Nelson, H., & Zerriffi, H. (2018). Firewood: Cause or consequence? Underlying drivers of firewood production in the south of Chile. *Energy for Sustainable Development, 42*, 97–108.
37. Nahuelhual, L., Saavedra, G., Jullian, C., Mellado, M. A., & Benra, F. (2019). Exploring traps in forest and marine socio-ecological systems of southern and austral Chile. In V. H. Marín & L. E. Delgado (Eds.), *Social-ecological systems of Latin America: Complexities and challenges* (pp. 323–345). Springer.
38. Reyes Gallardo, R. A., Blanco Wells, G., Lagarrigue Ibañez, A., & Rojas Marchini, F. (2017). *Ley de bosque nativo: desafíos socioculturales para su implementación*. https://biblio-tecadi gital.infor.cl/handle/20.500.12220/21352
39. Vergara, G., & Gayoso, J. (2004). Efecto de factores físico-sociales sobre la degradación del bosque nativo. *Bosque (Valdivia), 25*(1), 43–52.
40. Schueftan, A., Sommerhoff, J., & González, A. D. (2016). Firewood demand and energy policy in south-central Chile. *Energy for Sustainable Development, 33*, 26–35.
41. Jara, J. C., Palma, P., & Pantoja, R. (2006). El bosque ya no es matorral: mujeres rurales revalorizando el bosque a través de la avellana. In R. Catalán, P. Wilken, A. Kandzior, D. Tecklin, & H. Burschel (Eds.), *Bosques y comunidades del sur de Chile* (pp. 253–267). Editorial Universitaria.
42. Molina, M. (2010). *Caracterización y evolución de la oferta de leña nativa en las comunas de Los Muermos y Puerto Montt en las últimas dos décadas*. Thesis, Universidad Austral de Chile, Facultad de Ciencias Agrarias, Valdivia, Chile.
43. Fajardo, A. (2016). *Leña, contaminación y conservación del bosque nativo en Aysén*. Revista Lignum, Columna 11 Mayo, 2016. https://issuu.com/revistamch/docs/lignum_162
44. Altamirano, A., & Lara, A. (2010). Deforestación en ecosistemas templados de la pre-cordillera andina del centro-sur de Chile. *Bosque (Valdivia), 31*(1), 53–64.

45. Corporación Nacional Forestal. (2012). *Monitoreo de cambios, corrección cartográfica y actualización del catastro de recursos vegetacionales nativos de la Región de Aysén.* http://bibliotecadigital.ciren.cl/handle/123456789/26311

46. Graells, G., Corcoran, D., & Aravena, J. C. (2015). Invasion of North American beaver (*Castor canadensis*) in the province of Magallanes, southern Chile: Comparison between dating sites through interviews with the local community and dendrochronology. *Revista Chilena de Historia Natural, 88*(1), 3.

47. Alimonda, H. (2015). Mining in Latin America: Coloniality and degradation. In E. Bryan (Ed.), *The international handbook of political ecology* (pp. 149–161). Edward Elgar Publishing.

48. Antonelli, M. A. (2009). Minería transnacional y dispositivos de intervención en la cultura. La gestión del paradigma hegemónico de la "minería responsable y el desarrollo sustentable". In M. Svampa y M. Antonelli (Eds.), *Minería transnacional, narrativas del desarrollo y resistencias sociales* (pp. 51–100). Editorial Biblos.

49. Romero-Toledo, H. (2019). Extractivismo en Chile: la producción del territorio minero y las luchas del pueblo Aymara en el Norte Grande. *Colombia Internacional, 98*, 3–30.

50. Hernández, A. J., & Pastor, J. (2008). Relationship between plant biodiversity and heavy metal bioavailability in grasslands overlying an abandoned mine. *Environmental Geochemistry and Health, 30*(2), 127–133.

51. Palacios-Torres, Y., Jesus, D., & Olivero-Verbel, J. (2020). Trace elements in sediments and fish from Atrato River: An ecosystem with legal rights impacted by gold mining at the Colombian Pacific. *Environmental Pollution, 256*, 113290.

52. Siqueira-Gay, J., Sonter, L. J., & Sánchez, L. E. (2020). Exploring potential impacts of mining on forest loss and fragmentation within a biodiverse region of Brazil's northeastern Amazon. *Resources Policy, 67*, 101662.

53. Larondelle, N., & Haase, D. (2012). Valuing post-mining landscapes using an ecosystem services approach: An example from Germany. *Ecological Indicators, 18*, 567–574.

54. Comisión Chilena del Cobre (COCHILCO). (2017). *Anuario de estadísticas del cobre y otros minerales 1998–2017.* https://www.cochilco.cl/Paginas/Estadisticas/Publicaciones/Anuario.aspx

55. Mansilla, C. A., Domínguez, E., Mackenzie, R., Hoyos-Santillán, J., Henríquez, J. M., Aravena, J. C., & Villa-Martínez, R. (2023). *Peatlands in Chilean Patagonia: Distribution, biodiversity, ecosystem services, and conservation.* Springer.

56. Vega-Valdés, D., & Domínguez, E. (2015). Análisis espacial de la distribución geográfica de las turberas de *Sphagnum* en la Región de Magallanes y la Antártica chilena. In E. Domínguez & D. Vega-Valdéz (Eds.), *Funciones y servicios ecosistémicos de las turberas de Magallanes* (pp. 43–77). Colección de libros INIA N°33. Instituto de Investigaciones Agropecuarias Kampenaike. Centro Regional de Investigación.

57. Domínguez, E., Bahamonde, N., & Muñoz-Escobar, C. (2012). Efectos de la extracción de turba sobre la composición y estructura de una turbera de *Sphagnum* explotada y abandonada hace 20 años, Chile. *Anales Instituto Patagonia (Chile), 40*(2), 37–45.

58. Gajardo, P., Mondaca, E., & Santibáñez, P. (2017). La minería industrial como una nueva amenaza al espacio marinocostero de Chiloé: bahía de Cucao como caso de studio. Revista Iberoamericana de Viticultura. *Agroindustria y Ruralidad, 3*(10), 110–138.

59. Ministerio de Energía. (2017). *Política energética en Magallanes y la Antártica chilena.* https://energia.gob.cl/sites/default/files/energia_magallanes_2050.pdf

60. Bustos, B., Folchi, M., & Fragkou, M. (2017). Coal mining on pastureland in southern Chile: Challenging recognition and participation as guarantees for environmental justice. *Geoforum, 84*, 292–304.

61. Fundación Terram. (2016). *Fracking: fracturando el futuro energético de Chile.* https://www.terram.cl/wp-content/uploads/2017/02/APP-N%C2%B0-62-Fracking-Fracturando-el-futuro-energ%C3%A9tico-de-Chile.pdf

62. Sundriyal, S., Shridhar, V., Madhwal, S., Pandey, K., & Sharma, V. (2018). Impacts of tourism development on the physical environment of Mussoorie, a hill station in the lower Himalayan range of India. *Journal of Mountain Science, 15*(10), 2276–2291.

63. Ministerio de Obras Públicas [Ministerio de Economía, Fomento y Turismo, Subsecretaría de Turismo]. (2017). *Plan especial de infraestructura MOP de apoyo al turismo sustentable a 2030.* http://www.subturismo.gob.cl/wp-content/uploads/2017/12/00_Plan-Especial-de-Inf raestructura-MOP-de-Apoyo-al-Turismo-Sustentable-a-2030-RM.pdf

64. Instituto Nacional de Estadísticas. (2017). *Censo de población y vivienda.* http://www.ine.cl/ estadisticas/sociales/censos-de-poblacion-y-vivienda

65. Guala, C., Veloso, K., Farías, A., & Sariego, F. (2023). *Analysis of tourism development linked to protected areas in Chilean Patagonia.* Springer.

66. Román-Collado, R., Ordoñez, M., & Mundaca, L. (2018). Has electricity turned green or black in Chile? A structural decomposition analysis of energy consumption. *Energy, 162,* 282–298.

67. Hernando-Arrese, M., & Blanco, G. (2016). Territorio y energías renovables no convencionales: aprendizajes para la construcción de política pública a partir del caso de Rukatayo Alto, Región de Los Ríos, Chile. *Gestión y Política Pública, 25*(1), 165–202.

68. Ministerio de Energía. (2015). *Medición del consumo nacional de leña y otros combustibles sólidos derivados de la madera.* http://www.biblioteca.digital.gob.cl/handle/123456789/586

69. Villalobos, A. M., Barraza, F., Jorquera, H., & Schauer, J. J. (2017). Wood burning pollution in southern Chile: PM2. 5 source apportionment using CMB and molecular markers. *Environmental Pollution, 225,* 514–523.

70. Inodú. (2018). *Estudio de variables ambientales y sociales que deben abordarse para el cierre o reconversión programada y gradual de generación eléctrica a carbón.* Informe preparado para el Ministerio de Energía. Informe final - Licitación ID: 584105-9-LE18. https://energia. gob.cl/sites/default/files/12_2018_inodu_variables_ambientales_y_sociales.pdf

71. Comisión Nacional de Energía. (2015). *Informe de previsión de demanda 2015–2030 SIC - SING fijación de precios de nudo octubre 2015.* Santiago, Chile. https://www.cne.cl/wp-con tent/uploads/2015/11/Informe-de-Previsi%C3%B3n-de-Demanda-2015-2030-Oct-2015.pdf

72. Vega-Coloma, M., & Zaror, C. A. (2018). Environmental impact profile of electricity generation in Chile: A baseline study over two decades. *Renewable and Sustainable Energy Reviews, 94,* 154–167.

73. Ministerio de Energía. (2016). *El potencial hidroeléctrico de Chile: actualización.* http:// ernc.dgf.uchile.cl/Explorador/DAANC2017/info/datos/v2016/pdf/PotencialHidroelectrico2 016.pdf

74. Ministerio de Energía. (2017). *Energía 2050. Una política energética para Chile.* https://ene rgia.gob.cl/sites/default/files/energia_2050_-_politica_energetica_de_chile.pdf

75. Oficina de Estudios y Políticas Agrarias (ODEPA). (2018). *Región Magallanes y de la Antártica chilena información regional.* https://www.odepa.gob.cl/wp-content/uploads/2019/07/ Magallanes.pdf

76. Radic-Schilling, S., Corti, P., Muñoz, R., Butorovic, N., & Sánchez, L. (2023). *Steppe ecosystems in Chilean Patagonia: Distribution, climate, biodiversity, and threats to their sustainable management.* Springer.

77. Flores, J. P., Espinoza, M., Martínez, E., Henríquez, G., Avendaño, P., Torres, P., Ahumada, I., Retamal, M., Toledo, B., & Marín, M. L. (2010). *Determinación de la erosión actual y potencial de los suelos de Chile.* (Pub. CIREN N° 139). http://bibliotecadigital.ciren.cl/han dle/123456789/2016

78. Fauna Australis. (2007). *Informe técnico final proyecto. Evaluación del conflicto entre carnívoros silvestres y ganadería.* Pontificia Universidad Católica de Chile. https://www.sag. gob.cl/sites/default/files/INFORME%2520FINAL%2520PROYECTO2.pdf

79. Quintanilla, V., Cadiñanos, J., & Lozano, P. (2008). Degradaciones actuales en ecosistemas nord- patagónicos de Chile, derivadas de los incendios de bosques durante el siglo pasado. *Tiempo y Espacio, 21,* 6–24.

80. Instituto Nacional de Estadísticas. (1997). *Censo agropecuario.* http://www.ine.cl/estadisti cas/economia/agricultura-agroindustria-y-pesca/censos-agropecuarios
81. Instituto Nacional de Estadísticas. (2007). *Censo agropecuario.* http://www.ine.cl/estadisti cas/economia/agricultura-agroindustria-y-pesca/censos-agropecuarios
82. Oficina de Estudios y Políticas Agrarias (ODEPA). (2018). *Región Carlos Ibañez del Campo. Información regional.* https://www.odepa.gob.cl/wp-content/uploads/2019/07/Aysen.pdf
83. Castellaro, G., Morales, L., Rodrigo, P., & Fuentes, G. (2016). Carga ganadera y capacidad de carga de los pastizales naturales de la Patagonia chilena: estimación a nivel comunal. *Agro Sur, 44*(2), 11–23.
84. Molinet, C., & Niklitschek, E. J. (2023). *Fisheries and marine conservation in Chilean Patagonia.* Springer.
85. Subsecretaría de Pesca y Acuicultura. (2018). *El estado actual de las principales pesquerías chilenas.* http://www.subpesca.cl/portal/618/w3-article-107314.html
86. Nahuelhual, L., Saavedra, G., Blanco, G., Wesselink, E., Campos, G., & Vergara, X. (2018). On super fishers and black capture: Images of illegal fishing in artisanal fisheries of southern Chile. *Marine Policy, 95,* 36–45.
87. Nahuelhual, L., Saavedra, G., Mellado, M. A., Vergara, X. V., & Vallejos, T. (2020). A social-ecological trap perspective to explain the emergence and persistence of illegal fishing in small-scale fisheries. *Maritime Studies, 19,* 105–117.
88. Domenici, P., Torres, R., & Manríquez, P. H. (2017). Effects of elevated carbon dioxide and temperature on locomotion and the repeatability of lateralization in a keystone marine mollusk. *Journal of Experimental Biology, 220*(4), 667–676.
89. Torres, R., Manríquez, P. H., Duarte, C., Navarro, J. M., Lagos, N. A., Vargas, C. A., & Lardies, M. A. (2013). Evaluation of a semi-automatic system for long-term seawater carbonate chemistry manipulation. *Revista Chilena de Historia Natural, 86,* 4.
90. Vargas, C. A., de la Hoz, M., Aguilera, V., Martin, V. S., Manríquez, P. H., Navarro, J. M., Torres, R., Lardies, M., & Lagos, N. A. (2013). CO_2-driven ocean acidification reduces larval feeding efficiency and change food selectivity in the mollusk *Concholepas concholepas. Journal of Plankton Research,* 1–10.
91. Buschmann, A. H., Niklitschek, E. J., & Pereda, S. (2023). *Aquaculture and its impacts on the conservation of Chilean Patagonia.* Springer.
92. Lara, C., Saldías, G. S., Tapia, F. J., Iriarte, J. L., & Broitman, B. R. (2016). Interannual variability in temporal patterns of chlorophyll-a and their potential influence on the supply of mussel larvae to inner waters in northern Patagonia (41–44 S). *Journal of Marine Systems, 155,* 11–18.
93. Molina, M. R. (2017). La economía política del virus ISA: la crisis acuícola en Chile y Noruega. *Revista Enfoques: Ciencia Política y Administración Pública, 15*(27), 69–95.
94. Vike, S., Nylund, S., & Nylund, A. (2009). ISA virus in Chile: Evidence of vertical transmission. *Archives of Virology, 154*(1), 1–8.
95. Outeiro, L., & Villasante, S. (2013). Linking salmon aquaculture synergies and trade-offs on ecosystem services to human wellbeing constituents. *Ambio, 42*(8), 1022–1036.
96. Seto, K. C., Fragkias, M., Güneralp, B., & Reilly, M. K. (2011). A meta-analysis of global urban land expansion. *PLoS ONE, 6,* e23777.
97. Inostroza, L., Baur, R., & Csaplovics, E. (2013). Urban sprawl and fragmentation in Latin America: A dynamic quantification and characterization of spatial patterns. *Journal of Environmental Management, 115,* 87–97.
98. Romero, H., Molina, M., Moscoso, C., Sarricolea, P., Smith, P., & Vásquez, A. (2007). Caracterización de los cambios de usos y coberturas de suelos causados por la expansión urbana de Santiago, análisis estadístico de sus factores explicativos e inferencias ambientales. In C. de Mattos & R. Hidalgo (Eds.), *Santiago de Chile: movilidad espacial y reconfiguración metropolitana* (pp. 251–270). Colección Estudios Urbanos UC, Pontificia Universidad Católica de Chile.

99. Vásquez, A. E., Romero, H., Fuentes, C., López, C., & Sandoval, G. (2008). *Evaluación y simulación de los efectos ambientales del crecimiento urbano observado y propuesto en Santiago de Chile.* Actas del Congreso Nacional de Desarrollo Rural. Repositorio académico de la Universidad de Chile. http://repositorio.uchile.cl/handle/2250/118149

100. Ramírez, E., Modrego, F., Yáñez, R., & Macé, J. C. (2011). *Dinámicas territoriales de Chiloé, del crecimiento económico al desarrollo sostenible.* Documento de trabajo no. 86/Programa Dinámicas Territoriales Rurales. RIMISP-Centro Latinoamericano para el Desarrollo Rural. http://www.rimisp.org/wp-content/files_mf/1366295376N862011Ramirez ModregoYanezMaceDinamicasterritorialesChiloe.pdf

101. Díaz, G. I., Nahuelhual, L., Echeverría, C., & Marín, S. (2011). Drivers of land abandonment in southern Chile and implications for landscape planning. *Landscape and Urban Planning, 99*(3–4), 207–217.

102. Mancilla, M. (2010). *Caracterización del proceso de subdivisión predial, en la comuna de Ancud, entre los años 1999 y 2008.* Thesis, Universidad Austral de Chile, Facultad de Ciencias Agrarias, Valdivia, Chile. http://cybertesis.uach.cl/tesis/uach/2010/fad352a/doc/fad352a.pdf

103. González, D., and Viagniolo, J. R. (2004). Tendencias de la migración interna en Chile en los últimos 35 años: recuperación regional selectiva, desconcentración metropolitana y ruruban-ización. In: Congresso da Associação Latino Americana de População, ALAP (Eds.), *Anales Congreso Pobreza, desigualdad y exclusión en América Latina y el Caribe* (pp. 1–17). http://www.alapop.org/alap/images/PDF/ALAP2004_325.pdf

104. Aylwin, J., Arce, L., Guerra, F., Núñez, D., Álvarez, R., Mansilla P., Alday, D., Caro, L., Chiguay, C., & Huenucoy, C. (2023). *Conservation and indigenous people in Chilean Patagonia.* Springer.

105. Rockström, J., Steffen, W., Noone, K., Persson, Å., Chapin, F. S., Lambin, E. F., Lenton, T. M., Scheffer, M., Filke, C., Schellnhuber, H. J., Nykvist, B., Wit, C. A., Hughes, T., van der Leeuw, S., Rodhe, H., Sörling, S., Snyder, P. K., Constanza, R., Svedin, U., Falkenmark., M., Kalberg, L., Corell, R. W., Fabry, V. J., Hansen, J., Walker, B., Liverman, D., Richardson, K., Crutzen, P., & Foley, J. A. (2009). A safe operating space for humanity. *Nature, 461*(7263), 472–475.

106. Ward, J. D., Sutton, P. C., Werner, A. D., Costanza, R., Mohr, S. H., & Simmons, C. T. (2016). Is decoupling GDP growth from environmental impact possible? *PLoS ONE, 11*(10), e0164733.

107. Oficina de Estudios y Políticas Agrarias (ODEPA). (2014). *Sector pesquero: evolución de sus desembarques, uso y exportación en las últimas décadas.* http://www.odepa.cl/wp-con tent/files_mf/1392915533Sectorpesca201402.pdf

108. Baud, M., De Castro, F., & Hogenboom, B. (2011). Environmental governance in Latin America: Towards an integrative research agenda. *European Review of Latin American and Caribbean Studies/Revista Europea de Estudios Latinoamericanos y del Caribe,* 79–88.

109. De Castro, F., Hagenboom, B., & Baud, M. (2016). *Gobernanza ambiental en América Latina.* Buenos Aires: CLASCO. http://biblioteca.clacso.edu.ar/clacso/se/20150318053457/Gobern anzaAmbiental.pdf

110. Overbeek, H. W. (1993). *Restructuring hegemony in the global political economy: The rise of transnational neo-liberalism in the 1980s.* Routledge.

111. Peck, J. (2001). Neoliberalizing states: Thin policies/hard outcomes. *Progress in Human Geography, 25*(3), 445–455.

112. Bauer, C. J. (2015). Water conflicts and entrenched governance problems in Chile's market model. *Water Alternatives, 8*(2).

113. Bustos-Gallardo, B. (2013). The ISA crisis in Los Lagos Chile: A failure of neoliberal environmental governance? *Geoforum, 48,* 196–206.

114. Heilmayr, R., & Lambin, E. F. (2016). Impacts of non-state, market-driven governance on Chilean forests. *Proceedings of the National Academy of Sciences, 113*(11), 2910–2915.

115. Holmes, G. (2015). Markets, nature, neoliberalism, and conservation through private protected areas in southern Chile. *Environment and Planning A: Economy and Space, 47*(4), 850–866.

116. Barton, J. R., & Murray, W. E. (2009). Grounding geographies of economic globalization: Globalised spaces in Chile's non-traditional export sector, 1980–2005. *Tijdschrift Voor Economische en Sociale Geografie, 100*(1), 81–100.
117. Lee, W. H., & Abdullah, S. A. (2019). Framework to develop a consolidated index model to evaluate the conservation effectiveness of protected areas. *Ecological Indicators, 102*, 131–144.
118. Sierralta, L., Serrano, R., Rovira, J., & Cortés, C. (2011). *Las áreas protegidas de Chile, antecedentes, institucionalidad, estadísticas y desafíos.* Ministerio del Medio Ambiente. http://bibliotecadigital.ciren.cl/handle/123456789/6990
119. Tecklin, D., Farías, A., Peña, M. P., Gélvez, X. Castilla, J. C., Sepúlveda, M., Viddi, F. A., & Hucke-Gaete, R. (2023). *Coastal-marine protection in Chilean Patagonia: Historical progress, current situation, and challenges.* Springer.
120. Joppa, L., & Pfaff, A. (2010). Reassessing the forest impacts of protection: The challenge of nonrandom location and a corrective method. *Annals of the New York Academy of Sciences, 1185*(1), 135–149.
121. Petit, I., Campoy, A., Hevia, M., Gaymer, C. F., & Squeo, F. A. (2018). Protected areas in Chile: Are we managing them? *Revista Chilena de Historia Natural, 91*(1), 1–8.
122. Stolton, S., Redford, K. H., & Dudley, N. (2017). *The futures of privately protected areas.* https://portals.iucn.org/library/sites/library/files/documents/PATRS-001.pdf
123. Serenari, C., Peterson, M. N., Wallace, T., & Stowhas, P. (2017). Indigenous perspectives on private protected areas in Chile. *Natural Areas Journal, 37*, 98–107.
124. Zorondo-Rodríguez, F., Díaz, M., Simonetti-Grez, G., & Simonetti, J. A. (2019). Why would new protected areas be accepted or rejected by the public?: Lessons from an ex-ante evaluation of the new Patagonia park network in Chile. *Land Use Policy, 89*, 104248.
125. Gooden, J., & 't Sas-Rolfes, M. (2020). A review of critical perspectives on private land conservation in academic literature. *Ambio*, 1–16.
126. Andrade, B., Arenas, F., & Guijón, R. (2008). Revisión crítica del marco institucional y legal chileno de ordenamiento territorial: el caso de la zona costera. *Revista de Geografía Norte Grande, 41*, 23–48.
127. Cordero, E. (2011). Ordenamiento territorial, justicia ambiental y zonas costeras. *Revista de Derecho (Valparaíso), 36*, 209–249.
128. Cordero, E., & Vargas, I. (2016). Evaluación ambiental estratégica y planificación territorial. Análisis ante su regulación legal, reglamentaria y la jurisprudencia administrativa. *Revista Chilena de Derecho, 43*(3), 1031–1056.
129. Hoegh-Guldberg, O., & Bruno, J. F. (2010). The impact of climate change on the world's marine ecosystems. *Science, 328*(5985), 1523–1528.
130. Sumaila, U. R., Cheung, W. W., Lam, V. W., Pauly, D., & Herrick, S. (2011). Climate change impacts on the biophysics and economics of world fisheries. *Nature Climate Change, 1*(9), 449–456.
131. Halpern, B. S., Walbridge, S., Selkoe, K. A., Kappel, C. V., Micheli, F., D'Agrosa, C., Bruno, J. F., Casey, K., Elbert, C., Fox, H., Fujita, R., Heinemann, D., Lenihan, H., Madin, E., Perry, M., Selig, E., Spalding, M., Steneck, R., & Watson, R. (2008). A global map of human impact on marine ecosystems. *Science, 319*(5865), 948–952.
132. Ministerio del Medio Ambiente. (2019). *Quinto reporte del estado del medio ambiente.* https://sinia.mma.gob.cl/quinto-reporte-del-estado-del-medio-ambiente/
133. Inostroza, L., Zasada, I., & König, H. J. (2016). Last of the wild revisited: Assessing spatial patterns of human impact on landscapes in Southern Patagonia, Chile. *Regional Environmental Change, 16*, 2071–2085.
134. Ginocchio, R., Melo, O., Pliscoff, P., Camus, P., & Arellano, E. (2019). *Conflicto entre la intensificación de la agricultura y la conservación de la biodiversidad en Chile: alternativas para la conciliación.* Temas de Agenda Pública n°18. https://politicaspublicas.uc.cl/wp-content//uploads/2019/11/PAPER-N%C2%BA-118-VF.pdf

135. Hegland, T. J. (2017). Factors behind increasing ocean use: The IPAT equation and the marine environment. In M. Salomon & T. Markus (Eds.), *Handbook on marine environment protection* (pp. 533–542). Springer.
136. Andrade, B., Arenas, F., & Lagos, M. (2010). Territorial planning on the coast of central Chile: Incorporation of environmental fragility and risk criteria. *Revista de Geografía Norte Grande*, 5–20.
137. Belemmi, V. (2015). El ordenamiento territorial como catalizador de conflictos territoriales. [Thesis]. Facultad de Derecho, Universidad de Chile. http://repositorio.uchile.cl/handle/2250/138490
138. Datta, R. K. (2017). Practice-based interdisciplinary approach and environmental research. *Environments*, *4*(1), 22.
139. CONAF. (2006). Actualización del Catastro de los recursos vegetacionales de la región de Aysen, levantamiento de información a escala 1:50.000. Informe Final. Santiago, Chile.

Laura Nahuelhual Muñoz Agricultural Engineer, Universidad Austral de Chile. Master in Rural Development, Universidad Austral de Chile. Ph.D. in Agricultural and Resource Economics, University of Colorado, USA. Environmental economist.

Alejandra Carmona Engineer in Renewable Natural Resources, Universidad de Chile. Master in Rural Development, Universidad Austral de Chile. She has participated as a collaborator in research in the area of ecosystem services.

Analysis of Tourism Development Linked to Protected Areas in Chilean Patagonia

18

César Guala, Katerina Veloso, Aldo Farías, and Fernanda Sariego

Abstract

Tourism in protected areas has grown rapidly worldwide. Although tourism's contribution to biodiversity and local development has been the subject of research internationally, in Chile such research has not yet been carried out systematically or with a focus on specific territories. To address this gap, the chapter examines the role of the National System of State Wild Protected Areas (in Spanish SNASPE) in tourism development, with an emphasis on the relationship between tourism, conservation, and local development in Chilean Patagonia. The analysis includes a review of scientific and technical reports. The results show that public policies can enhance the role of the SNASPE for tourism and contribute to the growth of gateway communities with a symbiotic relationship between local people and PAs. The economic and environmental impacts of these policies are less clear, although there is evidence that tourism is a primary source of income for SNASPE, which has limitations and needs to be reviewed. Information on tourism impacts on biodiversity is also limited, but evidence shows that adopting better standards and planning in PAs is a critical requirement for tourism development.

Keywords

Patagonia · Chile · Tourism · Protected areas · Tourism supply · Demand and impacts

C. Guala (✉) · K. Veloso · A. Farías · F. Sariego
Austral Patagonia Program, School of Economic and Administrative Sciences, Institute of Tourism, Universidad Austral de Chile, Valdivia, Chile
e-mail: cesar.guala@uach.cl

© Pontificia Universidad Católica de Chile 2023
J. C. Castilla et al. (eds.), *Conservation in Chilean Patagonia*, Integrated Science 19,
https://doi.org/10.1007/978-3-031-39408-9_18

481

1   Introduction

Tourism in protected areas (PAs) has unique characteristics that allow it to contribute to biodiversity conservation, sustainability, and socioeconomic development of local communities, diversifying their economies and enhancing their cultural authenticity [1]. The opportunities and challenges of tourism in PAs center on maximizing these benefits and minimizing negative environmental and social impacts because of poor management [2]. Thus, tourism may be a key dimension for achieving conservation objectives and compliance with PA management standards [3]. Several authors have assessed the environmental impacts of tourism in PAs [4, 5] and suggested the application of management standards to address them [6]. Other authors point to the need to quantify the economic effects of tourism in PAs [7] and to estimate the impact on the well-being of local communities [1]; in particular for gateway communities, which are defined as those located near or at an entrance to a PA, and that provide cultural ecosystem goods and services, including tourism and recreation [8, 9].

The management of tourism within PAs has also been the subject of research, where two major models have been identified: one where the PA authority uses its own personnel and facilities to finance and provide tourism services within the area, and the other, where the PA authority contracts with one or several external entities that provide services under different legal arrangements (concessions, licenses, permits) [10]. The latter model requires a well-defined policy that ensures the benefits of tourism for conservation and local development [11].

Tourism in PAs in Chile has been evaluated regarding supply trends, demand [12], and the role of gateway communities for PA management and governance [13]. However, the evaluation of tourism impacts on natural ecosystems [14] and the estimation of its economic contribution have been less studied [15, 16].

Between Reloncaví Sound and the Diego Ramírez islands (Chilean Patagonia, 41° 42′ S 73° 02′ W; 56° 29′ S 68° 44′ W), different authors have identified: (i) conservation and enhancement of local heritage and opportunities for scientific tourism [17–21]; (ii) examined the impacts of tourism on biodiversity on PAs [14, 22]; (iii) developed public use planning models to minimize the different impacts of tourism [23]. The contribution of tourism to conservation and local welfare has been reviewed [24, 25], concluding that a territorial imaginary has emerged that values tourism and nature protection and is the result of centrally driven public policies [26]. The economic effects of tourism on PAs and gateway communities are much less studied, with some estimates based on the willingness to pay methodology [27, 28] and the economic contribution of cruise ship tourism [29].

Despite the above, a comprehensive analysis that addresses the relationship of tourism in PAs to conservation and gateway communities has not been published for Chile [30], except for the study by Muñoz and Torres [31], who examined aspects of connectivity, territorial openness, and the formation of a nature tourism destination for the Aysén Region. However, there is a lack of analysis with a systematic approach to the impacts of tourism on Chilean Patagonia, including variables such as supply, demand, and promotion of the sector, which would reveal the real role of the areas of the SNASPE system in Patagonian tourism development.

2   Scope and Objectives

In this chapter we analyze the distribution of tourism supply and demand, examining the economic contributions, environmental effects, and trends in Chilean Patagonia in relation to the SNASPE, and make recommendations about the present and future role of these protected areas in Chile's Patagonian tourism.

3   Methods

The study area is Chilean Patagonia, in particular the National Protected Wild Areas (NPWA) that comprise the SNASPE, within the regions of Los Lagos (41° 28′ 18″ S; 72° 56′ 12″ W), Aysén (45° 34′ 12″ S; 72° 3′ 58″ W), Magallanes and Chilean Antarctica (53° 9′ 45″ S; 70° 55′ 21″ W). Information to conduct the analyses was obtained from official documents and technical reports (Table 1). Relevant publications in mainstream journals and gray literature were consulted for the analysis of current tourism development and trends. National and regional policy tools were reviewed to learn about public intervention to promote tourism, systematizing those that mentioned—with a variety of key concepts—tourism development in the NPWA.

To identify gateway communities linked to the NPWA, the distribution of tourism supply was analyzed based on the number of tourism service providers per locality and registered with the National Tourism Service (in Spanish Servicio Nacional de Turismo, SERNATUR). These services include lodging, food, travel agencies, adventure tourism, and tour guides. The service providers were grouped by locality and linked to the 12 tourist destinations in Chilean Patagonia and the NPWA associated with each destination (Table 3). The relationship between tourism development and the management of the NPWA was evaluated based on the concentration of supply per locality (number of providers), the level of tourism development of the destination, and the level of management of the areas.[1] The supply within the NPWA was studied based on data from the National

---

[1] Destinations are classified as: Potential, Emerging and Consolidated [34]. Levels of management in NPWAs are classified as: (i) Initial: no management plan, personnel or budget; (ii) Basic: outdated management instruments and a minimum number of park rangers and budget; (iii) Intermediate: management plan in force, ranger team and sufficient infrastructure and equipment to carry out basic management activities; (iv) Consolidated: specialized ranger team, specific planning instruments and sufficient infrastructure and equipment to carry out advanced management activities [46].

Table 1 Sources consulted for the analysis of the supply and demand of tourism services in Chilean Patagonia (own elaboration)

Name of document	Type of information		
	Tourist offer	Tourism demand	Territorial planning
Tourism Yearbook 2018 (Undersecretariat of Tourism, 2019)		X	
SNASPE visitation statistics 2009–2019 [32]		X	
Application form for ZOIT Puelo and Cochamó [33]		X	
Report on Intensity and Definition of Tourism Destinations [34]			X
Arrivals of foreign tourists to Chile according to nationality or border crossing. January, 2018–August, 2018 series [35]		X	
Public–Private Cooperation Model for Sustainable Tourism Development in State Wildlife Protected Areas and their surroundings [36]	X		
ZOIT Aysén Patagonia Queulat Action Plan [37]		X	
Chelenko ZOIT Action Plan [38]		X	
Glacier Province ZOIT Action Plan [39]		X	
ZOIT Futaleufú Action Plan [40]		X	
ZOIT Cabo de Hornos Action Plan [34]		X	
Action Plan ZOIT Torres del Paine destination [34]		X	
Aysén 2018–2022 [41]			X
Los Lagos Region Plan 2018–2022 [42]			X
Magallanes Region Plan 2018–2022 [43]			X
National Registry of Tourism Service Providers [44]	X		

(continued)

Table 1 (continued)

Name of document	Type of information		
	Tourist offer	Tourism demand	Territorial planning
Tourism—Annual Report 2008		X	
Sustainable Tourism in Cape Horn: Tourism Development Plan 2015–2020 [45]		X	

Forestry Corporation (in Spanish Corporación Nacional Forestal, CONAF), identifying ecotourism concessions and permits and the economic income generated by concession contracts and entrance fees.

There is national information on tourism demand; however, this information is limited at the regional level and is distributed across various technical documents (Table 1). To examine the characteristics of this demand, annual information was systematized regionally for the NPWA according to CONAF statistics. Finally, the economic, social, and environmental impacts of tourism in Patagonia's NPWA are presented and development trends are addressed.

The results presented below have some limitations, mainly due to the lack of a complete and systematic database. For example, the data on tourism supply are based on official information, without reflecting the existence of an informal supply. A similar situation occurs with demand in NPWA, whose analysis is based only on CONAF records. Despite the importance of private conservation in Chilean Patagonia, this article focuses only on NPWA, given the lack of available data for analysis.

Los Cuernos viewpoint route, Torres del Paine, Magallanes and Chilean Antarctica Region. Photograph by Jorge López

4   Results

Tourism has grown significantly in Chile during the last decades, driven in part by an increased interest of tourism demand in learning about natural attractions in the NPWA [12, 16]. About 50% of the territory of Chilean Patagonia is protected by the SNASPE [46] propitiating a wide range of tourism activities [17]. This has led to a doubling of the number of tourists that have visited these areas in a decade [12], making tourism in the NPWA a relevant economic sector with great growth potential [27, 47].

4.1   Public Intervention Framework for the Promotion of Nature-Based Tourism

The growth of tourism in the NPWA can be explained by the multiple state policies that have explicitly recognized the opportunities of these areas for tourism promotion and development (Table 2). The Undersecretariat of Tourism has developed national development plans and programs that address these opportunities, while for Patagonia, regional governments have prioritized special interest tourism and recognized the relevance of NPWA through regional innovation and tourism strategies and plans.

The analysis of development instruments positions tourism in the NPWA as an axis of public investment in the activity. This can be seen in Chilean Patagonia in the declaration of seven Zones of Tourist Interest (in Spanish ZOIT) structured around the NPWA, where public–private governance is formed for the execution of plans with public financing. This has also translated into development and investment strategies that, with financing from the National Regional Development Fund, Strategic Programs of the Development Corporation, and Special Development Plans for Extreme Zones, have promoted actions to improve supply, investment in enabling infrastructure in the NPWA and marketing.

4.2   Characterization of Tourism Supply

The development of tourism in Chilean Patagonia can be explained by a series of factors related to its isolation and the presence of urban centers that determine accessibility. The cities of Puerto Montt, Coyhaique, and Punta Arenas provide access and connectivity to the territory, acting as distribution centers and articulating tourist flows to and from Patagonia. This supply is mostly linked around the Southern Highway (in Spanish Carretera Austral), which acts as the backbone of tourism development, connecting small towns (hereinafter gateway communities) that concentrate a varied tourist offer (Fig. 1). There are also maritime routes for cruise ship tourism.

The characteristics and distribution of tourism offer are heterogeneous in Patagonia. There is greater spatial distribution in destinations in the regions of Los Lagos and Aysén, where services, which differ in number and concentration, are distributed among various localities. In contrast, the Magallanes Region concentrates the supply mostly in two tourist centers, associated with two main destinations: Torres del Paine and the Strait of Magellan. A detailed analysis of the relationship with the NPWA and tourism development is presented in Table 3.

The largest number of services are found in the Torres del Paine National Park (NP), Coyhaique-Puerto Aysén and Estrecho de Magallanes destinations, concentrated in the cities of Puerto Natales, Coyhaique, and Punta Arenas, respectively. Puerto Natales is a gateway community for the activities that take place in Torres del Paine NP, while Punta Arenas and Coyhaique act as distribution centers for tourism flows to the NPWA in their respective areas of influence. Something

Table 2 Tourism development instruments (own elaboration)

Territorial coverage		Territorial planning	Validity in	Key concepts related to tourism and NPWA						
				Tourism	Special Interest Tourism	Tourism and NPWA	Sustainable tourism development	Tourist concessions	Tourism management in NPWA	Tourism infrastructure in NPWA
National		National tourism strategy	2012–2020	X	X	X	X	X		X
		Action plan for sustainable tourism in state wildlife protected areas	2014–2018	X	X	X	X	X	X	X
		Bidding plan Ministry of National Assets	2019–2020	X			X	X		
Regional	Los Lagos	Regional government region plan	2018–2022	X		X				
		Regional development strategy for the lakes region	2009–2020	X	X	X				
		Regional innovation strategy Los Lagos Region	2014–2019	X			X			
	Aysén	Regional government region plan	2018–2022	X		X				
		Regional development strategy	2009	X	X	X				

(continued)

Table 2 (continued)

Territorial coverage	Territorial planning	Validity in	Key concepts related to tourism and NPWA						
			Tourism	Special Interest Tourism	Tourism and NPWA	Sustainable tourism development	Tourist concessions	Tourism management in NPWA	Tourism infrastructure in NPWA
	Regional innovation strategy	2014–2020	X						
	Extreme zones development plan	2014	X	X	X				X
	PER tourism	2015–2025	X		X			X	X
Magallanes	Regional government region plan	2018–2022	X		X	X			
	Regional development strategy	2012–2020	X	X	X				
	Regional science, technology and innovation policy	2010–2020	X	X					
	Extreme zones development plan	2014	X	X	X				X
	PER tourism	2012–2020	X	X	X			X	

Table 3 Tourism development and NPWA in Chilean Patagonia (Prepared by the authors based on Subsecretaría de Turismo, 2019 and Servicio Nacional de Turismo [48]; Tacón et al. [46])

Tourists destinations (development level)	NPWA management level	Main locations (number of service providers)	Number of suppliers by type of service and destination	
Puerto Montt–Calbuco–Maullín [a] (Consolidated)	Alerce Andino NP (Int)	Puerto Montt (363)–Chamiza (5)–Correntoso (5)–Lago Chapo (5)	A (160)–R (56) AV (53)–TA (12) GT (35)–T (56)–O (14)	Total 386
Carretera Austral North (Emerging)	Hornopirén NP (Bas)–Pumalín NP (N/D)	Hornopirén (67)–Cochamó (14)–Puelo (11)	A (58)–R (13) AV (4)–TA (31) GT (25)–T (5)–O (4)	Total 140
Carretera Austral South (Emerging)	Pumalín NP–Corcovado NP (Ini)–Lago Palena NR (Bas)–Futaleufú NR (Bas)	Futaleufú (103)–Chaitén (56)–Palena (18)	A (122)–R (24) AV (16)–TA (49) GT (32)–T (3)–O (8)	Total 254
Carretera Austral Queulat NP Section (Emerging)	Queulat NP (Med)–NP Corcovado (Ini)–Melimoyu NP–Lago Rosselot NR (Ini)–Lago Las Torres NR (Ini)–Isla Magdalena NP (Ini)–Lago Carlota NR (Bas)	La Junta (70)–Puyuhuapi (60)–Puerto Cisnes (50)	A (142)–R (26) AV (8)–TA (31) GT (4)–T (9)–O (3)	Total 223
Coyhaique and Puerto Aysén (Consolidated)	Río Simpson NR (Int)–Coyhaique NR (Int)–Trapananda NR (Bas)–Cerro Castillo NP (Int)	Coyhaique (527)–Puerto Aysén (86)–Villa Cerro Castillo (24)	A (259)–R (73) AV (89)–TA (103) GT (77)–T (93)–O (20)	Total 714
Chelenko (Consolidated)	Laguna San Rafael NP (Int)–Patagonia NP–Cerro Castillo NP	Puerto Río Tranquilo (138)–Chile Chico (116)–Puerto Guadal (49)	A (182)–R (40) AV (50)–TA (103) GT (11)–T (30)–O (4)	Total 420

(continued)

Table 3 (continued)

Tourists destinations (development level)	NPWA management level	Main locations (number of service providers)	Number of suppliers by type of service and destination	
Province of Los Glaciares (Emerging)	Bernardo O'Higgins NP (Bas)–Katalalixar NP (Ini)–NP Laguna San Rafael (Int)–Patagonia NP	Cochrane (79)–Tortel (37)–Villa O'Higgins (35)	A (88)–R (13) AV (7)–TA (39) GT (9)–T (9)–O (2)	Total 167
Puerto Edén (Potential)	Bernardo O'Higgins NP (Bas)	Puerto Edén (0)	A (0)–R (0) AV (0)–TA (0) GT (0)–T (0)–O (0)	Total 0
Torres del Paine NP (Consolidated)	Torres del Paine NP (Cons)–Cueva del Milodón NM (Int)–Kawésqar NP–Bernardo O'Higgins NP (Bas)	Puerto Natales (626)–Torres del Paine NP (44)–Río Serrano (17)	A (256)–R (37) AV (95)–TA (114) GT (154)–T (52)–O (20)	Total 728
Strait of Magellan (Consolidated)	Magallanes NR (Int)–Laguna Parrillar NR (Bas)–Los Pinguinos NM (Int)–Kawésqar NP–Pali Aike NP (Bas)	Punta Arenas (525)–Agua Fresca (6) – Río Verde (4)	A (153)–R (44) AV (77)–TA (36) GT (110)–T (76) O (49)	Total 545
Tierra del Fuego (Potential)	Yendegaia NP (Ini)–RN Laguna de los Cisnes (Ini)–Alberto de Agostini NP (Ini)	Porvenir (59)–Cerro Sombrero (10)–Punta Delgada (5)	A (55)–R (14) AV (5)–TA (4) GT (4)–T (3)–O (2)	Total 87
Cape Horn (Emerging)	Cabo de Hornos NP (Ini)–Yendegaia NP (Ini)–Alberto de Agostini (NP) (Ini)	Puerto Williams (61)–Caleta Mejillones (1)	A (13)–R (5) AV (9)–TA (19) GT (11)–T (4)–O (1)	Total 62

[a] Destination Puerto Montt: only from Puerto Montt to the south is considered. Development level: 1. Initial (Ini), 2. Basic (Bas), 3. Intermediate (Int), 4. Consolidate (Cons). Services typology: Lodging (A), Restaurants and similar (B), Travel agencies and tour operators (AV), Adventure tourism (TA), Tour guides (GT), Transportation (T), Others such as recreation and handicrafts (O)

similar occurs with the destination of Puerto Montt. Other important destinations whose gateway communities are located along the Carretera Austral are: Chelenko (Puerto Río Tranquilo and Chile Chico), Carretera Austral southern section (Futaleufú and Chaitén), Carretera Austral-Queulat, Los Glaciares, and Carretera Austral northern section.

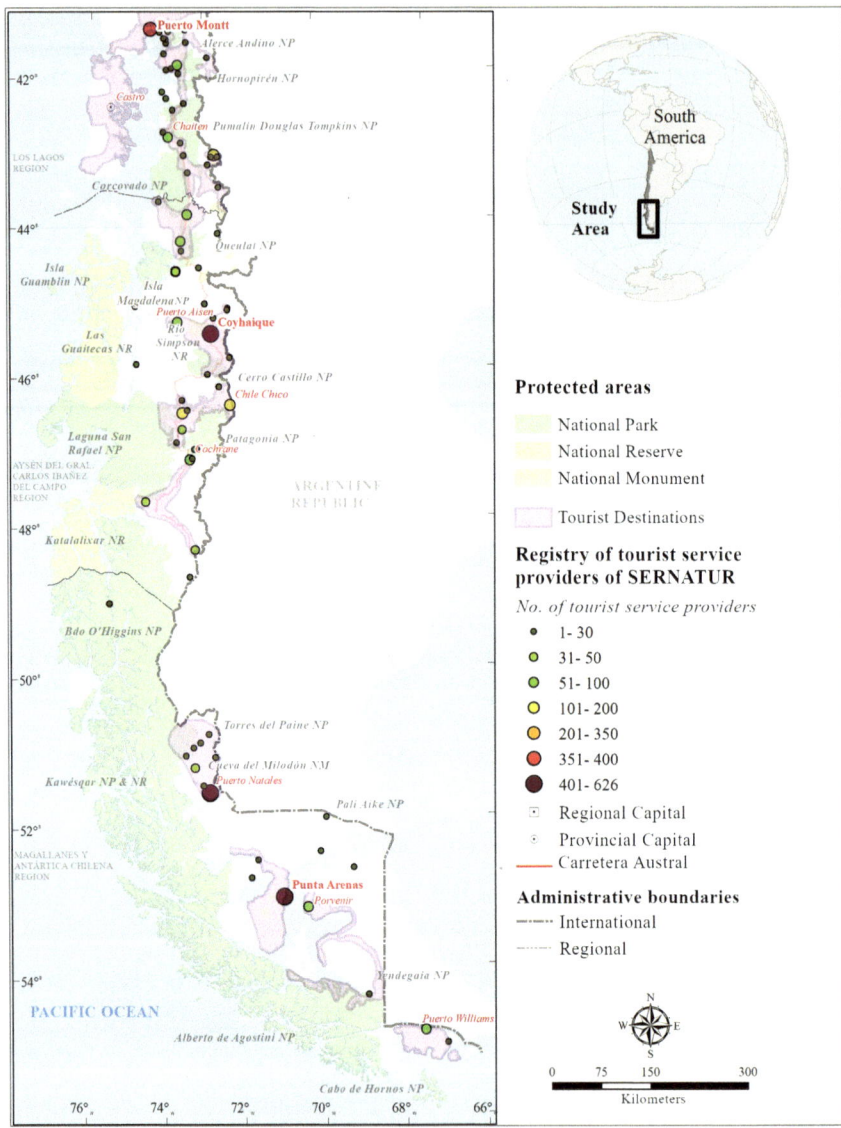

Fig. 1 Tourism development in Chilean Patagonia around NPWA (prepared by the authors based on: Subsecretary of Tourism, 2018 (in Spanish Subsecretaría de Turismo [49]) and National Tourism Service, 2019 (in Spanish Servicio Nacional de Turismo [48])

Lodging predominates the supply of services by destination, followed significantly by the collection of travel agencies, adventure tourism, and tour guides. The latter are of great relevance and show that outdoor activities in the NPWA are a central element of the supply. A more explicit relationship between level of tourism development and the NPWA emerges from the crossing of three variables that seem to be directly related: concentration of supply, level of tourism development of the destination, and the level of management of the NPWA (Table 3). One example is the gateway community of Puerto Natales, which concentrates the largest number of registered service providers in Chilean Patagonia, belonging to the best-positioned destination in the territory (level of development: Consolidated) and supporting the development of activities in Torres del Paine NP and Cueva del Milodón National Monument (NM), with Consolidated and Intermediate management levels, respectively.

At the other extreme, there are tourist destinations associated with the NPWA that are in an incipient condition, such as the gateway community of Puerto Eden (without official registration of service providers), with a Potential level of tourism development and close to Bernardo O'Higgins NP with a Basic management level. An additional important form of tourism is cruise ships and passenger maritime transport that travel through the SNASPE to visit its attractions as part of the navigation routes (Fig. 2).

Three types of routes are identified with little information and data available: (i) those operated by cruise ships with a long history (e.g. Laguna San Rafael, Skorpios); (ii) those of a mixed nature that serve to connect the territory and are used by tourists who travel independently (e.g. Tortel-Puerto Natales ferry); (iii) those operated by local agents who navigate the NPWA. Mixed routes predominate in northern Patagonia, connecting localities and NPWA as an alternative to Carretera Austral, while routes associated with cruise ships are mostly centered in Magallanes, with tourist routes associated with Alberto de Agostini NP and Cape Horn NP predominating.

4.3   Tourists Offer in the NPWA

Tourism development in Chilean Patagonia exists not only in the gateway communities near the NPWA, but also within the protected areas themselves. In some cases, this development is associated with the enhancement of value through the construction of enabling infrastructure for the development of activities (trails, viewpoints), while in others it has been accompanied by the operation of services granted to third parties through the system of tourism concessions (Table 4).

Thirty-seven tourism services operated by third parties were identified in the NPWA, concentrated in the regions of Aysén and Magallanes. Short-term operating permits are predominant for providing transportation and tour guide services in national parks in Aysén, while in Magallanes, long-term concessions are granted for the operation of lodging and gastronomic services. CONAF also identifies service providers that operate without permits or concessions.

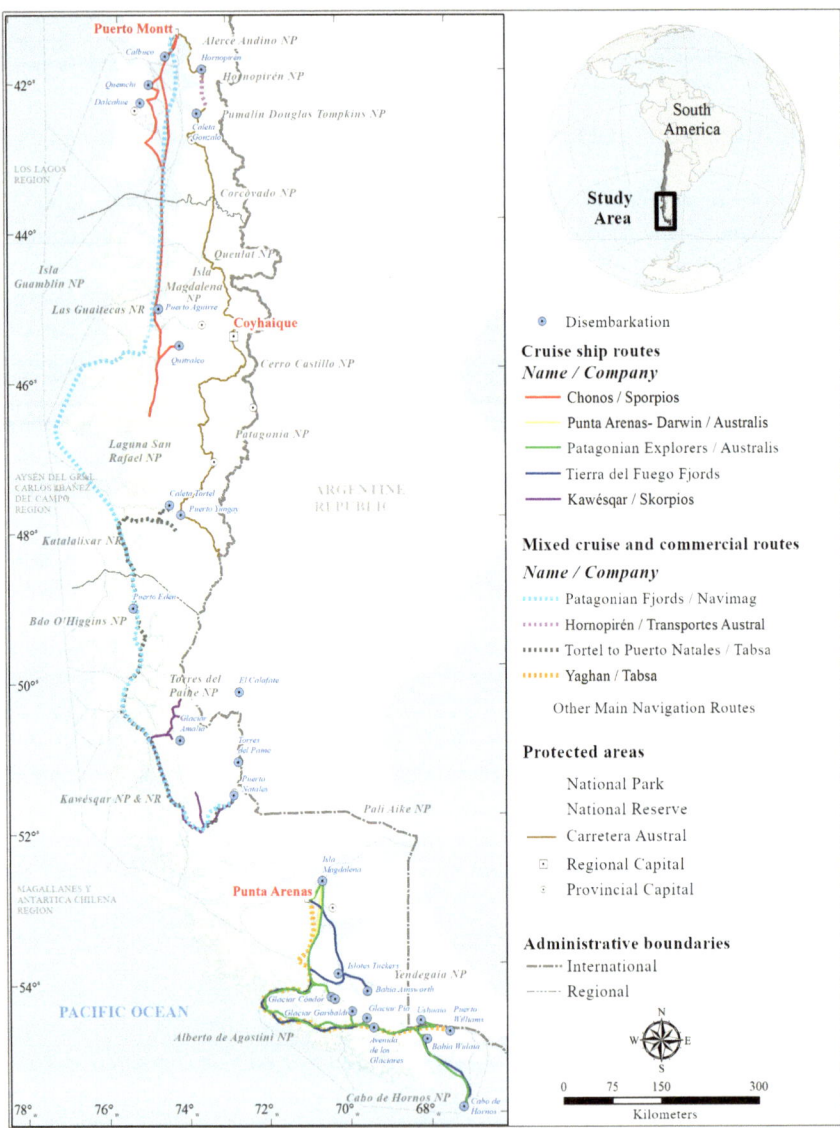

Fig. 2 Tourist navigation routes in Chilean Patagonia (own elaboration)

4.4   Tourism Demand in ASPE of Chilean Patagonia

Tourism demand in Chilean Patagonia has followed the national trend. The number of foreign tourists entering through border crossings alone doubled in the last 10 years, while the growth rate of visits to the NPWA increased by 9% annually for the same period. Patagonia's NPWA received a total of 751,000 visits (domestic

Table 4 Third-party tourism operations in NPWA in Chilean Patagonia (Prepared by authors based on CONAF personal communications[2])

Region	Unit name	Concession contract No. of providers	Ecotourism permit No. of providers	Total providers	Total regional
Los Lagos	Alerce Andino National Park	3	0	3	3
Aysén	Laguna San Rafael National Park	0	12	12	22
	Queulat National Park	1	0	1	
	Cerro Castillo National Park	2	1	3	
	Patagonia National Park	1	4	5	
	Simpson River National Reserve	1	0	1	
Magallanes	Torres del Paine National Park	11	0	11	12
	Cueva del Milodón Natural Monument	1	0	1	
	Total	20	17	37	

and foreign) in 2019, representing 20% of the national total of visits recorded by CONAF. The most visited areas in Patagonia were Torres del Paine NP and MN Cueva del Milodón (40 and 20%, respectively of the total visits to the Patagonian NPWA). The rest of the areas receive a considerably lower percentage of visits

---

[2] Personal communications: W. Rubilar, November, 5, 2019; M. Ruiz, October, 28, 2019; C. Hochstetter, 6 November, 2019.

(around 3% of the total), with Queulat NP, Alerce Andino and Los Pingüinos NP being slightly different (Fig. 3).

The records show that the tourist profile is mainly national, apart from the Magallanes Region where foreigners predominate, which is mainly explained by the influence of Torres del Paine NP (>60% international visitors). The age group varies between 25 and 50 years, with a length of stay in the destinations that

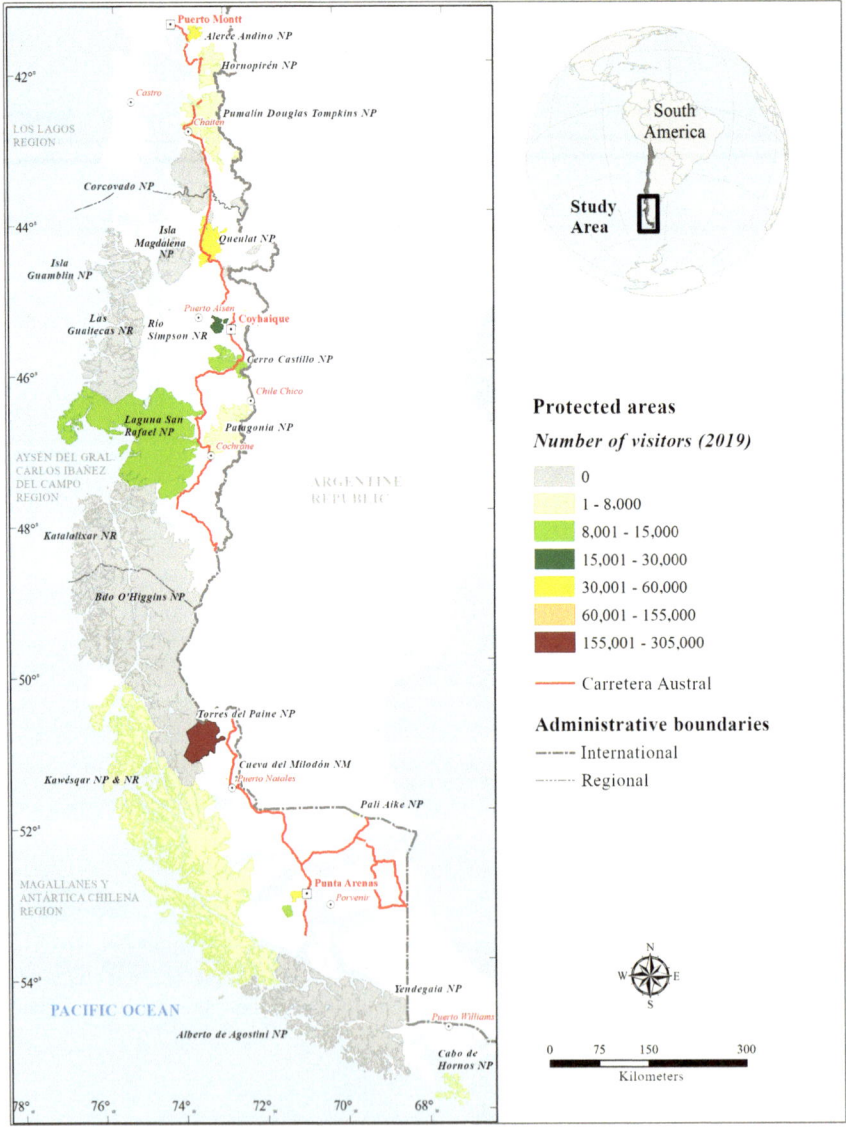

Fig. 3 Visitation in NPWA in Chilean Patagonia (own elaboration based on [50])

fluctuates between 1 and 5 days. The exception is Cape Horn, where the average stay is 7.1 days, due to the conditions of access to the destination. Tourist spending is also heterogeneous and fluctuates according to nationality and age group, with foreigners over 40 years of age spending the most and nationals under 30 years of age spending the least.

4.5 Impacts of Tourism Linked to NPWA in Chilean Patagonia

The economic contribution of tourism to the gateway communities and the SNASPE has not been quantified in depth for Chilean Patagonia. There are estimates based on the willingness to pay methodology [28], as well as a CONAF study to measure this contribution based on the Transbank system in Torres del Paine NP. The scope of the latter research cannot be analyzed, as it is not available. It is also difficult to analyze the SNASPE's income due to the lack of systematic data. However, recent CONAF reports show that in 2017 the Magallanes Region contributed about 4,400 million Chilean pesos (CLP) (*ca.* US$ 6.2 million[3]) to the SNASPE, which is equivalent to 40% of the total income of the system. Of this amount, 90% corresponds to income from entrance fees and 10% from concession contracts in Torres del Paine NP and Cueva del Milodón NM. The Aysén Region has shown a significant increase in revenues, from CLP 100 million in 2016 to CLP 200 million in 2017, which is mainly explained by the increase in visitors to Queulat NP. Thus, the contribution of this region to tourism is half of what Magallanes contributes from concessions alone (Fig. 2). Despite its growth potential, there is little data on demand and economic contribution to SNASPE from cruise tourism. Kirk et al. [29] studied the case of the Cabo de Hornos Biosphere Reserve and estimated a contribution of US$1 million annually.

The contribution to local development has also been addressed, although with rather qualitative approaches. Vela-Ruiz and Delgado [51], Rozzi et al. [20], Bourlon and Mao [19], Bourlon [18], and Bórquez et al. [17] investigated the role of tourism in local development and its possible contributions, while Núñez et al. [26], Blair et al. [24], and Zorondo-Rodríguez et al. [25], did this based on sociocultural impacts. Perception studies on the benefits of NPWA in Puerto Natales and Puerto Eden show recognition of the economic benefits for tourism, local development, and employment for women [52]. The seasonality of the activity has been identified as a negative impact of tourism [53], concluding that the construction of hotels within the NWPAs decreases the benefits for the community. In relation to environmental contributions, local inhabitants perceive a greater appreciation of the natural and cultural heritage of the environment that contributes to greater protection of nature [52]. However, the direct environmental impacts of tourism in NPWA in Patagonia have rarely been quantified. Most studies focus on Torres

---

[3] Average value of the US Dollar in 2019. Available at: http://www.sii.cl/valores_y_fechas/dolar/dolar2019.htm.

del Paine NP, where impacts generated by trails and informal camping areas have been identified [14, 54], with fires and soil erosion being important drivers of the loss of the structure of the landscape [14, 54] and native vegetation [55, 56]. This is more critical given the lack of planning instruments in NPWA, many of which do not have management and public use plans despite receiving tourists [23].

5   Discussion

Thousands of people have been motivated to visit Chilean Patagonia's NPWA in accordance with global trends in nature-based tourism development [1, 57]. These areas host a wide range of activities that motivate visitation, driving the emergence of tourism services in neighboring localities that act as gateway communities, located at the NPWA public access points (Fig. 1). This symbiotic relationship between NPWA, tourism and local communities can be illustrated in Torres del Paine NP, where there is a direct link between the concentration of supply in the gateway community of Puerto Natales, the destination's high level of tourism development and the park's higher level of management. However, it is not possible to establish with precision if the level of management of the NPWA is a consequence of the higher pressure from tourism use or if conversely, the tourism development of the gateway community is explained by the level of tourism development within the NPWA.

A relevant aspect is related to the heterogeneous distribution of gateway communities in Chilean Patagonia. While some areas receive a high level of visitation and have gateway communities that offer a significant number of services, there are others with similar levels of accessibility that have a low level of supply and visitation (Fig. 1), which could indicate the lack of a public policy to guide the development of tourism at the territorial level.

Tourism supply is also developed within the NPWA. In accordance with the models proposed by Spenceley et al. [10], two opposing models can be identified in Chilean Patagonia, which arise from the system of tourism concessions. This could account for the lack of a clearly defined policy that, coupled with CONAF's limited capacity to oversee unauthorized operations, constitute a risk to achieving tourism conservation objectives. Although there are no specific studies that have measured the impacts of the two models in Patagonia, the international trend suggests abandoning models of intensive development within the NPWA and proposes promoting the development of supply within gateway communities [10].

Although information on tourism in Chilean Patagonia is limited, recent data show a sustained growth in the number of tourists visiting the NPWA, accounting for nearly 20% of the national total, mostly in Torres del Paine NP. One type of tourism that is growing rapidly is cruises, where boats of different types and sizes travel through the NPWA to visit their attractions. Unfortunately, information on supply, demand, points of visitation, and impacts generated by these cruises in the NPWA is practically non-existent. This growth of the tourism sector increases the use pressures on NPWA, which in the absence of planning could increase impacts and reduce benefits [6]. The little research in Chilean Patagonia reports on these impacts and suggests paying more attention to the effects on soil, biodiversity loss, and hazard control, agreeing on the importance of measuring and monitoring the environmental impacts of tourism [2] and the need to adopt management standards [6]. This is particularly important because many NPWA do not have management and public use plans in place [23].

Although tourism emerges as a financial opportunity for the conservation and revitalization of economies in gateway communities [3], no studies quantifying these benefits have been identified in Chilean Patagonia. The data provided by CONAF in terms of entry fees and concessions in 2017 account for this contribution to SNASPE and demonstrate that the financing of the areas rests significantly on income derived from tourism. However, this could act as a perverse incentive to increase pressure on the areas. A similar situation occurs with the economic contributions to the gateway communities, where apart from a few specific cases, the effect of tourism on employment, income, and improvement in the quality of life of the inhabitants has not been estimated. Global studies have indicated that tourism could bring about changes in the socio-productive patterns of a community, generating economic dependence on the sector, which is why it is necessary to study these impacts further.

Finally, the growth of the tourism sector is framed within multiple sectorial policies that have placed emphasis on promoting tourism in the NPWA in Chilean Patagonia but have omitted essential aspects such as planning and mitigation of its impacts. Nature tourism must recognize that the main heritage of the sector is in the NPWA and their people, so contributing to and guaranteeing the objectives of conservation and local wellbeing is fundamental to the viability of the activity.

6 Conclusions and Recommendations

The synthesis presented in this chapter is the first systematic effort to examine the role of NPWA in tourism in Chilean Patagonia. Sectoral instruments have injected resources into the NPWA, bringing with them the urgent need to coordinate public–private investment to improve the areas' management, ensuring the sustainability of the natural and cultural heritage. The levels of tourism development associated with the NPWA appear to be mediated by the concentration of tourism services in gateway communities, by areas' level of management, and the stage of tourism development of the destination in which they are located. However, tourism development is currently concentrated in very few NPWA in Patagonia, reflecting the lack of a public policy to guide territorial tourism development.

There are two development models and several tourism concession systems operating in parallel within the areas, reflecting the lack of a clear policy to promote ecotourism in the SNASPE. Finally, given the growing tourism demand in Patagonia's NPWA, the analysis shows the need to quantify the direct economic and environmental impacts of tourism. This is critical, given that the effect of tourism on biodiversity in the NPWA is currently unknown or has not been measured. Based on the above, we present the following recommendations:

- Given the speed of tourism growth in Chilean Patagonia, it is recommended that progress be made in the adoption of management standards in the NPWA, guaranteeing the existence of management and public use plans as an enabling condition for tourism development. This is the only way to design development strategies focused on mitigating negative impacts and dispersing pressures, and to advance in a model that promotes economic benefits in gateway communities and strengthens the link between local inhabitants and the NPWA.
- The collection of visitor fees should be improved, and further analysis should be carried out of the concession systems currently operating in the SNASPE in order to improve their coordination. CONAF should have a well-defined policy for ecotourism development in the SNASPE, with better orientation of state and private investments. It is recommended that progress be made in defining a single concession mechanism based on a development model that benefits the gateway communities.
- There should be a focus on achieving a stable financing system for the SNASPE that creates appropriate incentives for its continuous improvement as a priority matter, together with the establishment of annual budgets that ensure a minimum floor for all NPWA and new revenue collection systems that include incentives for decentralized creative management, reducing the pressure for income derived from tourism.
- Investment in adapting and applying methodologies that allow for the systematic measurement and monitoring of the environmental impacts of tourism (e.g. biodiversity, ecosystem services) and the quantification of the contribution of visitation to NWPAs on local economies is a priority, the latter in order to have clarity regarding the return on state investment in each NPWA. Based on international experience, it is expected that investment in NPWA will be efficient and have a high impact in relation to other fiscal expenditures, and there are many validated methodologies for this purpose. However, limitations in local economic statistics present significant obstacles to their application. A joint investment by state actors and universities is recommended to generate a cost-effective methodology that can be replicated periodically for Chilean Patagonia.

Acknowledgements The authors thank CONAF and SERNATUR for providing information and acknowledge the important support of our colleague María Paz Peña and other members of the Austral Patagonia Program team at the Universidad Austral de Chile. We also express our most sincere thanks to the reviewer of the manuscript and to the editors, who kindly provided recommendations that strengthened the results and discussion presented. Finally, we thank The Pew Charitable Trusts for their financial support for the development of this work.

References

1. Yergeau, M. E. (2020). Tourism and local welfare: A multilevel analysis in Nepal's protected areas. *World Development, 127*, 104744.
2. Hawkins, D. E. (2004). A protected areas ecotourism competitive cluster approach to catalyze biodiversity conservation and economic growth in Bulgaria. *Journal of Sustainable Tourism, 12*(3), 219–244.
3. Leung, Y. F., Spenceley, A., Hvenegaard, G., & Buckley, R. (Eds.). (2019). *Gestión del turismo y de los visitantes en áreas protegidas: directrices para la sostenibilidad*. Serie Directrices sobre Buenas Prácticas en Áreas Protegidas no. 27. UICN. https://www.researchgate.net/publication/331412718_Gestion_del_turismo_y_de_los_visitan-tes_en_areas_protegidas_directrices_para_la_sostenibilidad
4. Balmford, A., Green, J. M. H., Anderson, M., Beresford, J., Huang, C., Naidoo, R., Walpole, M., & Manica, A. (2015). Walk on the wild side: Estimating the global magnitude of visits to protected areas. *PLoS Biology, 13*(2), 1–6.
5. Chung, M. G., Dietz, T., & Liu, J. (2018). Global relationships between biodiversity and nature- based tourism in protected areas. *Ecosystem Services, 34*(A), 11–23.
6. Eagles, P. F. J., Buteau-Duitschaever, W. C., Rattan, J., Havitz, M. E., Glover, T. D., Romagosa, F., & McCutcheon, B. (2012). Non-government organization members' perceptions of governance: A comparison between Ontario and British Columbia provincial parks management models. *Leisure/Loisir, 36*(3–4), 269–287.
7. Heagney, E. C., Rose, J. M., Ardeshiri, A., & Kovac, M. (2019). The economic value of tourism and recreation across a large protected area network. *Land Use Policy, 88*, 104084.
8. Bennett, N., Lemelin, R. H., Koster, R., & Budke, I. (2012). A capital assets framework for appraising and building capacity for tourism development in aboriginal protected area gateway communities. *Tourism Management, 33*(4), 752–766.
9. Frauman, E., & Banks, S. (2011). Gateway community resident perceptions of tourism development: Incorporating importance-performance analysis into a limits of acceptable change framework. *Tourism Management, 32*(1), 128–140.
10. Spenceley, A., Snyman, S., & Eagles, P. F. J. (2019). A decision framework on the choice of management models for park and protected area tourism services. *Journal of Outdoor Recreation and Tourism, 26*, 72–80.
11. Dinica, V. (2017). The environmental sustainability of protected area tourism: Towards a concession-related theory of regulation. *Journal of Sustainable Tourism, 26*(1), 146–164.
12. Rivas, H. (2018). Ecoturismo en Chile: Desafíos de una década de crecimiento en las áreas protegidas del Estado. L'écotourisme au Chili: les défis d'une décennie de croissance dans les zones protégées de l'Etat Ecotourism in Chile: Challenges of a Decade of Growth in the Pro. *Études Caribéennes, 41*, 0–17.
13. Mardones, G. (2017). Análisis de redes sociales para la gobernanza de un área protegida y su zona de amortiguación en el bosque templado del sur de Chile. *REDES. Revista Hispana para el Análisis de Redes Sociales, 28*, 1–61.
14. Barros, A., Monz, C., & Pickering, C. (2014). Is tourism damaging ecosystems in the Andes? Current knowledge and an agenda for future research. *Ambio, 44*(2), 82–98.
15. Cerda, C., Ponce, A., & Zappi, M. (2013). Using choice experiments to understand public demand for the conservation of nature: A case study in a protected area of Chile. *Journal for Nature Conservation, 21*(3), 143–153.
16. Outeiro, L., Villasante, S., & Oyarzo, H. (2018). The interplay between fish farming and nature based recreation-tourism in southern Chile: A perception approach. *Ecosystem Services, 32*, 90–100.
17. Bórquez, R., Bourlon, F., & Moreno, A. (2019). El turismo científico y su influencia en la comunidad local: el estudio de caso de la red de turismo científico en Aysén, Chile. *Turismo y Desarrollo, 12*(26).
18. Bourlon, F. (2017). La bio-geografía de Douglas Tompkins, una mirada comprensiva de la conservación privada en la Patagonia chilena. *Revista de Aysenología, 4*, 86–98.

19. Bourlon, F., & Mao, P. (2016). *La Patagonia chilena: Un nuevo El Dorado para el turismo científico*. Coyhaique, Chile: Ediciones Ñire Negro, Archipiélagos Patagónicos. https://www.researchgate.net/publication/315664139_La_Patagonia_Chilena_nuevo_El_Dorado_del_Turismo_Cientifico

20. Rozzi, R., Massardo, F., Cruz, F., Grenier, C., Muñoz, A., & Mueller, E. (2010). Galapagos and Cape Horn: Ecotourism or greenwashing in two iconic Latin American archipelagoes. *Environmental Philosophy, 7*(2), 1–32.

21. Bourlon, F., Osorio, M., Mao, P., & Gale, T. (Eds.). (2012). *Explorando las nuevas fronteras del turismo. Perspectivas de la investigación en turismo.* Centro de Investigación en Ecosistemas de.

22. Serenari, C., Peterson, M. N., Wallace, T., & Stowhas, P. (2017). Private protected areas, ecotourism development and impacts on local people's well-being: A review from case studies in southern Chile. *Journal of Sustainable Tourism, 25*(12), 1792–1810.

23. Gale, T., Adiego, A., & Ednie, A. (2018). A 360° approach to the conceptualization of protected area visitor use planning within the Aysén region of Chilean Patagonia. *Journal of Park and Recreation Administration, 36*(3), 22–46.

24. Blair, H., Bosak, K., & Gale, T. (2019). Protected areas, tourism, and rural transition in Aysén, Chile. *Sustainability (Switzerland), 11*(24), 1–22.

25. Zorondo-Rodríguez, F., Díaz, M., Simonetti-Grez, G., & Simonetti, J. A. (2019). Why would new protected areas be accepted or rejected by the public?: Lessons from an ex-ante evaluation of the new Patagonia park network in Chile. *Land Use Policy, 89*, 104248.

26. Núñez, A., Aliste, E., & Bello, A. (2014). El discurso del desarrollo en Patagonia-Aysén: la conservación y la protección de la naturaleza como dispositivos de una renovada colonización. Chile, siglos XX–XXI. *Scripta Nova: Revista Electrónica de Geografía y Ciencias Sociales, 18*(46), 1–13.

27. Nahuelhual, L., Carmona, A., Lozada, P., Jaramillo, A., & Aguayo, M. (2013). Mapping recreation and ecotourism as a cultural ecosystem service: An application at the local level in southern Chile. *Applied Geography, 40*, 71–82.

28. Nahuelhual, L., Vergara, X., Kusch, A., Campos, G., & Droguett, D. (2017). Mapping ecosystem services for marine spatial planning: Recreation opportunities in sub-Antarctic Chile. *Marine Policy, 81*, 211–218.

29. Kirk, C., Rozzi, R., & Gelcich, S. (2018). El turismo como una herramienta para la conservación del elefante marino del sur *(Mirounga leonina)* y sus habitats en Tierra del Fuego, Reserva de la Biósfera Cabo de Hornos, Chile. *Magallania (Punta Arenas), 46*(1), 65–78.

30. Pearce, D., Guala, C., Veloso, K., Llano, S., Negrete, J., Rovira, A., Gale, T., & Reis, A. (2017). Destination management in Chile: Objectives, actions and actors. *International Journal of Tourism Research, 19*(1), 50–67.

31. Muñoz, M., & Torres, R. (2010). Conectividad, apertura territorial y formación de un destino turístico de naturaleza: el caso de Aysén (Patagonia chilena). *Estudios y Perspectivas en Turismo, 19*(4), 447–470.

32. Corporación Nacional Forestal. (2020). Estadísticas de visitación SNASPE 2009-2019. https://www.conaf.cl/wpcontent/files_mf/1627054943VisitacionSNASPE_2020.pdf

33. Subsecretaría de Turismo. (2013). Formulario de Solicitud de Declaración ZOIT Puelo y Cochamó. https://pac.subturismo.gob.cl/wp-content/uploads/2021/07/Ficha-postulacion-ZOIT-Puelo-Cochamo-Hualaihue.pdf

34. (Undersecretariat of Tourism)Subsecretaría de Turismo. (2018). Plan de Acción ZOIT Cabo de Hornos. https://www.subturismo.gob.cl/wp-content/uploads/2021/06/plan-de-accion-zoit-20.pdf

35. Subsecretaria de Turismo. (2019). Anuario de Turismo. https://www.subturismo.gob.cl/wpcontent/uploads/2015/10/Anuario-de-Turismo-2018.pdf

36. Transforma Turismo y Fundación Sendero de Chile. (2017). Modelo de Cooperación público-privado para el desarrollo turístico sustentable en las Áreas Silvestres Protegidas del Estado y sus entornos. https://transformaturismo.cl/wpcontent/uploads/2018/03/transforma_turismo_memoria_2017.pdf

37. Servicio Nacional de Turismo. (2017). Plan de Acción ZOIT Provincia de los Glaciares https://www.subturismo.gob.cl/wp-content/uploads/2015/10/plan-acci%c3%93n-provincia-de-los-glaciares.pdf

38. Servicio Nacional de Turismo. (2017). Plan de Acción ZOIT Chelenko. https://www.subtur ismo.gob.cl/wpcontent/uploads/2015/10/Plan-Accion-ZOIT-Chelenko.pdf

39. Servicio Nacional de Turismo. (2017). Plan de Acción ZOIT Aysén Patagonia Queulat. https://www.subturismo.gob.cl/wp-content/uploads/2015/10/Plan-Accion-ZOIT-Patagonia-Queulat.pdf

40. (Undersecretariat of Tourism) Subsecretaría de Turismo. (2016). Ficha de actualización del Plan de Acción para Prórroga de Zona de Interés Turístico (ZOIT) "Futaleufú - Palena" https://www.subturismo.gob.cl/wpcontent/uploads/2021/06/plan-de-accion-zoit-16.pdf

41. Intendencia Regional, Gobierno de Chile. (2018). Plan Región de Aysén 2018–2022. http://www.intendenciaaysen.gov.cl/media/2019/09/Cuenta-P%C3%BAblica-Participativa-2018-2019.pdf

42. Intendencia Regional, Gobierno de Chile. (2018). Plan Región de Los Lagos 2018-2022. https://www.goreloslagos.cl/resources/descargas/acerca_de_gore/doc_gestion/Plan_Regi onal_Los_Lagos.pdf

43. Intendencia Regional, Gobierno de Chile. (2018). Plan Región de Los Lagos 2018–2022 . https://www.camara.cl/verdoc.aspx?prmid=169403&prmtipo=documentocomision

44. Servicio Nacional de Turismo y Instituto Nacional de Estadística. (2009). Informe Nacional de turismo. https://www.sernatur.cl/wp-content/uploads/2019/02/Turismo-Informe-Anual-2008.pdf

45. Ilustre Municipalidad de Cabo de Hornos. (2014). Turismo Sustentable en Cabo de Hornos: Plan de Desarrollo Turístico 2015–2020. http://www.imcabodehornos.cl/jsmallfib_top/pla detur.pdf

46. Tacón, A., Tecklin, D., Farías, A., Peña, M. P., & García, M. (2023). *Terrestrial protected areas In Chilean Patagonia: Characterization, historical evolution, and management.* Springer.

47. Silva, E. (2016). Patagonia, without dams! Lessons of a David *vs.* Goliath campaign. *Extractive Industries and Society, 3*(4), 947–957.

48. Servicio Nacional de Turismo. (2019). *Registro completo de los prestadores de servicios turísticos.* https://registro.sernatur.cl/descargas/

49. Subsecretaría de Turismo. (2018). *Informe de intensidad turística y definición de destinos turísticos.* http://www.subturismo.gob.cl/wp-content/uploads/2015/09/Informe-de-Intensidad-Turística-y-Definición-de-Destinos-Turísticos-2018-1.pdf

50. Corporación Nacional Forestal. (2019). *Estadísticas de visitación SNASPE.* https://www.conaf.cl/parques-nacionales/visitanos/estadisticas-de-visitacion/

51. Vela-Ruiz, G., & Delgado, M. M. (2010). Contribución del enfoque de desarrollo territorial rural a la comprensión de los procesos generados en torno a áreas protegidas en la Patagonia chilena 1. *Revista Chilena de Estudios Regionales, 2*(1), 83–96.

52. Vela-Ruiz, G. (2009). Contribución desde el enfoque de capitales a la comprensión de la inclusión de comunidades en los procesos generados por áreas protegidas, Región de Magallanes. In Fundación para la Superación de la Pobreza (Ed.), *Tesis País 2009. Piensa un país sin pobreza* (pp. 139–159). Fundación para la Superación de la Pobreza. http://www2.superacio npobreza.cl/wp-content/uploads/2019/06/02_Piensa-un-Pai%CC%81s-sin-Pobreza_2009.pdf

53. Marquet, P. A., Buschmann, A. H., Corcoran, D., Díaz, P. A., Fuentes-Castillo, T., Garreaud, R., Pliscoff, P., & Salazar, A. (2023). *Global change and acceleration of anthropic pressures on Patagonian ecosystems.* Springer.

54. Araya, P. (2007). El impacto del turismo en la conservación de una Reserva de la Biosfera y el desarrollo de su zona de influencia. El caso de la Reserva Torres del Paine. In G. Halffter, S. Guevara, & A. Melic (Eds.), *Hacia una cultura de conservación de la diversidad biológica* (Vol. 6, pp. 115–124). m3m: Monografías Tercer Milenio. http://sea-entomologia.org/PDF/PDFSM3MVOL6/Pdf13115124013Araya.pdf

55. Paula, S., & Labbé, D. L. (2018). Post-fire invasion in Torres del Paine Biosphere Reserve: The role of seed tolerance to heat. *International Journal of Wildland Fire, 28*(2), 160–166.

56. Vidal, O. J., Aguayo, M., Niculcar, R., Bahamonde, N., Radic, S., San Martín, C., Kusch, A., Latorre, J., & Félez, J. (2015). Plantas invasoras en el Parque Nacional Torres del Paine (Magallanes, Chile): estado del arte, distribución post-fuego e implicancias en restauración ecológica. *Anales del Instituto de la Patagonia (Chile), 43*(1), 75–96.
57. Eagles, P. F. J. (2002). Trends in park tourism: Economics, finance and management. *Journal of Sustainable Tourism, 10*(2), 132–153.

César Guala Catalán Master in Rural Development. Tourism Business Administrator, Universidad Austral de Chile. Ph.D. in Public Policy and Tourism, University of Victoria, Wellington, New Zealand. Director of the Austral Patagonia Program, Universidad Austral de Chile.

Katerina Veloso Santana Industrial Civil Engineer, Universidad Central de Chile. Master in Tourism Management and Innovation and Tourism Business Administrator, Universidad Austral de Chile. Coordinator of the Tourism and Conservation Line, Austral Patagonia Program, Universidad Austral de Chile.

Aldo Farías Forestry Engineer, Universidad Austral de Chile. Executive Coordinator of the Austral Patagonia Program, Universidad Austral de Chile, Valdivia.

Fernanda Sariego Bartsch Tourism Business Administrator, Universidad Austral de Chile, Programa Austral Patagonia-Universidad Austral de Chile, Valdivia.